S0-BWR-332

ANNUAL REVIEW OF CELL BIOLOGY

ANNUAL REVIEW OF CELL BIOLOGY

VOLUME 10, 1994

JAMES A. SPUDICH, *Editor*
Stanford University School of Medicine

STEVEN L. McKNIGHT, *Associate Editor*
TULARIK

RANDY SCHEKMAN, *Associate Editor*
University of California, Berkeley

ANNUAL REVIEWS INC. 4139 EL CAMINO WAY P.O. BOX 10139 PALO ALTO, CALIFORNIA 94303-0139

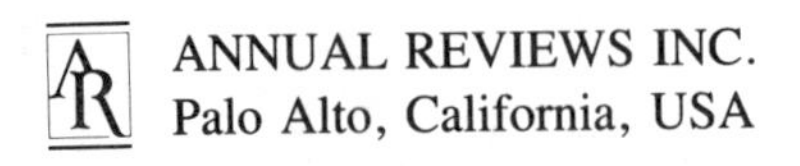

ANNUAL REVIEWS INC.
Palo Alto, California, USA

International Standard Serial Number: 0743–4634
International Standard Book Number: 0–8243–3110-9

♾ The paper used in this publication meets the minimum requirements of American National Standard for Information Sciences—Permanence of Paper for Printed Library Materials, ANSI Z39.48-1984.

Typesetting by Kachina Typesetting Inc., Tempe, Arizona; John Olson, President; Marty Mullins, Typesetting Coordinator; and by the Annual Reviews Inc. Editorial Staff

PRINTED AND BOUND IN THE UNITED STATES OF AMERICA

PREFACE

This tenth volume of the *Annual Review of Cell Biology* is dedicated to George Palade, who has retired as Editor. George established the *Annual Review of Cell Biology* and was the architect of its style in every sense. With his usual wisdom and scientific foresight, he based the series on the premise that modern cell biology consists of a continuous body of knowledge that encompasses all possible technologies, including biochemistry, genetics, and structural biology, to understand in molecular terms how cells carry out their diverse functions. By judiciously selecting subjects from all areas within cell biology and by demanding the highest quality, George has established one of the most important series that unifies a diverse and rapidly expanding field. Readers of the *Annual Review of Cell Biology* have benefited from and enjoyed the series immensely, and in these short nine years it has become one of the top journals of cell biology by any means of assessment. As Associate Editors, Bruce Alberts and I thank George for bringing us into this endeavor from the beginning and for giving us an opportunity to know him better and to grow scientifically as a result of this partnership.

Since Bruce Alberts, who also deserves to be acknowledged for his important contributions to this series, has taken on another important challenge as President of the National Academy of Sciences, the business of carrying forward this endeavor falls to me. The task will be made easier by having Steve McKnight, Randy Schekman, and John Gerhart as Associate Editors. The addition of John Gerhart marks a formal change acknowledging that over the years nearly a third of our reviews have encompassed developmental biology, which overlaps significantly with cell biology. The two continue to merge with every new discovery and, consequently, John joins at a time when we also change the name of the series to the *Annual Review of Cell and Developmental Biology* beginning with Volume 11. New members have also been added to the Editorial Committee to bring additional areas of expertise. Sharon Long and Hidde Ploegh will join us to represent the areas of plant biology and immunology. We are fortunate that Sandra Cooperman will continue to provide her expertise as Production Editor for the series. We all look forward to challenging but immensely satisfying work in the years to come to continue to provide the scientific community with a vigorous *Annual Review of Cell and Developmental Biology*. We trust that our readers will continue to enjoy and benefit from this effort.

JAMES A. SPUDICH
EDITOR

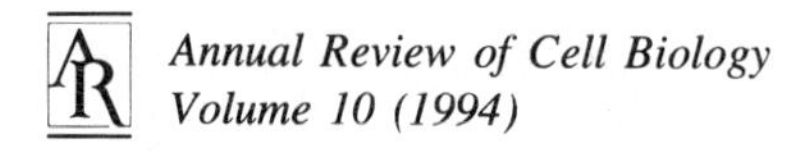
Annual Review of Cell Biology
Volume 10 (1994)

CONTENTS

OTHER REVIEWS OF INTEREST TO CELL BIOLOGISTS

From the *Annual Review of Biochemistry*, Volume 63 (1994):

A Molecular Description of Synaptic Vesicle Membrane Trafficking, Mark K. Bennett, Richard H. Scheller

Structure and Function of G Protein-Coupled Receptors, Catherine D. Strader, Tung Ming Fong, Micheal R. Tota, Dennis Underwood, Richard A. F. Dixon

The Retroviral Enzymes, Richard A. Katz, Anna Marie Skalka

Nitric Oxide: A Physiological Messenger Molecule, D. S. Bredt, S. H. Snyder

Structure, Function, Regulation, and Assembly of D-Ribulose-1,5 Bisphosphate Carboxylase/Oxygenase, Fred C. Hartman, Mark R. Harpel

Nitrogenase: A Nucleotide-Dependent Molecular Switch, James B. Howard, Douglas C. Rees

Role of Chromatin Structure in the Regulation of Transcription by RNA Polymerase II, Suman M. Paranjape, Rohinton T. Kamakaka, James T. Kadonaga

Quinoenzymes in Biology, Judith P. Klinman, David Mu

Intermediate Filaments: Structure, Dynamics, Function, and Disease, Elaine Fuchs, Klaus Weber

The Expression of Asymmetry During Caulobacter Cell Differentiation, Yves V. Brun, Greg Marczynski, Lucille Shapiro

Molecular Mechanisms of Action of Steroid/Thyroid Receptor Superfamily Members, Ming-Jer Tsai, Bert W. O'Malley

Homeodomain Proteins, Walter J. Gehring, Markus Affolter, Thomas Bürglin

The Biochemistry of Synaptic Regulation in the Central Nervous System, Mary B. Kennedy

Structures and Functions of Multiligand Lipoprotein Receptors: Macrophage Scavenger Receptors and LDL Receptor-Related Protein (LRP), Monty Krieger, Joachim Herz

The Centrosome and Cellular Organization, Douglas R. Kellogg, Michelle Moritz, Bruce M. Alberts

Energy Transduction by Cytochrome Complexes in Mitochondrial and Bacterial Respiration: The Enzymology of Coupling Electron Transfer Reactions to Transmembrane Proton Translocation, Bernard L. Trumpower, Robert B. Gennis

Regulation of Eukaryotic DNA Replication, Dawn Coverley, Ronald A. Laskey

Function and Structure Relationships in DNA Polymerases, Catherine M. Joyce, Thomas A. Steitz

Calcium Channel Diversity and Neurotransmitter Release: The ω-Conotoxins and ω-Agatoxins, Baldomero M. Olivera, George Miljanich, R. Ramachandran, Michael E. Adams

Genetic and Biochemical Studies of Protein N-Myristoylation, D. Russell Johnson, Rajiv S. Bhatnagar, Laura J. Knoll, Jeffrey I. Gordon

GTPases: Multifunctional Molecular Switches Regulating Vesicular Traffic, Claude Nuoffer, William E. Balch

From the *Annual Review of Biophysics and Biomolecular Structure,* Volume 23 (1994):

From the *Annual Review of Immunology,* Volume 12 (1994):

Structure of Peptides Associated with Class I and Class II MHC Molecules, Victor H. Engelhard

Roles of IGH and L Chains and of Surrogate H and L Chains in the Development of Cells of the Lymphocyte Lineage, Fritz Melchers, Dirk Haasner, Ulf Grawunder, Christian Kalberer, Hajime Karasuyama, Thomas Winkler, Antonius G. Rolink

Assembly, Transport, and Function of MHC Class II Molecules, Peter Cresswell

Tumor Antigens Recognized by T Lymphocytes, Thierry Boon, Jean-Charles Cerottini, Benoit den Van Eynde, Pierre van der Bruggen, Aline Van Pel

Lymphocyte Ontogeny and Activation in Gene Targeted Mutant Mice, K. Pfeffer, T. W. Mak

Signal Transduction by the B Cell Antigen Receptor and Its Coreceptors, J. C. Cambier, C. Pleiman, M. R. Clark

Thymus Independent T Cell Development and Selection in the Intestinal Epithelium, Philippe Poussier, Michael H. Julius

The Role of Protein Tyrosine Kinases and Protein Tyrosine Phosphatases in T Cell Antigen Receptor Signal Transduction, Andrew C. Chan, Dev M. Desai, Arthur Weiss

The Molecular Biology of Leukocyte Chemoattractant Receptors, Philip M. Murphy

Acquisition of Lymphokine-Producing Phenotype by CD4+ T Cells, William E. Paul, Robert A. Seder

Selective Events in T Cell Development, E. Robey, B. J. Fowlkes

Mechanisms of Transplantation Tolerance, Brett Charlton, Hugh Auchincloss, Jr., C. Garrison Fathman

The Binding and Lysis of Target Cells by Cytotoxic Lymphocytes: Molecular and Cellular Aspects, C. Berke

Antigen Presentation by Major Histocompatibility Complex Class I-B Molecules, Said M. Shawar, Jatin M. Vyas, John R. Rodgers, Robert R. Rich

From the *Annual Review of Microbiology,* Volume 48 (1994):

Comparative Molecular Biology of Lambdoid Phages, A. Campbell

The Molecular Phylogeny and Systematics of the Actinomycetes, T. M. Embley, E. Stackebrandt

Antisense RNA Control in Bacteria, Phages, and Plasmids, E. Gerhart, H. Wagner, Robert W. Simons

Genetic Controls for the Expression of Surface Antigens in African Trypanosomes, Etienne Pays, Luc Vanhamme, Magali Berberof

Biology and Genetics of Prion Diseases, Stanley B. Prusiner

From the *Annual Review of Neuroscience,* Volume 17 (1994):

Development of Motorneuronal Phenotype, Judith S. Eisen

Cloned Glutamate Receptors, Michael Hollmann, Stephen Heinemann

Hox *Genes and Regionalization of the Nervous System,* Roger Keynes, Robb Krumlauf

From the *Annual Review of Pharmacology and Toxicology,* Volume 34 (1994):

From the *Annual Review of Physiology,* Volume 56 (1994):

From the *Annual Review of Plant Physiology and Plant Molecular Biology,* Volume 45 (1994):

ANNUAL REVIEWS INC. is a nonprofit scientific publisher established to promote the advancement of the sciences. Beginning in 1932 with the *Annual Review of Biochemistry,* the Company has pursued as its principal function the publication of high-quality, reasonably priced *Annual Review* volumes. The volumes are organized by Editors and Editorial Committees who invite qualified authors to contribute critical articles reviewing significant developments within each major discipline. The Editor-in-Chief invites those interested in serving as future Editorial Committee members to communicate directly with him. Annual Reviews Inc. is administered by a Board of Directors, whose members serve without compensation.

Annu. Rev. Cell Biol. 1994. 10:1–29

THE RETINOBLASTOMA PROTEIN: More Than a Tumor Suppressor

Daniel J. Riley, Eva Y.-H.P. Lee, and Wen-Hwa Lee
Center for Molecular Medicine/Institute of Biotechnology, 15355 Lambda Drive, The University of Texas Health Science Center at San Antonio, San Antonio, Texas 78245

KEY WORDS: cell cycle, development, growth, p107, p130

CONTENTS

0743–4634/94/1115–0001$05.00

INTRODUCTION

The identification of the retinoblastoma gene (*rb*) has sparked intense investigation of tumor suppressor genes. These genes, like oncogenes, can encode transcriptional regulators, cytoplasmic signal tranducers, or molecules involved in DNA repair and cell adhesion. Studies of the *rb* gene product suggest that it may work as a fundamental regulator to coordinate pathways of cellular growth and differentiation. Some activities of Rb (the *rb* gene product) appear to be common to all mammalian cells, while others are specific to certain cell types and developmental stages.

In many different cultured tumor cells, replacement of a normal *rb* gene results in suppression of the cells' neoplastic properties. In humans, inactivating the germ line mutation of the *rb* gene leads to retinoblastomas, whereas in experimental mice, pituitary neuroendocrine tumors result. Thus the role of Rb in tumor predisposition is both tissue- and species-specific. In addition to suppressing tumor formation, the Rb protein apparently also has roles in normal development. In this review, we attempt to integrate studies regarding the role of Rb in cultured cells and experimental animals. Recent characterizations of Rb-associated proteins and proteins within the Rb family may provide some clues as to how Rb function may be differentially modulated in a cell type-specific manner.

FROM RECESSIVE HYPOTHESIS TO CLONING THE *RB* GENE

The existence of tumor suppressor genes was suggested years ago in experiments that fused normal cells with tumor cells and resulted in suppression of the neoplastic properties of the latter (Harris et al 1969). The typical growth and replication properties of the normal cells were dominant over the tumorigenic characteristics of cancer cells. This observation indicated that true neoplasia might require loss-of-function of a tumor suppressor. The first tumor suppressor system to be studied in humans was that governing the susceptibility to the childhood intraocular tumor, retinoblastoma. Familial cases of this tumor suggested germ line transmission of the susceptibility to retinoblastoma. By comparing the relative incidence and age at diagnosis of unilateral and bilateral retinoblastoma cases, Knudson proposed a "two-hit" hypothesis to explain the genetic data (Knudson 1971). Comings subsequently extended Knudson's hypothesis by proposing that the two mutations inactivate both alleles of a single gene responsible for suppressing tumor formation (Comings 1973). In familial cases, one such mutation was assumed to be transmitted in the germ line; the second stochastically occurred in somatic cells sometime during ontogeny. In sporadic cases, both alleles of a single tumor suppressor gene were predicted to suffer inactivating mutations during somatic differentiation.

Karyotypic examination of chromosomes from patients with hereditary retinoblastoma mapped the putative retinoblastoma susceptibility gene to chromosome 13q14 (Francke 1976; Strong et al 1981). When compared with normal somatic cells from the same patients, the majority of retinoblastoma cells have lost heterozygosity for one or more chromosome 13q14 markers. If one chromosome contains a germ line mutation, it is almost always the other normal chromosome that is lost in tumor cells (Cavenee et al 1983). Using a positional chromosome walking technique, three laboratories succeeded in cloning and characterizing the retinoblastoma gene (*rb-1*) (Friend et al 1986; Fung et al 1987; Lee W-H et al 1987a). The most compelling indication that the correct genetic locus has been cloned came from molecular studies comparing diseased and normal cells. The cloned *rb* gene was found to be mutated or deleted from all retinoblastoma samples tested. Normal tissues or other childhood brain tumors not associated with the retinoblastoma gene contained two normal copies of the encoding gene (Lee W-H 1987a). Subsequently, it was shown that inactivation of the *rb* gene also occurs with variable incidence in some other common tumors (reviewed by Bookstein & Lee 1991). Thus the significance of the *rb* gene and its product was broadened to include tumors other than retinoblastoma.

Additional evidence that *rb* is indeed a tumor suppressor or anti-oncogene comes from data demonstrating that protein Rb can be inactivated by certain oncogene products. The transforming viral proteins SV40 large T antigen, adenovirus E1A, and human papillomavirus E7 have been shown to associate tightly with Rb and inactivate its growth suppressive functions (Chellappan et al 1992; DeCaprio et al 1988; Dyson et al 1989b; Whyte et al 1988). In the normal cell cycle, activated Rb in G_1 phase serves as a gatekeeper to restrict access to subsequent phases of the cycle that commit the cell to proliferation. When Rb is complexed with an oncogene product, on the other hand, constraints to proliferation are removed and the unscheduled cellular proliferation characteristic of neoplasia can ensue.

RB GENE ORGANIZATION AND RB PROTEIN STRUCTURE

The rb Gene

The full-length human *rb-1* cDNA spans 4757 nucleotides. It has an open reading frame that encodes a protein of 928 amino acids, predicting a molecular mass of 106 kd. Using *rb* cDNA as a probe, the homologous human genomic DNA has been cloned and characterized (Hong et al 1989). The *rb* cDNAs from mouse and frog have also been cloned, and their predicted amino acid sequences are highly homologous to the human *rb* cDNA, especially within key regions that encode sites for binding of several important regulatory proteins (Bernards et al 1989; Destree et al 1992). RNA analysis of mouse and human tissue has revealed a 4.7 kb (4.2 kb in *Xenopus*) *rb* transcript in all tissues examined. Additional mRNA transcripts have been found in germ cells and in some embryonic tissues. For example, a 2.8 kb transcript is expressed in the testes of mice as spermatids mature (Bernards et al 1989).

The human *rb* transcription unit spans 200 kb and includes 27 separate exons, ranging from 31 to 1889 nucleotides. The largest intron is more than 60 kb in length and the smallest consists of only 80 base pairs. Exon 1 contains 5′ untranslated sequences, encodes the first methionine, and probably represents the most 5′ end of the mRNA. A 70 base pair promoter region located immediately upstream from the first exon is capable of activating transcription when fused to a heterologous gene (Hong et al 1989). The promoter is G+C-rich and has consensus sequences for the transcription factors E2F-1, ATF, and Sp-1 (Chellappan et al 1991; Kim et al 1992; Udvadia et al 1993). The functional importance of these regulating DNA sequences was demonstrated by the identification, in tumors, of inactive *rb* alleles bearing either deletions or point mutations within the promoter (Bookstein et al 1990a; Sakai et al 1992).

Rb Protein

The human *rb* gene product was first identified by antibodies raised against a TrypE-Rb fusion protein expressed in *E. coli* (Lee W-H 1987a). Antibodies recognized a protein, termed $p110^{Rb}$ (according to the nomenclature of Oncogene Products), that migrates as multiple closely-spaced bands between M_r 110 to 116 K when sized on denaturing polyacrylamide gels. The protein is absent from all the retinoblastoma lines examined thus far, which confirms that absence of Rb is important in the pathogenesis of the tumors (Lee W-H et al 1987a). Rb is a nuclear phosphoprotein that binds to double-stranded DNA nonspecifically (Lee W-H et al 1987b). The DNA-binding activity is intrinsic to a carboxyl-terminal region, consisting of about 300 amino acids (Wang et al 1990), which also contains a bipartite nuclear localization signal (Shew et al 1990; Zacksenhaus et al 1993).

Distinct domains of the Rb protein have been found to be important for Rb's biologic functions (Figure 1). Partial proteolytic digestion of purified human Rb protein has revealed four protease-resistant domains (Hensey et al 1994). Consistent with these results, computer-assisted analysis of the amino acid sequence of Rb protein predicts that its tertiary structure should contain several globular domains and a hydrophilic tail region. Two of the globular domains that bind viral oncoproteins have been mapped by deletion mutagenesis to the

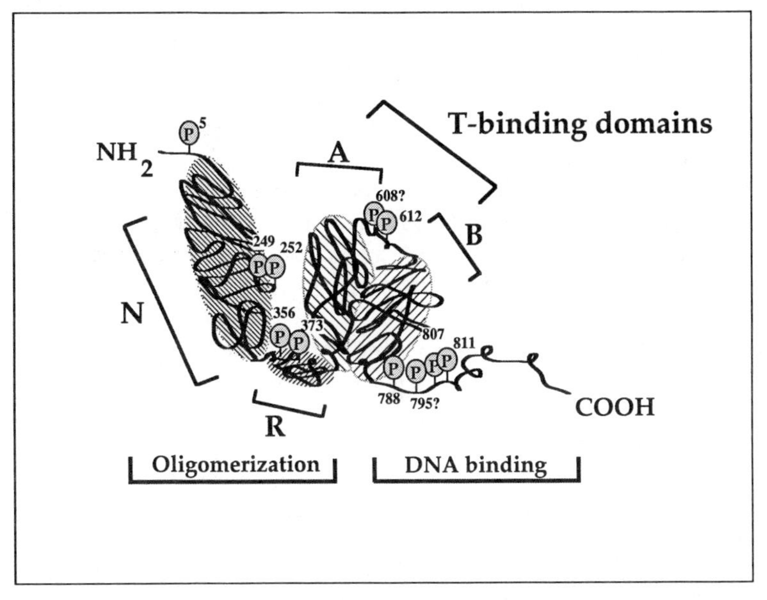

Figure 1 Schematic representation of the structure of the human Rb protein. Rb protein is organized into at least three discrete structural domains (*shaded areas*). The N-terminal domain (N) is important for oligomerization in vitro (Hensey et al 1994). Two other domains (A and B) are frequently altered in tumors and are required for the binding of E2F (Hu et al 1990; Huang et al 1990; Chittenden et al 1991) and several DNA tumor virus oncoproteins including SV40 large T antigen (Chellappan et al 1992). Nonspecific DNA binding is intrinsic to the carboxy-terminal portion of the protein (Wang et al 1990). Potential Cdc2 phosphorylation sites are marked P, as are the amino-acid numbers of the corresponding serine or threonine residues (Lin et al 1991; Lees et al 1991).

C-terminus (Hu et al 1990; Huang et al 1990; Y Qian et al 1992). These regions are critical for several of the activities of Rb, which include phosphorylation during G_1 phase of the cell cycle (Buchkovich et al 1989; Chen et al 1989), binding to transcription factors (Nevins 1992; Y Qian et al 1992), tethering to nuclear structures (Mittnacht & Weinberg 1991; Templeton, 1992), and growth suppression in cell culture (Goodrich et al 1991; X-Q Qian et al 1992). Mutations within these domains are precisely those found most frequently in human tumors (Hu et al 1990; Huang et al 1990).

Although the two N-terminal globular domains have no known biologic function, interesting clues to their potential function have emerged. For example, when intact Rb protein is sized on a nondenaturing electrophoretic gel, it migrates in the form of oligomers, a property not shared by $p56^{Rb}$, a derivative truncated at the N-terminus (Hensey et al 1994). Electron microscopy of purified $p110^{Rb}$ shows filamentous structures in vitro, which lends further support to the idea that $p110^{Rb}$ can form higher ordered structures (Hensey et al 1994). The structures appear to result from association between the N- and C-termini of the full-length Rb protein. The ability of Rb to form complex structures, when combined with its propensity to bind to nuclear matrix proteins (see below; Mancini et al 1994; Shan et al 1992) and several nuclear growth-promoting proteins (Chellappan et al 1991; Rustgi et al 1991), suggests a potential mode by which Rb functions in regulating cellular activity. Rb may keep growth-promoting proteins "corralled" into a nuclear subcompartment until signals are appropriate for such proteins to be released so that the cell can progress beyond G_1 (Lee et al 1991).

REGULATION OF RB DURING CELL CYCLE PROGRESSION AND DIFFERENTIATION

Abnormal proliferation of cells in cancer and other pathologic states is considered to be the consequence of deregulated progression through certain cell cycle checkpoints. Such checkpoints restrict progression as the cells cycle through phases of growth, DNA synthesis, and division. They might also regulate exit from the cell cycle and commitment to terminal differentiation (Marx 1994; Pardee 1989). In this framework, inactivation of proteins involved in the checkpoints, or activating them at inappropriate times, can lead to a number of problems. Inappropriate passage of cells normally halted at the G_1/S boundary might cause them to miss precisely timed signals to exit the cell cycle for differentiation or death. Alternatively, the cells might synthesize new DNA in S phase, from faulty templates, before scheduled DNA repair could be completed in G_1.

Overexpression of Rb in early G_1, for example, results in reversible G_1 arrest (Goodrich et al 1991; Hinds et al 1992; see more below). Since Rb protein is constitutively expressed in normal cells, has a half-life of at least 12 hr (Chen et al 1989), and is present in all mammalian cells tested to date (Lee W-H et al 1987b), regulation of Rb must be accomplished post-translationally. Rather than changes in expression and degradation leading to the regulation of Rb during phases of the cell cycle, existing Rb must be activated and deactivated at appropriate times. Work determining how Rb protein is turned "on" or "off," and how it is directed to its molecular sites of action, has focused on three general mechanisms: phosphorylation, compartmentation, and association with other cellular proteins whose functions are known or under investigation.

Modification of Rb by Phosphorylation

Phosphorylation and dephosphorylation of cellular proteins are recognized as important regulatory mechanisms controlling a variety of cellular events (Hunter 1987). The *rb* gene product is clearly a phosphoprotein as revealed by biochemical studies of Rb prepared from cells at different stages of the cell cycle (Chen et al 1989; Lee W-H 1987b). At least five distinct electrophoretic bands migrating from M_r 110 to 116 kd have been shown to correspond to different phosphorylated forms of $p110^{Rb}$ (Ludlow et al 1990; Shew et al 1989).

The phosphorylation status of the Rb protein oscillates regularly during the cell cycle (Buchkovich et al 1989; Chen et al 1989; DeCaprio et al 1989, 1992). Hypophosphorylated forms predominate in G_0 and G_1, while more highly phosphorylated forms exist in S, G_2, and M phases. Phosphopeptide mapping has shown that Rb protein is phosphorylated on serines and threonines in at least three distinct stages: in mid G_1, S, and near the G_2/M transition (Chen et al 1989; DeCaprio et al 1989, 1992; Furukawa et al 1990). Pulse-chase experiments were used to show that the phosphopeptide maps of phosphorylated $p110^{Rb}$ vary in different stages of the cell cycle by the addition or subtraction of phosphate residues. Specific susceptible sites are phosphorylated sequentially starting at G_1. Dephosphorylation begins during anaphase and also continues stepwise until completion in the ensuing G_1 phase (Ludlow et al 1993).

The primary biological function of underphosphorylated Rb is believed to mediate growth inhibition in G_0 and G_1. Phosphorylation in mid G_1 is thought to inactivate Rb, thereby overcoming its growth suppression activity. This phosphorylation allows progression past G_1 and commitment to DNA synthesis in the subsequent stage of the cell cycle. Such reasoning fits with observations that the transforming proteins of tumor viruses bind only to the hypophosphorylated form of Rb. Viral oncoproteins, including SV40 large T antigen, adenovirus E1A, and papillomavirus 16 E7, appear to function in a manner that mimics phosphorylation of Rb: they inactivate Rb at G_1 and allow cell cycle progression into S phase and beyond (Chellappan et al 1992; DeCaprio et al 1988; Dyson et al 1989a,b; Ludlow et al 1990; Whyte et al 1988).

G_1 Arrest by Overexpression of Rb

Direct evidence that Rb is involved in cell cycle regulation stems from single cell microinjection experiments. Injection into early G_1 cells of excess, purified, unphosphorylated Rb protein (either full-length $p110^{Rb}$ or $p56^{Rb}$) inhibits progression into S phase (Goodrich et al 1991). The reversible G_1 arrest is seen in normal monkey kidney CV-1 cells and in human osteosarcoma Saos-2 cells expressing inactivated Rb protein. Injection of similar amounts of Rb into cells arrested in late G_1 or early S phase has no effect on DNA synthesis. These observations reveal a specific restriction point in early G_1; passage through this point commits cells to DNA synthesis and cell division. Similar experiments, using transfection of *rb* gene constructs into cycling Saos-2 cells rather than microinjection of purified Rb protein, also demonstrate G_1 arrest (Hinds et al 1992).

Possible Roles for Rb in Other Phases of the Cell Cycle

Current hypotheses hold that only the unphosphorylated form of Rb, present in G_0 and G_1, is active. This conclusion is based on data showing G_1 arrest and on the binding of growth-regulating proteins specifically with only hypophosphorylated Rb. The stepwise and sequential phosphorylation of Rb from G_1 to S and G_2, however, suggests that Rb may act at other phases in addition to G_0 and G_1. The first series of phosphorylations might not be a master off switch for all the functions of Rb. A recent report demonstrated cell cycle arrest at G_2 by overproduction of Rb protein during S phase of the cell cycle. This study used a rapid, temperature-sensitive gene amplification system to overexpress murine Rb in monkey kidney BTS-1 cells specifically during S phase (Karantza et al 1993).

Identification of proteins that interact specifically with hyperphosphorylated Rb may help advance the idea that Rb functions at several stages of the cell cycle. Ongoing investigations have begun to reveal roles for Rb in M phase, based on the tight association of Rb with other proteins that are part of the mitotic apparatus (X Zhu, M Mancini, W-H Lee, unpublished data).

Interaction of Rb with Cyclin/Cdk Complexes and Protein Phosphatases

Regulators of the Rb protein, particularly kinases, have recently begun to be discovered and dissected. Cdc2 kinase, the crucial gatekeeper of G_1 to S and G_2 to M transitions (Myerson et al 1990; Wittenberg & Reed 1988), may regulate Rb function. Since Rb appears to be involved in restricting one or both of the same cell cycle transitions (Goodrich et al 1991; Hinds et al 1992; Karantza et al 1993), it was logical to consider Cdc2 kinase as a direct regulator of Rb function. Indeed, Cdc2 can efficiently phosphorylate the Rb protein in vitro on many of the same sites normally phosphorylated in vivo (Lee et al 1991; Lees et al 1991; Lin et al 1991). Human Cdc2 has also been shown to interact physically with Rb (Hu et al 1991; Lin et al 1991). However, a direct functional role for Cdc2 in phosphorylating of Rb in vivo has not been demonstrated conclusively.

Cdc2 is only one member of a family of cyclin-dependent kinases (Cdks). Other members of this family consist of one of several catalytic subunits, Cdks, and one of several regulatory subunits, or cyclins. Most Cdks are expressed with relative constancy throughout the cell cycle. Their activity, however, is modulated by cell cycle-specific changes in the concentrations of specific cyclins (Sherr 1993).

Using cotransfection of cyclin and *rb* genes, it has been shown that ectopic expression of cyclins A and E can overcome Rb-mediated phase cell cycle arrest (Hinds et al 1992). These experiments showed that Rb is hyperphosphorylated in cells overexpressing the specific cyclins and that phosphorylation of Rb is essential for cyclin A- and E-mediated rescue of Rb-blocked cells. Complexes of cyclins A and E with Cdk might be able to phosphorylate Rb and inactivate it. Although the G_1 cyclins D1, D2, and D3 can also form complexes with Rb, they do not function equivalently (Dowdy et al 1993; Ewen et al 1993). Cyclin D2-Cdk2 and cyclin D2-Cdk4 complexes, like cyclins A and E, can reverse the G_1 arrest caused by overexpression of Rb alone and lead to Rb phosphorylation (Ewen et al 1993). The sites on Rb that cyclin D2-Cdk4 phosphorylates in vitro are identical to sites phosphorylated in vivo (Kato et al 1993). Cyclins D1 and D3, on the other hand, do not lead to phosphorylation of Rb in cultured cells (Ewen et al 1993). Such observations indicate that their interaction with Rb may not be functionally significant. When interpreting these experiments, it is important to note that the data show only correlations between cyclin/Cdk complexes and Rb. Direct genetic evidence—i.e. creation of specific Rb or Cdk mutations that selectively interfere with Rb-cyclin/Cdk complex formation and/or phosphorylation of crucial residues—remains to be shown.

Dephosphorylation appears to be the primary mode of Rb reactivation immediately prior to G_1 (Ludlow et al 1990). Dephosphorylation of Rb protein may involve protein phosphatase 1 (PP1). The catalytic subunit of PP1 binds to Rb (Durfee et al 1993); moreover, dephosphorylation of Rb in M and G_1 phases by PP1 has been demonstrated by blocking PP1 activity using specific phosphatase inhibitors (Alberts et al 1993; Kim et al 1993; Ludlow et al 1993). It remains to be shown whether other phosphatases may also be involved in Rb dephosphorylation.

Importance of Phosphorylation and Dephosphorylation

The phosphorylation of Rb has specific and important cellular consequences. Hypophosphorylated Rb (*a*) binds well to DNA (Chen et al 1989; Lee W-H et al 1987b; Templeton 1992); (*b*) "tethers" to the nuclear structure when other cellular components are extracted by low salt/detergent solutions (Mittnacht & Weinberg 1991; Templeton 1992); and (*c*) binds specifically to many other nuclear proteins (Lee et al 1991; Shan et al 1992). Phosphorylation of crucial residues on Rb apparently abolishes or weakens the interactions of Rb with other protein structures thereby allowing Rb to dissociate from nuclear complexes and subcompartments at appropriate times during the cell cycle.

Proper phosphorylation of Rb, in turn, depends not only on having critical serine and threonine residues available for phosphorylation, but also on sites for the binding of kinases and phosphatases. Rb point mutations that specifically alter phosphorylation sites or groups of sites will be useful to determine which phosphorylation events are crucial for the associations of Rb with kinases and other proteins at various stages of the cell cycle. A specific point mutation of Rb at residue 706 (Cys to Phe), which prevents phosphorylation and binding to SV40 T antigen (Bignon et al 1990; Kaye et al 1990), helps illustrate the concept that Rb protein conformation is important for determination of protein kinase-binding sites. Since the Cys^{706} is not specifically a phosphorylation site, the Cys to Phe mutation might change the conformation of Rb and prevent it from serving as a substrate for phosphorylation.

Rb Phosphorylation and Cellular Differentiation

A role for the Rb protein in cell differentiation has been suggested from studies of several model systems. For example, in human fibroblasts, terminal withdrawal from the cell cycle during senescence has been associated with the loss of Rb phosphorylation (Stein et al 1990). It has also been shown that treatment of promyelocytic HL-60 or monoblastic U937 leukemia cells with phorbol esters or retinoic acid leads to terminal differentiation that is associated with marked dephosphorylation of Rb protein prior to arrest of cell growth (Chen et al 1989). Differentiation of U937 cells along a monocytic pathway is also associated with increased expression of the transcription factor NF/IL-6, which co-immunoprecipitates with Rb through interaction with its T-binding domains (P-L Chen & W-H Lee, unpublished data).

Muscle differentiation has provided another system for studies of Rb function. The Rb protein has been reported to bind directly to MyoD, a basic helix-loop-helix protein important in skeletal muscle differentiation and cell cycle suppression (Gu et al 1993). Inactivation of Rb was observed to inhibit differentiation of myoblasts to myotubes in culture. Rb inactivation also allowed myotubes, which are thought to be terminally differentiated, to reenter the cell cycle. This study did not account for other proteins that may have growth suppressive functions similar to those of Rb, however, and is inconsistent with the studies of mouse development (Jacks et al 1992; Lee et al 1992), which demonstrated that skeletal muscle develops normally in fetal mice expressing no Rb protein. The functional role of Rb in cell differentiation may vary according to the cell type-specific expression of different Rb-associated proteins as well as proteins that complement the functions of Rb.

TUMOR SUPPRESSION BY RB

In addition to retinoblastoma, inactivating mutations of *rb* have also been found in osteosarcomas, soft tissue sarcomas, leukemias, and small cell lung carcinoma, as well as carcinomas of lung epithelial cells, and breast, esophagus, prostate, and renal cells (reviewed by Bookstein & Lee 1991). Such correlative data strongly suggests that Rb inactivation is involved to a varying degree in tumor formation in a number of cell and tissue types.

Proof of the tumor suppressive properties of the *rb* gene product required supplementing *rb* in tumor cells bearing mutated endogenous genes. Retrovirus-mediated gene transfer of a copy of the wild-type *rb* gene into such tumor cells has been shown to suppress their neoplastic properties (Huang et al 1988). Compared to uninfected retinoblastoma or osteosarcoma cells, which lack normal Rb expression, *rb*-transduced cells were shown to be much less tumorigenic when injected subcutaneously into athymic nude mice. Although the bulk populations of cells expressing wild-type Rb initially grew more slowly than controls, this reduction in growth rate was eliminated when individual clones were selected. The variable effect on cell growth was attributed to different levels of wild-type Rb expression in individual clones. Those clones expressing the highest levels of Rb protein proliferated most slowly and therefore tended to be overgrown in bulk cell populations. Futhermore, in subsequent generations, *rb* gene copies were lost to varying degrees, which allowed eventual reversion to the neoplastic state (Chen et al 1992).

Several reports have now shown that replacement of a wild-type *rb* gene, and expression of wild-type Rb protein, suppresses tumorigenicity in other malignant cell lines (Table 1). Such studies begin to confirm the concept of tumor suppression by the *rb* gene. Thus far, all studies, except one that employed an inducible metallothionein promoter (Muncaster et al 1992), showed some degree of tumor suppression by the *rb* gene; effects on the duration of tumor suppression and on the suppression of cell growth in culture, however, were variable. Tumor suppression has been observed whether tumor cells are injected subcutaneously (Bookstein et al 1990b; Fung et al 1993; Goodrich et al 1992; Huang et al 1988; Sumegi et al 1990; Takahashi et al 1991; Wang et al 1993) or intraocularly (Fung et al 1993; Medraperla et al 1991; Xu et al 1991).

The ultimate test of cancer suppression by the retinoblastoma gene must be done in studies that do not involve xenografts, i.e. in tumors that arise de novo. The first step toward this important test of the two-hit model is to inactivate the endogenous wild-type *rb* gene in animals and score for tumor formation. Three separate groups succeeded in inactivating the *rb* gene in the germ line of mice (Clarke et al 1992; Jacks et al 1992; Lee et al 1992). They found that mice homozygous for the inactivating *rb-1* mutation do not survive. Animals with heterozygous *rb-1* mutations, however, do survive and develop tumors with nearly 100% penetrance (Hu et al 1994). Surprisingly, however, the tumors do not arise in the retina. They rise instead from a distinct cell type, the melanocorticotroph, located within the neuro-intermediate lobe of the pituitary gland (Hu et al 1994). Why is the tissue specificity of tumors arising from Rb-deficiency different in mice than in humans? Perhaps the mouse retina has fewer target cells or a different window of susceptibility to tumorigenesis than the human retina, so that genes in murine retinoblasts are not mutated with sufficient frequency to lead to the retinal tumor phenotype. Alternatively,

Table 1 Summary of studies using the *rb* gene to suppress cell growth and tumorigenicity

Tumor type	Cell line	Cell morphology changes	Growth rate in culture	Growth in soft agar	Tumorigenicity in nude mice	Reference
Retinoblastoma	WERI-27	Yes	Decreased	ND	Abolished	Huang et al 1988
	WERI-27	Yes	Decreased	ND	Reduced	Medraperla et al 1991; Xu et al 1991
	WERI-Rb1	No	Unchanged	Unchanged	Unchanged	Muncaster et al 1992[1]
	Y-79	No	Unchanged	Unchanged	Unchanged	Muncaster et al 1992[1]
Osteosarcoma	Saos-2	Yes	Decreased	ND	Abolished	Huang et al 1988
	(2 subpops.)	No	Unchanged	ND	Abolished	Huang et al 1988
Fibrosarcoma	HT 1080	No	Unchanged	ND	Unchanged	Fung et al 1993[2]
Prostate	DU-145	No	Unchanged	ND	Reduced	Bookstein et al 1990
Bladder	J82	No	Decreased	Decreased	Reduced	Goodrich et al 1992
	HT 1376	No	Unchanged	Unchanged	Reduced	Goodrich et al 1992
	TCC-SUP	No	Unchanged	Unchanged	Slightly reduced	Goodrich et al 1992
	HT 89	No	Decreased	Decreased	Reduced	Takahashi et al 1991
Breast	BT 549	+/−	Unchanged	Decreased	Reduced	Wang et al 1993
	MDA-MB 468	+/−	Unchanged	Decreased	Reduced	Wang et al 1993
	MDA-MB 468S4	No	Unchanged	Unchanged	Unchanged	Muncaster et al 1992[1]

In all but two studies[1,2], reintroduction of a wild-type *rb* gene into tumor cells that lacked it resulted in reduced tumorigenicity, as measured by decreased anchorage-independent growth in soft agar or decreased tumorigenecity in nude mice. Variable effects in cell morphology and cell growth rate in culture have been observed. ND = not done.

pituitary melanocorticotrophs in mice may be stimulated to grow and differentiate by a pathway related to one in human retinoblasts.

By crossing transgenic mice expressing different amounts of human Rb protein (Bignon et al 1993) with mice heterozygous for the endogenous *rb* gene, offspring were obtained that are heterozygous for the endogenous murine *rb* gene, but that also express varying amounts of human Rb encoded by a single *rb* allele (h$rb^{+/-}$, m$rb^{+/-}$) (Chang et al 1993). These mice with the genotype (h$rb^{+/-}$, m$rb^{+/-}$) do not develop tumors. Genetic rescue of the tumor phenotype can be accomplished by expression of relatively small amounts of wild-type human Rb. Only if functional Rb is expressed from more than one allele, however, is the tumor phenotype suppressed. Mice that express even supranormal amounts of human Rb protein are still prone to a tissue-specific cancer if they harbor mouse *rb* null mutations and express all functional Rb from a single human *rb* locus. These results further support the two-hit hypothesis.

To establish the concept of cancer suppression more firmly, the next step is directly to supplement intact tumors with the *rb* gene. The prediction is that *rb* gene replacement should be able to suppress tumor progression, if not completely reverse it. However, stable expression of the *rb* gene in 100% of the cells in intact tumors is more difficult than germ line transmission to prevent tumor formation. Trials of direct tumor suppression in vivo are ongoing, using viral and liposome vectors to deliver the *rb* gene. The most promising results will probably be obtained by direct injection of cells engineered continually to produce a retrovirus carrying the *rb* gene. The method takes advantage of the incorporation of retroviral DNA into rapidly proliferating tumor cells in preference to surrounding normal tissue and the relative isolation of the brain from some immune responses. A similar approach has been used to introduce a gene encoding susceptibility to an antiviral agent into glioma cells and successfully treat experimental brain tumors (Culver et al 1992). Although *rb* inactivation is undoubtedly not the only mutation leading to the pituitary tumors in mice with heterozygous germ line mutations of *rb-1*, it is reasonable to suggest that the role of Rb in tumor growth is fundamental and that restoration of normal Rb expression should suppress tumor growth in vivo to some degree.

RB AND DEVELOPMENT

Tumor suppression is not the only role of the *rb* gene. The *rb* gene product is also vital in development, but only in some cells at critical times. In *Xenopus, rb* mRNA is detected in the oocyte and throughout embryogenesis. In adult frogs, *rb* mRNA expression is greatest in the ovary and testis. Rb protein can be identified in the embryo shortly after fertilization of the ovum and prior to the mid-blastula stage; protein levels continue to increase thereafter. Although cells in *Xenopus* embryos, before the mid-blastula stage, do not have a G_1 phase, Rb protein is still clearly expressed. It is possible that Rb is present at levels lower than a critical threshold or sequestered from the nucleus such that it is not functional before the mid-blastula transition (Destree et al 1992).

In mice, *rb* mRNA is found at embryonic day 9.5, the earliest stage studied, and becomes more abundant in many tissues thereafter. The mRNA is present in greatest amounts in the brain and liver, where neurons and hematopoietic cells are differentiating (Bernards et al 1989). Rb protein, predominantly in its hyperphosphorylated form, is present by embryonic day 10.5. Only later in development is hypophosphorylated Rb found in significant amounts. Immunofluorescence staining of immunodeficient mouse fetuses has shown highest Rb protein expression in differentiating cells of the retina and other specialized tissues (Szekely et al 1992).

The consequences of inactivation of both copies of *rb* in the mouse are dire: death occurs in utero by embryonic day 16 and defects occur in the ability of cells in the nervous and hematopoietic systems to differentiate correctly (Clarke et al 1992; Jacks et al 1992; Lee et al 1992). Although embryonic lethality is caused by inactivation of Rb expression, the late timing of the fetal demise is surprising. Before the 16th day of gestation, the embryo and many of its organs are already formed. The notion that Rb must play an essential role in all cells is therefore untenable. Either Rb is not necessary or there are redundant mechanisms to substitute for functions of Rb in early cell proliferation and differentiation.

Rb-deficient Mouse Phenotype

Most major brain regions in Rb-deficient mouse embryos begin to develop normally, but cell death occurs throughout the central nervous system as early as embryonic day 11.5. The highest concentrations of apoptotic cells are found in the hindbrain, spinal cord, trigeminal ganglia, and dorsal root ganglia. Cell death is evident in the intermediate zone of the developing hindbrain and spinal cord, whereas the ventricular zone, where neurons and glia are first generated, is relatively normal. In addition to abnormal apoptosis, abnormally high mitotic indices occur in the intermediate zone, outside of the ventricular zone to which neuronal mitoses are usually confined (Lee et al 1992).

The timing of aberrant mitoses correlates well with the developmental schedule. Trigeminal ganglia are normally the first ganglia to differentiate in mouse embryos. During normal development, programmed cell death starts at embryonic day 11.5. By embryonic day 14.5, cells that express markers for neuronal differentiation begin to appear. In the trigeminal ganglia of Rb-deficient ($rb^{-/-}$) embryos, however, many ectopic mitoses are evident at embryonic day 11.5. These surviving cells differentiate abnormally; they do not form

Nissel bodies and neuron-specific tubulin is not expressed in them (E Lee et al, unpublished data).

Abnormalities in erythropoiesis in mouse embryos are found at about the same time as abnormalities in CNS development. The blood-forming liver in the mouse during mid and late gestation is undersized and nearly devoid of hepatocytes. In wild-type embryos at this stage, peripheral blood contains mostly enucleated erythrocytes. Blood from homozygous mutant embryos, in contrast, contains mostly immature, nucleated erythrocytes (Clarke et al 1992; Jacks et al 1992; Lee et al 1992). The abnormal erythropoiesis can be explained by postulating deregulated proliferation of immature cells or failure of erythroid precursors to differentiate in a proper and timely manner. Alternatively, it may be a deficiency in hepatocytes that creates a local environment to block proper differentiation of blood cells.

Since Rb-deficient embryos die before most hematopoietic lineages are present, other approaches have been applied to address the effect of Rb on the differentiation of specific blood cells. One approach monitors the effects of transplanted cells from Rb-deficient mouse embryos in normal recipient mice. In the recipients, donor erythrocytes never become post-mitotic, and the animals develop extensive extramedullary erythropoiesis (N Hu et al, unpublished results). Another approach uses Rb-deficient embryonic stem cells to complement blastocytes from mice deficient in other genes necessary for the differentiation of specific blood cell lineages. Results from these approaches suggest that Rb is not required for lymphocyte differentiation (Chen et al 1993), but may be required, directly or indirectly, to support erythrocyte differentiation.

The abnormal neuronal phenotype seems to show that Rb is necessary for slowing or stopping cell proliferation and for controlling terminal differentiation in specific cells. Similar roles for Rb in erythroid precursors are suggested by transplant studies. Why are only certain neurons and, perhaps, immature erythrocytes affected? The simplistic answer is that these cells are only the first to be affected in development. The dysfunction they cause in the organism could lead to fetal demise before abnormalities might eventually be seen in other cells and tissues. Further studies using tissue-specific promoters selectively to inactivate *rb* may help determine whether Rb protein has roles in other tissues in later stages of development.

Rb Dosage Determines Whether Mice Survive to Postnatal Stages

The dosage of Rb is crucial in allowing mice to develop past certain embryonic stages. Embryos heterozygous for the mutant *rb* gene express 50–60% of the amount of Rb protein that is expressed in wild-type littermates, and their development to adulthood is grossly normal (Chang et al 1993). By crossing these heterozygous mice (murine $rb^{+/-}$) with transgenic mice expressing varying amounts of human Rb protein from different copy numbers of the *rb* gene, it was discovered that amounts of Rb protein ≥ 50% of normal are necessary to rescue the lethal embryonic phenotype. Unlike rescue of the pituitary tumor phenotype in heterozygous mice by germ line transmission of a human *rb* transgene, rescue of the embryonic lethal mutant phenotype depends more on the level of Rb expression than on the number of alleles from which Rb protein is expressed (Chang et al 1993).

Overexpression of Rb in Germ Cells Results in Dwarf Mice

The dosage of Rb protein appears to have another important effect on mouse development: control of the ultimate size of the animal. Overexpression of Rb protein by germ line transmission of a human *rb* minitransgene results in growth retardation starting at mid-embryonic stages (Bignon et al 1993). The greater the amount of Rb protein expressed, up to a critical maximum, the smaller the animals become in relation to their nontransgenic littermates. Assuming that human Rb and murine Rb function equivalently in mice, expression of about two and one-half times the normal amount of Rb results in mice that are about 60% as large as wild-type littermates. Almost all structures in the dwarf mice are normal, simply smaller proportionally compared to wild-type littermates. Expression of total amounts of Rb protein greater than about two and one-half times normal in live adult animals is not seen. Animals expressing greater amounts are grossly runted and die in late embryonic or neonatal stages. Thus appropriately measured threshold amounts of Rb protein appear to be necessary for embryonic development. Minimal amounts may be required for differentiation of crucial tissues, whereas supramaximal amounts may prevent a sufficient number of cell divisions within the time of embryogenesis. Table 2 summarizes the effects of Rb dosage in mice.

The dwarf mouse phenotype resulting from Rb overexpression, in which mice do not exhibit catch-up growth later in life, resembles the phenotype of mice that lack functional insulin-like growth factor I (IGF-I) or IGF-I receptors (Baker et al 1993). During the period required for completion of embryonic development, proliferative events occurring during ontogeny generate fewer cells than in wild-type littermates (Baker et al 1993). Proliferative events correlate with body mass during embryogenesis (Enesco & LeBond 1962). Thus the dwarf phenotype observed in mice that either overexpress Rb or lack IGF-I probably results from retarded mitotic proliferation during embryogenesis. Since IGF-I signaling via IGF-I receptors causes positive regulation of the progression from G_1 to S phase of the cell cycle (Pardee 1989), and since Rb overexpression causes just the opposite effect at the same transition, it is not surprising that the phenotypes of mice lacking IGF-I and those expressing excessive amounts of Rb are similar. Whether Rb is a downstream mediator of IGF-I signals remains to be seen.

Table 2 Summary of the effects of RB dosage on mouse phenotypes

Genotype		Rb protein amount relative to normal (times)	Phenotype	
m*rb*	h*rb*		Development	Cancer
+/+	Rb3 + Rb1	~2.5	Super-mini	No
+/+	Rb3		Dwarf	No
+/+	Rb1-(4)		Dwarf	No
+/−	Rb3		Dwarf	No
+/−	Rb1-(4)	2	Dwarf	No
−/−	Rb3		Dwarf	Yes
−/−	Rb1-(4)		Dwarf	Yes
+/+	Rb4		Dwarf	No
+/+	Rb2		Dwarf	No
+/+	Rb1-(1)		Dwarf	No
+/+	−/−	1	Normal	No
+/−	Rb4		Normal	No
+/−	Rb2		Normal	No
+/−	Rb1-(1)		Normal	No
+/−	−/−	0.5	Larger	Yes
−/−	Rb4		Death in utero with defects in	—
−/−	Rb2		neurogenesis and hemato-	—
−/−	Rb1-(1)		poiesis	—
−/−	−/−	0		—

Mice expressing human Rb protein from a human *rb* minitransgene construct (Bignon et al 1993) were crossed to obtain offspring that express different amounts of Rb protein from endogenous mouse (m*rb*) and exogenous human (h*rb*) genes. Wild-type mice (genotype $mrb^{+/+}$, $hrb^{-/-}$) represent the normal amount of Rb protein. The amount of total Rb protein expressed in other individual mouse lines varies from zero to two-and-a-half times normal. Rb1, Rb2, Rb3, and Rb4 refer to mice generated from four different founder transgenic mice, each of which carries a different number of copies of human *rb* cDNA and expresses a different amount of human Rb protein. Note that mice in the Rb1-(1) and Rb1-(4) lines contain one or four copies of human *rb* cDNA, respectively, and express different amounts of human Rb protein from transgenes integrated into the mouse genome at identical sites. Increasing levels of total Rb expression during development, up to a maximum of about two-and-a-half times normal, results in increasing degrees of dwarfism. Cancer suppression, in contrast, seems to depend more directly on expression of Rb from more than one allele (Chang et al 1993).

INTERACTION OF RB WITH ITS ASSOCIATED PROTEINS

One of the possible ways by which the effects of Rb protein are restricted with respect to tumorigenesis and embryonic development is through the interaction of Rb with other proteins that are expressed in a tissue- or time-dependent manner. Studies have demonstrated that exogenous oncoproteins form complexes with Rb (Chellappan et al 1992). These oncoproteins bind to the same regions of Rb protein that are often mutated in spontaneous tumors (Hu et al 1990; Huang et al 1990). A logical and prevailing theory for explaining how these oncoproteins function to transform cells is by binding to and inactivating Rb or similar growth-restraining proteins, in effect mimicking inactivating mutations of a ubiquitous tumor suppressor gene (Whyte et al 1988).

The demonstration of direct links between positively acting oncogenes and a negatively acting tumor suppressor started the effort to find cellular proteins that have molecular functions similar to those of exogenous oncoproteins. Several methods—passing cell extracts over an Rb-affinity column (Kaelin et al 1991; Lee et al 1991), screening lambda expression libraries using purified Rb protein as a probe (Defeo-Jones et al 1991; Shan et al 1992), and modifications of the yeast two-hybrid system (Durfee et al 1993; Fields & Song 1989)—have been used to identify more than thirty separate cellular proteins that bind to Rb.

The proteins characterized thus far represent a diverse group including transcription factors (Fattaey et al 1993; Helin et al 1992; Kim et al 1992; Shan et al 1992), growth regulators (Rustgi et al 1991; Qian et al 1993), protein kinases (Hu et al 1992; Kato et al 1993; Lin et al 1991), protein phosphatases (Durfee et al 1993), and nuclear matrix proteins (Mancini et al 1994). Novel Rb-associated proteins apparently involved in chromosome segregation during M phase are still being characterized (X Zhu & W-H Lee, unpublished data).

Transcription Factors

The best characterized cellular Rb-associated protein is the transcription factor E2F-1 (Bandara & LaThangue 1991; Chellappan et al 1991; Chittenden et al 1991). E2F binds to the same region of Rb as do the DNA tumor virus oncoproteins, and E2F-Rb complexes can be dissociated by these oncoproteins (Chittenden et al 1991). The E2F-Rb interaction has been used as a paradigm to demonstrate how Rb restrains cell cycle progression (Nevins 1992). E2F has been shown to activate transcription of the dihydrofolate reductase (DHFR) gene, which is active during the DNA synthetic phase of the cell cycle (Blake & Azizkhan 1989). The importance of Rb in this activation has been demonstrated by showing that DHFR expression is upregulated in mouse embryonic fibroblasts that express no Rb, but normal in similar cells that express functional Rb and are otherwise isogenic (G Wahl, personal communication).

Since there are E2F recognition sites in the promoters of several other growth-promoting genes, it is suspected that E2F can also transactivate these genes, which include c-*myc*, N-*myc*, DNA polymerase α, and thymidilate kinase (Dou et al 1992; Nevins 1992; Pearson et al 1991; Rustgi et al 1991). A recent report suggests that E2F also regulates *rb* gene expression in vivo

(Shan et al 1994). The *rb* gene promoter contains an E2F-1 recognition site, and overexpression of Rb suppresses E2F-1-mediated activation of *rb* transcription. E2F-1 thus participates in a feedback loop for regulation of Rb expression.

From these findings it appears that one function of Rb is to block transcriptional activation by E2F and other transcription factors that play important roles in the G_1/S transition (Y Chen & W-H Lee, unpublished data). When E2F and other transcription factors are free, they can activate transcription of genes required for S phase and beyond. When these transcription factors are complexed with Rb and inactivated, transcription is inhibited and cell cycle progression is blocked.

Under other circumstances, Rb might act as a positive regulator of transcription. An example of positive regulation of a transcription factor by Rb involves the transcription of transforming growth factor beta (TGF-β). TGF-β isoforms 1 and 2 inhibit growth of many cell types by blocking cell cycle progression from G_1 to S phase, similar to the effect of Rb overexpression (Laiho et al 1990). Rb remains unphosphorylated during TGF-β-induced growth suppression and may function to regulate TGF-β expression by an autocrine or paracrine loop. In mink lung epithelial cells, Rb expression has been reported to activate TGF-β2 gene expression, apparently through binding the transcription factor ATF-2, which itself binds to a high-affinity promoter element on the TGF-β2 gene (Kim et al 1992). This study suggests that Rb might constrain cell proliferation in part by activating expression of the inhibitory growth factor TGF-β2. If so, TGF-β2, when synthesized and secreted, could have a similar effect on neighboring cells, causing G_1 arrest. This explanation does not apply to all cells; many Rb-overexpressing cells do not interfere with the growth of neighboring normal cells in culture.

Intracellular Mitogens and Signaling Proteins

Rb protein also associates, at least in vitro, with the cellular growth promoters c-myc and N-myc (Rustgi et al 1991). These proteins, when activated, are oncogenic and therefore normally can function to antagonize Rb in some cells. Indeed, co-injection of c-myc with Rb has been shown to abrogate the G_1 phase arrest induced by Rb protein (Goodrich & Lee 1992). C-myc has not been shown to associate with Rb in vivo, however, so the ability of c-myc to antagonize Rb may be indirect.

Rb may also associate with intracellular signaling proteins. An Rb-associated protein, p48, which is a human homologue of yeast MSI1, was recently identified (Qian et al 1993). MSI1 is a putative negative regulator of Ras in yeast. Although functional data implicating p48 in a mammalian cell Ras pathway have not yet emerged, a possible antagonistic interaction between Rb and Ras in human cells is being pursued.

The Nuclear Matrix

The viral oncoproteins that bind Rb (adenovirus E1A, SV40 T antigen, papillomavirus E7) are known to associate with the nuclear matrix (Chatterjee & Flint 1986; Deppert & von der Weth 1990; Greenfield et al 1991). This association is similar to the interaction of Rb protein with components of the nuclear matrix (Mancini et al 1994; Shan et al 1992). The insoluble matrix is a predictable player in the regulation of cell growth and metabolism. The matrix is a chromatin-free complex that provides structure to the nucleus. It can serve as a docking site for protein complexes at critical sites of transcription, replication, and RNA processing. These sites have been characterized as complex assemblies (Wan et al 1994) and are known to contain multiple components of machinery for RNA processing and DNA replication (Hozak et al 1993). Using immunolabeling with confocal microscopy and electron microscopy, a specific association in G_1 phase between Rb and the nuclear matrix has been shown (Mancini et al 1994). Rb localizes not only within the assemblies, but also at the nuclear lamina and the nucleolar remnant. The latter locations are consistent with the identification of lamin A/C and UBF, a ribosomal transcription factor, as Rb-associated proteins (Mancini et al 1994; Shan et al 1992). In cells expressing mutant Rb that does not bind T antigen, there is no association between Rb and the nuclear matrix.

Another nuclear matrix protein, p84 (according to its sizing by SDS-PAGE), associates specifically with the N-terminal portion of hypophosphorylated Rb during G_1 phase of the cell cycle. The p84 protein is novel and still being characterized. It is constitutively expressed and colocalizes with splicing centers by confocal immunofluorescence (Durfee et al 1994).

The Mitotic Apparatus

Two other Rb-associated proteins seem to have their major functions during mitosis. One 90-kd Rb-associated protein has 60% sequence identity with the nuc2 protein of fission yeast and the bimA protein of *Aspergillus* (P-L Chen & W-H Lee, unpublished data), proteins that are apparently important in metaphase spindle elongation (Hirano et al 1988; O'Donnell et al 1991). The human protein has therefore been named H-nuc (for human Nuc).

A second Rb-associated protein, mitosin, is a 350-kd phosphoprotein expressed during S and M phases of the cell cycle, but not in G_1 phase. It is phosphorylated from G_2 through M phases and degraded at the end of mitosis. During mitosis, it colocalizes with centromeres/kinetichores. Preliminary studies in monkey CV1 cells indicate that overexpression of mitosin results in delayed exit from G_2/M (X Zhu & W-H Lee, unpublished results).

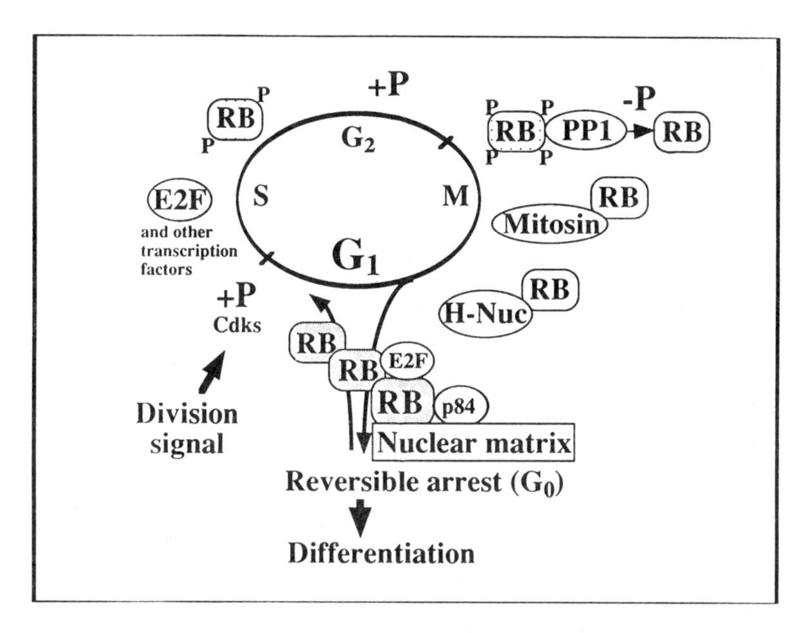

Figure 2 Rb and its associated proteins regulate progression through the cell cycle. Unphosphorylated Rb protein (RB) is active in G_1 phase of the cell cycle. In this unphosphorylated form, it binds to nuclear matrix-associated proteins, including p84, as well as to E2F and other transcription factors. Rb may also form oligomers that corral growth-promoting proteins. Phosphorylation of Rb in mid G_1, possibly by specific Cdk/cyclin complexes, leads to release of E2F and other transcription factors. The cell is then committed to progression through the rest of the cell cycle, during which Rb is further phosphorylated in S and G_2. In M phase, a portion of Rb that remains unphosphorylated associates with two novel proteins, mitosin and H-nuc. Hyperphosphorylated Rb associates with protein phosphatase 1 (PP1), and dephosphorylation leads to reactivation of Rb in G_1. When signals are appropriate, the cell can exit the cell cycle, either temporarily or terminally.

The localization of PP1 specifically to chromosomes during mitosis (Fernandez et al 1992) raises the possibility that dephosphorylation of Rb may occur regionally to facilitate interaction with mitotic proteins. The significance of the interactions of Rb with protein phosphatase 1, H-nuc, and mitosin during mitosis, when most of the Rb is highly phosphorylated, will require further study. Such exploration will be important in discovering functions for Rb other than growth arrest and regulation of transcription factors in G_1. A subfraction of Rb may remain unphosphorylated during mitosis, and it may be this subfraction that binds mitosin and other mitotic proteins.

Figure 2 is a schematic diagram showing some of the modifications of Rb and its proposed interactions with other proteins at different stages of the cell cycle. The array of Rb-associated proteins, from transcription factors being regulated in G_1, to kinetichore proteins involved in the mitotic apparatus, adds another layer of complexity to the functions of Rb during the cell cycle. If the interactions between Rb and some of its associated proteins prove to be significant in vivo, the former dogma that states only the hypophosphorylated form of Rb is active may need to be reevaluated. Paradoxically, only by adding to the complexity of the interactions of Rb with its associated proteins can we productively continue to understand the functions of Rb in growth suppression, differentiation and, perhaps, even in mitosis. The task is daunting, and it is only just beginning.

RB-RELATED PROTEINS AND OTHER TUMOR SUPPRESSORS

The search for activities of Rb protein in different cells at different times has been rewarded by another layer of complexity, i.e. the discovery a family of proteins that share structural similarity with $p110^{Rb}$. Two other proteins, p107 and p130, have been isolated by their interaction with adenovirus E1A (Harlow et al 1986). A similar region of E1A is required for binding with all three of these proteins, which suggests a structural similarity between them. Amino acid sequence comparisons show a high degree of identity (53%) between p107 and p130 and a significant degree of identity (30%) between p107 or p130 and $p110^{Rb}$ (Ewen et al 1991; Mayol et al 1993). The sequence similarity shared among the three proteins is greatest in the regions that correspond to the SV40 T antigen-binding domains in Rb. Based on these similarities, the three proteins are considered to be members of an Rb family and to have some overlapping biologic functions. Indeed, p107 has been shown to share some properties with Rb. Both proteins inhibit E2F-mediated transcriptional activation and both can inhibit progression through G_1 (Zhu et al 1993). However, the biologic functions of p107 certainly are not identical to those of Rb. Transient growth arrest by Rb and p107 can be rescued differentially by various regulators of the cell cycle (E1A, E2F-1, and cyclins A and E for Rb; only E1A and E2F-1 for p107). Furthermore, growth arrest in cervical carcinoma cells was achieved by p107, but not by Rb. To date, no mutation of p107 has been found in human tumors. The precise role of p107 in growth and tumor suppression therefore remains to be determined.

No functional comparisons of p130 with Rb have been reported yet, but p130 associates with E2F and some of the same cyclins and kinases as Rb (Cobrinik et al 1993; Hannon et al 1993; Li et al 1993). Allelic loss of the region of chromosome 16 to which p130 maps is seen in several common tumors, and inactivating mutations of p130 may eventually be found to be essential in some of them (Hannon et al 1993; Li et al 1993).

It is attractive to think that p107 and p130 are similar enough to Rb that the three proteins have many functions that are redundant in vivo. Such a theory would help to explain the tissue specificity seen in tumor predisposition caused by inactivation of the *rb* gene. Patients with heterozygous germ line inactivation of the *rb-1* gene, for example, are predisposed to tumors of the retina and soft tissues, but not other tumors with any great incidence (Abramson et al 1984; Horowitz et al 1990). Perhaps p107 and p130 substitute for Rb and act as tumor suppressors in other cells. The differences in the respective protein structures could allow differential association with other crucial proteins. Studies comparing Rb, p107, and p130 will undoubtedly discover more differences in their interactions with growth-promoting proteins than have already been established.

PERSPECTIVES

Studies exploring the role of the retinoblastoma gene product in cancer have begun to show that fundamental pathways for normal cell growth and differentiation intersect with pathways for oncogenesis. Factors controlling the growth and maintenance of a mammal are undoubtedly complex. Subtle changes such as premature progression through G_1 can lead to unscheduled DNA synthesis, in turn generating damaged DNA or chromosomes otherwise unready for mitosis. This abnormal progression could, by leading to increased mutational frequency, either activate protooncogenes or inactivate other tumor suppressor genes. The end result might be the generation of the highly deregulated clonal populations of cells that form cancers. Several other pathways involving growth factors, intracellular signals, cyclins, and newly identified Cdk-interacting proteins are converging on roads that involve Rb, p53, and other tumors suppressor gene products (Harper et al 1994; Hunter 1993; Marx 1994).

One challenge of future research focusing on the retinoblastoma gene product and Rb-related proteins will be to explore the complexity of the proteins' respective functions and to determine how tissue and temporal specificity are achieved. Why, for example, are tumors of particular neuroendocrine cells the only ones that develop spontaneously in mice bearing germ line inactivation of one *rb* allele? Why is it that retinoblastomas do not develop, as they do in humans?

The tissue specificity of tumors arising or progressing due to tumor suppressor inactivation is probably achieved, at least in part, by the differential expression of Rb-associated proteins and proteins whose functions partially overlap with those of Rb. By its association with and activation of some regulatory proteins and by its function to corral others into an inactive subcompartment within the nucleus, Rb function can be differentially controlled. In cells that are cycling under different influences, the same regulatory proteins that can associate with Rb might associate instead with Rb-related proteins to achieve alternate activities. Thus humans and mice bearing inactivations of the Rb-related proteins p107 or p130 might be expected to reveal phenotypes quite different from those bearing inactivations of the *rb* gene itself. When one considers that Rb and Rb-related proteins can interact with dozens of other proteins involved in the regulation of growth and differentiation, it is easy to see that an extraordinarily diverse array of possible cellular responses is possible using only a relatively small number of interacting players. By finding cellular proteins that interact strongly with several known or putative tumor suppressors, the convergence of growth regulation pathways will be made more convincing and sharply defined.

Although the functions of Rb and Rb-related proteins in the basic cell cycle machinery will continue to be explored in a variety of cells in culture, much future research on the functions of these proteins in normal cell growth and differentiation will have to be extrapolated to cells in the whole organism. Animal studies might show which mammalian cell types should receive the most concentrated effort in the artificial environment of the petri dish. In particular, developmental studies in animal models will be useful, especially if tissue-specific expression of Rb and related proteins can be achieved.

Complexity is the hallmark of advancing evolution. It is achieved only at the cost of having more potential points where mistakes can be made. The convergence of studies on normal cell growth and division of those of the cancerous cell is useful and satisfying. As is the case with most pathologic states, only by understanding normal cellular physiology can we hope to comprehend abnormalities with enough confidence to intervene and correct them. The restoration of neoplastic cells to a normal state, a goal envisioned long ago by cell fusion studies, is coming even closer as we recognize the molecular mechanisms that control cell proliferation.

ACKNOWLEDGMENTS

We thank ZD Sharp, M Mancini, and S McKnight for critical reading of the manuscript, as well as all the members of our labs and G Wahl for helpful discussions. Work performed in the authors' laboratories was supported by grants from the National Institutes of Health, the Council for Tobacco Research, and the AP McDermott Endowment Fund. DJR is the recipient of a Physician's Training Research Award from the American Cancer Society.

Literature Cited

Abramson DH, Ellsworth RM, Kitchin D, Tung G. 1984. Second nonocular tumors in retinoblastoma survivors. *Ophthalmology* 91: 1351–55

Alberts AS, Thorburn AM, Shenolikar S, Mumby MC, Feramisco JR. 1993. Regulation of cell cycle progression and nuclear affinity of the retinoblastoma protein by protein phosphatases. *Proc. Natl. Acad. Sci. USA* 90:388–92

Baker J, Lui J-P, Robertson EJ, Efstratiadis A. 1993. Role of insulin-like growth factors in embryonic and postnatal growth. *Cell* 75:73–82

Bandara LR, LaThangue NB. 1991. Adenovirus E1a prevents the retinoblastoma gene product from complexing with a cellular transcription factor. *Nature* 351:494–97

Bernards R, Schackleford GM, Gerber MR, Horowitz JM, Friend SH, et al. 1989. Structure and function of the murine retinoblastoma gene and characterization of its encoded protein. *Proc. Natl. Acad. Sci. USA* 86:6474–78

Bignon Y-J. 1993. Expression of a retinoblastoma transgene results in dwarf mice. *Genes Dev.* 7:1654–62

Bignon Y-J, Shew J-Y, Rappolee D, Naylor SL, Lee EY-HP, et al. 1990. A single Cys^{706} to Phe substitution in the retinoblastoma protein causes loss of binding to the SV40 T antigen. *Cell Growth Differ.* 1:647–51

Blake MC, Azizkhan JC. 1989. Transcription factor E2F is required for efficient expression of the hamster dihydrofolate reductase gene in vitro and in vivo. *Mol. Cell. Biol.* 9:4994–5002

Bookstein R, Lee W-H. 1991. Molecular genetics of the retinoblastoma suppressor gene. *CRC Crit. Rev. Oncogenesis* 2:211–27

Bookstein R, Rio P, Medraperla SA, Hong F, Allred C, et al. 1990a. Promoter deletion and loss of retinoblastoma gene expression in human prostate carcinoma. *Proc. Natl. Acad. Sci. USA* 87:7762–66

Bookstein R, Shew J-Y, Chen P-L, Scully P, Lee W-H. 1990b. Suppression of tumorigenicity of human prostate carcinoma cells by replacing a mutated RB gene. *Science* 247:712–15

Buchkovich K, Duffy LA, Harlow E. 1989. The retinoblastoma protein is phosphorylated during specific phases of the cell cycle. *Cell* 58:1097–105

Cavenee WK, Dryja TP, Phillips RA, Benedict WF, Godbout R, et al. 1983. Expression of recessive alleles by chromosomal mechanisms in retinoblastoma. *Nature* 305:779–84

Chang C-Y, Riley DJ, Lee EY-HP, Lee W-H. 1993. Quantitative effects of the retinoblastoma gene on mouse development and tissue-specific tumorigenesis. *Cell Growth Differ.* 4:1057–64

Chatterjee PK, Flint SJ. 1986. Partition of E1A proteins between soluble and structural fractions of adenovirus-infected and -transformed cells. *J. Virol.* 60:1018–26

Chellappan SP, Heibert S, Mudryj M, Horowitz JM, Nevins JR. 1991. The E2F transcription factor is a cellular target for the RB protein. *Cell* 65:1053–61

Chellappan S, Kraus VB, Kroger B, Munger K, Howley PM, et al. 1992. Adenovirus E1A, simian virus 40 tumor antigen, and human papillomavirus E7 protein share the capacity to disrupt the interaction between transcription factor E2F and the retinoblastoma gene product. *Proc. Natl. Acad. Sci. USA* 89: 4549–53

Chen J, Gorman JR, Stewart V, Williams B, Jacks T, Alt FW. 1993. Generation of normal lymphocyte populations by Rb-deficient embryonic stem cells. *Curr. Biol.* 3:405–13

Chen P-L, Chen Y, Shan B, Bookstein R, Lee W-H. 1992. Stability of RB gene expression determines tumorigenicity of reconstituted retinoblastoma cells. *Cell Growth Differ.* 3:119–25

Chen P-L, Scully P, Shew J-Y, Wang JYJ, Lee W-H. 1989. Phosphorylation of the retinoblastoma gene product is modulated during the cell cycle and cellular differentiation. *Cell* 58:1193–98

Chittenden T, Livingston DM, Kaelin WG Jr. 1991. The T/E1A-binding domain of the retinoblastoma product can interact selectively with a sequence-specific DNA-binding protein. *Cell* 65:1073–82

Clarke AR, Mandaag ER, van Roon M, van der Lugt NMT, van der Valk M, et al. 1992. Requirement for a functional Rb-1 gene in murine development. *Nature* 359:328–30

Cobrinik D, Whyte P, Peeper DS, Jacks T, Weinberg RA. 1993. Cell cycle-specific association of E2F with the p130 E1A-binding protein. *Genes Dev.* 7:2392–402

Comings DE. 1973. A general theory of carcinogenesis. *Proc. Natl. Acad. Sci. USA* 70:3324–28

Culver KW, Ram Z, Wallbridge S, Ishii H, Oldfield EH, Blaese RM. 1992. In vivo gene transfer with retroviral vector-producer cells for treatment of experimental brain tumors. *Science* 256:1550–52

DeCaprio JA, Furukawa Y, Ajchenbaum F, Griffin JD, Livingston DM. 1992. The retinoblastoma-susceptibility gene product becomes phosphorylated in multiple stages during cell cycle entry and progression. *Proc. Natl. Acad. Sci. USA* 89:1795–98

DeCaprio JA, Ludlow JW, Figge J, Shew J-Y,

Huang C-M, et al. 1988. SV40 large tumor antigen forms a specific complex with the product of the retinoblastoma susceptibility gene. *Cell* 54:275–83

DeCaprio JA, Ludlow JW, Lynch D, Furukawa Y, Griffin J, et al. 1989. The product of the retinoblastoma susceptibility gene has properties of a cell cycle regulating element. *Cell* 58:1085–95

Defeo-Jones D, Huang PS, Jones RE, Haskell KM, Vuocolo GA, et al. 1991. Cloning of cDNAs for cellular proteins that bind to the retinoblastoma gene product. *Nature* 352: 251–54

Deppert W, von der Weth A. 1990. Functional interaction of nuclear transport-defective simian virus 40 large T antigen with chromatin and nuclear matrix. *J. Virol.* 64838–46

Destree OHJ, Lam KT, Peterson-Maduro LJ, Eizema K, Diller L, et al. 1992. Structure and expression of the *Xenopus* retinoblastoma gene. *Dev. Biol.* 153:141–49

Donehower LA, Harvey M, Slagle B, McArthur MJ, Montgomery CA, et al. 1992. Mice deficient for p53 are developmentally normal but are susceptible to spontaneous tumours. *Nature* 356:215–21

Dou Q-P, Markell PJ, Pardee AB. 1992. Thymidine kinase transcription is regulated at the G_1/S phase by a complex that contains retinoblastoma-like protein and a cdc2 kinase. *Proc. Natl. Acad. Sci. USA* 89:3256–60

Dowdy SF, Hinds PW, Louie K, Reed SI, Arnold A, Weinberg RA. 1993. Physical interaction of the retinoblastoma protein with human D cyclins. *Cell* 73:499–511

Durfee T, Becherer K, Chen P-L, Yeh S-H, Yang Y, et al. 1993. The retinoblastoma protein associates with the protein phosphatase type 1 catalytic subunit. *Genes Dev.* 7:555–69

Dyson N, Buchkovich K, Whyte P, Harlow E. 1989a. The cellular 107 kD protein that binds to adenovirus E1A also associates with the large T antigens of SV40 and JC virus. *Cell* 58:249–55

Dyson N, Howley PM, Munger K, Harlow E. 1989b. The human papilloma virus-16 E7 oncoprotein is able to bind to the retinoblastoma gene product. *Science* 243:934–37

Enesco M, Leblond CP. 1962. Increase in cell number as a factor in the growth of organs and tissues of the young male rat. *J. Embryol. Exp. Morphol.* 10:530–63

Ewen ME, Sluss HK, Sherr CJ, Matsushime H, Kato J-y, Livingston DM. 1993. Functional interactions of the retinoblastoma protein with mammalian D-type cyclins. *Cell* 73: 487–97

Ewen ME, Xing Y, Lawrence JB, Livingston DM. 1991. Molecular cloning, chromosomal mapping, and expression of the cDNA for p107, a retinoblastoma gene product-related protein. *Cell* 66:1155–64

Fattaey AR, Helin K, Dembski MS, Dyson N, Harlow E, et al. 1993. Characterization of the retinoblastoma binding proteins RBP1 and RBP2. *Oncogene* 8:3149–56

Fernandez A, Brautigan DL, Lamb NJC. 1992. Protein phosphatase type 1 in mammalian cell mitosis: chromosomal localization and involvement in mitosis. *J. Cell Biol.* 116: 1421–30

Fields S, Song O. 1989. A novel genetic system to detect protein-protein interactions. *Nature* 340:245–46

Francke U. 1976. Retinoblastoma and chromosome 13. *Birth Defects* 12:131–39

Friend SH, Bernards R, Rogelj S, Weinberg RA, Rapaport JM, et al. 1986. A human DNA segment with properties of the gene that predisposes to retinoblastoma and osteosarcoma. *Nature* 323:643–46

Fung Y-KT, Murphee AL, T'Ang A, Qian J, Hinrichs SH, Benedict WF. 1987. Structural evidence for the authenticity of the human retinoblastoma gene. *Science* 236:1659–61

Fung Y-KT, T'Ang A, Murphee AL, Zhang F-H, Qui W-R, et al. 1993. The Rb gene suppresses the growth of normal cells. *Oncogene* 8:2659–72

Furukawa Y, DeCaprio JA, Freedman A, Kanakura Y, Nakamura M, et al. 1990. Expression and state of phosphorylation of the retinoblastoma susceptibility gene product in cycling and non-cycling human hematopoietic cells. *Proc. Natl. Acad. Sci. USA* 87: 2770–74

Goodrich DW, Chen Y, Scully P, Lee W-H. 1992. Expression of the retinoblastoma gene product in bladder carcinoma cells associates with a low frequency of tumor formation. *Cancer Res.* 52:1968–73

Goodrich DW, Lee W-H. 1992. Abrogation by c-myc of G_1 phase arrest induced by RB protein but not by p53. *Nature* 360:177–79

Goodrich DW, Wang NP, Qian Y-W, Lee EY-HP, Lee W-H. 1991. The retinoblastoma gene product regulates progression through the G1 phase of the cell cycle. *Cell* 67:293–302

Greenfield I, Nickerson J, Penman S, Stanley M. 1991. Human papillomavirus 16 E7 protein is associated with the nuclear matrix. *Proc. Natl. Acad. Sci. USA* 88:11217–21

Gu W, Schneider JW, Condorelli G, Kaushal S, Mahdavi V, Nadal-Ginard B. 1993. Interaction of myogenic factors and the retinoblastoma protein mediates muscle cell commitment and differentiation. *Cell* 72: 309–24

Hannon GJ, Demetrick D, Beach D. 1993. Isolation of the Rb-related p130 through its interaction with CDK2 and cyclins. *Genes Dev.* 7:2378–91

Harper JW, Adami GR, Wei N, Keyomarsi K, Elledge SJ. 1994. The p21 cdk-interacting protein cip1 is a potent inhibitor of G1-cyclin dependent kinases. *Cell* 75:805–16

Harlow E, Whyte P, Franza BR, Schley C. 1986. Association of adenovirus early-region 1A region proteins with cellular polypeptides. *Mol. Cell. Biol.* 6:1579–89

Harris H, Miller OJ, Klein G, Worst P, Tachibana T. 1969. Suppression of malignancy by cell fusion. *Nature* 223:363–68

Helin K, Lees JA, Vidal M, Dyson N, Harlow E, Fattaey A. 1992. A cDNA encoding an Rb-binding protein with properties of the transcription factor E2F. *Cell* 70:337:50

Hensey CE, Hong F, Durfee T, Qian Y-W, Lee EY-HP, Lee W-H. 1994. Identification of discrete structural domains in the retinoblastoma protein: amino-teminal domain is required for its oligomerization. *J. Biol. Chem.* 269:1380–87

Hinds PW, Mittnacht S, Dulic V, Arnold A, Reed SI, Weinberg RA. 1992. Regulation of retinoblastoma protein functions by ectopic expression of human cyclins. *Cell* 70:993–1006

Hirano T, Hiraoka Y, Yanagida M. 1988. A temperature-sensitive mutation of the *Schizosaccharomyces pombe* gene *nuc2+* that encodes a nuclear scaffold-like protein blocks spindle elongation in mitotic anaphase. *J. Cell Biol.* 106:1171–83

Hong FD, Huang H-JS, To H, Young L-JS, Oro A, et al. 1989. Structure of the human retinoblastoma gene. *Proc. Natl. Acad. Sci. USA* 86:5502–6

Horowitz JM, Park S-H, Bogenmann E, Cheng J-C, Yandell DW, et al. 1990. Frequent inactivation of the retinoblastoma anti-oncogene is restricted to a subset of human tumor cells. *Proc. Natl. Acad. Sci. USA* 87: 2775–79

Hozak P, Hassan AB, Jackson DA, Cook PR. 1993. Visualization of replication factors attached to a nucleoskeleton. *Cell* 73:361–73

Hu N, Gutsmann A, Herbert DC, Bradley A, Lee W-H, Lee EY-HP. 1994. Heterozygous *Rb*-$1^{\Delta 20/+}$ mice are predisposed to tumors of the pituitary gland with nearly complete penetrance. *Oncogene* 9:1021–27

Hu Q, Bautista C, Edwards GM, Defeo-Jones D, Jones RE, Harlow E. 1991. Antibodies specific for the human retinoblastoma protein family identify a family of related polypeptides. *Mol. Cell. Biol.* 11:5792–99

Hu Q, Dyson N, Harlow E. 1990. The regions of the retinoblastoma protein needed for binding to adenovirus E1A or SV40 large T antigen are common sites for mutations. *EMBO J.* 9:1147–55

Hu QJ, Lees JA, Buchkovich KJ, Harlow E. 1992. The retinoblastoma protein physically associates with the human cdc2 kinase. *Mol. Cell. Biol.* 12:971–80

Huang H-JS, Yee J-K, Shew J-Y, Chen P-L, Bookstein R, et al. 1988. Suppression of the neoplastic phenotype by replacement of the RB gene in human cancer cells. *Science* 242: 1563–66

Huang S, Wang NP, Tseng BY, Lee W-H, Lee EY-HP. 1990. Two distinct and frequently mutated regions of the retinoblastoma protein are required for binding to SV40 T antigen. *EMBO J.* 9:1815–22

Hunter T. 1987. A thousand and one protein kinases. *Cell* 50:823–29

Hunter T. 1993. Braking the cycle. *Cell* 75: 839–41

Jacks T, Fazeli A, Schmitt EM, Bronson RT, Goodell MA, Weinberg RA. 1992. Effects of an Rb mutation in the mouse. *Nature* 359: 295–300

Kaelin WGJ, Pallas DC, DeCaprio JA, Kaye FJ, Livingston DM. 1991. Identification of cellular proteins that can interact specifically with the T/E1A-binding region of the retinoblastoma gene product. *Cell* 64:521–32

Karantza V, Maroo A, Fay D, Sedivy JM. 1993. Overproduction of Rb protein after the G_1/S boundary causes G_2 arrest. *Mol. Cell. Biol.* 13:6640–52

Kato J-y, Matsushime H, Hiebert SW, Ewen ME, Sherr CJ. 1993. Direct binding of cyclin D to the retinoblastoma gene product (pRb) and pRb phosphorylation by the cyclin D-dependent kinase CDK4. *Genes Dev.* 7:331–42

Kaye FJ, Kratzke RA, Gerster JL, Horowitz JM. 1990. A single amino acid substitution results in a retinoblastoma protein defective in phosphorylation and oncoprotein binding. *Proc. Natl. Acad. Sci. USA* 87:6922–26

Kim S-J, Wagner S, Lui F, O'Reilly MA, Robbins PD, Green MR. 1992. Retinoblastoma gene product activates expression of the human TGF-β2 gene through transcription factor ATF-2. *Nature* 358:331–34

Kim T-A, Velasquez BR, Wenner LE. 1993. Okadaic acid regulation of the retinoblastoma gene product is correlated with inhibition of growth factor-induced cell proliferation in mouse fibroblasts. *Proc. Natl. Acad. Sci. USA* 90:5460–63

Knudson AG Jr. 1971. Mutation and cancer: statistical study of cancer. *Proc. Natl. Acad. Sci. USA* 68:820–23

Laiho M, DeCaprio JA, Ludlow JW, Livingston DM, Massagué J. 1990. Growth inhibition by TGF-β linked to suppression of retinoblastoma protein phosphorylation. *Cell* 62:175–85

Lee EY-HP, Chang C-Y, Hu N, Wang Y-CJ, Lai C-C, et al. 1992. Mice deficient for Rb are nonviable and show defects in neurogenesis and haematopoiesis. *Nature* 359: 288–94

Lee W-H, Bookstein R, Hong F, Young L-J, Shew J-Y, Lee EY-HP. 1987a. Human retinoblastoma gene: cloning, identification, and sequence. *Science* 235:1394–99

Lee W-H, Hollingworth RE Jr, Qian Y-W, Chen P-L, Hong F, Lee EY-HP. 1991. RB protein as a cellular "corral" for growth promoting proteins. *Cold Spring Harbor Symp. Quant. Biol.* 56:211–17

Lee W-H, Shew J-Y, Hong FD, Sery TW, Donoso LA, et al. 1987b. The retinoblastoma susceptibility gene encodes a nuclear phosphoprotein associated with DNA binding activity. *Nature* 329:642–45

Lees JA, Buchkovich KJ, Marshak DR, Anderson CW, Harlow E. 1991. The retinoblastoma protein is phosphorylated on multiple sites by human cdc2. *EMBO J.* 10:4279–90

Li Y, Graham C, Lacy S, Duncan AMV, Whyte P. 1993. The adenovirus E1A-associated 130-kD protein is encoded by a member of the retinoblastoma gene family and physically interacts with cyclins A and E. *Genes Dev.* 7:2366–77

Lin BT-Y, Gruenwald S, Morla AO, Lee W-H, Wang JYJ. 1991. Retinoblastoma cancer suppressor gene product is a substrate of the cell cycle regulator cdc2 kinase. *EMBO J.* 10:857–64

Ludlow JW, Glendening CL, Livingston DM, DeCaprio JA. 1993. Specific enzymatic dephosphorylation of the retinoblastoma protein. *Mol. Cell. Biol.* 13:367–72

Ludlow JW, Shen J, Pipas JM, Livingston DM, DeCaprio JA. 1990. The retinoblastoma susceptibility gene product undergoes cell cycle-dependent dephosphorylation and binding to and release from SV40 large T. *Cell* 60:387–96

Mancini M, Shan B, Nickerson J, Penman S, Lee W-H. 1994. The retinoblastoma gene product is a cell-cycle dependent, nuclear-matrix associated protein. *Proc. Natl. Acad. Sci. USA* 91:418–22

Marx J. 1994. Research news: How cells cycle toward cancer. *Science* 263:319–21

Mayol X, Grana X, Baldi A, Sang N, Hu Q, Giordano A. 1993. Cloning of a new member of the retinoblastoma gene family (pRB2) which binds to the E1A transforming domain. *Oncogene* 8:2561–66

Medraperla SA, Whittum-Hudson JA, Prendergast RA, Chen P-L, Lee W-H. 1991. Intraocular tumor suppression of retinoblastoma gene-reconstituted retinoblastoma cells. *Cancer Res.* 61:6381–84

Mittnacht S, Weinberg RA. 1991. Phosphorylation of the retinoblastoma protein is associated with an altered affinity for the nuclear compartment. *Cell* 65:381–93

Muncaster MM, Cohen BL, Phillips RA, Gallie BL. 1992. Failure of RB1 to reverse the malignant phenotype of human tumor cell lines. *Cancer Res.* 52:654–61

Myerson M, Enders GH, Wu C-L, Ju L-K, Gorka C, et al. 1990. The human cdc2 kinase family. *EMBO J.* 11:2909–18

Nevins JR. 1992. E2F: a link between the Rb tumor suppressor protein and viral oncoproteins. *Science* 258:424–29

O'Donnell KL, Osmani AH, Osmani SA, Morris NR. 1991. bimA encodes a member of the tetratricopeptide repeat family of proteins and is required for the completion of mitosis in *Aspergillus nidulans*. *J. Cell Sci.* 99:711–19

Pardee AB. 1989. G_1 events and regulation of cell proliferation. *Science* 246:603–8

Pearson BE, Nasheuer H-P, Wang TS-F. 1991. Human DNA polymerase α gene: sequences controlling expression in cycling and serum-stimulated cells. *Mol. Cell. Biol.* 11:2081–95

Qian X-Q, Chittenden T, Livingston DM, Kaelin WG Jr. 1992. Identification of a growth suppression domain within the retinoblastoma gene product. *Genes Dev.* 6: 953–74

Qian Y, Luckey C, Horton L, Esser M, Templeton DJ. 1992. Biological function of the retinoblastoma protein requires domains for hyperphosphorylation and transcription factor binding. *Mol. Cell. Biol.* 12:5363–72

Qian Y-W, Wang Y-CJ, Hollingsworth RE Jr, Jones D, Ling N, Lee EY-HP. 1993. A retinoblastoma-binding protein related to a negative regulator of Ras in yeast. *Nature* 364:648–52

Rustgi AK, Dyson N, Bernards R. 1991. Amino-terminal domains of c-myc and N-myc proteins mediate binding to the retinoblastoma gene product. *Nature* 352:541–44

Sakai T, Ohtani N, McGee TL, Robbins PD, Dryja TP. 1992. Oncogenic germ-line mutations in Sp1 and ATF sites in the human retinoblastoma gene. *Nature* 353:83–86

Shan B, Chang C-Y, Lee W-H. 1994. The transcription factor E2F-1 mediates autoregulation of RB expression. *Mol. Cell. Biol.* 14: 299–309

Shan B, Zhu X, Chen P-L, Durfee T, Yang Y, Sharp D, Lee W-H. 1992. Molecular cloning of the cellular genes encoding retinoblastoma-associated proteins: identification of a gene with properties of the transcription factor E2F. *Mol. Cell. Biol.* 12:5620–31

Sherr C. 1993. Mammalian G_1 cyclins. *Cell* 73:1059–65

Shew J-Y, Lin BT-Y, Chen P-L, Tseng BY, Yang-Feng TL, Lee W-H. 1990. C-terminal truncation of the retinoblastoma gene product leads to functional inactivation. *Proc. Natl. Acad. Sci. USA* 87:6–10

Shew J-Y, Ling N, Yang X, Fodstad O, Lee W-H. 1989. Antibodies detecting abnormalities of the retinoblastoma susceptibility gene product (pp110RB) in osteosarcomas and synovial sarcomas. *Oncogene Res.* 1:205–14

Stein GH, Beeson M, Gordon L. 1990. Failure to phosphorylate the retinoblastoma gene product in senescent human fibroblasts. *Science* 249:666–68

Strong LC, Riccardi VM, Ferrell RE, Sparkes RS. 1981. Familial retinoblastoma and chromosome 13 deletion transmitted via an insertional translocation. *Science* 213:1501–5

Sumegi J, Uzvolgyi E, Klein G. 1990. Expression of the RB gene under control of MuLV-LTR suppresses tumorigenicity of WERI-RB-27 retinoblastoma cells in immunodeficient mice. *Cell Growth Differ.* 1: 247–50

Szekely L, Jiang W-Q, Bulic-Jakus F, Rosen A, Ringertz N, et al. 1992. Cell type and differentiation dependent heterogeneity in retinoblastoma protein expression in SCID mouse fetuses. *Cell Growth Differ.* 3:149–56

Takahashi R, Hashimoto T, Xu H-J, Hu S-X, Matsui T, et al. 1991. The retinoblastoma gene functions as a growth and tumor suppressor in human bladder carcinoma cells. *Proc. Natl. Acad. Sci. USA* 88:5257–61

Templeton DJ. 1992. Nuclear binding of purified retinoblastoma gene product is determined by cell cycle-regulated phosphorylation. *Mol. Cell. Biol.* 12:435–43

Udvadia AJ, Rogers KT, Higgins PDR, Murata Y, Martin KH, et al. 1993. Sp-1 binds promoter elements regulated by the RB protein and Sp-1-mediated transcription is stimulated by RB coexpression. *Proc. Natl. Acad. Sci. USA* 90:3265–69

Wan KM, Nickerson JA, Krockmalnic G, Penman S. 1994. The B1C8 protein is in the dense assemblies of the nuclear matrix and relocates to the spindle and pericentriolar filaments at mitosis. *Proc. Natl. Acad. Sci. USA* 91:594–98

Wang NP, Chen P-L, Huang S-H, Donosos LA, Lee W-H, Lee EY-HP. 1990. DNA binding activity of retinoblastoma protein is intrinsic to its carboxyl-terminal region. *Cell Growth Differ.* 1:233–39

Wang NP, To H, Lee W-H, Lee EY-HP. 1993. Tumor suppressor activity of RB and p53 genes in human breast carcinoma cells. *Oncogene* 8:279–88

Whyte P, Buchkovich KJ, Horowitz JM, Friend SH, Raybuck M, et al. 1988. Association between an oncogene and an anti-oncogene: the adenovirus E1A protein binds to the retinoblastoma gene product. *Nature* 334:124–29

Wittenberg C, Reed SI. 1988. Control of the yeast cell cycle is associated with assembly/disassembly of the cdc28 protein kinase complex. *Cell* 54:1068–72

Xu H-J, Sumegi J, Hu S-X, Banerjee A, Uzvolgyi E, et al. 1991. Intraocular tumor formation of RB reconstituted retinoblastoma cells. *Cancer Res.* 51:4481–85

Zacksenhaus E, Bremner R, Phillips RA, Gallie BL. 1993. A bipartite nuclear localization signal in the retinoblastoma gene product and its importance for biologic activity. *Mol. Cell. Biol.* 13:4588–99

Zhu L, van der Heuvel S, Helin K, Fattaey A, Ewen M, et al. 1993. Inhibition of cell proliferation by p107, a relative of the retinoblastoma protein. *Genes Dev.* 7: 1111–25

ADDED IN PROOF

Durfee T, Mancini M, Jones D, Elledge SJ, Lee W-H. 1994. The amino-terminal region of the retinoblastoma gene product binds a novel nuclear matrix protein that localizes to RNA processing centers. *J. Cell Biol.* In press

Annu. Rev. Cell Biol. 1994. 10:31–54

SMALL GTP-BINDING PROTEINS AND THE REGULATION OF THE ACTIN CYTOSKELETON

Alan Hall

MRC Laboratory For Molecular Cell Biology and Department of Biochemistry, University College London, Gower Street, London WC1E 6BT, England

KEY WORDS: rho, rac, focal adhesion, integrins, membrane ruffling

---

CONTENTS

INTRODUCTION

The actin cytoskeleton is a highly dynamic structure that is reshaped and reformed during the cell cycle and in response to extracellular signals. Its biological functions are diverse; in combination with myosin it forms the contractile ring required for cytokinesis in animal cells; in association with

0743–4634/94/1115–0031$05.00

cell adhesion molecules it affects the nature of cell/cell and cell/substrate interactions; while at the plasma membrane, it provides the driving force for cell movement and surface remodeling. To perform these biological tasks, the spatial and structural properties of the actin cytoskeleton and its ability to be rapidly assembled and disassembled must be harnessed. Since physiological conditions favor the spontaneous assembly of polymerized (F) actin from monomeric (G) actin, this is achieved by sequestering actin monomers and by capping actin filaments, in such a way that they can be rapidly mobilized when required. A second and equally important problem, the spatial organization of polymerized actin, appears to be taken care of by targeting polymerization to discrete sites (nucleation sites), often on the plasma membrane.

A complete description of how these processes are regulated in animal cells is still a way off, but extracellular factors such as matrix proteins, cell/cell contacts, or soluble agonists that interact with plasma membrane receptors are known to have a major influence on the rate of formation and spatial organization of polymerized actin. Although many intracellular signaling molecules have been identified that could account for these responses, there is now compelling evidence that, as with other signal transduction pathways, GTP-binding proteins (GTPases) play a key regulatory role.

GTP-binding proteins function as molecular switches, cycling between an inactive GDP-bound conformation and an active GTP-bound conformation. Thus far, all known GTPases are activated by guanine nucleotide exchange and inactivated through an intrinsic GTPase activity (although this simple idea has been elaborated upon during evolution) so that different kinds of GTP-binding protein can carry out diverse regulatory functions (Bourne et al 1991). Heterotrimeric G proteins, for example, couple transmembrane receptors to classical intracellular signaling molecules such as adenylyl cyclase and phospholipase C, and second messengers derived from these signal transduction pathways can affect the actin cytoskeleton. In addition, although the main function of the small GTP-binding protein ras is to control cell growth and differentiation, it too can influence the organization of the actin cytoskeleton. However, a major breakthrough during the last few years has been the discovery of a subfamily of small GTP-binding proteins, the rho-related GTPases, whose function is specifically to regulate the organization of the actin cytoskeleton in eukaryotic cells.

ACTIN POLYMERIZATION—THE PROBLEM

Actin is a highly abundant protein found in all eukaryotic cells. It has a strong tendency to self-assemble into helical filaments and, in fact, under physiological conditions of temperature and ionic strength, more than 99% of monomeric actin would be expected to spontaneously polymerize. In practice, however, it is generally well below 50%. To maintain this unfavorable thermodynamic state, actin monomers must be sequestered and actin filaments must be capped. In addition, sequestered monomeric actin and capped filaments must be capable of being released in an organized fashion in response to specific signals.

Monomer Sequestration and Filament Capping

One of the best-characterized actin monomer-binding proteins is profilin, which binds to actin in a 1:1 complex and with a relatively low dissociation constant (Pollard & Cooper 1986; Theriot & Mitchison 1993). There has been a great deal of interest in profilin as a potential key regulator of actin polymerization since it can also interact with the plasma membrane lipid phosphatidyl-4,5-bis-phosphate (PIP_2). PIP_2 is rapidly turned over and resynthesized after activation of a wide variety of receptors under conditions where actin polymerization is induced; an interaction between profilin and this lipid could lead to the release of actin monomers (Lassing & Lindberg 1988). In addition, profilin has other activities that suggest a regulatory role; it can stimulate the conversion of actin•ADP to the faster polymerizing form, actin•ATP (Goldschmidt-Clermont et al 1991), and profilin:actin complexes can bind directly to the barbed ends of actin filaments which, if followed by release of profilin, would result in actin polymerization (Pring et al 1992; Panatoloni & Carlier 1993).

Profilin cannot, however, be the whole story, and several other actin monomer-binding proteins with similar properties have been identified. These include actin depolymerizing factor (ADF), cofilin, and perhaps the most likely candidate to account for the bulk of actin sequestering activity, thymosin-β4, a small, 5-kd peptide that is found in all cell types at extremely high concentrations (Safer 1992). It is now thought that thymosin-β4 acts as a buffer for actin•ADP monomers, and that through mass action, these can be captured by profilin, converted into the ATP form, and rapidly incorporated into filaments (Goldschmidt-Clermont et al 1992; Theriot & Mitchison 1993).

A number of proteins have been identified that can bind to the barbed (plus) or pointed (minus) ends of actin filaments, and agonist-stimulated release of these proteins, particularly at the barbed end, could be a mechanism for promoting localized actin assembly onto pre-existing filaments. For example, villin, an epithelial-specific protein, and gelsolin, in addition to binding actin monomers, can cap the barbed ends of filaments (Matsudaira & Janmey 1988). Like profilin, each has a binding site for phosphoinositides, and it is possible that local changes in these lipids at the membrane release free filament ends.

Actin Nucleation

Actin polymerization does not occur uniformly throughout the cytoplasm but instead is found at discrete sites referred to as actin nucleation sites. Although the nature of these sites is not clear, there is much evidence to suggest they are located at the plasma membrane.

Microinjection of fibroblasts with biotinylated or fluorescently labeled monomeric actin reveals its rapid incorporation into motile peripheral regions of the cell such as lamellipodia and microspikes. Within one minute it can be seen at the distal ends of fibers, i.e. at the plasma membrane, and by five minutes, actin fibers in lamellipodia are uniformly labeled (Okabe & Hirokawa 1989; Symons & Mitchison 1991). Incorporation of actin monomers can be inhibited by coinjection of a barbed-end capping protein, CapZ, which suggests that the release of sequestered monomers is not rate limiting but that nucleation is dependent on the generation of free barbed ends adjacent to the plasma membrane.

A number of proteins have been identified at the tips of lamellipodia and microspikes, including α-actinin, fimbrin, and talin, and it is likely that a multimolecular complex is responsible for organizing the membrane structures and for controlling the rapid rates of localized actin polymerization (see Figure 1). Interestingly, talin is capable of nucleating actin polymerization in vitro, and a careful study of talin localization in cells showed high levels at the outer edge of lamellipodia and, with further concentration, in nodes, or tips coincident with the ends of actin fibers (DePasquale & Izzard 1991; Kaufmann et al 1992).

More stable plasma membrane structures associated with actin filaments are seen at focal adhesions and at adherens junctions. Focal adhesions are clusters of integrins attached to the extracellular matrix, associated with intracellular proteins, and forming the sites of attachment of actin stress fibers (Burridge et al 1988; Gumbiner 1993). A dozen or so proteins, listed in Figure 1, have been localized to these structures including talin, vinculin, α-actinin, paxillin, tensin, focal adhesion kinase (FAK), as well as actin (Turner & Burridge 1991). Adherens junctions are clusters of cadherins that mediate cell/cell interactions and again a large number of proteins, in addition to actin, are found associated at the cytoplasmic side of the junction, including vinculin, α-actinin, and radixin (Geiger et al 1990; Sato et al 1992).

There has been much interest in the interaction of bacteria with mammalian cells because they often induce major changes in the actin cytoskeleton (for review see Falkow et al 1992). *Listeria monocytogenes* is a particularly interesting case because after invading mammalian cells it migrates through the cytoplasm by inducing unidirectional actin polymerization. Nucleation occurs at the rear end of the bacterium, which becomes associated with barbed ends of filaments, and this is thought to provide the driving force for movement (Tilney et al 1992). Clearly the bacteria can hijack the host's machinery for controlling actin polymerization, and the bacterial protein responsible for this,

Actin-associated membrane complexes	Component proteins
focal adhesions	integrin / vinculin / talin / α-actinin / tensin / zyxin / focal adhesion kinase / P-Tyr / paxillin
adherens junctions	cadherin / catenin / ERM = ezrin, radixin, moesin / vinculin / filamin / α-actinin / P-Tyr / (not talin, not paxillin)
lamellipodia, ruffles, leading-edge	integrin / talin / P-Tyr / ERM / fimbrin / α-actinin / (not vinculin)
cortical actin cytoskeleton	fodrin-spectrin / ankyrin / transporters / ion channels / adhesion molecules
contractile ring	CD43 (leukosialin) / ERM / filamin / α-actinin / (not vinculin, not talin)

Figure 1 Actin-associated membrane complexes found in mammalian cells. In general, these complexes cannot be isolated biochemically, and the identification of protein components has been done mostly by immunofluorescence analysis.

the product of the *actA* gene, has some sequence homology with vinculin (Domann et al 1992). Interestingly, ActA, like vinculin, has proline-rich motifs that suggest possible interactions with SH3 domain-containing proteins or perhaps profilin (Ren et al 1993).

Two major problems emerge from this discussion: How does the cell regulate the assembly of cytoskeletal-associated multimolecular complexes at the plasma membrane, and which components of the complex are responsible for actin nucleation?

REGULATION OF ACTIN POLYMERIZATION BY SECOND MESSENGERS

An early response of many cells to extracellular factors is a rapid reorganization of the actin cytoskeleton, and this can be reflected in changes in cell shape, in adherence to a substrate or to other cells, in membrane organization to produce ruffles and lamellipodia, in pinocytosis, or in cell motility. The intracellular signals that couple membrane receptors to these effects must be capable of recruiting an actin nucleation activity to a spatially defined site on the plasma membrane (for review see Stossel 1993).

G Protein-Coupled Receptors

One of the best studied areas of receptor-mediated actin reorganization is chemotaxis (for review see Devreotes & Zigmond 1988). Addition of chemoattractants such as the N-formylated peptide FMLP to polymorphonuclear leukocytes induces cytochalasin-sensitive directed movement toward the stimulant. Activation of the G protein-coupled FMLP receptor stimulates a phospholipase, PLCβ, to produce two second messengers, DAG and inositol trisphosphate (IP_3) from PIP_2 hydrolysis, but whether this is responsible for the rapid and dramatic increase in actin polymerization is not known (Thomas et al 1990). IP_3 releases Ca^{2+} from intracellular stores, and although this would be unlikely to produce a localized signal, many actin-binding proteins are known to be Ca^{2+}-sensitive (Vandekerkhove 1990). Profilin and gelsolin, on the other hand, can be dissociated from actin by PIP_2, and it has been suggested that despite an initial reduction, the localized, steady state levels of PIP_2 might actually increase (Stossel 1989).

FMLP is, in addition, a potent stimulator of PI 3-kinase, which converts PIP_2 into the putative second messenger, PIP_3; and, in fact, the time course for PIP_3 formation, unlike that for PIP_2, directly correlates with actin polymerization (Eberle et al 1990). Thrombin, also a potent activator of PI 3-kinase, causes its translocation to a detergent-insoluble fraction along with actin, talin, and vinculin during platelet aggregation (Zhang et al 1992). It has been suggested, although not directly demonstrated, that these proteins exist in a cytoskeletal-associated membrane complex. Whether PIP_3 plays a direct role in actin polymerization, i.e. can liberate sequestered actin monomers or capped filaments, is not known, but a potent inhibitor of PI 3-kinase, wortmannin, has been described that should help define the role of PIP_3 (Yano et al 1993).

It is now widely accepted that G protein-coupled receptors can activate tyrosine kinases as well as classical second messenger pathways. Bombesin, for example, independently of PKC or Ca^{2+}, can rapidly induce tyrosine phosphorylation of several proteins, including p125FAK, originally identified as a src tyrosine kinase substrate present in focal adhesions (Sinnett-Smith et al 1993; Schaller et al 1992). Tyrosine phosphorylation of p125FAK is induced by a variety of extracellular stimuli, including tumor necrosis factor, antibody-mediated integrin crosslinking, or attachment of cells to extracellular matrix. Since p125FAK is itself a tyrosine kinase, it has been speculated that it is a key regulator of focal adhesion and actin assembly and perhaps other signals generated by activated integrins (Kornberg et al 1992; Fuortes et al 1993; Burridge et al 1993; Damsky & Werb 1992).

Tyrosine Kinase Receptors

The signal transduction pathways downstream of tyrosine kinase receptors have been examined in great detail (Ullrich & Schlessinger 1990). Both platelet-derived growth factor (PDGF) and epidermal growth factor (EGF) receptors can activate phospholipase (PLCγ) and PI 3-kinase and, as outlined above, it is possible that the second messengers derived from these enzymes could lead to actin polymerization. In fact, work with mutated PDGF receptors strongly suggests that its interaction with PI 3-kinase is essential for the PDGF-induced ruffling response (Wennstrom et al 1994). In A431 cells, EGF induces membrane ruffling and the disassembly of actin stress fibers, although this does not correlate with changes in phosphoinositide levels but instead seems to depend on the activation of phospholipase A2 (Dadabay et al 1991; Peppelenbosch et al 1993). It has been proposed that products of lipoxygenase metabolism are specifically required for the ruffling effect, whereas the dissolution of stress fibers is dependent on cyclooxygenase metabolites.

A major step forward in understanding cell signaling came with the realization that tyrosine kinase receptors and G protein-coupled receptors activate small GTP-binding proteins. EGF-induced activation of ras, for example, occurs through an SH2/SH3-containing adaptor molecule, grb2, and a ras guanine nucleotide exchange factor, sos, which leads to the activation of the MAP kinase cascade (see Figure 2) (Egan & Weinberg 1993). This signal transduction pathway has been shown to be absolutely essential for growth factor-induced DNA synthesis. Some years ago it was shown that microinjection of activated mutant forms of ras into fibroblasts could induce rapid membrane ruffling in the absence of growth factors, and it was widely believed, therefore, that ras was a key signal for receptor-induced actin polymerization (Bar-Sagi & Feramisco 1986). It is now clear, however, that the stimulation of actin polymerization in fibroblasts by growth factors or by microinjection of ras is mediated by rac, a distinct member of the rho subfamily of ras-related GTP-binding proteins (Ridley et al 1992). The analysis of this subfamily of small GTPases has brought new insights into how actin polymerization is controlled in eukaryotic cells.

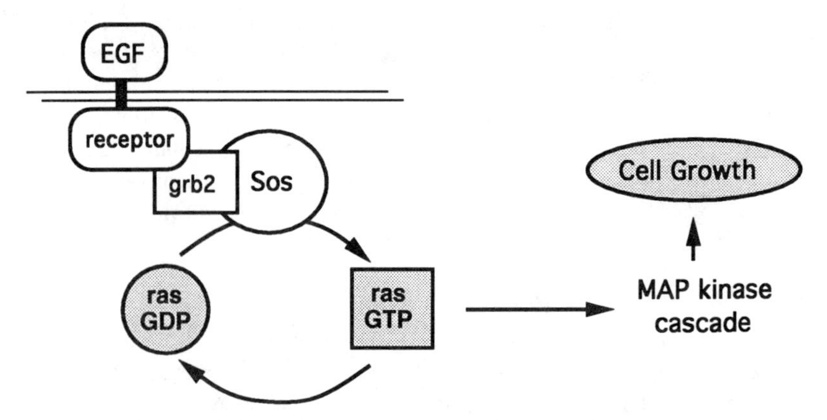

Figure 2 Cell growth induced by epidermal growth factor (EGF) is mediated by a signal transduction pathway controlled by the ras GTP-binding protein.

SMALL GTP-BINDING PROTEINS—THE SOLUTION?

The Ras Superfamily

Ras is an 189 amino acid, ubiquitously expressed GTP-binding protein that is a key regulator of eukaryotic cell growth and differentiation (for review see Bollag & McCormick 1991). Its discovery and the subsequent identification of its mechanism of action (Figure 2) have increased our understanding of growth control and differentiation. The discovery of a large superfamily of ras-related GTP-binding proteins is having an equally profound effect on understanding of other fundamental biological processes, particularly the organization of the actin cytoskeleton (Hall 1990a).

Over 50 mammalian ras-related proteins have been described so far, and these are listed in Figure 3. Based on sequence homologies, they can be grouped into five subfamilies: ras, rho, rab, ARF, and Ran, and evidence is accumulating that individual members of these subfamilies have some functional similarities (Hall 1990a; Boguski & McCormick 1993). The rab and ARF subfamilies have evolved to control different aspects of vesicular transport (for review see Pryer et al 1992), while the ras- and the rho-related proteins regulate signal transduction pathways linking plasma membrane receptors to biological responses. To date most work has centered around ras because of its role in growth control, but recent breakthroughs demonstrating a role for rho-related proteins in controlling the organization of the actin cytoskeleton have now stimulated much interest in these proteins.

subfamily	members	biological functions
Ras	H-ras/Ki-ras/N-ras; R-ras; rap1A/B; rap2A/B; ralA/B; TC21	Growth Differentiation
Rho	rho A/rhoB/rhoC; rac 1/2; CDC42/G25K; rhoG; TC10	Integrin activity Actin cytoskeleton NADPH oxidase
Rab	rab1 .... rab26	Vesicle transport
ARF	ARF1 ..... ARF6	Vesicle transport Stimulation of PLD
Ran	Ran1	Nuclear protein import

Figure 3 Members of the mammalian ras superfamily of small GTP-binding proteins.

The Rho Subfamily

The mammalian rho subfamily, shown in Figure 3, currently consists of nine proteins, each having 50–55% homology with each other and around 30% homology to ras (Ridley & Hall 1993). Rho was first isolated serendipitously from an *Aplysia* cDNA library, but subsequently three mammalian *rho* genes (*rhoA, rhoB,* and *rhoC*) were cloned (Madaule & Axel 1985; Yeramian et al 1987; Chardin et al 1988). Later two rac proteins (1 and 2) differing at 15 amino acids, and two CDC42-like proteins (CDC42Hs and G25K) differing at 9 amino acids, were identified (Didsbury et al 1989; Munemitsu et al 1990; Shinjo et al 1990). Finally, two further cDNAs, TC10 and *rhoG*, have been reported (Drivas et al 1990; Vincent et al 1992). The rho-related proteins appear to be widely expressed, only TC10 thus far shows significant tissue-specific expression, but interestingly both *rhoG* and *rhoB* are serum-inducible genes, at least in fibroblasts (Vincent et al 1992; Jahner & Hunter 1991). Five rho subfamily members have been identified in *Saccharomyces cerevisiae* (RHO 1–4 and CDC42) (Madaule et al 1987; Matsui & Toh-e 1992; Johnson & Pringle 1990)

Function of Rho

The first suggestion that rho function might be linked to the actin cytoskeleton came from studies using C3 transferase, an exoenzyme produced by the bacterium *Clostridium botulinum.* This 25-kd protein, which can be introduced into cells by microinjection, or by addition to tissue culture medium, ADP-ribosylates rho proteins on an asparagine residue at codon 41 (Aktories et al 1989; Sekine et al 1989). Its effects on Vero cells or on fibroblasts are dramatic; within 15 min, cell rounding occurs accompanied by a complete loss of actin stress fibers (Chardin et al 1989; Paterson et al 1990). Paterson et al were later able to show that ribosylation by C3 transferase inactivates rho, consistent with ribosylation occurring in a region that, by analogy with ras, would be expected to be required for interaction with downstream targets (Paterson et al 1990). In complementary experiments, constitutively activated recombinant rho protein microinjected into fibroblasts or into the epithelial cell line MDCK induced changes in cell shape and stimulated the formation of actin stress fibers (Paterson et al 1990).

As previously mentioned, the organization of the actin cytoskeleton in many cells is critically dependent on extracellular factors, and Swiss 3T3 fibroblasts have been found to be particular sensitive. Overnight serum-starvation, for example, results in almost complete dissolution of the actin cytoskeleton and disassembly of focal adhesions, which leaves only a faint cortical ring of polymerized actin visible with fluorescently labeled phalloidin, shown in Figure 4a (Ridley & Hall 1992). Readdition of serum (the active constituent of which is lysophosphatidic acid, LPA) induces the rapid reformation of actin stress fibers (Figure 4c) and the assembly of focal adhesions (Figure 4d); this action can be completely blocked if quiescent cells are first microinjected with C3 transferase to inactivate endogenous rho proteins (Ridley & Hall 1992). A rho-regulated signal transduction pathway is then established in Swiss 3T3 cells that links the LPA receptor (and the bombesin receptor) to the assembly of focal adhesions and actin stress fibers.

The function of rho has been probed in other cell types using C3 transferase. In a lymphocyte aggregation assay, cell/cell interactions mediated by the

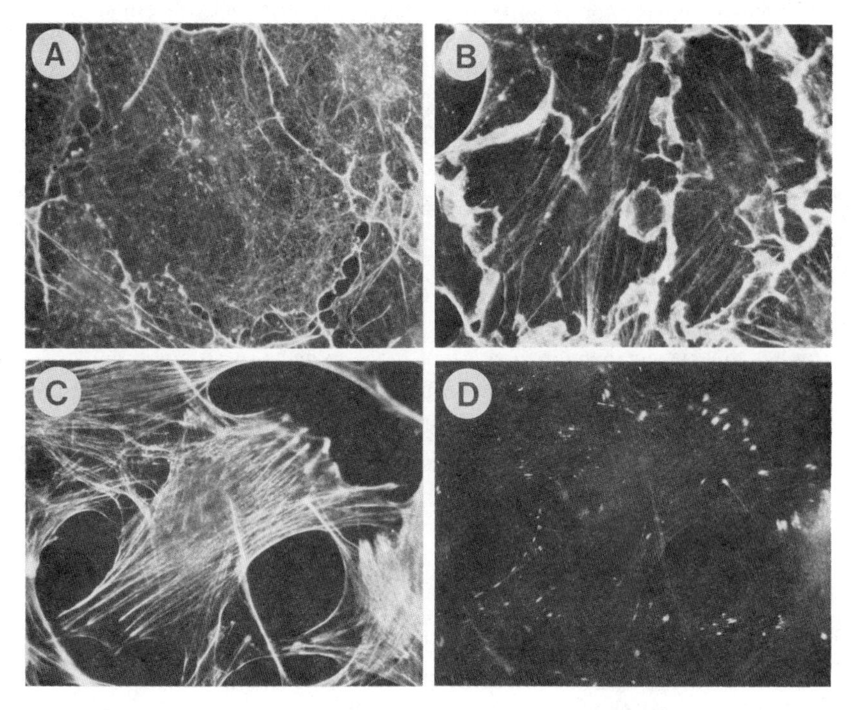

Figure 4 Rho and rac control different aspects of the actin cytoskeleton in Swiss 3T3 fibroblasts. (*a*) Polymerized actin in quiescent cells, (*b*) polymerized actin in PDGF-treated cells, (*c*) polymerized actin in LPA-treated cells, (*d*) focal adhesions in LPA-treated cells. Polymerized actin is visualized using fluorescently labeled phalloidin, and focal adhesions are visualized with a fluorescently labeled antibody to vinculin.

activated leukocyte integrin LFA-1 were inhibited by C3 transferase, but not by cytochalasin, an inhibitor of actin polymerization (Tominaga et al 1993). This strongly supports the idea expressed previously that the primary role of rho is to regulate the assembly of functional integrin complexes and that the formation of actin stress fibers seen in Swiss 3T3 cells is a secondary consequence of this activity (Hall 1992a). One possibility is that rho, under the control of extracellular factors, promotes the assembly of certain key components of the focal adhesion and that one or more of these is capable of acting as a nucleation site for actin polymerization. Interestingly, however, immunofluorescence studies suggest that rho is itself not a stable component of the adhesion complex or of the actin cytoskeleton (Adamson et al 1992).

In platelets, C3 transferase blocks thrombin-induced aggregation mediated by the platelet integrin glycoprotein IIb-IIIa, which suggests that rho may control not only intracellular events such as the organization of the actin cytoskeleton, but also the avidity of the platelet integrin for extracellular matrix proteins (Morii et al 1992). C3 transferase also blocks the function of natural killer cells, although in this case it has not been established whether inhibition is due to a block in the attachment of killer cells to the target cell or to a block in the delivery of cytotoxic granules (Lang et al 1992).

An alternative approach that has been used to inactivate cellular rho is microinjection of rhoGDI, a protein that forms a 1:1 complex with all members of the rho subfamily. In this way, it has been shown that rho function is essential for cell motility of Swiss 3T3 fibroblasts (Takaishi et al 1993). Some care has to be taken in interpreting these observations since rhoGDI (and C3 transferase if used in excess) will block rac and CDC42 function. However, the result implies that rho-mediated formation of focal contacts (presumably in combination with localized actin polymerization to produce lamellipodia) is required for cell motility. A similar conclusion has been reached in neutrophils since chemotaxis, but not superoxide production or degranulation, is blocked by C3 transferase (Stasia et al 1991).

RHO AND SECOND MESSENGERS The relationship between rho function and the second messengers described above has not been studied extensively. In Swiss 3T3 fibroblasts, the classical second messengers PIP_2, DAG, Ca^{2+}, and cAMP, generally assumed to be important for actin polymerization, are not required in the rho-mediated signal transduction pathway linking the LPA or bombesin receptors to focal adhesion or actin stress fiber formation (Ridley & Hall 1994). There is evidence, however, that a genestein-sensitive tyrosine kinase is required.

The role of PI 3-kinase in the rho pathway, if any, is not clear. It has been shown that activation of PI 3-kinase in Swiss 3T3 cells and in a platelet cytosolic extract is dependent on rho (Zhang et al 1993; Kumagi et al 1993). This would suggest that the enzyme lies downstream of rho, but does not address whether it is necessary for the rho-mediated effects on actin polymerization.

Function of Rac

Microinjection of recombinant rac protein into serum-starved Swiss 3T3 cells has a dramatic effect on the actin cytoskeleton that is quite distinct from rho-induced effects; it stimulates the rapid polymerization of actin at the plasma membrane to produce lamellipodia and membrane ruffles (Ridley et al 1992). These responses are similar to those seen after the addition of a variety of growth factors such as PDGF, EGF, or insulin, as shown in Figure 4b. Microinjection of a dominant negative mutant rac protein prior to addition of these growth factors completely inhibits ruffling and lamellipodia formation, although it has no effect on LPA-induced stress fiber formation. Thus a distinct rac-regulated signal transduction pathway exists in Swiss 3T3 cells linking plasma membrane receptors to the polymerization of actin at the plasma membrane.

Rac is a very poor substrate for C3 transferase, and its role in other cell types has not been easy to address (Ridley et al 1992). However, ruffling and lamellipodia are common early effects observed in many cell types challenged with agonists, and it is likely that rac will prove to be involved in mediating these responses (Kadowaki et al 1986; Downey et al 1992). The biological significance of membrane ruffling is not known, but it involves localized polymerization of actin to produce short, highly cross-linked filaments underneath the plasma membrane (Bretscher 1991; Mitchison & Kirschner 1988). There is evidence that pinocytosis occurs through membrane ruffling, and cells overexpressing rac have very high levels of pinocytosis (Haigler et al 1979; Bar-Sagi & Feramisco 1986; Ridley et al 1992). Lamellipodia, on the other hand, are key features of motile cells and are particularly apparent during chemotaxis. It is expected, therefore, that rac will play a major role in different kinds of cell locomotion, but this has not been directly demonstrated.

The interaction of a number of types of bacteria with mammalian cells can lead to membrane ruffling. *Salmonella typhimurium,* for example, will induce ruffles on many different cell types including fibroblasts and epithelial cells (Francis et al 1993). *Salmonella* also induces actin polymerization and ruffles in quiescent Swiss 3T3 cells, but in contrast to growth factor addition, this is not blocked by a dominant negative rac protein (Jones et al 1993). It appears that binding of *Salmonella* to its surface receptor (currently unknown) can bypass the rac-dependent step required for actin polymerization.

The biochemical mechanism of the action of rac is unclear at this time, although there has been one report that rac1 and CDC42 interact with a brain serine/threonine kinase, $p65^{PAK}$, which is related in sequence to the STE20 kinase of *S. cerevisiae* (Manser et al 1994). There is good evidence that actin polymerization occurs at discrete foci on the plasma membrane to induce ruffles and lamellipodia, and it is possible that rac regulates the formation of multimolecular protein complexes at these sites (Bretscher 1991; Symons & Mitchison 1991; DePasquale & Izzard 1991). Unlike focal adhesions, however, which are centered around clusters of integrins, the structure of these complexes has not been clearly defined.

A SPECIALIZED FUNCTION FOR RAC IN PHAGOCYTES Several groups have reported that activation of an NADPH oxidase in phagocytic cells of the immune system is dependent on rac1 or rac2 (Abo et al 1991; Knaus et al 1991). Activation of this enzyme, leading to the production of superoxide radicals, occurs in vivo in response to microbial infection, and it has been known for some time that this response depends upon the formation of a complex between the two subunits, membrane-bound cytochrome b (gp91phox and p21phox), and two proteins (p67-phox and p47-phox) recruited from the cytosol. The absence of any one of these four components blocks oxidase activation, which results in the life-threatening disease chronic granulomatous disease (CGD) (for review see Segal & Abo 1993). Further biochemical analysis of the NADPH oxidase activity in vitro has allowed the identification of a fifth component that is essential for activity, namely rac. An active oxidase has now been reconstituted in vitro using just five proteins: homogeneously purified gp91 and p21, recombinant p47 and p67, and recombinant rac in the GTP-bound form (Abo et al 1993). This is the first in vitro assay for a ras-related GTPase where all the components are known, and the role of rac is currently under investigation. One possibility is that rac interacts with one or more of the other proteins to induce a conformational change that permits self-assembly of the complex, while an alternative possibility is that the complex can form in the absence of rac but that rac acts as an allosteric regulator of enzymatic activity.

It is interesting to note that NADPH oxidase activation occurs concomitantly with phagocytosis after microbial infection. Phagocytosis is a membrane actin-driven event that bears some similarities to membrane ruffling, and the possibility that rac coordinately regulates the oxidase and phagocytosis has been suggested but not tested (Greenberg et al 1991; Hall 1992a).

Function of CDC42

The biological function of CDC42 has not been examined in mammalian cells, although in vitro the protein can bind to a brain-specific, nonreceptor tyrosine kinase, ACK (Manser et al 1993). In *S. cerevisiae* genetic analysis has identified CDC42 as essential for the assembly of components at the bud site of dividing yeast (Adams et al 1990; Johnson & Pringle 1990). Bud formation in *S. cerevisiae* is genetically determined and defines both the site at which the daughter cell is formed and the polarity and organization of the actin cytoskeleton (Drubin 1991). Mutations in *CDC42* lead to a general enlargement of the yeast cell (i.e. no buds) and disruption of the actin cytoskeleton.

Other studies have revealed that the site of bud formation is determined by another ras-related protein, RSR1 (homologue of mammalian rap1), and that mutations in the *RHO3* and *RHO4* genes also affect bud growth (Bender & Pringle 1989; Matsui & Toh-e 1992). It appears, therefore, that a hierarchy of small GTPases encoded by *RSR1, CDC42*, and *RHO* control the positioning and assembly of the bud during yeast cell division. Both rho and CDC42 have been implicated in the formation of the contractile ring during cytokinesis in animal cells; just how related this process will be to yeast cell division is too early to tell (Hart et al 1993; Kishi et al 1993; A Hall & S Brill, unpublished results).

Regulation of Rho-Related Proteins

The activation of ras by the EGF receptor is mediated by a ras guanine nucleotide exchange factor, sos, coupled to the receptor through an adapter protein, grb2, but the mechanism of activation of rho and rac by growth factor receptors is not known (Buday & Downward 1993). There has been one report that antibodies to grb2 injected into Swiss 3T3 cells inhibit membrane ruffling and cell growth but not stress fiber formation, which suggests that grb2 couples growth factor receptors to a rac nucleotide exchange factor (Matuoka et al 1993). However, these experiments are complicated by the possibility that antibodies raised against grb2 might cross-react with other SH2/SH3-containing proteins.

Another possibility is that other known signaling molecules activated by growth factor receptors such as PLCγ and PI 3-kinase lead to activation of rac or rho. It is well known, for example, that activators of protein kinase C stimulate many rho- and rac-dependent processes (e.g. lymphocyte aggregation, NADPH oxidase activation, membrane ruffling), and since their effects are blocked by inhibiting rho or rac, this suggests that PKC may lie upstream of these GTPases (Tominaga et al 1993; Ridley et al 1992). There is good evidence now that PI 3-kinase acts somewhere upstream of rac, since an inhibitor of the kinase, wortmannin, blocks PDGF but not microinjected rac-induced ruffling (Wennstrom et al 1994; Yano et al 1993; C Nobes & A Hall, unpublished results). In general, it seems that a variety of classical second messengers are able to activate rho and rac, presumably by regulating guanine nucleotide exchange factors.

The mechanism of activation of small GTP-binding proteins by nontyrosine kinase receptors such as those for bombesin and LPA is also not clear. LPA, which probably acts via a G protein-coupled 7 pass transmembrane receptor, has been shown to activate ras through a Gi-like protein since it can be blocked by pertussis toxin (van der Bend 1992; van Corven et al 1993). However, a tyrosine kinase is required in addition (van Corven et al 1993). In contrast, the activation of rho by LPA or by bombesin in Swiss 3T3 cells is not blocked by pertussis toxin, nor can rho be activated by addition of PMA, calcium ionophores, or by affecting cAMP levels (Ridley & Hall 1994). It remains a strong possibility that tyrosine kinases play a key role in the activation of rho and rac by G protein-coupled receptors.

NUCLEOTIDE EXCHANGE FACTORS The only candidate exchange factors for the rho subfamily so far described have emerged from the observation that the product of the *dbl* oncogene catalyzes nucleotide exchange on CDC42 (Hart et al 1991). A clue that dbl might be an exchange factor came from the observation that the product of the *CDC24* gene in *S. cerevisiae* lies genetically upstream of *CDC42* (thus suggesting that it might be a nucleotide exchange factor) and the CDC24 protein has 29% sequence identity to dbl (Ron et al 1991). Although expression of normal dbl is restricted to the brain, a family of proteins harboring a dbl-like domain has been identified, including the ubiquitously expressed breakpoint cluster region protein (bcr), two oncoproteins (ect2 and vav), and the amino terminal end of a brain-specific ras nucleotide exchange factor, ras GRF (Adams et al 1992; Miki et al 1993; Shou et al 1992). Ect2 forms a tight complex with rho, rac, and CDC42 but has no measurable exchange activity in vitro (Miki et al 1993). Whether these proteins provide a link between receptors and rho-like GTPases remains to be seen; an unexpected report that vav can catalyze nucleotide exchange on ras, but not on rho, suggests that this family of proteins deserves closer attention (Gulbins et al 1993; Boguski & McCormick 1993).

One other potential exchange factor, smgGDS, has been characterized and cloned, but this appears not to be specific for the rho subfamily since it is also active on Ki-ras and rap1 (Kikuchi et al 1992).

GTPASE-ACTIVATING PROTEINS Work on ras has revealed another class of regulatory proteins that can affect small GTPases, namely GTPase-activating proteins or GAPs (Boguski & McCormick 1993). Although GAPs are clearly potential downregulators of GTPase function, there is evidence that they can act as downstream effectors and contribute to signal transduction (Hall 1990b). A GAP for rho, rhoGAP, was first identified in human spleen, but since then four other proteins having GAP activity have been identified in mammalian cells (bcr, chimerin, p190, and abr) and one in *S. cerevisiae,* Bem3 (Garrett et al 1991; Diekmann et al 1991; Settleman et al 1992; Barford et al 1993; Lancaster et al 1994; Heisterkamp et al 1993; Zheng et al 1993). In addition, the p85 subunit of PI 3-kinase and a partially characterized clone, 3BP-1, have significant sequence homology to the catalytic domain of rhoGAP (Otsu et al 1991; Cicchetti et al 1992). The function of these multidomained proteins is not known, although many of them have been implicated in signal transduction pathways (Hall 1992b). It is possible that they are downregulators of rho GTPases acting at different cellular locations or in different molecular complexes. Alternatively, the rhoGAP domain could function as a signal for recruitment into molecular complexes by rho-related proteins.

In vitro, the GAPs interact with multiple members of the rho family, although there are some differences in relative activities (Barford et al 1993; Lancaster et al 1994). Microinjection experiments with isolated GAP domains also suggest that they might have some intrinsic specificity; p190 and rhoGAP, for example, interfere with rho signaling but not with rac in Swiss 3T3 cells, while the reverse is true for bcr (Ridley et al 1993).

P190GAP is of particular interest since it was first identified as a rasGAP-binding protein, which raised the possibility that the ras and rho signaling pathways are coordinately controlled (Settleman et al 1992). Further support for this idea has come from the observations that the noncatalytic (amino terminal) domain of rasGAP, when expressed in cells, interacts constitutively with p190 and that the actin cytokeleton becomes hypersensitive to serum deprivation (McGlade et al 1993).

RHOGDI The ubiquitous protein, rhoGDI, plays a critical role in regulating the activity of rho-like proteins and forms a 1:1 complex with all members of the subfamily (Ohga et al 1989; Hart et al 1992). In resting neutrophils, all the cellular rac is found complexed to rhoGDI, and upon stimulation, around 10% dissociates and translocates to the membrane (Quinn et al 1993). How this translocation step is coordinated with GDP/GTP exchange and what the signal is for GDI release are not known. In vitro, rhoGDI will solubilize rho-like GTPases from a membrane-bound form, and originally this action was thought to be specific for the GDP-bound form of the GTPase (Isomura et al 1991). This led to a cycling model, similar to that proposed for rab proteins, where the activated GTPase dissociates from rhoGDI and moves to its site of action at the plasma membrane. GTP is then hydrolzed to GDP, and the inactivated GTPase is removed from the membrane by rhoGDI thus completing the cycle. However, recent work has shown that rhoGDI can also interact with and solubilize the GTP form of CDC42 from membranes, at least in vitro, an observation that is hard to reconcile with the simple cycling model (Leonard et al 1992).

Rho-like GTPases and Cell Growth

Many observations point to an intimate link between the cell cycle and the organization of the actin cytoskeleton. Growing fibroblasts have well-defined actin stress fibers and focal adhesions throughout G1 and S phase but round up and lose their stress fibers during mitosis (Jackson & Bellett 1989). In addition, many cell types must be attached to a solid support during the G1 phase of the cell cycle in order to enter S (Guadagno & Assoian 1991). Interfering with integrin/cytoskeletal function on the other hand, using inhibitors of actin polymerization such as cytochalasin, or using C3 to inhibit rho, blocks the cell cycle in G1 (Maness & Walsh 1982; Yamamoto et al 1993). These observations support the idea that there is a signal transduction pathway from integrins and/or the actin cytoskeleton which, in combination with signals from growth factors (acting through ras), controls the ability of cells to traverse G1 of the cell cycle.

The nature of integrin-mediated signals is, however, unclear at this time but has been the subject of much speculation (for reviews see Damsky & Werb 1992; Juliano & Haskill 1993). A recent paper suggests that a key element of the anchorage-dependent signaling system might be cyclin A and its associated kinase activity, which are expressed in an adhesion-dependent manner during late G1 (Guadagno et al 1993). In agreement with this proposal, constitutive expression of cyclin A in fibroblasts results in anchorage-independent growth.

Although the biochemical details of integrin signals have not been elucidated, there is evidence to suggest that the integrin and ras-mediated signal transduction pathways are coordinately regulated. For example, a component of the ras pathway, p120rasGAP, interacts directly with p190rhoGAP, a likely key player in the rho-mediated assembly of integrin complexes (Settleman et al 1992). Furthermore, cell transformation with ras leads to anchorage-independent growth and changes in the organization of the actin cytoskeleton, while candidate exchange factors for the rho subfamily (related to dbl) can themselves act as oncogenes (Lombardi et al 1990; Eva & Aaronson 1985).

Further support for a link between ras and the actin cytoskeleton comes from *S. cerevisiae* with the identification of a protein, CAP, that is essential for ras-mediated activation of adenylyl cyclase. It turns out that only the N-terminal domain of CAP is required in the ras pathway, but deletion of its C-terminal end leads to disruption of the actin cytoskeleton, a defect that can be rescued by profilin (Vojtek et al 1991). It appears that in yeast, CAP is responsible for coordinating ras-induced signals with those that control the actin cytoskeleton. A mammalian counterpart of CAP has been purified as an actin-binding protein from pig platelets, and human and mouse cDNAs have been cloned (Gieselman & Mann 1992; Matviw et al 1992; Vojtek & Cooper 1993). The protein seems to localize to the leading edge of motile cells, and it will be of great interest to see what the role of CAP might be in mammalian signal transduction pathways (Vojtek & Cooper 1993).

CONCLUSIONS

There have been many reports that classical second messengers such as PIP_2, DAG, Ca^{2+}, cAMP, and more recently PIP_3 control the organization of the actin cytoskeleton in response to extracellular signals, but their sites and mechanisms of action have remained unresolved. It has now been established that the rho subfamily of ras-related, GTP-binding proteins are key regulators of the organized polymerization of actin in all eukaryotic cells, and it appears increasingly likely that at least some of the effects ascribed to second messengers are due to their effects on rho and rac.

Once activated, rho and rac promote the assembly of clustered multimolecular complexes at the plasma membrane, components of which act as nucleation sites for actin polymerization. Some of these ideas for rho are summarized in Figure 5, where experiments carried out so far have linked its activity to integrin function both at focal adhesions and at cell-cell contacts. The mechanism by which rho controls the assembly of functional integrin complexes and the

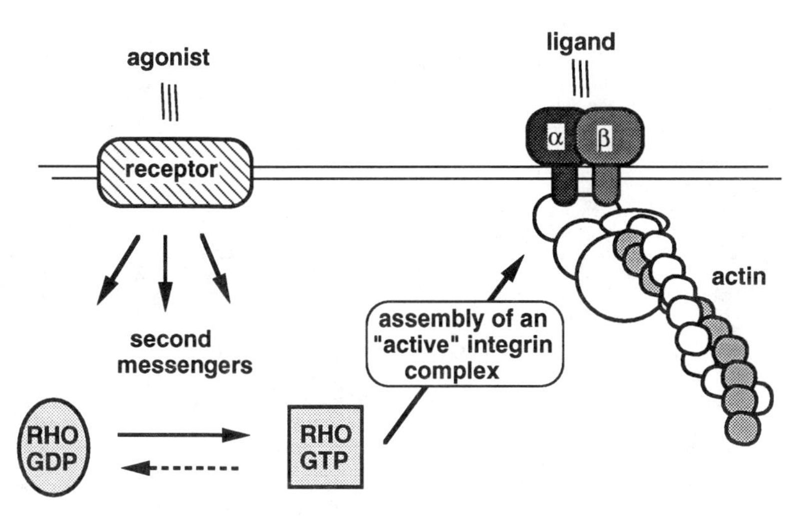

Figure 5 Schematic representation of rho function. The activity of rho is controlled by extracellular agonists acting on plasma membrane receptors. The biochemical mechanism by which rho is activated is unknown, but may be through second messengers acting on nucleotide exchange factors. In the presence of integrin ligands (extracellular matrix or cell surface proteins) active, GTP-bound rho promotes the formation of clusters of integrins associated with a variety of cytoplasmic proteins and polymerized actin. There is evidence that these complexes can themselves generate signals that affect cell behavior.

subsequent polymerization of actin is not known at this time: it could promote protein:protein interactions by inducing conformational changes in key components or, alternatively, it could regulate an enzymatic activity such as a kinase that subsequently results in the formation of protein complexes. It is clearly important to identify protein targets for rho in order to distinguish these possibilities. Rac's role in actin polymerization also appears to be confined to regulating the formation of multimolecular structures located at the plasma membrane. In this case, actin polymerization is qualitatively different from that induced by rho and leads to membrane restructuring to produce lamellipodia and ruffles.

The next few years should see real progress in elucidating the signal transduction pathways downstream of rho, rac, and other related GTPases, which in turn should go a long way toward clarifying the molecular basis of numerous biological processes that depend upon the intrinsic ability of actin to polymerize.

Literature Cited

Abo A, Boyhan A, West I, Thrasher AJ, Segal AW. 1993. Reconstitution of neutrophil NADPH oxidase in the cell free system by four components: p67-phox, p47-phox, p21rac1 and cytochrome b. *J. Biol. Chem.* 267:16767–70

Abo A, Pick E, Hall A, Totty N, Teahan CG, Segal AW. 1991. Activation of the NADPH oxidase involves the small GTP-binding protein p21rac1. *Nature* 353:668–69

Adams AEM, Johnson DI, Longnecker RM, Sloat BF, Pringle JR. 1990. *CDC42* and *CDC43*, two additional genes involved in budding and the establishment of cell polarity in the yeast *Saccharomyces cerevisiae. J. Cell Biol.* 111:131–42

Adams JM, Houston H, Allen J, Lints T, Harvey R. 1992. The haematopoietically expressed vav proto-oncogene shares homology with the dbl GDP/GTP exchange factor, the *bcr* gene and a yeast gene *CDC42* involved in cytoskeletal organization. *Oncogene* 7:611–18

Adamson P, Paterson H, Hall A. 1992. Intracellular localization of the p21rho proteins. *J. Cell Biol.* 119:617–27

Aktories K, Braun U, Rosener S, Just I, Hall A. 1989. The rho gene product expressed in *E. coli* is a substrate of botulinum ADP-ribosyltransferase C3. *Biochem. Biophys. Res. Comm.* 158:209:13

Barford ET, Zheng Y, Kuang WJ, Hart MJ, Evans T, et al. 1993. Cloning and expression of a human CDC42 GTPase activating protein reveals a functional SH3 binding domain. *J. Biol. Chem.* 268:26059–62

Bar-Sagi D, Feramisco JR. 1986. Induction of membrane ruffling and fluid phase pinocytosis in quiescent fibroblasts by ras proteins. *Science* 233:1061–68

Bender A, Pringle JR. 1989. Multicopy suppression of cdc24 budding defect in yeast by *CDC42* and three newly identified genes including ras-related *RSR1* gene. *Proc. Natl. Acad. Sci. USA* 86:9976–80

Boguski MS, McCormick F. 1993. Proteins regulating ras and its relatives. *Nature* 366:643–54

Bollag G, McCormick F. 1991. Regulators and effectors of ras proteins. *Annu. Rev. Cell Biol.* 7:601–32

Bourne HR, Sanders DA, McCormick F. 1991. The GTPase superfamily: conserved structure and molecular mechanism. *Nature* 349: 117–26

Bretscher A. 1991. Microfilament structure and function in the cortical cytoskeleton. *Annu. Rev. Cell Biol.* 7:337–74

Buday L, Downward J. 1993. Epidermal growth factor regulates p21ras through the formation of a complex of receptor, grb2 adapter protein and sos nucleotide exchange factor. *Cell* 73:611–20

Burridge K, Fath K, Kellt T, Nuckolls G, Turner C. 1988. Focal adhesions: transmembrane junctions between the extracellular matrix and the cytoskeleton. *Annu. Rev. Cell Biol.* 4:487–523

Burridge K, Turner CE, Romer LH. 1993. Tyrosine phosphorylation of paxillin and pp125FAK accompanies cell adhesion to extracellular matrix: A role in cytoskeletal assembly. *J. Cell Biol.* 119:893–903

Chardin P, Boquet P, Madaule P, Popoff MR, Rubin EJ, Gill DM. 1989. The mammalian G protein rhoC is ADP-ribosylated by *Clostridium botulinum* exoenzyme C3 and affects actin microfilaments in Vero cells. *EMBO J.* 8:1087–92

Chardin P, Madaule P, Tavitian A. 1988. Coding sequence of human rho cDNAs clone 6 and clone 9. *Nucleic Acids Res.* 16:2717

Cicchetti P, Mayer BJ, Thiel G, Baltimore D. 1992. Identification of a protein that binds to the SH3 region of abl and is similar to bcr and GAP-rho. *Science* 257:803–6

Dadabay CY, Patton E, Cooper JA, Pike LJ. 1991. Lack of correlation between changes in polyphosphoinositide levels and actin/gelsolin complexes in A431 cells treated with epidermal growth factor. *J. Cell Biol.* 112:1151–56

Damsky CH, Werb Z. 1992. Signal transduction by integrin receptors for extracellular matrix: cooperative processing of extracellular information. *Curr. Biol.* 4:772–81

DePasquale JA, Izzard CS. 1991. Accumulation of talin in nodes at the edge of the lamellipodium and separate incorporation into adhesion plaques at focal contacts in fibroblasts. *J. Cell Biol.* 113:1351–59

Devreotes PN, Zigmond SH. 1988. Chemotaxis in eukaryotic cells. *Annu. Rev. Cell Biol.* 4: 649–86

Didsbury J, Weber Rf, Bokoch GM, Evans T, Snyderman R. 1989. Rac, a novel ras-related family of proteins that are botulinum toxin substrates. *J. Biol. Chem.* 264:16378–82

Diekamnn D, Brill S, Garrett MD, Totty N, Hsuan J, et al. 1991. Bcr encodes a GTPase activating protein for p21rac. *Nature* 351: 400–2

Domann E, Wehland J, Rohed M, Pistor S, Hartl M, et al. 1992. A novel bacterial virulence gene in *Listeria monocytogenes* required for host cell microfilament interaction with homology to the proline-rich region of vinculin. *EMBO J.* 11:1981–90

Downey GP, Chan CK, Lea P, Takai A, Grinstein S. 1992. Phorbol ester-induced actin assembly in neutrophils: role of protein kinase C. *J. Cell Biol.* 116:695–706

Drivas GT, Shih A, Coutavas E, Rush MG, D'Eustachio P. 1990. Characterization of four novel ras-like genes expressed in a human teratocarcinoma cell line. *Mol. Cell. Biol.* 10:793–1798

Drubin DG. 1991. Development of cell polarity in budding yeast. *Cell* 65:1093–96

Eberle M, Traynor-Kaplan AE, Sklar LA, Norgauer J. 1990. Is there a relationship between phosphatidylinositol trisphosphate and F-actin polymerization in human neutrophils? *J. Biol. Chem.* 265:16725–28

Egan SE, Weinberg RA. 1993. The pathway to signal achievement. *Nature* 365:781–83

Eva A, Aaronson SA. 1985. Isolation of a new human oncogene from diffuse B-cell lymphoma. *Nature* 316:273–75

Falkow S, Isberg RR, Portnoy DA. 1992. The interaction of bacteria with mammalian cells. *Annu. Rev. Cell Biol.* 8:333–63

Francis CL, Ryan TA, Jones BD, Smith SJ, Falkow S. 1993. Ruffles induced by *Salmonella* and other stimuli direct macropinocytosis of bacteria. *Nature* 364:639–42

Fuortes M, Jin WW, Nathan C. 1993. Adhesion-dependent protein tyrosine phosphorylation in neutrophils with tumor necrosis factor. *J. Cell Biol.* 120:777–84

Garrett MD, Major GN, Totty N, Hall A. 1991. Purification and N-terminal sequence of the p21rho GTPase activating protein. *Biochem. J.* 276:833–36

Geiger B, Ginsberg D, Salomon D, Volberg T. 1990. The molecular basis for the assembly and modulation of adherens-type junctions. *Cell Diff. Dev.* 32:343–53

Gieselman R, Mann K. 1992. ASP-56, a new actin sequestering protein from pig platelets with homology to CAP, an adenylyl cyclase associated protein from yeast. *FEBS Lett.* 298:149–53

Goldschmidt-Clermont PJ, Machesky LM, Doberstein SK, Pollard TD. 1991. Mechanism of the interaction of human platelet profilin with actin. *J. Cell Biol.* 113:1081–89

Goldschmidt-Clermont PJ, Furman MI, Wachsstock D, Safer D, Nachmias VT, Pollard TD. 1992. The control of actin nucleotide exchange by thymosin b4 and profilin. A potential regulatory mechanism for actin polymerization in cells. *Mol. Biol. Cell.* 3: 1015–20

Greenberg S, El Khoury J, DiVirgilio F, Kaplan EM, Silverstein SC. 1991. Ca^{2+} independent F-actin assembly and disassembly during Fc receptor-mediated phagocytosis in mouse macrophages. *J. Cell Biol.* 113:747–67

Guadagno TM, Assoian RK. 1991. G1/S control of anchorage independent growth in the fibroblast cell cycle. *J. Cell Biol.* 115:1419–25

Guadagno TM, Ohtsubo M, Roberts JM, Assoian RK. 1993. A link between cyclin A expression and adhesion-dependent cell cycle progression. *Science* 262:1572–75

Gulbins E, Coggeshall KM, Baier G, Katzav S, Burn P, Altman A. 1993. Tyrosine kinase-stimulated guanine nucleotide exchange activity of vav in T-cell activation. *Science* 260:822–25

Gumbiner BM. 1993. Proteins associated with the cytoplasmic surface of adhesion molecules. *Neuron* 11:551–64

Haigler HT, McKanna JA, Cohen S. 1979. Rapid stimulation of pinocytosis in human carcinoma cells A431 by epidermal growth factor. *J. Cell Biol.* 83:82–90

Hall A. 1990a. The cellular function of small GTP-binding proteins. *Science* 249:635–40

Hall A. 1990b. ras and GAP - who's controlling whom. *Cell* 61:921–23

Hall A. 1992a. Ras-related GTPases and the cytoskeleton. *Mol. Biol. Cell* 3:475–79

Hall A. 1992b. Signal transduction through small GTPases - A tale of two GAPs. *Cell* 60:389–91

Hart MJ, Eva A, Evans T, Aaronson SA, Cerione RA. 1991. Catalysis of guanine nucleotide exchange on the CDC42Hs protein by the dbl oncogene product. *Nature* 354: 311–14

Hart MJ, Leonard D, Zheng Y, Shinjo K, Evans T, Cerione RA. 1993. The mammalian homolog of the yeast cell division cycle protein CDC42. In *GTPases in Biology,* ed BF Dickey, L Birnbaumer, 1:579–95. Berlin: Springer-Verlag. 697 pp.

Hart MJ, Maru Y, Leonard D, Witte ON, Evans T, Cerione RA. 1992. A GDP dissociation inhibitor that serves as a GTPase inhibitor for the ras-like protein CDC42Hs. *Science* 258:812–15

Heisterkamp N, Kaartinen V, van Soest S, Bokoch G, Groffen J. 1993. Human abr encodes a protein with GAPrac activity and homology to the dbl nucleotide exchange factor domain. *J. Biol. Chem.* 268:16903–6

Isomura M, Kikuchi A, Ohga N, Takai Y. 1991. Regulation of binding of rhoB p20 to membranes by its specific regulatory protein, GDP dissociation inhibitor. *Oncogene* 6: 119–24

Jackson P, Bellett AJD. 1989. Relationship between organization of the actin cytoskeleton and the cell cycle in normal and adenovirus-infected rat cells. *J. Virol.* 63:311–18

Jahner D, Hunter T. 1991. The ras-related gene *rhoB* is an immediate early gene inducible by v-fps, epidermal growth factor and platelet derived growth factor in rat fibroblasts. *Mol. Cell. Biol.* 11:3682–90

Johnson DI, Pringle JR. 1990. Molecular characterization of *CDC42*, a *Saccharomyces cerevisiae* gene involved in the development of cell polarity. *J. Cell Biol.* 111:143–52

Jones BD, Paterson HF, Hall A, Falkow S. 1993. *Salmonella typhimurium* induces membrane ruffling by a growth factor receptor independent mechanism. *Proc. Natl. Acad. Sci. USA* 90:10390–94

Juliano RL, Haskill S. 1993. Signal transduction from the extracellular matrix. *J. Cell Biol.* 120:577–85

Kadowaki T, Koyasu S, Nishida E, Sakai H, Takaku F, et al. 1986. Insulin-like growth factors, insulin, and epidermal growth factor cause rapid cytoskeletal reorganization in KB cells. *J. Biol. Chem.* 261:16141–47

Kaufmann S, Kas J, Goldmann WH, Sackman E, Isenberg G. 1992. Talin anchors and nucleates actin filaments at lipid membranes. A direct demonstration. *FEBS Lett.* 314:203–5

Kikuchi A, Kuroda S, Sasaki T, Kotani K, Hirata K, et al. 1992. Functional interactions of stimulatory and inhibitory GDP/GTP exchange proteins and their common substrate small GTP-binding protein. *J. Biol. Chem.* 267:14611–15

Kishi K, Sasaki T, Kuroda S, Itoh T, Takai Y. 1993. Regulation of cytoplasmic division of *Xenopus* embryo by p21rho and its inhibitory GDP/GTP exchange protein (rhoGDI). *J. Cell Biol.* 120:1187–95

Knaus UG, Heyworth PG, Evans T, Curnutte JT, Bokoch GM. 1991. Regulation of phagocyte oxygen radical production by the GTP-binding protein rac2. *Science* 254:1512–15

Kornberg L, Earp HS, Parsons JT, Schaller M, Juliano RL. 1992. Cell adhesion or integrin clustering increases phosphorylation of a focal adhesion-associated tyrosine kinase. *J. Biol. Chem.* 267:23439–42

Kumagi N, Morii N, Fujisawa K, Nemoto Y, Narumiya S. 1993. ADP-ribosylation of rho p21 inhibits lysophosphatidic acid induced protein tyrosine phosphorylation and phosphatidylinositol 3-kinase activation in cultured Swiss 3T3 cells. *J. Biol. Chem.* 268: 24535–38

Lancaster CA, Taylor-Harris P, Self A, Brill S, Van Erp HE, Hall A. 1994. Characterization of rhoGAP: A GTPase activating protein for rho-related small GTPases. *J. Biol. Chem.* 269:1137–42

Lang P, Guizani L, Mony IV, Stancou R, Dorseuil O, et al. 1992. ADP-ribosylation of the ras-related GTP-binding protein rhoA inhibits lymphocyte mediated cytotoxicity. *J. Biol. Chem.* 267:11677–80

Lassing I, Lindberg U. 1988. Specificity of the interaction between phosphatidylinositol 4,5 bisphosphate and the profilin:actin complex. *J. Cell Biol.* 37:255–67

Leonard D, Hart MJ, Platko JV, Eva A, Henzel W, et al. 1992. The identification and characterization of a GDP-dissociation inhibitor (GDI) for the CDC42Hs protein. *J. Biol. Chem.* 267:22860–68

Lombardi L, Ballinari D, Bongarzone I, Migliari M, Mondellini P, et al. 1990. Ultrastructural cytoskeleton alterations and modification of actin expression in the NIH 3T3 cell line after transformation with Ha-ras activated oncogene. *Cell Motil. Cytoskeleton* 15:220–29

Madaule P, Axel R. 1985. A novel ras-related gene family. *Cell* 42:31–40

Madaule P, Axel R, Myers AM. 1987. Characterization of two members of the *rho* gene family from the yeast *Saccharomyces cerevisiae. Proc. Natl. Acad. Sci. USA* 84: 779–83

Maness PF, Walsh RC. 1982. Dihydrocytochalasin B disorganizes actin cytoarchitecture and inhibits initiation of DNA synthesis in 3T3 cells. *Cell* 30:253–62

Manser E, Leung T, Salihuddin H, Tan L, Lim L. 1993. A non-receptor tyrosine kinase that inhibits the GTPase activity of p21cdc42. *Nature* 363:364–67

Manser E, Leung T, Salihuddin H, Zhao Z, Lim L. 1994. A brain serine/threonine protein kinase activated by Cdc42 and rac1. *Nature* 367:40–46

Matsudaira PT, Janmey PA. 1988. Pieces in the actin-severing protein puzzle. *Cell* 54:139–40

Matsui Y, Toh-e A. 1992. Yeast *RHO3* and *RHO4* ras superfamily genes are necessary for bud growth and their defect is suppressed by a high dose of bud formation genes *CDC42* and *BEM1*. *Mol. Cell. Biol.* 12:5690–99

Matuoka K, Shibasaki F, Shibata M, Takenawa T. 1993. Ash/grb2, a SH2/SH3-containing protein couples to signaling for mitogenesis and cytoskeletal reorganization by EGF and PDGF. *EMBO J.* 12:3467–73

Matviw H, Yu G, Young D. 1992. Identification of a human cDNA encoding a protein that is structurally and functionally related to the yeast adenylyl cyclase-associated CAP proteins. *Mol. Cell. Biol.* 12:5033–40

McGlade J, Brunkhorst B, Anderson D, Mbamalu G, Settleman J, et al. 1993. The N-terminal region of GAP regulates cytoskeletal structure and cell adhesion. *EMBO J.* 12:3073–81

Miki T, Smith CL, Long JE, Eva A, Fleming TP. 1993. Oncogene ect2 is related to regulators of small GTP-binding proteins. *Nature* 362:462–65

Mitchison T, Kirschner M. 1988. Cytoskeletal dynamics and nerve growth. *Neuron* 1:761–72

Morii N, Teru-uchi T, Tominaga T, Kumagai N, Kozaki S, et al. 1992. A rho gene product in human platelets. *J. Biol. Chem.* 267: 20921–26

Munemitsu S, Innis MA, Clark R, McCormick F, Ulrich A, Polakis P. 1990. Molecular cloning and expression of a G25K cDNA; the human homolog of the yeast cell cycle gene *CDC42*. *Mol. Cell. Biol.* 10:5977–82

Ohga N, Kikuchi A, Ueda T, Yamamoto J, Takai Y. 1989. Rabbit intestine contains a protein that inhibits the dissociation of GDP from and the subsequent binding of GTP to rhoB p20, a ras p21-like GTP-binding protein. *Biochem. Biophys. Res. Comm.* 163: 1523–33

Okabe S, Hirokawa N. 1989. Incorporation and turnover of biotin-labelled actin microinjected into fibroblastic cells: an immunoelectron microscope study. *J. Cell Biol.* 109:1581–95

Otsu M, Hiles I, Gout I, Fry M, Ruiz-Larrea F, et al. 1991. Characterization of two 85 kDa proteins that associate with receptor tyrosine kinases, middle T, pp60src complexes and PI 3-kinase. *Cell* 64:91–104

Panataloni D, Carlier MF. 1993. How profilin promotes actin filament assembly in the presence of thymosin b4. *Cell* 75:1007–14

Paterson HF, Self AJ, Garrett MD, Just I, Aktories K, Hall A. 1990. Microinjection of recombinant p21rho induces rapid changes in cell morphology. *J. Cell Biol.* 111:1001–7

Peppelenbosch MP, Tertoolen LGJ, Hage WJ, de Laat SW. 1993. Epidermal growth factor-induced actin remodelling is regulated by 5-lipoxygenase and cyclooxygenase products. *Cell* 74:565–75

Pollard TD, Cooper JA. 1986. Actin and actin-binding proteins: a critical evaluation of mechanisms and functions. *Annu. Rev. Biochem.* 55:987–1035

Pring M, Weber A, Bubb MR. 1992. Profilin-actin complexes directly elongate actin filaments at the barbed end. *Biochemistry* 31: 1827–36

Pryer NK, Wuestehube LJ, Schekman R. 1992. Vesicle mediated protein sorting. *Annu. Rev. Biochem.* 61:471–516

Quinn MT, Evans T, Loetterle LR, Jesaitis AJ, Bokoch GM. 1993. Translocation of rac correlates with NADPH oxidase activation. *J. Biol. Chem.* 268:20983–87

Ren R, Mayer BJ, Cicchetti P, Baltimore D. 1993. Identification of a ten-amino acid proline-rich SH3 binding site. *Science* 259: 1157–61

Ridley AJ, Hall A. 1992. The small GTP-binding protein rho regulates the assembly of focal adhesions and actin stress fibres in response to growth factors. *Cell* 70:389–99

Ridley AJ, Hall A. 1993. Rho and rho-like proteins. See Hart el al 1993, pp. 563–77

Ridley AJ, Hall A. 1994. Signal transduction pathways regulating rho-mediated stress fibre formation: requirement for a tyrosine kinase. *EMBO J.* In press

Ridley AJ, Paterson HF, Johnston CL, Diekmann D, Hall A. 1992. The small GTP-binding protein rac regulates growth factor induced membrane ruffling. *Cell* 40: 401–10

Ridley AJ, Self AJ, Kasmi F, Paterson HF, Hall A, et al. 1993. Rho family GTPase activating protein p190, bcr and rhoGAP show distinct specificities in vitro and in vivo. *EMBO J.* 12:5151–60

Ron D, Zannini M, Lewis M, Wickner RB, Hunt LT, et al. 1991. A region of proto-dbl essential for its transforming activity shows sequence similarity to a yeast cell cycle gene, *CDC24* and the human breakpoint cluster region gene, *bcr*. *New Biol.* 3:372–79

Safer D. 1992. The interaction of actin with thymosin b4. *J. Musc. Res. Cell Motil.* 13: 269–71

Sato N, Funayama N, Nagafuchi A, Yonemura S, Tsukita S, Tsukita S. 1992. A gene family consisting of ezrin, radixin and moesin. *J. Cell Sci.* 103:131–43

Schaller MD, Borgman CA, Cobb BS, Vines RR, Reynolds AB, Parsons JT. 1992. pp125FAK, a structurally distinctive protein tyrosine kinase associated with focal adhesions. *Proc. Natl. Acad. Sci. USA* 89:5192–96

Segal AW, Abo A. 1993. The biochemical basis of the NADPH oxidase of phagocytes. *Trends Biochem. Sci.* 18:43–47

Sekine A, Fujiwara M, Narumiya S. 1989, Asparagine residue in the rho gene product is the modification site for botulinum ADP-ribosyltransferase. *J. Biol. Chem.* 264:8602–5

Settleman J, Albright CE, Foster LC, Weinberg RA. 1992. Association between GTPase activators for rho and ras families. *Nature* 359: 153–54

Shinjo K, Koland JG, Hart MT, Narasimhan V, Johnson DI, et al. 1990. Molecular cloning of the gene for the human placental GTP-binding protein Gp (G25K): Identification of this GTP-binding protein as the human homolog of the yeast cell division cycle protein CDC42. *Proc. Natl. Acad. Sci. USA* 87: 9853–57

Shou C, Farnsworth CL, Neel BG, Feig LA. 1992. Molecular cloning of cDNAs encoding a guanine nucleotide-releasing factor for ras p21. *Nature* 358:351–54

Sinnett-Smith J, Zachary I, Valverde AM, Rozengurt E. 1993. Bombesin stimulation of p125 focal adhesion kinase tyrosine phosphorylation. *J. Biol. Chem.* 268:14261–68

Stasia MJ, Jouan A, Bourmeyster N, Boquet P, Vignais PV. 1991. ADP ribosylation of a small size GTP-binding protein in bovine neutrophils by the C3 exoenzyme of *Clostridium botulinum* and effect on the cell motility. *Biochem. Biophys. Res. Comm.* 180: 615–22

Stossel TP. 1989. From signal to pseudopod. *J. Biol. Chem.* 264:18261–64

Stossel TP. 1993. On the crawling of animal cells. *Science* 260:1086–94

Symons MH, Mitchison TJ. 1991. Control of actin polymerization in live and permeabilized fibroblasts. *J. Cell Biol.* 114:503–13

Takaishi K, Kikuchi A, Kuroda S, Kotani K, Sasaki T, Takai Y. 1993. Involvement of rho p21 and its regulatory GDP/GTP exchange protein rhoGDI in cell motility. *Mol. Cell. Biol.* 13:72–79

Theriot JA, Mitchison TJ. 1993. The three faces of profilin. *Cell* 75:835–38

Thomas KM, Pyun HY, Navarro J. 1990. Molecular cloning of the fMet-Leu-Phe receptor from neutrophils. *J. Biol. Chem.* 265:20061–64

Tilney LG, DeRosier DJ, Weber A, Tilney MS. 1992. How Listeria exploits host actin to form its own cytoskeleton. II. Nucleation, actin filament polarity, filament assembly and evidence for pointed end capper. *J. Cell Biol.* 118:83–93

Tominaga T, Sugie K, Hirata M, Morii N, Fukata J, et al. 1993. Inhibition of PMA-induced LFA-1-dependent lymphocyte aggregation by ADP ribosylation of the small molecular weight GTP-binding protein, rho. *J. Cell Biol.* 120:1529–37

Turner CE, Burridge K. 1991. Transmembrane molecular assemblies in cell-extracellular matrix interactions. *Curr. Opin. Cell Biol.* 3:849–53

Ullrich A, Schlessinger J. 1990. Signal transduction by receptors with tyrosine kinase activity. *Cell* 61:200–12

van Corven EJ, Hordijk H, Hedema RH, Bos JL, Moolenaar WH. 1993. Pertussis toxin sensitive activation of p21ras by G protein coupled receptor agonists in fibroblasts. *Proc. Natl. Acad. Sci. USA* 90:1257–61

Vandekerckhove J. 1990. Actin binding proteins. *Curr. Opin. Cell Biol.* 2:41–50

van der Bend R, Brunner J, Jalink K, van Corven EJ, Moolenaar WH, Blitterswijk WJ. 1992. Identification of a putative membrane receptor for the bioactive phospholipid, lysophosphatidic acid. *EMBO J.* 11:2495–501

Vincent S, Jeanteur P, Fort P. 1992. Growth regulated expression of rhoG, a new member of the ras homolog gene family. *Mol. Cell. Biol.* 12:3138–48

Vojtek A, Haarer B, Field J, Gerst J, Pollard TD, et al. 1991. Evidence for a functional link between profilin and CAP in the yeast S. cerevisiae. *Cell* 66:497–505

Vojtek A, Cooper JA. 1993. Identification and characterization of a cDNA encoding mouse CAP: a homolog of the yeast adenylyl cyclase associated protein. *J. Cell Sci.* 105:777–85

Wennstrom S, Siegbahn A, Yokote K, Arvidsson AK, Heldin CH, et al. 1994. Membrane ruffling and chemotaxis transduced by the PDGF b-receptor require the binding site for phosphatidylinositol 3-kinase. *Oncogene* 9: 651–60

Yamamoto M, Maru N, Sakai T, Morii N, Kozaki S, et al. 1993. ADP-ribosylation of the rhoA gene product by botulinum C3 exoenzyme causes Swiss 3T3 cells to accumulate in the G1 phase of the cell cycle. *Oncogene* 8:1449–53

Yano H, Nakanishi S, Kimura K, Hanai N, Saitoh Y, et al. 1993. Inhibition of histamine secretion by wortmannin through the blockade of phosphatidyl 3-kinase in RBL-2H3 cells. *J. Biol. Chem.* 268:25846–56

Yeramian P, Chardin P, Madaule P, Tavitian A. 1987. Nucleotide sequence of human rho cDNA clone 12. *Nucleic Acids Res.* 15:1869

Zhang J, Fry MJ, Waterfield MD, Jaken S, Liao L, et al. 1992. Activated phosphatidylinositide 3-kinase associates with membrane skeleton in thrombin-exposed platelets. *J. Biol. Chem.* 267:4686–92

Zhang J, King WC, Dillon S, Hall A, Feig L, Rittenhouse SE, et al. 1993. Activation of platelet phosphatidylinositide 3-kinase requires the small GTP-binding protein rho. *J. Biol. Chem.* 268:22251–54

Zheng Y, Hart MJ, Shinjo K, Evans T, Bender A, Cerione RA. 1993. Biochemical comparisons of the *Saccharomyces cerevisiae* Bem2 and Bem3 proteins. *J. Biol. Chem.* 268: 24629–34

Annu. Rev. Cell Biol. 1994. 10:55–86

PROTEIN SERINE/THREONINE PHOSPHATASES—New Avenues for Cell Regulation

Shirish Shenolikar
Department of Pharmacology, Duke University Medical Center, Durham, North Carolina 27710

KEY WORDS: phosphoprotein phosphatases, gene expression, cell growth, signal transduction, ion channels, receptors

---

CONTENTS

INTRODUCTION

Protein phosphorylation is now well recognized as a mechanism widely utilized by plant and animal cells to regulate their functions. The covalent modification of regulatory proteins coordinates diverse processes within a cell to provide tight control of the cell's physiology. This mechanism is used by external stimuli such as hormones, growth factors, and neurotransmitters to transduce their signals within the cell. The phosphorylation state of key regulatory proteins defines both the strength and duration of a physiological response to hormones. A number of protein kinases are directly activated by intracellular second messengers, which led to the proposal that protein kinases were the key mediators of the hormonal response. Several protein kinases have also been harnessed together in a sequence of phosphorylation reactions. These

0743–4634/94/1115–0055$05.00

cascades provide tremendous amplification to the initial physiological stimulus and enhance the sensitivity of the cell to hormones.

Inherent in this mode of regulation is its reversibility, which allows cells to reset prior to each new stimulus. Protein phosphatases reverse the actions of protein kinases and are therefore seen as negative regulators of hormone response. Earlier biochemical studies suggested that protein phosphatases might represent a much smaller group of enzymes than protein kinases. Moreover, protein phosphatases were originally found to display a broad in vitro substrate specificity and many of them did not change in activity following hormonal stimulation. These factors contributed to the view that the diversity and specificity of hormonal responses primarily reflected the complement of protein kinases and their substrates that were present in cells. By comparison, protein phosphatases were thought to act constitutively and oppose hormonally regulated protein kinases.

During the last four years, we have gained information on the primary structure of all the major protein (serine/threonine) phosphatases. Molecular cloning of cDNAs encoding phosphatase catalytic subunits from different species indicates that protein (serine/threonine) phosphatases are among the most highly conserved proteins in evolution. This implies that protein phosphatases perform functions that are essential to all eukaryotic cells. Consequently, genetic studies in lower eukaryotes have begun to be exploited to gain new insight into the physiological functions of phosphatases.

Protein kinases that phosphorylate either tyrosine or serine/threonine residues share easily recognizable amino acid sequence motifs. These motifs specify common secondary and tertiary folds and indicate that protein kinases evolved from a single ancestral gene. These conserved sequences have become hallmarks for these enzymes and have proved useful in identifying new protein kinases. In contrast, protein (serine/threonine) phosphatases share no structural homology with the protein (tyrosine) phosphatases (Charbonneau & Tonks 1992) and therefore probably evolved along separate evolutionary pathways. However, several different protein (serine/threonine) phosphatases share blocks of sequence homology that might comprise their catalytic domain (Shenolikar & Nairn 1991). These sequences have been used to clone many new protein (serine/threonine) phosphatases (MX Chen et al 1992). However, at least one protein serine/threonine phosphatase, PP2C, lacks these conserved sequences, thereby demonstrating that they are not essential to defining a phosphatase. This raises the possibility that all existing phosphatases may not be identified by homology screening or PCR. While the newly identified cDNAs have increased the number of phosphatases found in eukaryotic cells, the total number of protein phosphatases still lags behind the protein kinases. It is noteworthy that all the phosphatase catalytic subunits exist as multigene families (Table 1). Moreover, the number of phosphatase regulatory subunits

Table 1 Human protein serine/theronine phosphatase genes

Catalytic subunits	
PPP1CA	Protein phosphatase 1, catalytic subunit, α isoform
PPP1CB	Protein phosphatase 1, catalytic subunit, β isoform
PPP1CC	Protein phosphatase 1, catalytic subunit, γ isoform
PPP2CA	Protein phosphatase 2A, catalytic subunit, α isoform
PPP2CB	Protein phosphatase 2A, catalytic subunit, β isoform
PPP3CA	Protein phosphatase 2B, catalytic subunit, α isoform
PPP3CB	Protein phosphatase 2B, catalytic subunit, β isoform
PPP3CC	Protein phosphatase 2B, catalytic subunit, γ
PPP4C	Protein phosphatase X, catalytic subunit
Regulatory subunits	
PPP1R1A	Protein phosphatase 1, inhibitor-1
PPP1R1B	Protein phosphatase 1, DARPP-32
PPP1R2	Protein phosphatase 1, inhibitor-2
PPP1R3	Protein phosphatase 1, Glycogen and sarcoplasmic reticulum binding subunit (skeletal muscle)
PPP1R4	Protein phosphatase 1, Myofibrillar subunit (skeletal muscle)
PPP2R1A	Protein phosphatase 2A, A subunit (PR 65), α isoform
PPP2R1B	Protein phosphatase 2A, A subunit (PR 65), β isoform
PPP2R2A	Protein phosphatase 2A, B subunit (PR 52), α isoform
PPP2R2B	Protein phosphatase 2A, B subunit (PR 52), β isoform
PPP2R2C	Protein phosphatase 2A, B subunit (PR 52), γ isoform
PPP2R3	Protein phosphatase 2A, C (previously B'') subunit (PR 72), α isoform
	Protein phosphatase 2A, C (previously B'') subunit (PR 130), β isoform (an alternate spliced product)
PPP2R4	Protein phosphatase 2A, B' subunit (PR 53)
PPP3R1	Protein phosphatase 2B, 19 kd B subunit, α isoform
PPP3R2	Protein phosphatase 2B, 19 kd B subunit, β isoform
Protein phosphatases unrelated to the PPP family of genes	
PPM1A	Protein phosphatase 2C, α isoform
PPM1B	Protein phosphatase 2C, β isoform

The numbering scheme presented in this table was recommended by a Nomenclature Committee formed at FASEB Summer Research Conference on Protein Phosphatases in 1992. When the nearly complete primary structure or gene location points to a novel human protein phosphatase, it can be assigned a new number by contacting Dr. Patricia T. W. Cohen, MRC Protein Phosphorylation Unit, Department of Biochemistry, University of Dundee, Dundee DD1 4HN, Scotland (FAX: 44 382-23778), or Dr. Phyllis McAlpine (nomenclature editor of the human genome database), Department of Human Genetics, The University of Manitoba, Faculty of Medicine, T250-770 Bannatyne Avenue, Winnipeg, Manitoba, Canada K3E OW3 (FAX: 204 786-8712).

far exceeds the number of catalytic subunits, which suggests that the functional diversity of these enzymes arises from association of the different catalytic subunits with one or more regulatory subunits.

Protein phosphatases have recently been shown to be the targets for many environmental toxins that inhibit enzyme activity. Selected protein phosphatases are also inhibited by natural products that have been developed as immunosuppressive agents. Many of these compounds are cell permeable and can inhibit phosphatases in the intact cell, which has opened the way for analyzing phosphatase functions in cells.

The biochemical properties of the major protein (serine/threonine) phosphatases have been discussed in several recent reviews (Shenolikar & Nairn 1991; Cohen 1989; Bollen & Stalmans 1992). The goal of this review is to discuss how current pharmacological and molecular genetic approaches have combined to provide a better understanding of the structure, function, and regulation of protein (serine/threonine) phosphatases. This discussion focuses primarily on data published within the last year to illustrate the remarkable progress that has been made in understanding the physiological role of phosphatases. These recent studies show that protein phosphatases are in fact highly regulated enzymes that play critical roles in controlling cell function. Moreover, protein phosphatases act as positive regulators of many hormonal responses. Increasing evidence suggests that extensive crosstalk between kinases and phosphatases is essential in exercising tight control over cell physiology. With the use of newly identified phosphatase inhibitors, as well as genetic manipulations to modulate cellular phosphatase activity, the physiological functions of individual protein (serine/threonine) phosphatases are finally being realized.

STRUCTURE, FUNCTION, AND REGULATION OF PROTEIN PHOSPHATASES

Early biochemical studies of protein phosphatases developed the criteria for their classification based on substrate specificity and sensitivity to endogenous inhibitors (Shenolikar & Nairn 1991; Cohen 1989). These properties divided protein (serine/threonine) phosphatases into two broad groups, termed type-1 and type-2. The type-2 enzymes were further separated into three subgroups, 2A, 2B, and 2C, because of their unique subunit structure, spectrum of substrates, and regulation by divalent cations. Recent molecular cloning of these enzymes confirms that they represent distinct gene products (Table 1).

Okadaic acid (OA), a marine toxin, inhibits both of the major cellular protein (serine/threonine) phosphatases, PP1 and PP2A (Shenolikar & Nairn 1991; Cohen 1989). OA inhibits PP2A (IC_{50} 0.1–1.0 nM) at concentrations 10- to 100-times lower than those required to inhibit PP1. PP2B can also be inhibited by OA, but only at high micromolar concentrations. In contrast, PP2C activity is unaffected by OA, thus providing a convenient assay for this enzyme in the absence of the other major phosphatase.

The membrane-permeability of OA has made it a useful reagent for examining the role of protein phosphorylation in many cellular processes. However, in vitro sensitivity of PP2A to OA is decreased by high protein concentrations. Moreover, radiolabeled OA predominantly accumulates in the membranes, where PP1 is more abundant than PP2A. Thus it may be difficult to utilize solely the OA sensitivity of a particular process in the intact cell to identify the phosphatase involved.

This problem may be partially circumvented by the discovery of phosphatase inhibitors with differing chemical structures and specificity for protein (serine/threonine) phosphatases (Luu et al 1993). For example, calyculin A is structurally unrelated to OA and shows a tenfold higher potency for the inhibition of PP1 than of PP2A. By comparing the biological effects of this cell-permeable inhibitor with those of OA, we can gain initial insight into the differing actions of PP1 and PP2A in the intact cell. Several bacterial toxins, tautomycin, microcystin, and nodularin, also inhibit PP1 and PP2A. However, some of these compounds are limited by their lower potency as phosphatase inhibitors and the inability of these toxins to enter cells. The insect toxin cantharidin and structurally related herbicides known as endothals also inhibit PP1 and PP2A (Li & Casida 1992; Honkanen 1993). While the toxicity of these compounds suggests that they enter cells, their lower potency as inhibitors (IC_{50} of cantharidin of 160 nM for PP2A and 1.7 μM for PP1) compared with OA and calyculin A may limit their use. Several newly identified protein phosphatases (e.g. PP3 and PPX) are also inhibited by OA, and additional experiments are essential to identify the OA-sensitive phosphatase.

OA and other cell-permeable phosphatase inhibitors often lead to a preferential increase in the phosphothreonine content of metabolically labeled proteins (Bu et al 1993; Turner et al 1993). This may indicate that phosphothreonine is more rapidly turned over than phosphoserine in cells. Consequently, phosphatase inhibitors may be helpful tools to visualize the phosphorylation of threonines. Studies of hormone-induced phosphorylation in the absence of phosphatase inhibitors may have underestimated the regulatory importance of threonine phosphorylation.

Phosphorylation of serine and threonine residues accounts for more than 97% of the protein-bound phosphate in stimulated cells. Nevertheless, tyrosine phosphorylation clearly plays an important role in growth and differentiation of eukaryotic cells. Some serine/threonine phosphatases, e.g. PP2A and PP2B, possess in vitro phosphotyrosine phosphatase (PTPase) activity, and cellular factors have been identified that can enhance the PTPase activity of PP2A

(Cayla et al 1990). However, the abundance of competing phosphoserine- and phosphothreonine-containing substrates in the cell questions the physiological relevance of these findings. Identification of numerous PTPases with significantly higher specific activity towards phosphotyrosine-containing substrates also suggests that different enzymes reverse the covalent modifications of tyrosines and serine/threonines. On the other hand, the discovery of dual-specificity protein phosphatases that dephosphorylate both phosphotyrosine and phosphoserine/threonine (Guan et al 1991) raises questions as to the unique dedication of protein phosphatases to one or the other category.

Eukaryotic proteins also contain phosphohistidine. Phosphohistidine is unstable to heating at low pH, a treatment that is used to analyze other phospho amino acids. Thus the presence of phosphohistidine in many cellular proteins may have been missed. The phosphohistidine content of some proteins is modulated by hormones and oncogenes (Hegde & Das 1990). Protein kinases that specifically phosphorylate histidines have been identified (Huang et al 1991). Hence the stage is set for demonstrating the role for this covalent modification. The protein (serine/threonine) phosphatases PP1, PP2A, and PP2C possess in vitro phosphohistidine phosphatase activity (Kim et al 1993). Moreover, dephosphorylation of phosphohistidine-containing proteins in yeast extracts was partially inhibited by OA, which indicates that PP1 or PP2A contributes to phosphohistidine dephosphorylation. These studies also demonstrated the presence of OA-insensitive phosphohistidine phosphatases in yeast.

Protein Phosphatase-1

Multiple mammalian PP1 cDNAs have been identified (Cohen 1988; Shima et al 1993b; Cohen et al 1990). PP1α and PP1δ are expressed in many different rat tissues including brain, lung, liver, skeletal muscle, small intestine, and testes (Shima et al 1993b). In contrast, PP1γ1 is predominantly expressed in rat brain tissue. A fourth isoform, PP1γ2, is almost exclusively expressed in rat testes (Shima et al 1993a) and has been localized to nuclei of meiotic cells in the seminiferous tubules. Loss of one of four PP1 genes (Dombradi et al 1993) in *Drosophila* inhibits chromosome separation, which indicates a selective role for this PP1 isoform in mitosis (Axton et al 1990). Based on primary sequences conserved in the mammalian PP1s, six plant cDNAs were isolated by PCR (Smith & Walker 1993). The presence of multiple genes, their differential expression in tissues, as well as differential subcellular localization of selected isoforms have led to the hypothesis that the PP1 catalytic subunits have unique physiological functions.

In mammalian cells, PP1 is associated with many subcellular compartments (Shenolikar & Nairn 1991; Cohen 1989; Bollen & Stalmans 1992). This association is mediated by targeting proteins (Hubbard & Cohen 1993). The

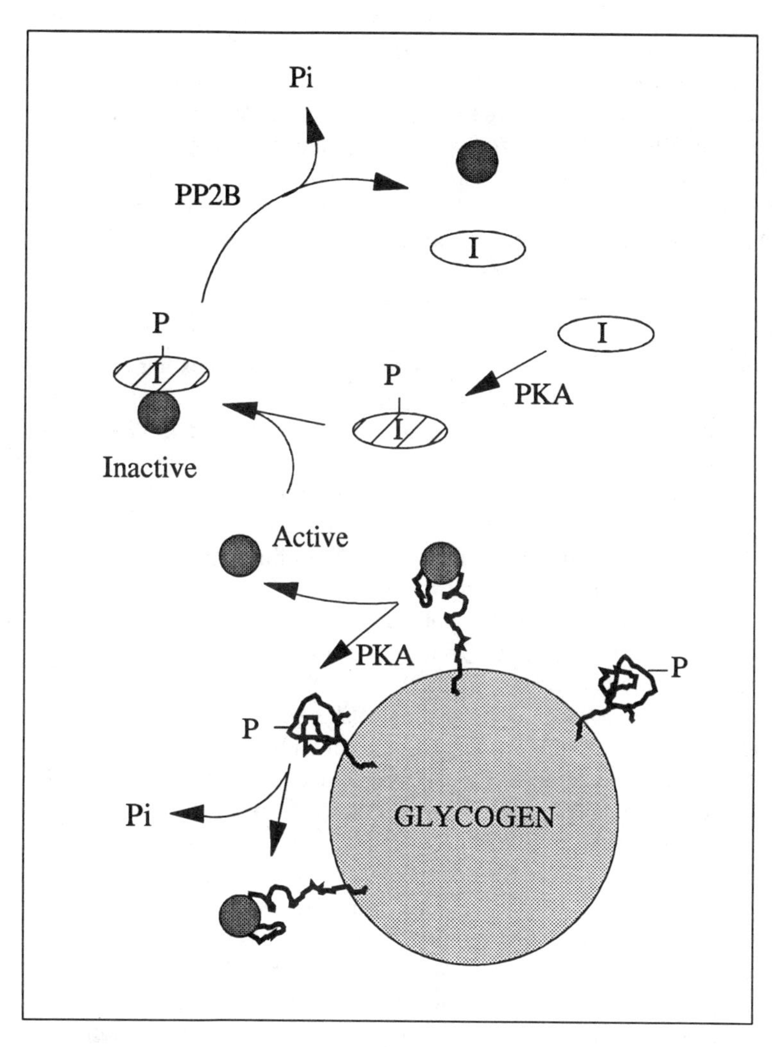

Figure 1 Hormonal regulation of protein phosphatase-1. PP1 catalytic subunit (*filled circles*) associates with glycogen via the G-subunit and regulates phosphoproteins involved in glycogen metabolism. PKA-mediated phosphorylation of the G-subunit translocates PP1 to the cytosol and decreases its activity against the glycogen enzymes. PKA also phosphorylates and activates cytosolic I-1 (I), which leads to rapid inactivation of soluble PP1. PP2B catalyzes calcium-dependent dephosphorylation of I-1. The dephosphorylation of I-1 and G-subunit permits reassociation of PP1 with glycogen.

best-characterized targeting protein is the glycogen-binding or G-subunit, which confers PP1 association with skeletal muscle glycogen. The G-subunit also enhances the dephosphorylation of phosphoproteins found in glycogen particles. The 161-kd G-subunit is only expressed in adult skeletal muscle (Tang et al 1991), which suggests that a different targeting protein may direct PP1 association with glycogen in the liver (Bollen & Stalmans 1992). Hormones that elevate intracellular cAMP promote phosphorylation of the G-subunit by cAMP-dependent protein kinase (PKA) and lead to the displacement of PP1 from glycogen (Figure 1). Translocation of PP1 to cytosol decreases its activity towards the enzymes of glycogen metabolism. PKA also phosphorylates and activates the cytosolic PP1 inhibitor, inhibitor-1 (I-1), in skeletal muscle. This should lead to rapid inactivation of PP1. Thus hormone-induced translocation of PP1 is coordinated with I-1 activation to regulate protein dephosphorylation in the mammalian skeletal muscle. The G-subunit also mediates PP1 association with sarcoplasmic reticulum in skeletal and cardiac muscle, where the enzyme regulates the phosphorylation state of phospholamban and calcium uptake (Steenaart et al 1992).

PP1 binds to smooth muscle myosin, and this binding is mediated through two polypeptides, one of 130 and the other 25 kd, that comprise the M-complex. Association with the M-complex enhances the dephosphorylation of smooth muscle myosin by PP1, but has no effect on skeletal muscle myosin dephosphorylation (Alessi et al 1992). Thus distinct PP1-targeting protein(s) seem to regulate the dephosphorylation of skeletal muscle myosin (Dent et al 1992). High concentrations of arachidonic acid inhibit myosin dephosphorylation and potentiate smooth muscle contraction (Gong et al 1992). Whether inhibition of myosin phosphatase is associated with the displacement of PP1 from smooth muscle myosin is not yet clear.

High molecular weight PP1 complexes are present in all cell extracts. In extracts of fission yeast, the PP1 isoform, *dis*2, is present as multiple high molecular weight complexes ranging in size from 80 to 200 kd (Kinoshita et al 1990). However, only the 80 kd complex possesses phosphatase activity, which suggests that the larger complexes probably also contain inhibitors. The *GLC7* gene in *Saccharomyces cerevisiae* encodes the PP1 catalytic subunit and is identical to *dis*2*s*1, the cDNA cloned by hybridization with the fission yeast cDNA. Genetic studies in budding yeast have identified a PP1 mutation, *glc7•1,* that diminishes glycogen accumulation (Feng et al 1991). An independent mutation in the *GAC1* gene also fails to accumulate glycogen and is associated with the inability to activate glycogen synthase. The protein product of the *GAC1* gene, Gac1p, shows some sequence homology with rabbit skeletal muscle G-subunit (Francois et al 1992). Gac1p levels increase during growth and correlate with glycogen accumulation in yeast.

Gac1p physically associates with PP1 as seen by co-immunoprecipitation and the two-hybrid assay (Stuart et al 1994). A point mutation in *GLC7* (arginine-73 to cysteine) prevented PP1 from associating with Gac1p. However, increasing the dosage of the *GAC1* gene partially suppresses the *glc7-1* defect, which suggests that the *GAC1* mutation reduces the affinity of the yeast glycogen-binding subunit for PP1. Mutations that eliminate key phosphorylations of glycogen synthase suppress the glycogen accumulation defect in *gac1* and *glc7⁻1* strains, which suggests that PP1 association with Gac1p regulates the phosphorylation state of glycogen synthase. Gac1p is phosphorylated in vivo and hyperphosphorylated in the *glc7⁻1* strain (Stuart et al 1994). PP1 preferentially associates with the underphosphorylated form of Gac1p. Thus, as in skeletal muscle, the phosphorylation of the yeast G-subunit may regulate its association with PP1. Conversely, PP1 might maintain Gac1p in its dephosphorylated state. The *GLC7* gene also suppresses mutations in GCN2, the protein kinase that phosphorylated the yeast initiation factor eIF2α. This suggests that PP1 also regulates protein synthesis in the budding yeast (Wek et al 1992).

The recently developed yeast two-hybrid assay provides a means of analyzing protein-protein interactions and might be used to identify PP1-binding proteins. However, to identify multisubunit complexes like the M-complex, where more than one polypeptide is required for high affinity binding to PP1, a different approach is required. The two-hybrid assay has demonstrated the interaction of PP1 with the 100-kd protein product of the human retinoblastoma tumor suppressor gene (Durfee et al 1993). Purified PP1 preferentially binds to unphosphorylated RB and requires C-terminal sequences that are also essential for RB's tumor suppressor activity. Co-immunoprecipitation assay has shown that PP1 associates with RB during G_0/G_1 and mid G_1, but this association is lost as cells progress through S and G_2 when RB becomes hyperphosphorylated (Durfee et al 1993). PP1 reassociates with RB in M-phase extracts. The RB phosphatase in M-phase extracts was biochemically characterized as PP1 (Ludlow et al 1993). PP1 association may prevent the untimely phosphorylation of RB. How PP1 is displaced from RB and its relationship to RB phosphorylation during late G_1 and S-phase are not clear. One might predict that PP1 mutations that prevent binding to RB might promote RB phosphorylation and lead to cell transformation. It is interesting to note that the chromosomal location of human PP1α gene at 11q13 has been linked with some cancers (Barker et al 1993).

An essential gene, *sds*22$^+$, genetically interacts with two genes, *dis*2$^+$ and *sds*21$^+$, that encode PP1 isoforms in *Schizosaccharomyces pombe* (Stone et al 1993). The *sds*22 protein is localized within the nucleus and contains 11 leucine-rich repeats, each 22 amino acids in length. The central repeats are necessary for its interaction with PP1. Association with *sds*22 has been shown to modify the substrate specificity for PP1 so that it dephosphorylates histone

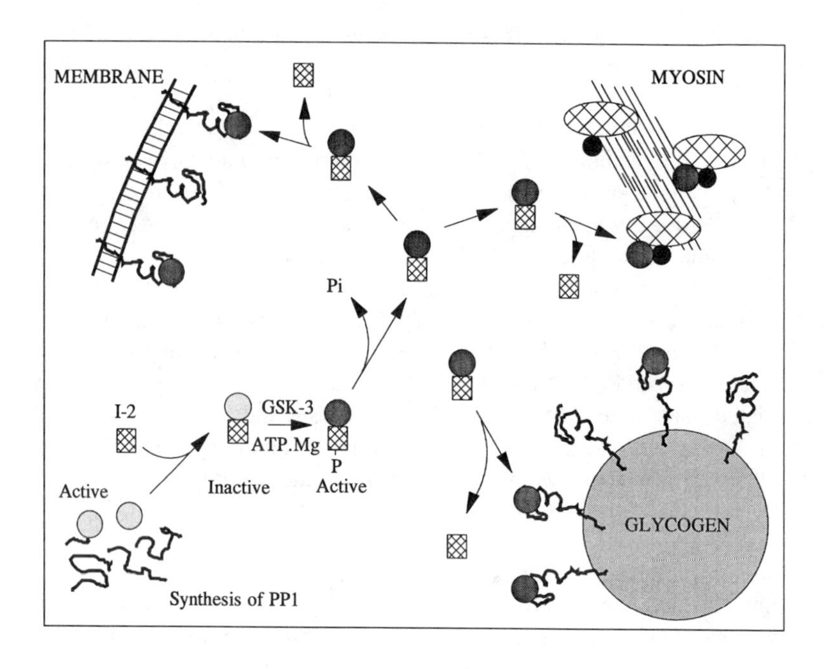

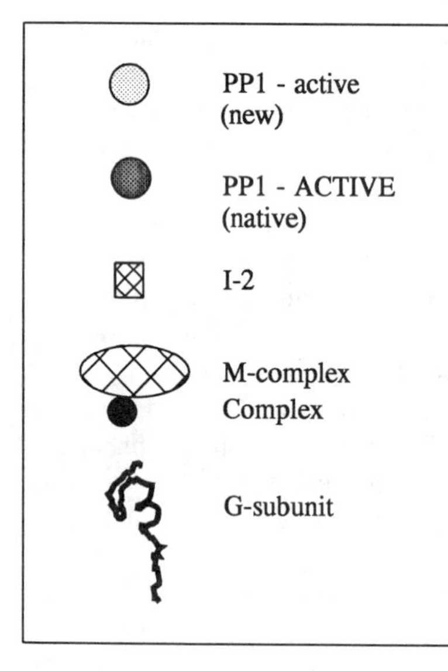

Figure 2 Delivery of newly synthesized protein phosphatase-1 to subcellular compartments. Newly synthesized PP1 catalytic subunit (*open circles*) is essentially inactive under physiological conditions and is unable to interact with many substrates and regulators. However, I-2 (*hatched squares*) binds to the newly synthesized PP1 with high affinity and, through its phosphorylation by GSK-3, refolds PP1 to its fully active conformation (*filled circles*). Autodephosphorylation produces an active complex that delivers PP1 to subcellular compartments such as membranes and glycogen, where I-2 is displaced by the G-subunit, or at the myofibrils, where it is displaced by the myosin complex.

H1, which is phosphorylated by *cdc*2 kinase. This *sds*22 association showed negligible effect on PP1 activity against phosphorylase *a*, the classical PP1 substrate. PP1/*sds*22 complex was inhibited by I-1 and OA to the same degree as the free PP1 catalytic subunits. In vivo, phosphorylation of the *sds*22 protein modulates its association with PP1 (Stone et al 1993). Thus *sds*22 shows many of the properties of a potential nuclear targeting protein that could regulate PP1 localization in the nucleus and enhance phosphatase activity against nuclear phosphoproteins.

Individual mammalian PP1 isoforms have been expressed in bacteria (Zhang et al 1993). Such recombinant PP1s show similar activity against several substrates. However, unlike the enzyme purified from tissues, recombinant PP1 requires Mn^{2+} ions for activity. Moreover, the recombinant PP1s also show increased activity towards histone H1 phosphorylated by *cdc*2 relative to phosphorylase *a* when compared to native PP1.

Bacterially expressed PP1 isoenzymes are inhibited by OA and inhibitor-2 (I-2) in a manner similar to that of native PP1 (Alessi et al 1993b). They also bind to the G-subunit and are translocated to glycogen. However, they demonstrate significantly decreased sensitivity to I-1 compared to the native PP1 catalytic subunit. Like native PP1, the recombinant isoenzymes form inactive or latent complexes with I-2 that can be reactivated by incubation with glycogen synthase kinase-3 (GSK-3) and ATP-Mg (Shenolikar & Nairn 1991; Cohen 1989; Bollen & Stalmans 1992). Following reactivation, these enzymes no longer require Mn^{2+} ions for activity and possess properties indistinguishable from those of the native PP1. These observations suggest that the interaction with I-2 induces the native conformation in bacterially expressed PP1 catalytic subunits.

To test this hypothesis, native PP1 was converted to a Mn^{2+}-dependent form by incubation with the inhibitor NaF. This enzyme was restored to its previous Mn^{2+}-independent state by incubation with I-2 and GSK-3/ATP-Mg. Moreover, PP1 denatured by guanidine hydrochloride was rapidly reactivated by I-2/GSK-3. These data support the hypothesis that I-2 acts like a chaperon to refold the PP1 catalytic subunits to their native conformation. Following reactivation, PP1 catalytic subunits can be displaced from the PP1/I-2 complex by G-subunit or the M-complex (Figure 2). Thus the PP1/I-2 complex may also deliver PP1 to different subcellular compartments.

As PP1 catalytic subunits are highly conserved, one might predict that I-2 or a functionally related protein is necessary for PP1 synthesis in all eukaryotic organisms. Thus far there is no evidence for the presence of a PP1 inhibitor like I-2 in plants and fungi. These studies also suggest that structural differences in the PP1 isoforms do not contribute significantly to their activity or regulation. Thus the physiological relevance of the PP1 isoenzymes remains unknown.

A newly developing theme in protein serine/threonine phosphatase regulation involves the direct covalent modification of the catalytic subunits. For example, PP1 phosphorylation in vitro by viral tyrosine kinases inhibits its activity (Shenolikar & Nairn 1991; Villa-Moruzzi et al 1991). In contrast, in vitro phosphorylation by a serine/threonine kinase, *cdc*2, activates PP1 (Villa-Moruzzi 1992). However, there is currently no evidence for the phosphorylation of PP1 in mammalian cells.

Protein Phosphatase-2A

Native PP2A enzymes are heterotrimers of two regulatory subunits (A and B) and a catalytic subunit (Cohen 1989; Shenolikar & Nairn 1991). Multiple genes encode the catalytic and regulatory subunits of PP2A in plants and animals (Table 1). Evidence indicates that more PP2A regulatory subunits remain to be cloned (Hendrix et al 1993). If each catalytic subunit can associate with all A and B regulatory subunits, up to 40 different PP2A complexes could exist in mammalian cells. Biochemical studies have shown that different regulatory subunits differentially inhibit PP2A activity against selected substrates. Thus the complement of regulatory subunits defines the substrate specificity of an individual PP2A complex.

Two genes (*ppa*1 and *ppa*2) encode PP2A catalytic subunits in fission yeast. Lethality is observed when both PP2A-encoding genes are disrupted, which suggests that PP2A is essential for growth in *S. pombe*, but that the two catalytic subunits share overlapping functions (Kinoshita et al 1990). Disruption of *ppa*2 alone reduces cell size, consistent with premature entry of yeast cells into mitosis. By comparison, loss of both PP2A catalytic subunits (PPH21 and PPH22) impairs growth in budding yeast, but is not lethal. Inactivation of another protein phosphatase gene, PPH3, eliminates residual growth of *pph*21*pph*22 double mutants, which suggests a functional interaction between PPH3 and PP2A (Ronne et al 1991).

In *S. cerevisiae*, the TPD3 gene encodes the A-subunit of PP2A (Van Zyl et al 1992) and antibodies against Tpd3p precipitate PP2A activity from yeast extracts (Van Zyl et al 1992). Temperature-sensitive *tpd*3 strains were defective in cytokinesis, becoming multinucleated and multibudded at nonpermissive temperatures. This phenotype is also observed following overexpression of a PP2A catalytic subunit or the loss of CDC55 gene, which encodes the *S. cerevisiae* B-subunit. Thus the loss of the A- and B-subunits is functionally similar to increasing the cellular content of PP2A catalytic subunit. However, differences in the *cdc*55 and *tpd*3 strains argue that each regulatory subunit has unique effects on PP2A activity. For instance, PP2A inhibits the entry of yeast cells into mitosis (Kinoshita et al 1993). Thus by inhibiting PP2A, OA promotes *cdc*2 activation and induces entry into mitosis (Shenolikar & Nairn 1991; Bollen & Stalmans 1992). In contrast, temperature-sensitive *tpd*3 strains do not arrest at mitosis, which suggests that the *tpd*3-associated PP2A is not a key regulator of mitosis. On the other hand, the A-subunit of PP2A, encoded by *tpd*3, is required for cytokinesis in yeast.

The *tpd*3 strains are defective in transcription of tRNA genes in vivo and in vitro. Loss of the A-subunit is unlikely to "increase" PP2A activity as OA does not restore tRNA transcription in the *tpd*3 extracts. Thus the loss of function of the A-subunit most likely modifies the substrate specificity of PP2A, which leads to sustained phosphorylation and inactivation of RNA polymerase III (Van Zyl et al 1992). Transcription is restored by the addition of exogenous RNA polymerase III. RNA polymerase II activity is also diminished in *tpd*3 strains, perhaps indicating an overlap in function between PP2A and the SIT4 phosphatase, which also regulates RNA polymerase II in *S. cerevisiae* and cell cycle progression from G_1 to S (Sutton et al 1991). Lethality results from disruption of both *sit*4 and *tpd*3, which is indicative of a functional overlap of the two gene products.

The A-subunit also modifies PP2A activity in mammalian cells. Intracellular injection of either the PP2A catalytic subunit or an expression plasmid encoding the PP2A cDNA leads to enhanced transcriptional activity to the nuclear protooncogene, *c-Jun* (Alberts et al 1993a). Co-transfection of the A-subunit inhibits PP2A's ability to induce *c-Jun* transactivation. Similarly, microinjection of PP2A catalytic subunit, but not the AC complex, inhibits the expression of a gene controlled by the serum response element from the *c-fos* gene (Schonthal et al 1991). As discussed above, RB is a nuclear phosphoprotein that, in its underphosphorylated state, is tightly associated with the nucleus. Microinjection of the PP2A catalytic subunit, but not the AC complex, increases the nuclear affinity of RB (Alberts et al 1993b).

In *Drosophila melanogaster,* there appears to be only one gene for each of the three PP2A subunits. The *Drosophila* PP2A catalytic subunit is 94% identical to the mammalian enzyme, perhaps reflecting one of the most remarkable degrees of evolutionary conservation in any known protein. Whereas PP2A catalytic subunit and the A-subunit (PR65) are present at similar levels throughout development, there are high maternal stores of the B-subunit (PR55) mRNA in embryos and, subsequently, a dramatic increase in the abundance of this mRNA during development (Mayer-Jaekel et al 1993). *Drosophila* imaginal discs are excellent models for studying cell specialization and differentiation from initially homogenous cell masses. High levels of PR55 mRNA are found in *Drosophila* imaginal discs, which suggests a link with pattern formation (Uemura et al 1993). Loss of the *twins* gene, which encodes the PP2A B-subunit, results in the formation of an extra wing blade anlagen, thus implicating both PP2A regulatory subunits in pattern formation in *Drosophila.* Lack of defects in other imaginal discs such as the eye-antenna disc indicates that the B-subunit may function in a restricted way to regulate PP2A activity in different discs. The *twins* locus maps close to the *aar* mutation in

Drosophila. Wild-type *twins*, when introduced into *Drosophila* by P-element-mediated transformation, rescues the *aar* defect, which shows abnormal mitotic figures (Gomes et al 1993). Increased mitotic indices in *aar* mutants could result from premature entry into mitosis, consistent with PP2A's proposed role as an inhibitor of $p34^{cdc2}$ activation. Abnormal mitotic figures are seen in the *twins* neuroblasts. However, there are also differences in the phenotypes associated with these mutations, which suggests that neither the *aar* nor the *twins* mutation represents a complete loss of function of the B-subunit.

In budding yeast, CDC55 encodes the B-subunit of PP2A (Healy et al 1991). The CDC55 deletion strain shows a cold-sensitive phenotype, with defective growth at the restrictive temperature. Recent studies suggest that CDC55 regulates morphogenesis in yeast. Three ELM genes were identified on the basis that mutations in these genes induced cell elongation or pseudohyphal growth of *S. cerevisiae* (Blacketer et al 1993), similar to that seen in wild-type yeast following nitrogen starvation. Mutation in the ELM1 gene, encoding a protein kinase, combined with the *cdc55* mutation, results in a synthetic lethal phenotype. These data suggest that the Elm1p kinase and PP2A act together to regulate morphogenesis and cell division in the budding yeast.

One of the most exciting recent developments in protein phosphatase research is the discovery that DNA tumor viruses utilize PP2A to regulate viral replication and cell transformation (Walter & Mumby 1993). T-antigens produced by both polyoma and SV40 viruses associate with the AC complex. SV40 small-t binds to both free A-subunit and the PP2A complex. B-subunit inhibits the interaction of small-t with the A-subunit, which suggests a common binding site. Interestingly, small-t exchange is restricted to only a subset of B-subunits of the heterotrimeric PP2A complex, such as the 55-kd Bα subunit from bovine brain, but not to the 54-kd B′-subunit from cardiac tissue.

Polyoma middle-T and SV40 small-t antigens exchange with the B-subunit to produce PP2A enzymes with altered substrate specificity. High affinity binding of small-t to the AC complex inhibits dephosphorylation of some substrates and enhances other dephosphorylation reactions (Yang et al 1991). Small-t has no effect on protein dephosphorylation by the PP2A catalytic subunit or the ABC complex. This establishes that the predominant interaction of small-t is with the A-subunit at or near the binding site for the B-subunit. The polyoma middle-T-associated PP2A shows an enhanced activity as a protein tyrosine phosphatase compared with native PP2A complexes (Cayla et al 1993). The physiological significance of this finding is unclear.

PP2A dephosphorylates the phosphoserines that regulate large-T binding at the SV40 origin of replication (Scheidtmann et al 1991). Association of small-t with PP2A prevents in vitro dephosphorylation of SV40 large-T (Scheidtmann et al 1991). Elevations of the phosphorylation state of large-T and other nuclear phosphoproteins such as p53 may represent important mechanisms by which small-t collaborates with large-T to enhance viral transcription and cell transformation (Mumby & Walter 1991). Interaction of small-t with PP2A also inhibits in vitro dephosphorylation and inactivation of ERK1 and MEK1, two protein kinases involved in mitogenic signaling in mammalian cells. Transfection of small-t into CV-1 cells activates MEK and ERK-2 (Sontag et al 1993), but does not affect the activity of Raf-1 kinase, a preferred substrate for PP1 (Kovacina et al 1990). Small-t is as effective as serum in stimulating the growth of quiescent CV-1 cells. Co-expression of a dominant negative ERK2 abolishes cell proliferation induced by small-t, which emphasizes the involvement of the MAP kinase pathway in the proliferative effects of small-t. Thus small-t can inhibit PP2A in intact CV-1 cells and, in comparison to other phosphatase inhibitors, may be a more specific tool to modulate PP2A activity in cells. Overexpression of small-t regulates transcription of selected hormone-sensitive genes in NIH 3T3 cells (Alberts et al 1994). The activation of MAP kinases and transcription of specific immediate early genes could represent additional mechanisms contributing to cell transformation by DNA tumor viruses.

The mechanisms responsible for regulating PP2A activity in normal cells are still poorly understood. Activation of many G protein-linked receptors promotes their transient phosphorylation and attendant desensitization. Occupation of the cholecystokinin (CCK) receptor by agonist activates both a receptor kinase and a phosphatase (Lutz et al 1993). PP2A, the predominant CCK receptor phosphatase in pancreatic acinar cells, has been shown to be transiently activated in response to CCK (Lutz et al 1993). However, the phosphorylase phosphatase activity was unchanged. Thus the ability to show hormonal regulation of PP2A could depend on the use of an appropriate substrate. Somatostatin (White et al 1991) and atrial natriuretic peptides (White et al 1993) in pituitary cells utilize a protein phosphatase activity to regulate hormone secretion. These hormones act via distinct second messenger pathways to activate a PP2A-like enzyme. The presence of phorbol esters, activators of protein kinase C (PKC), led to the dephosphorylation of negative regulatory sites in c-Jun, a component of the AP-1 transcription complex (Boyle et al 1991). How PKC activation is translated into the dephosphorylation of sites that are the preferred substrates for PP2A, or how different receptor-linked pathways regulate PP2A activity, remain to be addressed.

Tumor necrosis factor-α (TNF-α) induces sphingomyelin hydrolysis and ceramide production in U937 leukemia cells. Moreover, ceramide mimics the actions of TNF-α to inhibit cell growth, down-regulate *c-myc* mRNA, and induce differentiation. These physiological effects of ceramide are attenuated by OA, which indicates the involvement of an OA-sensitive phosphatase. The ceramide-activated protein phosphatase has been identified as PP2A (Dobrowsky et al 1993). Ceramide activates heterotrimeric PP2A in vitro, yet has no effect on the activity of PP2A catalytic subunit or the AC complex, which demonstrates an essential role for the B-subunit in ceramide effects (Dobrowsky et al 1993). Ceramide may dissociate the B-subunit or perhaps promote a conformational change that activates PP2A.

Direct phosphorylation of the PP2A catalytic subunit in vitro by receptor and non-receptor tyrosine kinases inactivates PP2A (J Chen et al 1992; Chen et al 1994). Transformation of 10T cells by *v-src* increases tyrosine phosphorylation of the PP2A catalytic subunit (Chen et al 1994). Serum also promotes PP2A phosphorylation on tyrosine. Thus activation of growth factor receptor tyrosine kinases leads to transient phosphorylation and inhibition of PP2A. This may prohibit PP2A from reversing key steps in the MAP kinase cascade that are also activated by growth factors and are essential for the mitogenic response. In vitro phosphorylation of the PP2A catalytic subunit on threonine by an autophosphorylation-activated protein kinase (AK) also reduces its activity (Damuni & Guo 1993; Guo et al 1993). The catalytic subunit is phosphorylated whether present as a free catalytic subunit, or in the AC and ABC complexes. In the heterotrimeric complex, both the catalytic subunit and B-subunit are phosphorylated by AK. However, the in vivo phosphorylation of the PP2A subunits on either serine or threonine has not been demonstrated.

The major methylated protein in bovine brain cytosol has been identified as the PP2A catalytic subunit (Lee & Stock 1993). Methylation occurs on the C-terminal leucine that is conserved in most if not all PP2A enzymes. The presence of cellular methyl esterases suggests that this modification is reversible and could play a role in regulating PP2A function.

Protein Phosphatase-2B

Multiple genes encode PP2B (also known as calcineurin) catalytic A-subunit and the regulatory B-subunits in mammalian cells (Kincaid 1993). The mRNA derived from a third gene encoding a testis-specific A-subunit (α3) increases during hormone-regulated spermatogenesis (Muramatsu et al 1992). PP2B may also regulate sperm motility (Shenolikar & Nairn 1991).

The B-subunits belong to the EF-hand family of calcium-binding proteins (Stemmer & Klee 1991). To examine the role of different B-subunits in PP2B function, the two murine B-subunits, β1 from brain and β2 from testes, have been expressed in bacteria (Ueki & Kincaid 1993). Both B-subunits bind to the mammalian A-subunit isoforms, α1 from brain and α3 from testes, as well as to the *Neurospora crassa* A-subunit, and thereby increase their calcium/calmodulin-dependent phosphatase activity. The optimal time for association of the A- and B-subunits at 4°C was 1 hr, substantially less than that required to activate the phosphatase (> 10 hr). This suggests that the B-subunits induce a slow change in the A-subunit to its fully active conformation. The mammalian B-subunits are myristoylated at their N-termini. This modification does not occur in bacteria, apparently indicating that myristoylation is not essential for functional interaction between the two PP2B subunits. The highly conserved region in the mammalian and fungal A-subunits, located between the putative catalytic core and calmodulin-binding domain, may represent the binding site for the B-subunit.

Two PP2B A-subunits (CNA1 and CNA2) have been cloned from *S. cerevisiae.* Yeast strains containing disruption of both genes are viable (Cyert et al 1991). However, these double mutants are highly sensitive to growth arrest induced by the yeast mating pheromone, alpha factor, and fail to resume growth during continued exposure to alpha factor. Apparently, PP2B antagonizes the mating response pathway, which is homologous to the mammalian MAP kinase pathway (Pelech 1993). Subsequent studies (Cyert & Thorner 1992) have shown that yeast strains containing mutations in the B-subunit (the product of the CNB1 gene) are also defective in recovery from alpha factor arrest, which emphasizes that PP2B is required in this signaling pathway. Surprisingly, the loss of the B-subunit completely eliminates PP2B activity in yeast extracts. Thus in contrast to the active A-subunits expressed in bacteria (Ueki & Kincaid 1993), the A-subunits expressed in yeast appear to require the interaction with B-subunit to exhibit phosphatase activity.

One of the most exciting recent discoveries in the protein phosphatase field is that PP2B is the major cellular target of immunosuppressive drugs (Shenolikar 1992) that elicit their pharmacological actions by inhibiting phosphatase activity. The drugs, cyclosporin A (CsA), FK506, and rapamycin, bind to intracellular receptors known as immunophilins. Cellular levels of immunophilins greatly exceed the concentration of drug required for immunosuppression, which suggests that the drug-immunophilin complex mediates PP2B inhibition. Overexpression of the A-subunits of PP2B in T cells enhances cytokine gene transcription and increases cellular resistance to the aforementioned immunosuppressive drugs. These studies suggest that the PP2B catalytic subunit is the primary target of the drug-immunophilin complex (Li & Handschumacher 1993). Co-transfection of the B-subunit synergizes with that of the A-subunit to increase drug resistance. Thus the B-subunit appears to increase the affinity of the drug only for the A-subunit. However, recent biochemical studies show that ^{125}I-labeled cyclophilin (CyP) cross-links solely to B-subunit in the presence of CsA. The presence of A-subunit is essential because no cross-linking is seen with the purified B-subunit. The B-subunit is also the target for cross-linking with ^{125}I-labeled FKBP (FK506-binding protein) in the presence of FK-506. The CsA-CyP complex competes with FK506-FKBP for B-subunit binding. Rapamycin, which also binds to FKBP, does not promote the FKBP cross-linking to PP2B, consistent with studies that show that the rapamycin-FKBP complex inhibits T cell activation by a different mechanism (Shenolikar 1992). CyP-CsA cross-linking to the intact PP2B enzyme is Ca^{2+}/calmodulin dependent. These data suggest that the physiological target of drug-immunophilin complexes is the Ca^{2+}/calmodulin-activated form of PP2B.

The immunophilins are highly conserved from yeast to man, thereby opening the way for genetic studies of drug-phosphatase interactions in lower eukaryotes. The immunosuppressive drugs inhibit the recovery of budding yeast from alpha-factor arrest, a response similar to that produced by deletion of PP2B genes and confirms that the drugs indeed inhibit the yeast PP2B activity (Foor et al 1992). Interestingly, FK506 is a better inhibitor of PP2B in vitro when associated with the yeast FKBP12 than the human homologue (Rotonda et al 1993). Moreover, L-685,818, a structural derivative of FK506 and a well known antagonist of FK506-mediated immunosuppression, does not inhibit PP2B activity when combined with human FKBP12. Surprisingly, this antagonist is a potent PP2B inhibitor when bound to yeast FKBP12. Comparison of the three-dimensional structures of drug-immunophilin complexes formed by yeast and human FKBP12 showed only 10 amino acid differences that might represent surface contacts for PP2B. The targeted expression of yeast FKBP12 in animal tissues, combined with the use of L-685,818, may provide a unique opportunity for examining the physiological role of PP2B.

Multiple mammalian cyclophilins (CyP) and FKBPs have been identified. Expression of CyP (A or B) or FKBP12 increased T-cell sensitivity to CsA and FK506, respectively (Bram et al 1993). In contrast, expression of CyPC, FKBP13, or FKBP15 had no effect on drug responses in transfected T cells. In vitro studies with the drug-immunophilin-complexes have showed that PP2B inhibition is necessary but insufficient to explain differences in the in vivo effects of the selected immunophilins. Subcellular localization may also play an important part in determining the functions of these drug-receptor complexes. Transfer of C-terminal ER localization sequence from an active immunophilin, CyPB, to the inactive immunophilin, CyPC, provided a gain of function to CypC, which indicates that ER localization does not prevent the interaction of drug-immunophilin complexes with PP2B. In contrast, transfer of C-terminal sequences from the inactive immunophilin, CyPC, impaired the ability of CyPB to mediate CsA-dependent inhibition of cytokine gene expression. These studies provide the first insight into structural elements that are present in a subset of immunophilins that mediate immunosuppression. They also suggest that components of the drug-sensitive signal transduction pathway regulated by PP2B may be spatially restricted.

Increasing evidence suggests that selected immunophilins interact with proteins other than PP2B. For instance, the chicken progesterone receptor complexes contain 50–54-kd FK506-binding proteins (Smith et al 1993). Moreover, chicken progesterone receptors can be purified on a rapamycin-affinity matrix. Other steroid receptors, such as the glucocorticoid receptor, also bind FKBP52 in the presence of the drug FK506. Although the role of immunophilins in the control of steroid hormone function is still unclear, it raises the possibility that proteins other than PP2B may also be targeted by immunosuppressive drugs.

Protein Phosphatase-2C

PP2C, a Mg^{2+}-dependent protein phosphatase, is also a member of a multigene family and is structurally different from all other protein serine/threonine phosphatases (Shenolikar & Nairn 1991). In vitro studies show that PP2C has unusually high activity towards the enzymes of cholesterol metabolism. However, in vivo studies designed to examine the role of PP2C in this metabolic pathway have not been carried out. One would predict that the unique insensitivity of PP2C to OA and other phosphatase inhibitors might be exploited to decipher its physiological functions. For instance, OA concentrations that abolish PP1 and PP2A activity enhance phosphorylation of Ca^{2+}/calmodulin-dependent protein kinase II (CaMK II) in rat cerebellar granule cells (Fukunaga et al 1993), thus confirming the ability of PP1 and PP2A to dephosphorylate this enzyme in vitro and in vivo. However, the OA-induced phosphorylation of CaMK II in the granule cells is transient, returning to basal level within 10 min. This observation suggests that another OA-insensitive phosphatase also reversed CaMK II phosphorylation in cells. One such phosphatase could be PP2C, which dephosphorylates CaMK II in vitro at the key regulatory sites that generate the calcium-independent form of this kinase.

Insight into the physiological role of PP2C can also be gained from genetic studies in budding yeast. Disruption of two *S. cerevisiae* protein tyrosine phosphatase genes (PTP1 and PTP2) produced no apparent phenotype. Disruption of the PTC1 gene, which encodes PP2C, also failed to affect either the sexual or asexual growth of the yeast. However, loss of function of *ptp*2, combined with disruption of PTC1 gene, does inhibit cell growth (Maeda et al 1993). Overexpression of PTP1 or PTP2 suppresses the growth defect associated with the *ptc*1*ptp*2 double mutants. These data may suggest that the PTPases and PP2C share a substrate whose dephosphorylation at either tyrosine or serine/threonine causes a similar functional consequence, as seen with MAP kinases and *cdc*2 kinases. Alternately, unique substrates for these protein phosphatases may be functionally redundant. Interestingly, PCR cloning has identified two other yeast PP2C genes (PTC2 and PTC3), but their expression in yeast is insufficient to overcome the loss of PTC1 function.

Whereas PP2C shares no structural homology with PP1, PP2A, or PP2B, some sequence homology was observed with the mitochondrial enzyme pyruvate dehydrogenase phosphatase (Lawson et al 1993). Biochemical properties of a protein phosphatase isolated from archaebacteria also resemble those of PP2C (Kennelly et al 1993). This might suggest that the PP2C-like phosphatases arose earlier in evolution than the other serine/threonine phosphatases.

New Protein Serine/Threonine Phosphatases

Protein serine/threonine phosphatases that differ in their biochemical properties from PP1, 2A, 2B, and 2C have been isolated from mammalian tissues. However, currently we have insufficient structural information to classify them as new enzymes. For example, PP3, an OA-sensitive phosphatase, has been extracted by detergents from a particulate fraction of bovine brain (Honkanen et al 1991). This putative membrane-bound enzyme demonstrates substrate specificity similar to that of PP1, yet is not inhibited by I-2. While several peptide sequences from PP3 were present in PP1, unique peptides not represented in any PP1 isoform were also obtained. This suggests that PP3 may be a novel protein phosphatase. A new myosin phosphatase has been isolated from avian smooth muscle (Tulloch & Pato 1991). This enzyme, SMP-III, is inhibited by OA, yet shows very low activity against phosphorylase *a*, a common substrate for the well-known OA-sensitive phosphatases, PP1 and PP2A. Moreover, unlike all other serine/threonine phosphatases, SMP-III does not dephosphorylate phosphorylase kinase. Thus SMP-III may also represent a novel phosphatase.

Identification of novel cDNAs that encode protein serine/threonine phosphatases from plants and animals suggests that many more protein phosphatases have yet to be discovered. However, most of these enzymes still have not been biochemically characterized. For example, a novel *Drosophila* phosphatase catalytic subunit (*rdg*C) contains a calcium-binding domain fused to the catalytic domain (Steele et al 1992). Defects in the *rdg*C gene result in retinal degeneration, which implies a role for *rdg*C phosphatase in phototransduction. However, the inability of this enzyme to dephosphorylate available substrates has made it difficult to analyze the function or regulation of the *rdg*C protein.

PPX, a novel mammalian phosphatase, shows 65% primary sequence identity with PP2A. Moreover, PPX expressed in insect cells (Brewis et al 1993) exhibits in vitro substrate specificity and inhibitor sensitivity of PP2A. However, unlike PP2A, PPX does not bind the A-subunit and shows a preferential localization at centrosomes in interphase and during mitosis. Thus PPX may play a specific role in microtubule nucleation. PCR cloning from *S. cerevisiae* has identified a protein phosphatase, PPG (Posas et al 1993), that shares 60% sequence identity with PPX. Disruption of the PPG gene has shown that it is not essential for growth, yet causes a decreased accumulation of glycogen. Unlike the disruption of PP2A genes and loss of function in PP1 that prevents glycogen synthase activation and the accumulation of glycogen, PPG disruptants regulate glycogen synthase normally, but show greatly decreased levels of this enzyme. Thus regulation of glycogen metabolism in budding yeast may require three different protein phosphatases.

A murine immediate early gene, *3CH134*, that is induced by many mitogenic stimuli (Duff et al 1993) encodes a phosphatase with structural homology to the PTPase represented by the VH1 gene of *Vaccinia* virus (Guan et al 1991). The 3CH134-encoded phosphatase represents a dual-specificity enzyme that dephosphorylates both phosphotyrosine and phosphothreonine on MAP kinases, which results in the inactivation of these enzymes (Charles et al 1993; Zheng & Guan 1993). The human homologue CL100 has been expressed in bacteria and shown to dephosphorylate phosphothreonine and phosphotyrosine in MAP kinases 8-20-fold more rapidly than either a serine/threonine phosphatase, PP2A, or a PTPase, CD45 (Alessi et al 1993a). CL100 does not dephosphorylate a variety of serine/threonine or tyrosine-containing proteins. Moreover, the mutation of a cysteine conserved in all PTPases abolished CL100 phosphatase activity.

Growth factors and oncogenes activate a complex network of kinases (Crews & Erikson 1993; Pelech 1993) and phosphatases (Figure 3) to control many cellular processes ranging from metabolism to gene expression. Overexpression of 3CH134 selectively dephosphorylates $p42^{MAPK}$, but has no effect on most other phosphotyrosyl-containing proteins found in COS cells (Sun et al 1993). Overexpression of receptor or non-receptor PTPases CD45 and PTP1B in COS cells does not dephosphorylate and inactivate MAP kinases, which also argues that PTPases do not regulate MAP kinases. In contrast, addition of the human homologue CL100 to *Xenopus* extracts suppresses oncogenic *ras*-mediated activation of MAP kinase (Alessi et al 1993a). Expression of 3CH134 also prevents phosphorylation and activation of $p42^{MAPK}$ induced by serum as well as oncogenes such as *ras* and *raf*. Surprisingly, the inactive phosphatase 3CH134CS (cysteine-258 to serine mutant) augments MAP kinase phosphorylation. Overexpression of 3CH134CS activates MAP kinase in the absence of serum. The mutant phosphatase physically associates with phosphorylated $p42^{MAPK}$ and inhibits its dephosphorylation by the endogenous enzyme. The use of such dominant negative mutations may provide an important new approach to analyzing phosphatase function.

MAP kinases translocate to the nucleus following mitogenic stimuli (Pelech 1993) where they regulate a variety of transcription factors. MAP kinases in the nucleus may be regulated by nuclear dual-specificity phosphatases such as PAC-1 (Rohan et al 1993). The dual specificity of phosphatases, such as *cdc*25 (Millar & Russell 1992), 3CH134, and CL100, that dephosphorylate adjacent tyrosine and threonine residues in their target substrates provides a very high level specificity to these enzymes. Thus tight control can be exercised over key phosphoproteins, such as the *cdc*2 kinases and MAP kinases, that control cell growth, and inappropriate regulation by other phosphatases can be avoided. The dual-specificity phosphatases show two different modes for regulation, one being reversible phosphorylation of enzymes like *cdc*25 (Millar & Russell 1992) and the other being transient expression of the VH1-like enzymes (e.g. 3CH134, CL100, PAC-1, etc). VH1-like protein phosphatases

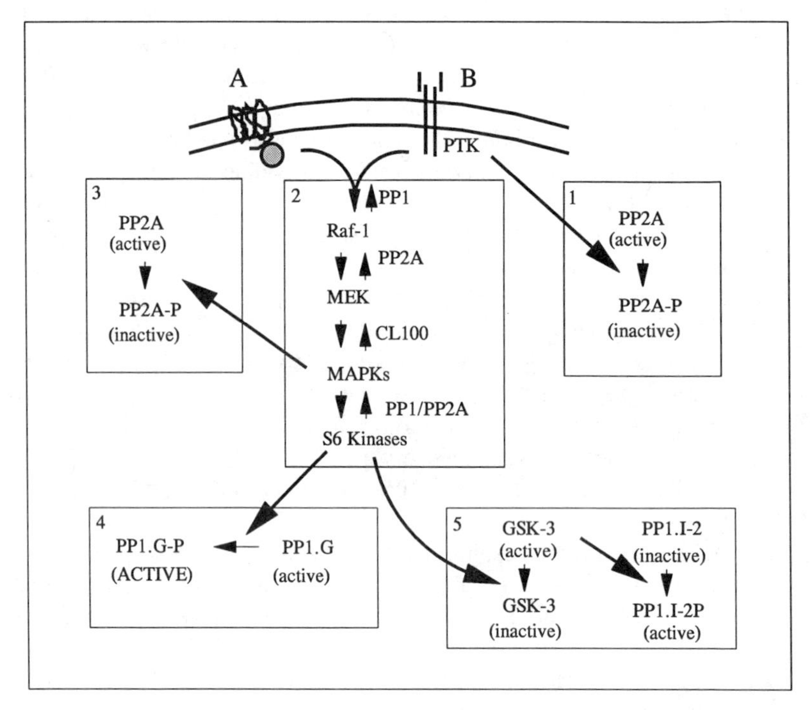

Figure 3 Integration of protein kinases and phosphatases during mitogenic signaling. Cell proliferation is stimulated by growth factors and hormones. (*Panel 1*) Activation of growth factor receptor kinases either directly or indirectly through soluble kinases leads to the transient tyrosine phosphorylation and inactivation of PP2A. This could facilitate the growth signals mediated by MAP kinases. (*Panel 2*) Mitogenic signals initiated by many different receptors are transduced by phosphorylation and activation of four protein kinases that constitute the MAP kinase cascade. Three protein serine/threonine phosphatases are required to reverse this pathway, thereby providing greater control over the mitogenic stimulus. (*Panel 3*) During the activation of the MAP kinase cascade, serine/threonine phosphorylation of PP2A catalytic and regulatory subunits may lead to a second phase of phosphatase inactivation that maintains the growth signals to promote expression of immediate early genes and other events essential for growth. (*Panel 4*) Insulin and growth factors enhance the phosphorylation of G-subunit by the p90 ribosomal S6 kinase and promote the dephosphorylation of glycogen synthase. Growth and metabolism are coordinated by the MAP kinase cascade. (*Panel 5*) The MAP kinase cascade also leads to the phosphorylation and inactivation of GSK-3β (Wang et al 1994), the primary activator of the latent PP1/I-2 complex in the cytosol (Welsh & Proud 1993). Thus growth factors can produce opposing changes in different PP1 pools within the cell.

are the key virulence factors for many poxviruses (Hakes et al 1993). The unregulated expression of viral enzymes may intervene in the normal functions of cellular dual-specificity phosphatases and promote cell death.

In *S. cerevisiae,* PKC activates the MAP kinase cascade. Loss of function of the PKC1 gene results in cell lysis with the inability to remodel the cell wall during growth. Two protein serine/threonine phosphatases (PPZ1 and PPZ2) suppress growth defects associated with *pkc*1 mutations (Lee et al 1993). Deletions of PPZ1 and PPZ2 induce cell lysis, which is similar to the effect of mutations in genes encoding PKC (PKC1), MAP kinase kinase (MKK1), and MAP kinase (MPK1). Overexpression of any one of these kinases suppresses *ppz*1*ppz*2 double disruptants. This suggests that PPZ1 and PPZ2 functions overlap with those of the MAP kinase cascade. In considering how a protein kinase and a phosphatase might elicit similar physiological effects, one could examine phos- phoproteins (e.g. *c-src* and c-Jun) that show positive and negative regulation by phosphorylation. In these proteins, phosphorylation of the positive regulatory site or dephosphorylation of the negative site would enhance biological activity. Alternately, the different targets for the kinase and the phosphatase may be functionally redundant.

The SIT4 protein phosphatase in *S. cerevisiae* is required for accumulation of the late G_1 cyclins and progression through S phase (Fernandez-Sara et al 1992). The *Drosophila* protein phosphatase PPV shares 63% sequence identity with SIT4 and rescues yeast SIT4 mutants, which allows growth at non-permissive temperatures (Mann et al 1993). In contrast, another phosphatase, *Drosophila* PP1 (40% sequence identity with PPV), fails to complement the SIT4 mutation (Mann et al 1993). However, when the PPV N-terminus was fused to PP1, the chimeric enzyme rescued the SIT4 mutations, which indicates that the N-terminal domain plays a key role such as localization or substrate recognition in PPV functions. Transient increase of PPV mRNA and protein in *Drosophila* occurs in late syncytial and early cellular blastoderm embryos, which suggests that the principal role for PPV in this organism is in a later phase of the cell cycle, namely transition through G_2.

These experiments illustrate how an heterologous system can be used to analyze structure-activity relationships for protein phosphatases. The *bim*G11 mutation in the PP1 catalytic subunit of *Aspergillus nidulans* results in a temperature-sensitive block in mitosis. This defect was rescued by expression of the mammalian PP1α in *Aspergillus* (Doonan et al 1991). Similarly, the cold-sensitive mutation *dis*2–11 in the fission yeast represents a mutation in the PP1 catalytic subunit that prevents separation of sister chromatids during cell division and was complemented by a plant PP1 (Nitschke et al 1992). This shows that structurally related protein phosphatases from different species are functionally interchangeable, thereby allowing the analysis of animal and plant enzymes in fungi.

FUTURE DIRECTIONS

Although much is known about the biochemistry of the major protein serine/threonine phosphatases (Shenolikar & Nairn 1991; Cohen 1989; Bollen &

Stalmans 1992), recent incorporation of molecular biology and genetics has not only identified many new enzymes, but has also begun to elucidate their role in cell regulation. We now have a wide range of new tools, including toxins and tumor promoters, that can decrease or even abolish the activity of the major phosphatases, PP1 and PP2A, in the intact cell. These reagents demonstrate an essential role for phosphatases in complex events such as cell growth and differentiation. OA and calyculin A inhibit PP1 and PP2A and interrupt gene expression in many organisms. For example, the use of phosphatase inhibitors suggests that PP1 activates multiple transcription factors involved in light-induced gene expression in plants (Sheen 1993). The use of these inhibitors has also revealed new regulatory pathways for steroid-inducible genes such that expression of hormone-sensitive target genes can be achieved in the absence of steroid hormone (Zhang et al 1994). Indeed, OA-induced phosphorylation and activation of "orphan" receptors that are members of the steroid receptor superfamily may identify their target genes in the absence of available ligands for these receptors.

Hormones stimulate gene transcription in a time frame ranging from minutes to hours. In contrast, the same hormones can modulate cellular metabolism within seconds or minutes. The cellular events that account for temporal differences in physiological responses are just beginning to be understood. Recent studies show that activation as well as translocation of kinases (Hagiwara et al 1993) and phosphatases (Fernandez et al 1992) into the nucleus may coordinate the amplitude and duration of the physiological response. Targeting proteins (Hubbard & Cohen 1993) and endogenous inhibitors, such as the newly discovered nuclear protein NIPP-1 (Beullens et al 1992, 1993), may regulate the localization and activity of protein phosphatases in the nucleus to control gene transcription.

Studies of hormone-regulated gene expression show that cellular responsiveness is controlled not only by phosphorylation and desensitization of cell surface receptors (Clark et al 1993), but also by phosphorylation-dephosphorylation of regulatory proteins such as transcription factors (Hagiwara et al 1992). As the number of transcription factors subject to phosphorylation continues to grow (Hunter & Karin 1992), examples of genes that are activated by dephosphorylation of specific nuclear factors will also increase. For example, phosphorylation of myogenin occurs within a sequence that is highly conserved in all myogenic determining factors and abolishes its DNA binding (Olson 1993). Thus a protein serine/threonine phosphatase may play a critical role in the coordinated activation of myogenic factors during muscle differentiation.

As immunophilins are expressed in all eukaryotic cells, the use of CsA and FK506 has revealed new roles for PP2B in both plant (Luan et al 1993) and animal cells. PP2B activation in T cells is essential for dephosphorylation and activation of a number of different transcription factors (Schreiber & Crabtree 1992). A major challenge for the future is to elucidate the mechanisms utilized by cell surface receptors to communicate with protein phosphatases that control processes such as gene expression.

Identification of regulatory proteins that inhibit or redirect the normal functions of protein phosphatases has provided new tools for the analysis of phosphatases. For instance, the overexpression of an activated I-1 peptide (Alberts et al 1994; Hagiwara et al 1992) or small-t (Sontag et al 1993; Alberts et al 1994) now allows us to examine the specific roles of PP1 and PP2A in cell function. An added layer of complexity may arise from the finding that virtually all the catalytic subunits and many regulatory subunits of protein serine/threonine phosphatases can be phosphorylated in vitro by many different protein kinases. Moreover, recent findings that protein serine/threonine phosphatases are subject to phosphorylation on tyrosine residues and that PTPases are phosphorylated on serine/threonine (Schievella et al 1993; Clarke et al 1993) in vivo increases the potential crosstalk between signaling pathways that utilize these enzymes. Deciphering the physiological relevance of this cross-communication will be another key challenge for the future.

The complete three-dimensional structure of several protein phosphatases will soon be available. These studies, in conjunction with structure-function analysis of the protein phosphatases, should identify residues that are critical for catalytic activity. Mutations of some of these residues should generate dominant negative enzymes as new tools for investigating protein phosphatase functions. For example, all the major protein serine/threonine phosphatases require cysteine for activity (Nemani & Lee 1993). As with 3CH134 (Sun et al 1993), one might predict that substitutions at this cysteine will generate dominant negative mutants of PP1, PP2A, and PP2B. However, the specific cysteine required for activity of these protein serine/threonine phosphatases has not been identified. The cold-sensitive mutation *dis*2-11 in the fission yeast results from the single substitution of arginine-245 with glutamine (Kinoshita et al 1990). This arginine is conserved in the major protein serine/threonine phosphatases. Introduction of cold-sensitive *dis*2 gene in high copy number into wild-type cells inhibits growth. Identical mutation in yeast PP2A also yields a growth inhibitor of otherwise normal yeast cells. Screening for a synthetic lethal phenotype in such conditional phosphatase mutations in yeast should provide insight into the functional interactions of these phosphatases and other components of cell signaling pathways.

In *S. cerevisiae,* the GLC7-1 mutation represents a substitution of arginine-73 by cysteine in the PP1 catalytic subunit. This mutation does not eliminate phosphatase activity, but prevents PP1 association with at least one targeting protein, the yeast G-subunit, and inhibits glycogen synthesis (Stuart et al 1994). Subtle mutations that do not prevent substrate recognition but compromise phosphatase activity or localization may yield useful reagents for sorting out the normal physiological functions of protein phosphatases.

Recent studies have shown that protein phosphatases regulate many aspects of neuronal function (Nairn & Shenolikar 1992). For instance, PP1 may mediate the reduction of synaptic transmission known as long-term depression in the mammalian brain (Mulkey et al 1993, 1994). These studies suggest that receptor-mediated events in specific hippocampal neurons are translated into activation of PP1 during LTD. Moreover, this signaling pathway could represent an example of a protein phosphatase cascade whereby activation of PP2B leads to dephosphorylation of I-1 and consequent increase in PP1 activity and induction of LTD.

Both physiological (Shenolikar & Endo 1993) and genetic studies (Asztalos et al 1993) in invertebrates point to PP1 as a key regulator of learning and memory. A mutation in the *Drosophila* chromosome at the locus of a PP1 gene impairs visual conditioning and associative olfactory learning (Asztalos et al 1993). Protein phosphorylation also plays a role in the transduction of odor signals to electrical responses in mammalian olfactory cilia (Boehoff & Breer 1992). Termination of olfactory signals within milliseconds may reflect rapid phosphorylation and desensitization of odorant receptors. Such studies are beginning to dispel a long-held view that protein phosphorylation induced by diffusible messengers cannot occur in milliseconds and emphasize that coordinated regulation of kinases and phosphatases can lead to rapid control of physiological processes. In conclusion, tremendous progress has been made in our understanding of the physiological functions of protein phosphatases in the last two years. With many new reagents and experimental approaches available, even greater progress towards establishing the physiological role of protein serine/threonine phosphatases is anticipated in the years to come.

ACKNOWLEDGMENTS

With the explosion of new information on protein phosphatases, it is a daunting task to write a truly comprehensive review on this topic. By arbitrarily focusing on publications that appeared during the last year, we identified 545 manuscripts that discussed important aspects of structure, function, and regulation of protein serine/threonine phosphatases. However, space limitation has forced us to further restrict the scope of this review, and many areas have received only cursory mention. I apologize to my many friends whose contributions to the recent advances in protein phosphatases are not fully recognized in this review. I would also like to thank my colleagues Shogo Endo and Dylan Steer for their helpful and sometimes lengthy debates on the recent literature.

Literature Cited

Alberts AS, Deng T, Lin A, Meinkoth JL, Schonthal A, Mumby MC, et al. 1993a. Protein phosphatase 2A potentiates the activity of promoters containing AP-1-binding elements. *Mol. Cell. Biol.* 13:2104–12

Alberts AS, Montminy M, Shenolikar S, Feramisco JR. 1994. Expression of a peptide inhibitor of protein phosphatase-1 increases phosphorylation and activity of CREB in NIH-3T3 fibroblasts. *Mol. Cell. Biol.* In press

Alberts AS, Thorburn AM, Shenolikar S, Mumby MC, Feramisco JR. 1993b. Regulation of cell cycle progression and nuclear affinity of the retinoblastoma protein by protein phosphatases. *Proc. Natl. Acad. Sci. USA* 90:388–92

Alessi D, MacDougall LK, Sola MM, Ikebe M, Cohen P. 1992. The control of protein phosphatase-1 by targeting subunits. The major myosin phosphatase in avian smooth muscle is a novel form of protein phosphatase-1. *Eur. J. Biochem.* 210:1023–35

Alessi DR, Smythe C, Keyse SM. 1993a. The human CL100 gene encodes a Tyr/Thr-protein phosphatase which potently and specifically inactivates MAP kinase and suppresses its activation by oncogenic ras in *Xenopus* oocyte extracts. *Oncogene* 8:2015–20

Alessi DR, Street AJ, Cohen P, Cohen PTW. 1993b. Inhibitor-2 functions like a chaperone to fold three expressed isoforms of mammalian protein phosphatase-1 into a conformation with the specificity and regulatory properties of the native enzyme. *Eur. J. Biochem.* 213:1055–66

Asztalos Z, Von Wegerer J, Wustman G, Dombradi V, Gausz J, et al. 1993. Protein phosphatase 1-deficient mutant *Drosophila* is affected in habituation and associative learning. *J. Neurosci.* 13:924–30

Axton JM, Dombradi V, Cohen PTW, Glover DM. 1990. One of the protein phosphatase 1 isoenzymes in Drosophila is essential for mitosis. *Cell* 63:33–46

Barker HM, Jones TA, de Cruz e Silva EF, Spurr NK, Sheer D, Cohen PTW. 1990. Localization of the gene encoding type 1 phosphatase catalytic subunit to human chromosome band 11q13. *Genomics* 7:159–66

Beullens M, Eynde AV, Bollen M, Stalmans W. 1993. Inactivation of nuclear inhibitory polypeptides of protein phosphatase-1 (NIPP-1). *J. Biol. Chem.* 268:13172–77

Beullens M, Eynde AV, Stalmans W, Bollen M. 1992. The isolation of novel inhibitory polypeptides of protein phosphatase 1 from bovine thymus nuclei. *J. Biol. Chem.* 267: 16538–44

Blacketer MJ, Koehler CM, Coats SG, Myers AM, Madaule P. 1993. Regulation of dimorphism in *Saccharomyces cerevisiae*: involvement of the novel protein kinase homolog Elm1p and protein phosphatase 2A. *Mol. Cell. Biol.* 13:5567–81

Boehoff I, Breer H. 1992. Termination of second messenger signaling in olfaction. *Proc. Natl. Acad. Sci. USA* 89:4171–74

Bollen M, Stalmans W. 1992. The structure, role and regulation of type-1 protein phosphatases. *Crit. Rev. Biochem. Mol. Biol.* 27: 227–81

Boyle WJ, Smeal T, Defize LHK, Angel P, Woodgett JR, et al. 1991. Activation of protein kinase C decreases phosphorylation of c-Jun at sites that negatively regulate its DNA-binding activity. *Cell* 64:573–84

Bram RJ, Hung DT, Martin PK, Schreiber SL, Crabtree GR. 1993. Identification of the immunophilins capable of mediating inhibition of signal transduction by cyclosporin A and FK506: roles of calcineurin binding and cellular location. *Mol. Cell. Biol.* 13:4760–69

Brewis ND, Street AJ, Prescott AR, Cohen PT. 1993. PPX, a novel protein serine/threonine phosphatase localized to centrosomes. *EMBO J.* 12:987–96

Bu X, Hass DW, Hagedorn CH. 1993. Novel phosphorylation sites of eukaryotic initiation factor-4F and evidence that phosphorylation stabilizes interactions with the p25 and p220 subunits. *J. Biol. Chem.* 268:4975–78

Cayla X, Ballmer-Hofer K, Merlevede W, Goris J. 1993. Protein phosphatase 2A associated with polyomavirus small-T or middle-T antigen is an okadaic acid-sensitive tyrosyl phosphatase. *Eur. J. Biochem.* 15: 281–86

Cayla X, Goris J, Hermann J, Hendrix P, Ozon R, Merlevede W. 1990. Isolation and characterization of a tyrosine phosphatase activator from rabbit skeletal muscle and *Xenopus laevis* oocytes. *Biochemistry* 29:658–67

Charbonneau H, Tonks NK. 1992. 1002 protein phosphatases? *Annu. Rev. Cell. Biol.* 8:463–93

Charles CH, Sun H, Lau LF, Tonks NK. 1993. The growth factor inducible immediate early

gene *3CH134* encodes a protein tyrosine phosphatase. *Proc. Natl. Acad. Sci. USA* 90: 5292–96
Chen J, Martin BL, Brautigan DL. 1992. Regulation of protein serine/threonine phosphatase type-2A by tyrosine phosphorylation. *Science* 257:1261–64
Chen J, Parsons S, Brautigan DL. 1994. Tyrosine phosphorylation of protein phosphatase 2A in response to growth stimulation and v-src transformation of fibroblasts. *J. Biol. Chem.* 269:7957–62
Chen MX, Chen YH, Cohen PT. 1992. Polymerase chain reactions using *Saccharomyces, Drosophila* and human DNA predict a large family of protein serine/threonine phosphatases. *FEBS Lett.* 306:54–59
Clark RB, Friedman J, Kunkel MW, January BG, Shenolikar S. 1993. Okadaic acid, a phosphatase inhibitor, induces both an augmentation and an inhibition of β_2-adrenergic stimulation of cAMP accumulation in S49 lymphoma cells. *J. Biol. Chem.* 268:3245–50
Clarke PR, Hoffman I, Draetta G, Karsenti E. 1993. Dephosphorylation of *cdc*25-C by a type-2A protein phosphatase: specific regulation during the cell cycle in *Xenopus* egg extracts. *Mol. Biol. Cell.* 4:397–411
Cohen P. 1989. Structure and regulation of protein phosphatases. *Annu. Rev. Biochem.* 58: 453–508
Cohen PT. 1988. Two isoforms of protein phosphatase-1 may be produced from the same gene. *FEBS Lett.* 232:17–23
Cohen PTW, Brewis ND, Hughes V, Mann D. 1990. Protein serine/threonine phosphatases: an expanding family. *FEBS Lett.* 268:355–59
Crews CM, Erikson RL. 1993. Extracellular signals and reversible protein phosphorylation: what to Mek of it all. *Cell* 74:215–17
Cyert MS, Kunisawa R, Kaim D, Thorner J. 1991. Yeast has homologs (CNA1 and CNA2 gene products) of mammalian calcineurin, a calmodulin-regulated phosphoprotein phosphatase. *Proc. Natl. Acad. Sci. USA* 88: 7376–80
Cyert MS, Thorner J. 1992. Regulatory subunit (CNB1 gene product) of yeast Ca^{2+}/calmodulin-dependent phosphoprotein phosphatases is required for adaptation to pheromone. *Mol. Cell. Biol.* 12:3460–69
Damuni Z, Guo H. 1993. Autophosphorylation-activated protein kinase phosphorylates and inactivates protein phosphatase 2A. *Proc. Natl. Acad. Sci. USA* 90:2500–4
Dent P, MacDougall LK, MacKintosh C, Campbell DG, Cohen P. 1992. A myofibrillar protein phosphatase from rabbit skeletal muscle contains the beta isoform of protein phosphatase complexed to a regulatory subunit which greatly enhances the dephosphorylation of myosin. *Eur. J. Biochem.* 210: 1037–44
Dobrowsky RT, Kamibayashi C, Mumby MC, Hannun YA. 1993. Ceramide activates heterotrimeric protein phosphatase 2A. *J. Biol. Chem.* 268:15523–30
Dombradi V, Mann DJ, Saunders RD, Cohen PT. 1993. Cloning of the fourth functional gene for protein phosphatase 1 in *Drosophila melanogaster* from its chromosomal location. *Eur. J. Biochem.* 212:177–83
Doonan JH, MacKintosh C, Osmani S, Cohen P, Bai G, et al. 1991. A cDNA encoding rabbit skeletal muscle protein phosphatase 1 alpha complements the *Aspergillus* cell cycle mutation, bimG11. *J. Biol. Chem.* 266: 18889–94
Duff JL, Marrero MB, Paxton WG, Charles CH, Bernstein KE, Berk BC. 1993. Angiotensin II induces 3CH134, a protein tyrosine phosphatase in vascular smooth muscle cells. *J. Biol. Chem.* 268:26037–40
Durfee T, Becherer K, Chen P-L, Yeh S-H, Yang Y, et al. 1993. The retinoblastoma protein associates with the protein phosphatase type 1 catalytic subunit. *Genes Dev.* 7:555–69
Feng ZH, Wilson SE, Peng ZY, Schlender KK, Reimann EM, Trumbly RJ. 1991. The yeast *GLC7* gene required for glycogen accumulation encodes a type 1 protein phosphatase. *J. Biol. Chem.* 266:23796–801
Fernandez A, Brautigan DL, Lamb NJ. 1992. Protein phosphatase type 1 in mammalian cell mitosis: chromosomal localization and involvement in mitotic exit. *J. Cell Biol.* 116: 1421–30
Fernandez-Sara MS, Sutton A, Zhong T, Arndt KT. 1992. SIT4 protein phosphatase is required for the normal accumulation of *SW14, CLN1, CLN2* and *HCS26* RNAs during late G_1. *Genes Dev.* 6:2417
Foor F, Parent SA, Morin N, Dahl AM, Ramadan N, et al. 1992. Calcineurin mediates inhibition by FK506 and cyclosporin of recovery from alpha-factor arrest in yeast. *Nature* 360:682–84
Francois JM, Thompson-Jaeger S, Skroch J, Zellenka U, Spevak W, Tatchell K. 1992. GAC1 may encode a regulatory subunit for protein phosphatase type 1 in *Saccharomyces cerevisiae. EMBO J.* 11:87–96
Fukunaga K, Kobayashi T, Tamura S, Miyamoto E. 1993. Dephosphorylation of autophosphorylated Ca^{2+}/calmodulin-dependent protein kinase II by protein phosphatase 2C. *J. Biol. Chem.* 268:133–37
Gomes R, Karess RE, Ohkura H, Glover DM, Sunkel CE. 1993. Abnormal anaphase resolution (*aar*): a locus required for progression through mitosis in *Drosophila. J. Cell Sci.* 104:1–11
Gong MC, Fuglsang A, Alessi D, Kobayashi S, Cohen P, et al. 1992. Arachidonic acid inhibits myosin light chain phosphatase and sen-

sitizes smooth muscle to calcium. *J. Biol. Chem.* 267:21492–98

Guan K, Broyles SS, Dixon JE. 1991. A tyr/Ser protein phosphatase encoded by *Vaccinia* virus. *Nature* 350:359–62

Guo H, Reddy SAG, Damuni Z. 1993. Purification and characterization of a distinct autophosphorylation-activated protein kinase that phosphorylates and inactivates protein phosphatase 2A. *J. Biol. Chem.* 268:11193–98

Hagiwara M, Brindle P, Alberts A, Meinkoth J, Feramisco J, et al. 1992. Transcriptional attenuation following cAMP induction requires PP-1 mediated dephosphorylation of CREB. *Cell* 70:105–13

Hagiwara M, Brindle P, Harootunian A, Armstrong R, Rivier J, et al. 1993. Coupling of hormonal stimulation and transcription via the cAMP-responsive factor CREB is rate limited by nuclear entry of protein kinase A. *Mol. Cell. Biol.* 13:4852–59

Hakes DJ, Martell KJ, Zhao WG, Massung RF, Esposito JJ, Dixon JE. 1993. A protein phosphatase related to the *Vaccinia* virus VH1 is encoded in the genomes of several orthopoxviruses and a baculovirus. *Proc. Natl. Acad. Sci. USA* 90:4017–21

Healy AM, Zolnierowicz S, Stapleton AE, Goebl M, DePaoli-Roach AA, Pringle JR. 1991. *CDC55*, a *Saccharomyces cerevisiae* gene involved in cellular morphogenesis: identification, characterization, and homology to the B subunit of mammalian type 2A protein phosphatase. *Mol. Cell. Biol.* 11: 5767–80

Hegde AN, Das MR. 1990. Glucagon and p21ras enhance the phosphorylation of the same 38-kilodalton membrane protein from rat liver cells. *Mol. Cell. Biol.* 10:2468–74

Hendrix P, Mayer-Jaekel RE, Cron P, Goris J, Hofsteenge J, et al. 1993. Structure and regulation of a 72-kDa regulatory subunit of protein phosphatase 2A—evidence for different forms produced by alternative splicing. *J. Biol. Chem.* 268:15267–76

Honkanen RE. 1993. Cantharidin, another natural toxin that inhibits the activity of serine/threonine phosphatases type 1 and 2A. *FEBS Lett.* 330:283–86

Honkanen RE, Zwiller J, Daily SI, Khatra BS, Dukelow M, Boynton AL. 1991. Identification, purification and characterization of a novel serine/threonine protein phosphatase from bovine brain. *J. Biol. Chem.* 266:6614–19

Huang J, Wei Y, Kim Y, Osterberg L, Matthews HR. 1991. Purification of a protein kinase from the yeast *Saccharomyces cerevisiae*. The first member of this class of protein kinases. *J. Biol. Chem.* 266:9023–31

Hubbard MJ, Cohen P. 1993. On target with a new mechanism for the regulation of protein phosphorylation. *Trends Biochem. Sci.* 18: 172–77

Hunter T, Karin M. 1992. The regulation of transcription by phosphorylation. *Cell* 70: 375–87

Kennelly PJ, Oxenrider KA, Leng J, Cantwell JS, Zhao N. 1993. Identification of a serine/threonine-specific protein phosphatase from the archaebacterium *Supholobus solfataricus*. *J. Biol. Chem.* 268:6505–10

Kim Y, Huang J, Cohen P, Matthews HR. 1993. Protein phosphatases 1, 2A and 2C are protein histidine phosphatases. *J. Biol. Chem.* 268:18513–18

Kincaid RL. 1993. Calmodulin-dependent protein phosphatases from microorganism to man: a study in structural conservatism and biological diversity. *Adv. Second Messenger Phosphoprotein Res.* 23:1–25

Kinoshita N, Ohkura H, Yanagida M. 1990. Distinct roles of type 1 and 2A protein phosphatases in the control of the fission yeast cell division cycle. *Cell* 63:405–15

Kinoshita N, Yamano H, Niwa H, Yoshida T, Yanagida M. 1993. Negative regulation of mitosis by the fission yeast protein phosphatase ppa2. *Genes Dev.* 7:1059–71

Kovacina KS, Yonezawa K, Brautigan DL, Tonks NK, Rapp UR, Roth RA. 1990. Insulin activates the kinase activity of Raf-1 protooncogene by increasing its serine phosphorylation. *J. Biol. Chem.* 265:12115–18

Lawson JE, Niu XD, Browning KS, Trang HL, Yan J, Reed LJ. 1993. Molecular cloning and expression of the catalytic subunit of bovine pyruvate dehydrogenase phosphatase and sequence similarity with protein phosphatase 2C. *Biochemistry* 32:8987–93

Lee J, Stock J. 1993. Protein phosphatase 2A catalytic subunit is methyl-esterified at its carboxyl terminus by a novel methyltransferase. *J. Biol. Chem.* 268: 19192–95

Lee KS, Hines LK, Levin DE. 1993. A pair of functionally redundant yeast genes (*PPZ1* and *PPZ2*) encoding type-1-related protein phosphatases function within the PKC-mediated pathway. *Mol. Cell. Biol.* 13:5843–53

Li W, Handschumacher E. 1993. Specific interaction of the cyclophilin-cyclosporin complex with the B-subunit of calcineurin. *J. Biol. Chem.* 268:14040–44

Li YM, Casida JE. 1992. Cantharidin-binding protein: identification as protein phosphatase 2A. *Proc. Natl. Acad. Sci. USA* 89: 11867–70

Luan S, Li W, Rusnak F, Assmann SM, Schreiber SL. 1993. Immunosuppressants implicate protein phosphatase regulation of K^+ channels in guard cells. *Proc. Natl. Acad. Sci. USA* 90:2202–6

Ludlow JW, Glendenning CL, Livingston DM, DeCaprio JA. 1993. Specific enzymatic dephosphorylation of the retinoblastoma protein. *Mol. Cell. Biol.* 13:367–72

Lutz MP, Gates LK, Pinon DI, Shenolikar S, Miller LJ. 1993. Control of cholecystokinin receptor dephosphorylation in pancreatic acinar cells. *J. Biol. Chem.* 268: 12136–42

Luu HA, Chen DZ, Magoon J, Worms J, Smith J, Holmes CF. 1993. Quantification of diarrhetic toxins and identification of novel protein phosphatase inhibitors in marine phytoplankton and mussels. *Toxicon* 31:75–83

Maeda T, Tsai AYM, Saito H. 1993. Mutations in protein tyrosine phosphatase gene (PTP2) and a protein serine/threonine phosphatase gene (PTC1) cause a synthetic growth defect in *Saccharomyces cerevisiae. Mol. Cell. Biol.* 13:5408–17

Mann DJ, Dombradi V, Cohen PTW. 1993. *Drosophila* protein phosphatase V functionally complements a SIT4 mutant in *Saccharomyces cerevisiae* and its amino-terminal region can confer this complementation to a heterologous phosphatase catalytic domain. *EMBO J.* 12:4833–42

Mayer-Jaekel RE, Ohkura H, Gomes R, Sunkel CE, Baumgartner S, et al. 1993. The 55 kDa regulatory subunit of Drosophila protein phosphatase 2A is required for anaphase. *Cell* 72:621–33

Millar JB, Russell P. 1992 The cdc25 M-phase inducer: an unconventional protein phosphatase. *Cell* 68:407–10

Mulkey RM, Endo S, Shenolikar S, Malenka, RC. 1994. Involvement of a calcineurin inhibitor-1 phosphatase cascade in hippocampal long-term depression. *Nature*. In press

Mulkey RM, Herron CE, Malenka RC. 1993. An essential role for protein phosphatases in hippocampal long-term depression. *Science* 261:1051–55

Mumby MC, Walter G. 1991. Protein phosphatases and DNA tumor viruses: transformation through the back door. *Cell Regul.* 2:589–98

Muramatsu T, Giri PR, Higuchi S, Kincaid RL. 1992. Molecular cloning of a calmodulin-dependent protein phosphatase from murine testis: identification of a developmentally expressed nonneural isoenzyme. *Proc. Natl. Acad. Sci. USA* 89:529–33

Nairn AC, Shenolikar S. 1992. Role of protein phosphatases in synaptic transmission, plasticity and neuronal development. *Curr. Top. Neurobiol.* 2:296–30

Nemani R, Lee EY. 1993. Reactivity of sulphydryl groups of the catalytic subunits of rabbit skeletal muscle protein phosphatases 1 and 2A. *Arch. Biochem. Biophys.* 300:24–29

Nitschke K, Fleig U, Schell J, Palme K. 1992. Complementation of the cs dis2–11 cell cycle mutant of *Schizosaccharomyces pombe* by a protein phosphatase from *Arabidopsis thaliana. EMBO J.* 11:1327–33

Olson EN. 1993. Signal transduction pathways that regulate skeletal muscle gene expression. *Mol. Endocrinol.* 7:1369–78

Pelech SL. 1993. Networking with protein kinases. *Curr. Biol.* 3:513–15

Posas F, Clotete J, Muns MT, Corominas J, Casamayor A, Arino J. 1993. The gene *PPG* encodes a novel protein phosphatase involved in glycogen metabolism. *J. Biol. Chem.* 268:1349–54

Rohan PJ, Davis P, Moskaluk CA, Kearns M, Krutzsch H, et al. 1993. PAC-1: a mitogen-induced nuclear protein tyrosine phosphatase. *Science* 259:1763–66

Ronne H, Carlberg M, Hu GZ, Nehlin JO. 1991. Protein phosphatase 2A in *Saccharomyces cerevisiae*: effects on growth and bud morphogenesis. *Mol. Cell. Biol.* 11:4876–84

Rotonda J, Burbaum JJ, Chan K, Marcy AI, Becker JW. 1993. Improved calcineurin inhibition by yeast FKBP12-drug complexes—crystallographic and functional analysis. *J. Biol. Chem.* 268:7607–9

Scheidtmann KH, Mumby MC, Rundell K, Walter G. 1991. Dephosphorylation of simian virus 40 large T antigen and p53 protein by protein phosphatase 2A: inhibition by small-t antigen. *Mol. Cell. Biol.* 11:1996–2003

Schievella AR, Paige LA, Johnson KA, Hill DE, Erikson RE. 1993. Protein tyrosine phosphatase 1B undergoes mitosis-specific phosphorylation on serine. *Cell Growth Diff.* 4:239–46

Schonthal A, Alberts AS, Rahman A, Meinkoth J, Mumby M, Feramisco JR. 1991. Involvement of serine/threonine protein phosphatases in signal transduction pathways that regulate gene expression and cell growth. In *Recent Advances in Cellular and Molecular Biology,* ed. RJ Wegmann, MA Wegmann, pp 67–74. Leuven: Peeters

Schreiber SL, Crabtree GR. 1992. The mechanism of action of cyclosporin A and FK506. *Immunol. Today* 13:136–42

Sheen, J. 1993. Protein phosphatase activity is required for light-inducible gene expression in maize. *EMBO J.* 12:3497–505

Shenolikar S. 1992. A window opens on immunosuppression. *Curr. Biol.* 2:549–51

Shenolikar S, Endo S. 1993. Protein phosphatases and memory channels. *Neurosci. Facts* 4:11–12

Shenolikar S, Nairn A. 1991. Protein phosphatases: recent progress. *Adv. Second Messenger Phosphoprotein Res.* 23:1–121

Shima H, Haneji T, Hatano Y, Kasugai I, Sugimura T, Nagao M. 1993a. Protein phosphatase 1γ2 is associated with nuclei of meiotic cells in rat testis. *Biochem. Biophys. Res. Commun.* 194:930–37

Shima H, Hatano Y, Chun Y-S, Sugimura T, Zhang Z, et al. 1993b. Identification of PP1 catalytic subunit isotypes PP1γ, PP1δ and PP1α in various rat tissues. *Biochem. Biophys. Res. Commun.* 192:1289–96

Smith DF, Baggenstoss BA, Marion TN, Rimerman RA. 1993. Two FKBP-related proteins are associated with progesterone receptor complexes. *J. Biol. Chem.* 268:18365–71

Smith RD, Walker JC. 1993. Expression of multiple type 1 protein phosphatases in *Arabidopsis thaliana*. *Plant Mol. Biol.* 21: 307–16

Sontag E, Fedorov S, Kamibayashi C, Robbins D, Cobb M, Mumby M. 1993. The interaction of SV40 small tumor antigen with protein phosphatase 2A stimulates the MAP kinase pathway and induces cell proliferation. *Cell* 75:887–97

Steele FR, Washburn T, Rieger R, O'Tousa JE. 1992. Drosophila retinal degeneration C (*rdgC*) encodes a novel serine/threonine protein phosphatase. *Cell* 69:669–76

Steenaart NA, Ganim JR, DiSalvo J, Kranias EG. 1992. The phospholamban phosphatase associated with cardiac sarcoplasmic reticulum is a type 1 enzyme. *Arch. Biochem. Biophys.* 293:17–24

Stemmer P, Klee CB. 1991. Serine/threonine phosphatases in the nervous system. *Curr. Opin. Neurobiol.* 1:53–64

Stone EM, Yamano H, Kinoshita N, Yanagida M. 1993. Mitotic regulation of protein phosphatases by fission yeast *sds22* protein. *Curr. Biol.* 3:13–26

Stuart JK, Frederick DL, Varner CM, Tatchell K. 1994. The mutant type 1 protein phosphatase encoded by glc7–1 from *Saccharomyces cerevisiae* fails to interact productively with the GAC1-encoded regulatory subunit. *Mol. Cell. Biol.* 14:896–905

Sun H, Charles CH, Lau LF, Tonks NK. 1993. MKP-1 (3CH134), an immediate early gene product, is a dual specificity phosphatase that dephosphorylates MAP kinase in vivo. *Cell* 75:487

Sutton AD, Immanuel D, Arndt KT. 1991. The SIT4 protein phosphatase functions in late G_1 for progression into S phase. *Mol. Cell. Biol.* 11:2133–48

Tang PM, Bondor JA, Swiderek KM, DePaoli-Roach AA. 1991. Molecular cloning and expression of the regulatory (RG1) subunit of the glycogen-associated protein phosphatase. *J. Biol. Chem.* 266:15782–89

Tulloch AG, Pato MD. 1991 Turkey gizzard smooth muscle myosin phosphatase-III is a novel protein phosphatase. *J. Biol. Chem.* 266:20168–74

Turner BC, Tonks NK, Rapp UR, Reed JC. 1993. Interleukin 2 regulates Raf-1 kinase activity through a tyrosine phosphorylation-dependent mechanism in a T-cell line. *Proc. Natl. Acad. Sci. USA* 90:5544–48

Ueki K, Kincaid RL. 1993. Interchangeable associations of calcineurin regulatory subunit isoforms with mammalian and fungal catalytic subunits. *J. Biol. Chem.* 268:6554–59

Uemura T, Shiomi K, Togashi S, Takeichi M. 1993. Mutation of twins encoding a regulator of protein phosphatase 2A leads to pattern duplication in *Drosophila* imaginal discs. *Genes Dev.* 7:429–40

Van Zyl W, Huang W, Sneddon AA, Stark M, Camier S, et al. 1992. Inactivation of the protein phosphatase 2A regulatory subunit A results in morphological and transcriptional defects in *Saccharomyces cerevisiae*. *Mol. Cell. Biol.* 12:4946–59

Villa-Moruzzi E. 1992. Activation of type-1 protein phosphatase by *cdc*2 kinase. *FEBS Lett.* 304:211–15

Villa-Moruzzi E, Dalta Zonca P, Crabb JW. 1991. Phosphorylation of the catalytic subunit of type-1 protein phosphatase by the *v-abl* tyrosine kinase. *FEBS Lett.* 293:67–71

Walter G, Mumby M. 1993. Protein serine/threonine phosphatases and cell transformation. *Biochim. Biophys. Acta* 1155:207–26

Wang QM, Park IK, Fiol CJ, Roach PJ, DePaoli-Roach AA. 1994. Isoform differences in substrate recognition by glycogen synthase kinase 3α and 3β in the phosphorylation of phosphatase inhibitor 2. *Biochemistry* 33:143–47

Wek RC, Cannon JF, Dever TE, Hinnebusch AG. 1992. Truncated protein phosphatase GLC7 restores translational activation of GCN4 expression in yeast mutants defective for eIF-2 alpha kinase GCN2. *Mol. Cell. Biol.* 12:5700–10

Welsh GI, Proud CG. 1993. Glycogen synthase kinase-3 is rapidly inactivated in response to insulin and phosphorylates eukaryotic initiation factor eIF-2B. *Biochem. J.* 294:625–29

White RE, Lee AB, Scherbatko AD, Lincoln TM, Schonbrunn A, Armstrong DL. 1993. Potassium channel stimulation by natriuretic peptides through cGMP-dependent dephosphorylation. *Nature* 361:263–66

White RE, Schonbrunn A, Armstrong DL. 1991. Somatostatin stimulates Ca^{2+}-activated K^+ channels through protein dephosphorylation. *Nature* 351:570–73

Yang SI, Licktieg RL, Estes R, Rundell K, Walter G, Mumby MC. 1991. Control of protein phosphatase 2A by simian virus 40 small-t antigen. *Mol. Cell. Biol.* 11:1988–95

Zhang Y, Bai W, Laagood VE, Weigel NL. 1994. Multiple signalling pathways activate the chicken progesterone receptor. *Mol. Endocrinol.* 8:577–84

Zhang Z, Bai G, Shima M, Zhao S, Nagao M, Lee EYC. 1993. Expression and characterization of rat protein phosphatase-1α, -1γ1, -1γ2 and -1γ. *Arch. Biochem. Biophys.* 303: 402–6

Zheng CF, Guan KL. 1993. Dephosphorylation and inactivation of the mitogen-activated protein kinase by a mitogen-induced thr/tyr protein phosphatase. *J. Biol. Chem.* 268:16116–19

Annu. Rev. Cell Biol. 1994. 10:87–119

SIGNAL SEQUENCE RECOGNITION AND PROTEIN TARGETING TO THE ENDOPLASMIC RETICULUM MEMBRANE

Peter Walter
Department of Biochemistry and Biophysics, University of California, San Francisco, California 94143–0448

Arthur E. Johnson
Department of Chemistry and Biochemistry, University of Oklahoma, Norman, Oklahoma 73019

KEY WORDS: signal recognition particle, SRP, SRP receptor

CONTENTS

0743–4634/94/1115–0087$05.00

INTRODUCTION

In this review, we attempt a timely survey of issues concerning protein translocation across the endoplasmic reticulum membrane, with a strong focus on the initial events that lead to the selection and proper delivery of proteins to this membrane system. Many new questions are raised by recent discoveries: it is now known that targeting can occur by multiple pathways and that the molecular machines that catalyze targeting and translocation are conserved in all cell types examined (from bacteria to mammalian cells). It is desirable to integrate the information from these different organisms into a coherent picture because we feel that the similarities—as well as the differences—found over such vast evolutionary distances will illuminate the fundamental principles that govern the inner workings of these components. Although this philosophy may cause an occasional oversimplification, we believe that it sets a useful conceptual framework to guide future experimental investigation. We focus here on recent developments and open questions, and do not intend this review to be comprehensive. Where appropriate, reference to more detailed reviews is given in the text.

OVERVIEW OF SIGNAL RECOGNITION PARTICLE-DEPENDENT PROTEIN TARGETING

Translocation of soluble proteins across the endoplasmic reticulum (ER) membrane or integration of membrane proteins into the ER membrane are the first steps in the processes that deliver proteins to the secretory pathway and thereby initiate their journey to the outside of the cell, to the plasma membrane, or to the intracellular organelles that comprise the endomembrane system (Palade 1975). In mammalian cells, the synthesis of these proteins takes place on ribosomes that are bound to the rough ER membrane, and protein translocation and integration occur simultaneously with ongoing protein synthesis, i.e. co-translationally. Because all other proteins are thought to be synthesized on ribosomes that are free in the cytosol, a mechanism must exist that mediates the selective attachment to the ER membrane of the ribosomes that synthesize proteins destined for secretion or for integration (Blobel & Dobberstein 1975).

In vitro assays have identified the signal recognition particle (SRP) and the SRP receptor (also known as the docking protein) as components required to target ribosomes to the ER membrane, and a detailed model describing their function has been proposed (Walter et al 1984; Walter & Lingappa 1986; Nunnari & Walter 1992; Gilmore 1993; Rapoport 1992; Sanders & Schekman 1992) (see Figure 1). The process is initiated when a signal sequence in the nascent protein chain emerges from the ribosome and is recognized by the SRP. This interaction causes the SRP to bind tightly to the ribosome, which

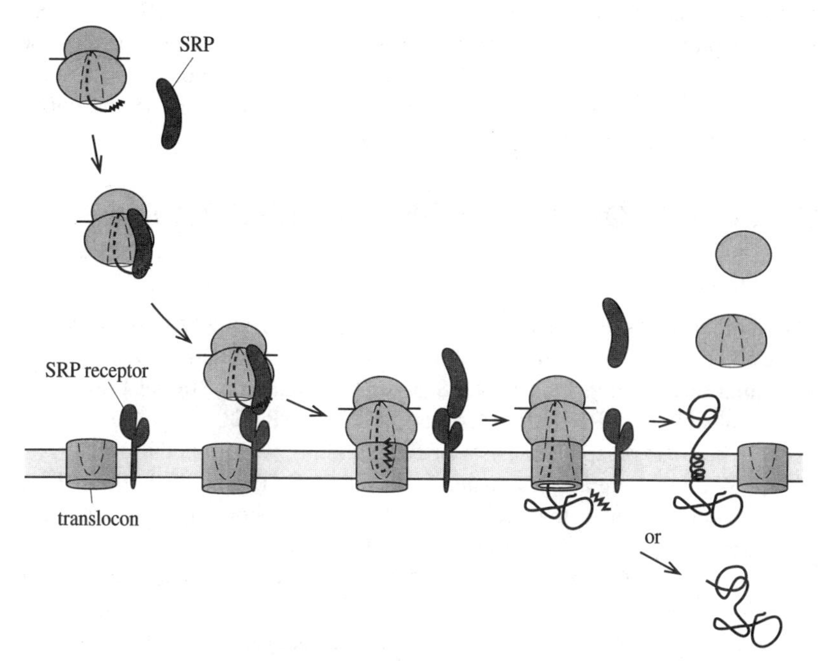

Figure 1 SRP-dependent protein targeting. SRP binds to the signal sequence of a nascent polypeptide emerging from the ribosome to form the targeting complex in which elongation is arrested. SRP in the targeting complex then binds to the SRP receptor in the ER membrane. This interaction leads to the formation of the ribosome-translocon junction and translation elongation resumes. SRP and SRP receptor then dissociate from each other and can engage in another round of targeting. After targeting, the signal sequence is initially in an aqueous compartment, formed by the ribosome and translocon components, that is sealed off from both the cytosol and the ER lumen. As the nascent chain grows, the translocation pore opens and allows the passage of the nascent protein across the membrane. In most cases, the signal sequence is removed on the lumenal side of the ER membrane by signal peptidase (not shown). Soluble proteins are released into the lumen of the ER. Transmembrane segments of nascent membrane proteins function as "stop-transfer" sequences. They must be recognized in the translocation pore, stop the translocation process, and then trigger the pore to open on one side to release the membrane protein laterally into the membrane.

effects a pause ("elongation arrest") in the translation of the nascent protein. The resulting complex consists of the ribosome, the nascent chain with its signal sequence, and the SRP, and it is herein referred to as the targeting complex. Interaction of the SRP in the targeting complex with the SRP receptor, an ER membrane protein, releases the SRP from the ribosome and signal sequence and allows translation to continue. Concomitantly, the ribosome becomes bound to other components in the ER membrane. These components, collectively termed a translocon (Walter & Lingappa 1986), catalyze the transfer of the growing protein chain across the membrane, presumably through a gated aqueous pore (Crowley et al 1994; Simon & Blobel 1991). SRP and SRP receptor act catalytically in this process; they are not part of the ribosome-translocon junction that mediates the transfer of the protein chain across the lipid bilayer. Rather, they function to direct the ribosome to the correct intracellular membrane and are then released from the ribosome.

STRUCTURE OF THE SRP AND SRP RECEPTOR

Nomenclature

Mammalian SRP contains one 7S RNA molecule, originally termed the 7SL RNA and referred to herein as SRP RNA, and six different polypeptides with molecular masses of 9, 14, 19, 54, 68, and 72 kd (Walter & Blobel 1980, 1982) that are designated SRP9, SRP14, SRP19, SRP54, SRP68, and SRP72, respectively. The SRP receptor consists of two subunits with molecular masses of 72 and 30 kd that are are designated SRα and SRβ (Tajima et al 1986). There is now good evidence that SRP and SRP receptor homologues exist in all organisms (see, e.g. Dobberstein 1994; Althoff et al 1994). We refer to the RNA in each SRP species as SRP RNA and identify the SRP proteins in different organisms by the name of the corresponding mammalian protein wherever there is sequence and functional homology, even if the molecular mass of the non-mammalian protein differs from that of its mammalian homologue. The *Escherichia coli* SRP54 is also known as Ffh or p48, and the *E. coli* SRP RNA is known as 4.5S RNA.

SRP RNA

SRP RNA is the central component of SRP: functionally it may mediate SRP's association with the ribosome and the SRP receptor, and structurally it provides the backbone onto which the SRP proteins assemble to form the SRP. The predicted secondary structure of the 300-nucleotide mammalian SRP RNA suggests an elongated conformation formed by extensively base-paired helices (Figure 2) (Ullu et al 1982). Two different ways are used to divide the SRP RNA structure into domains. First, four domains, designated domains I-IV (Figure 2A), are defined based on the distinct elements of the secondary structure: domain I is a variable structure at the 5′ end of the molecule, while domain II is the main stem that ends by bifurcating into two stem-loop structures defined as domains III and IV (Poritz et al 1988). The second division is based on the finding that mammalian SRP RNA is homologous for about 100 nucleotides from its 5′ end and for about 50 nucleotides from its 3′ end with the highly repetitive Alu DNA family (Ullu et al 1982). Although they are positioned on either end of the SRP RNA sequence, the 5′ and 3′ Alu sequences form a structurally contiguous domain ("Alu domain") in the folded

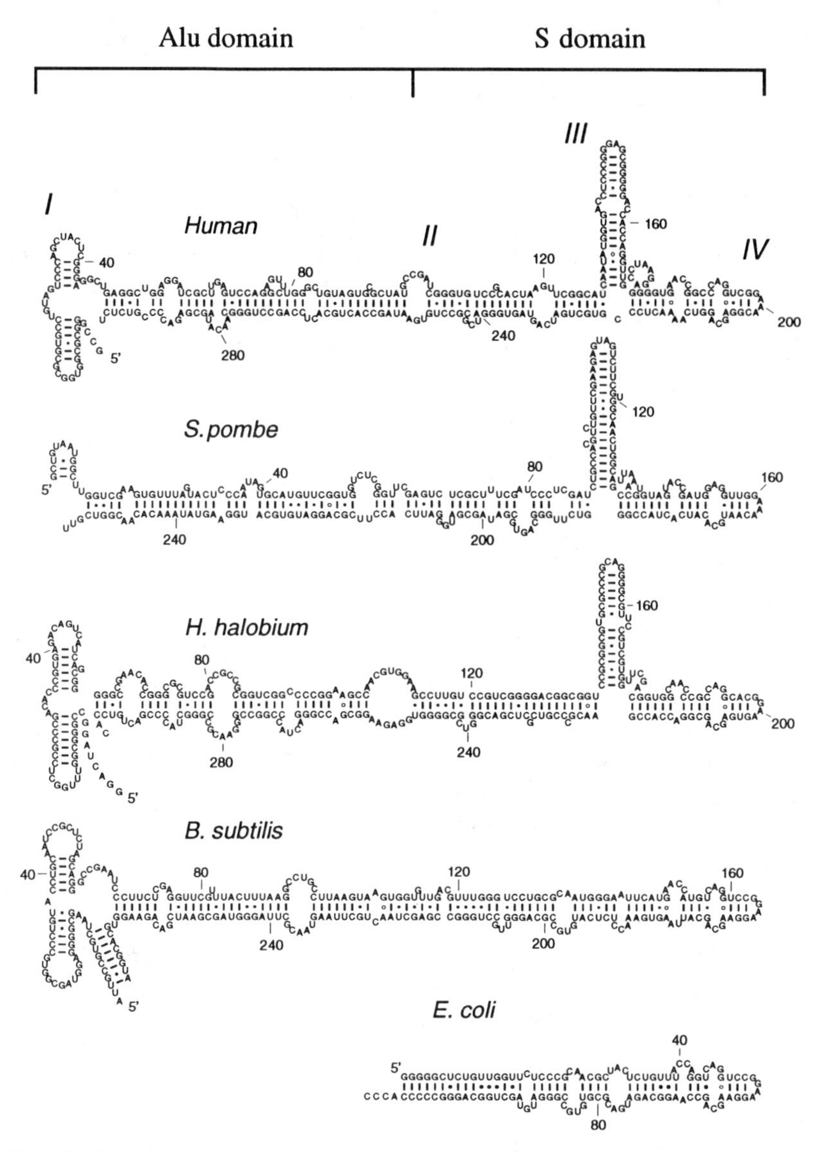

Figure 2 Secondary structures of SRP RNAs. RNAs from all three kingdoms are shown. Note the similarities both in the position and in the primary sequence of the bulges in domain IV. For a complete listing of sequences and secondary structures refer to a databank on SRP RNAs compiled by Larsen & Zwieb (1993).

SRP RNA. The Alu domain therefore consists of domain I and about half of domain II. Interestingly, SRP RNA is thought to be the evolutionay progenitor of Alu sequences, which may have arisen by reverse transcription from SRP RNA (Ullu & Tschudi 1984). In contrast, the central 150 nucleotides of SRP RNA comprise a unique sequence. The structural domain formed by this central RNA portion is termed the S domain (Siegel & Walter 1986; Ullu et al 1982). As discussed below, Alu and S domains define structurally and functionally distinct parts of SRP.

The secondary structure of SRP RNAs has been confirmed by phylogenetic comparison of SRP RNA sequences from a wide variety of organisms (Larsen & Zwieb 1991, 1993) and by experimental approaches that include nuclease digestion (Gundelfinger et al 1984), chemical modification (Andreazzoli & Gerbi 1991), and mutagenesis (Selinger et al 1993a; Zwieb 1991). The overall shape and dimensions of the secondary structure are conserved in eukaryotes and archeae despite a remarkable evolutionary drift in the sequences of the SRP RNAs. There are only very short regions of sequence conservation, and an unexpectedly large number of mutations in SRP RNA do not affect its function in vivo (Liao et al 1992). Therefore, it appears that the secondary and most likely tertiary structural features of SRP RNA are important for SRP assembly and function. The only known SRP RNA that has not yet been folded to fit the consensus structure is from the yeast *Saccharomyces cerevisiae.* This SRP RNA is about twice as long as the mammalian RNA (Felici et al 1989; Hann & Walter 1991), and it remains to be determined experimentally which portions of the sequence correspond to the known domains. Eubacterial SRP RNAs can be thought of as truncated versions of eukaryotic SRP RNA that lack domain III in the case of the *Bacillus subtilis* RNA and that lack domain III and the Alu-domain in the case of the *E. coli* RNA (Poritz et al 1988; Struck et al 1988) (Figure 2).

The most conserved sequence motif of SRP RNA is found in domain IV (Figure 2) and consists of a tetra loop and two bulges that are found in conserved positions and that contain short stretches of highly conserved nucleotide sequences. This motif was originally recognized as the most characteristic feature of all SRP RNAs (Poritz et al 1988; Struck et al 1988) and led to the discovery that 4.5S RNA is the *E. coli* SRP RNA homologue. The two conserved bulges in domain IV provide the binding site for SRP54 (Samuelsson 1992; Selinger et al 1993a; Wood et al 1992; L-S Kahng & P Walter, unpublished), whereas the conserved tetra loop at the end of domain IV can be replaced by a different sequence without affecting SRP function (Selinger et al 1993b). Hence, despite the conservation of its sequence, the tetra loop does not seem to be a target for base-specific contacts. A short conserved sequence motif in domain I has been implicated in the binding of SRP9 and SRP14 (Strub et al 1991).

SRP Structure and Assembly

The overall shape of mammalian SRP resembles an elongated rod, 240 Å long × 60 Å wide, in electron micrographs (Andrews et al 1985), and electron-dense SRP RNA appears to extend throughout the length of the rod (Andrews et al 1987). Such an extended structure may reflect a requirement for SRP to span a considerable physical distance in order to perform its multiple functional roles; signal sequence recognition occurs near the nascent chain exit site in the large ribosomal subunit, while elongation arrest is likely to be mediated via interactions about 160 Å away near the peptidyltransferase center of the ribosome (Berneabeau et al 1983). As discussed below, the signal recognition and elongation arrest activities of SRP map to SRP54 and to the Alu domain of SRP RNA, respectively, and are predicted to reside on opposite ends of the particle.

The protein subunit binding sites on the SRP RNA were determined by footprinting experiments (Siegel & Walter 1988b; Strub et al 1991) and are schematically depicted in Figure 3. These studies took advantage of the finding that SRP can be dissociated into its individual protein and RNA subunits when EDTA is added to remove magnesium ions. SRP subunits can then be purified and reassembled under the appropriate conditions to form an active SRP (Siegel & Walter 1985; Walter & Blobel 1983). SRP68 and SRP72 bind to SRP RNA as a stable heterodimer (designated SRP68/72), as do SRP9 and SRP14 (SRP9/14) (Siegel & Walter 1985; Walter & Blobel 1983). Both SRP54 and SRP19 bind to the SRP RNA individually, but SRP54 will not form a salt-stable complex with the SRP RNA unless SRP19 is also added (Poritz et al 1990; Walter & Blobel 1983; Miller et al 1993). A requirement for SRP19 to stabilize SRP54 binding has also been observed in vivo; a mutation in SRP19 in the yeast *S. cerevisiae* causes SRP54 to fall off the particle (Hann et al 1992). While the details of this intriguing interaction remain to be determined, it appears that binding of SRP19 to SRP RNA removes a destabilizing influence of domain III on the SRP54/domain IV interaction. In agreement with this hypothesis, mammalian SRP54 and its *E. coli* homologue form a stable complex with *E. coli* SRP RNA, which lacks domain III, in the absence of any additional protein (Römisch et al 1990; Zopf et al 1990; L-S Kahng & P Walter, unpublished).

The assembly of multicomponent ribonucleoprotein particles requires a number of specific, and probably ordered, protein-RNA and protein-protein interactions to obtain a complete functional particle. This has been amply demonstrated, for example, in many studies of ribosomal subunit assembly, where the association of ribosomal proteins with rRNA occurs in a specific order and the binding of proteins to the complex is often cooperative (e.g. Nomura et al 1969). An understanding of such interactions not only provides

A Mammalian SRP

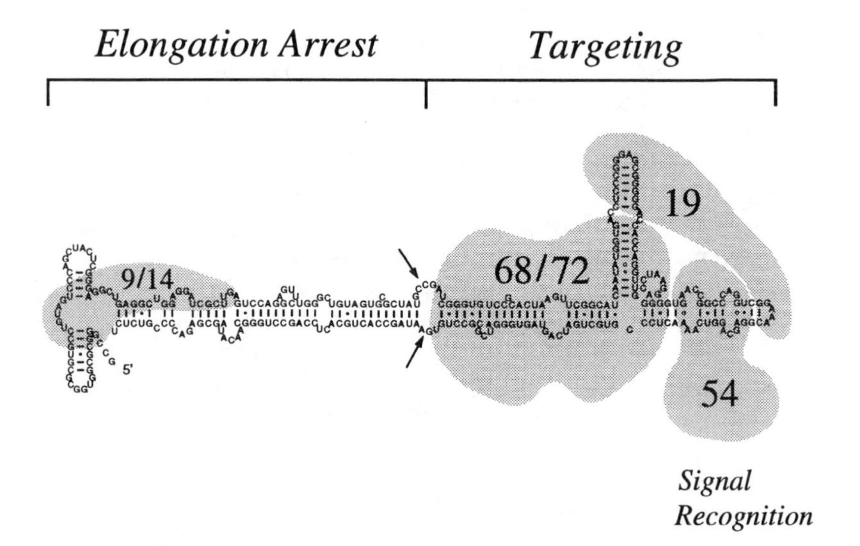

B *E. coli* SRP

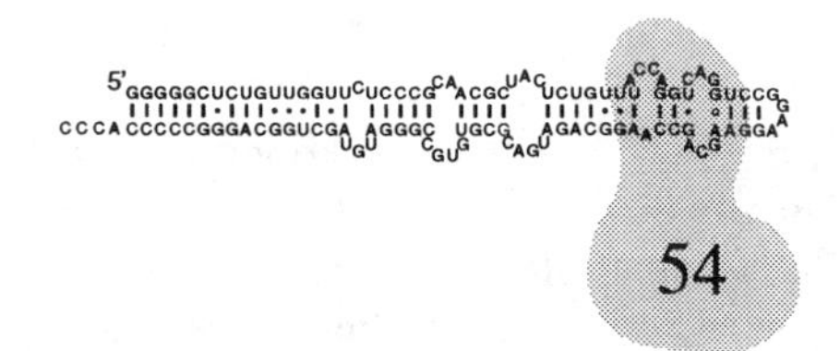

Figure 3 Domain structure of SRP. (*A*) The approximate positions of the SRP protein subunits on mammalian SRP RNA are shown. The binding sites of SRP19 and SRP68/72 were determined by enzymatic footprinting using α-sarcin, a nuclease that cleaves both single- and double-stranded RNA (Siegel & Walter 1988b). The SRP19 binding site was confirmed by SRP RNA fragmentation studies (Zwieb 1991). The binding site for the SRP9/14 heterodimer was determined using chemical footprinting (Strub et al 1991). These locations also agree with the enzymatic splitting of the SRP RNA into two halves in which the Alu domain co-purifies with SRP9/14 and the S domain is found with SRP68/72, SRP54, and SRP19 (Siegel & Walter 1986). The position of SRP54 is inferred from the binding site of its prokaryotic homologue on *E. coli* SRP RNA (*B*), which was determined by mutagenesis (Wood et al 1992), footprinting, and SELEX (L-S Kahng et al, unpublished) experiments and from mutagenesis studies in the yeast *S. pombe* (Selinger et al 1993a).

a recipe for particle formation, but may also provide clues about molecular interactions within the particle during its functional cycle.

The binding of SRP subunits may also occur in an ordered fashion. SRP proteins bind to the SRP RNA with a high affinity: the K_d for the SRP RNA complexes with SRP9/14 and with SRP68/72 are < 0.1 and 7 nM, respectively (Janiak et al 1992), and the K_d for *E. coli* SRP54/SRP RNA complex is 5 nM (BS Watson et al, unpublished) as determined by equilibrium binding assays using fluorescently labeled RNAs. The large difference in these K_d values may be important for SRP assembly. Because SRP9/14 binds with high affinity to the 5′ end of the SRP RNA, this association may serve to nucleate the folding of the SRP RNA and thereby ensure that SRP assembly is initiated properly. This is consistent with what has been observed during the assembly of the much larger bacterial ribosomal subunits, where the ribosomal proteins that nucleate their assembly also bind near the 5′ ends of the 16S and 23S rRNAs presumably because this portion of the rRNA is exposed first during RNA synthesis (Nowotny & Nierhaus 1988).

It also appears that SRP assembly is cooperative. Active SRP assembles even in the presence of an excess of naked SRP RNA (Walter & Blobel 1983), and the addition of SRP19 and SRP54 stabilizes the binding of SRP68/72 to the SRP RNA (J-C Chen & A Johnson, unpublished). In contrast, the binding affinity of SRP68/72 for a SRP RNA was unaffected by the presence of SRP9/14, which indicates that the binding of the heterodimers is noncooperative in the absence of SRP54 and SRP19 (Janiak et al 1992). SRP68/72 and SRP9/14 therefore associate randomly and independently with SRP RNA to form non-interacting domains in the particle, consistent with the substantial separation of their binding sites shown in Figure 3. It remains to be shown whether the assembly of the entire particle is cooperative in the presence of SRP19 and SRP54 or whether the cooperativity is restricted to the assembly of proteins on the S domain of SRP RNA.

In interpreting this cartoon, two main caveats must be kept in mind. First, the indicated limits of the protein binding sites in the S domain are not precise because the footprinting data obtained enzymatically do not have as high a resolution as those obtained chemically for SRP9/14. Moreover, the binding site for mammalian SRP54 has not been delineated experimentally, but is inferred from studies of the yeast and bacterial homologues. Second, it is possible that the conformation of the SRP RNA is sensitive to the presence of one or more of the SRP proteins, the SRP receptor, or other components of the system. Such conformational changes upon protein binding have been observed using both chemical modification (Andreazzoli & Gerbi 1991) and fluorescence (Janiak et al 1992) techniques.

The structural arrangement has been dissected in terms of domains with specific functions (Siegel & Walter 1988d). SRP54, SRP68/72, and SRP9/14 are involved in the three primary SRP functions: signal sequence recognition, targeting, and elongation arrest, respectively, as discussed in the text. The nucleolytic sites that allowed the isolation of an SRP subparticle containing only the S domain of SRP RNA are indicated by the arrows; this subfragment is active in signal recognition and targeting.

SRP Protein Subunits

The sequences of all mammalian SRP proteins have been determined (Bernstein et al 1989; Herz et al 1990; Lingelbach et al 1988; Lütcke et al 1993; Römisch et al 1989; Strub & Walter 1989, 1990). All six proteins have a high abundance of basic amino acids and in this aspect resemble ribosomal proteins. Otherwise, with the notable exception of SRP54 (see below), they have no characteristic sequence motifs or significant sequence similarity to other known proteins. Homologues to mammalian SRP subunits have also been identified in the yeast *S. cerevisiae* (Amaya et al 1990; Hann et al 1989; Stirling & Hewitt 1992; J Brown & P Walter, unpublished). Based on their sequence, SRP72, SRP68, SRP19, and SRP14 are only loosely conserved over this evolutionary distance (about 20–30% sequence identity), but are still clearly recognizable as homologues (Stirling & Hewitt 1992; Brown & Walter, unpublished). In contrast, SRP54 is highly conserved (47% sequence identity) (Hann et al 1989), which suggests that this protein plays a central role in SRP function that results in many more constraints on its structure. SRP54 is also the only known protein subunit of the *E. coli* SRP, and it is remarkably 31% identical to its mammalian counterpart.

No procedure is yet available to separate purified heterodimeric SRP9/14 and SRP68/72 protein complexes into their individual proteins without causing irreversible denaturation. With the availability of cDNA clones for mammalian SRP9, SRP14, SRP68, and SRP72, however, it became possible to produce these subunits individually by in vitro translation and to assess their RNA-binding properties and domain structures. Thus it was determined that dimerization of SRP9 and SRP14 is strictly required because neither protein will bind to SRP RNA in the absence of the other (Strub & Walter 1990). In contrast, SRP68 can bind weakly to SRP RNA in the absence of SRP72 (Lütcke et al 1993). SRP72 association with SRP68 significantly increases the binding affinity, but no stable binding of SRP72 alone to SRP RNA was observed. Thus SRP72 may associate with the SRP RNA via a protein-protein interaction with SRP68.

In contrast to the other SRP subunits, the sequence of SRP54 provides a wealth of information about its structure and function. As shown in the schematic alignment in Figure 4, SRP54 contains a central GTPase domain that is characterized by short sequence stretches that are conserved between most known GTPases and that are known from the X-ray structures of H-ras and EF-Tu to form juxtaposed loops on the surface of the protein that directly contact the bound nucleotide (Bourne et al 1991). The GTPase domain of SRP54 is most closely related to a GTPase domain contained in the SRP receptor subunit SRα, but is more distantly related to other known GTPases. Thus SRP54 and SRα together define a separate subfamily in the superfamily

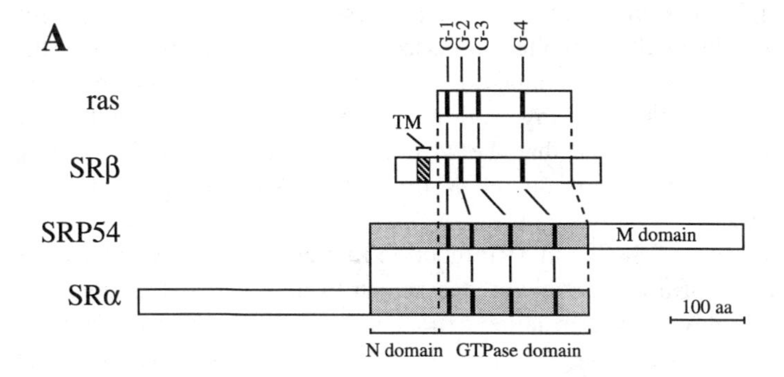

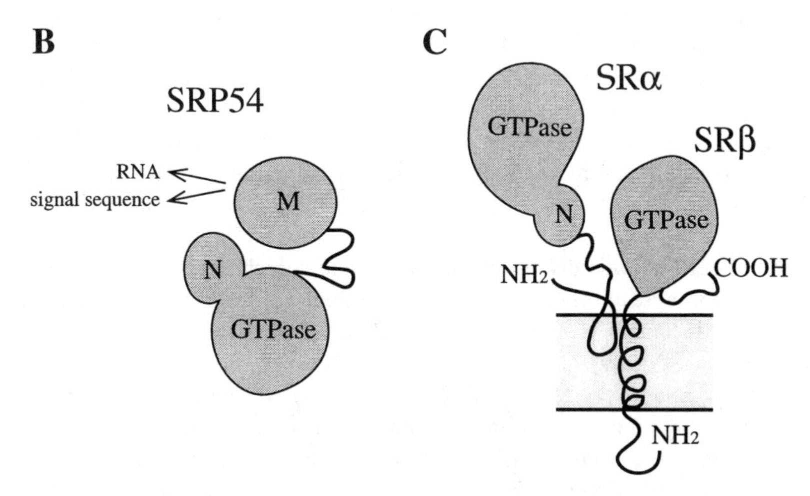

Figure 4 Domain structure of the three GTPases in SRP and SRP receptor. (*A*) The ras-like GTPase domain is characterized by four conserved sequence motifs (G-1 to G-4) as defined by Bourne et al (1991). Note that the GTPase domains of SRP54 and SRα are related to one another (*shaded box*) and that this homology extends through the N domain. The transmembrane region of SRβ is indicated (*TM*). (*B*) The M domain of SRP54 is linked to the N/G domains by a protease-sensitive hinge region. The N and the M domains are likely to be in close spatial proximity because the N and C termini of ras are in close proximity in the folded protein. (*C*) The heterodimeric SRP receptor is likely to be anchored in the membrane by the single transmembrane segment of SRβ. The N/G domains of SRα are linked to SRβ by a protease-sensitive hinge region. It is not known how the N-terminal region of SRα contacts SRβ nor whether it also contacts the hydrophobic core of the membrane.

of GTPases. The GTPasc domain is flanked on its N-terminus by the N domain, and on its C-terminus by the M domain, which is characterized by an unusually high abundance of Met residues (about 12% for mammalian SRP54) (Bernstein et al 1989). The M domain is connected to the rest of

SRP54 by a flexible hinge region that is protease-sensitive (Figure 4B). This has been experimentally exploited to dissect SRP54 into two fragments, the M domain and a fragment containing both N and GTPase domains, here referred to as the N/G domains. Following proteolytic cleavage within the hinge region, it was shown that the M domain contains the RNA-binding site that attaches SRP54 to domain IV of SRP RNA (Römisch et al 1990; Zopf et al 1990).

The central role of SRP54 in SRP function first became apparent when the signal sequence of a nascent chain was shown to photo cross-link solely to this SRP subunit in crude translation extracts (Krieg et al 1986; Kurzchalia et al 1986). This strongly suggests that SRP54 contains the signal sequence binding site of SRP. Specific cross-linking to SRP54 has now been shown for a variety of different signal sequences including signal-anchor sequences of integral membrane proteins (High et al 1991b). Isolated SRP54 can also be cross-linked to signal sequences, which indicates that the remainder of SRP is not required for this association (Lütcke et al 1992; Zopf et al 1993). The site of signal sequence cross-linking maps to the M domain (Römisch et al 1990; Zopf et al 1990), and more specifically its C-terminal 6-kd fragment (High & Dobberstein 1991), as concluded from cross-linking combined with proteolysis.

E. coli SRP54 was discovered by data bank searches as an open reading frame encoding a protein of unknown function that is highly similar in sequence and domain structure to mammalian SRP54 (Bernstein et al 1989; Römisch et al 1989). This unanticipated discovery was the key step that allowed the characterization of SRP54 homologues from other species. Comparison of the bacterial and mammalian sequences identified highly similar regions that were used in PCR-based approaches to identify SRP54 homologues from all cells analyzed thus far. To date nine different SRP54 sequences are known including those of mammals (Bernstein et al 1989; Römisch et al 1989), yeasts (Amaya et al 1990; Hann et al 1989), prokaryotes (Bernstein et al 1989; Römisch et al 1989; Samuelsson 1992), and even chloroplasts (Franklin & Hoffman 1993). In each case, the SRP54 is part of a soluble ribonucleoprotein complex that also contains an SRP RNA with the characteristic domain IV motif (in the case of chloroplast SRP54 this is not known, however).

SRP54 is the only known protein subunit of the *E. coli* SRP. As discussed above, SRP54 binding to *E. coli* SRP RNA does not require an SRP19, and it is possible that the other SRP subunits are also dispensible in *E. coli*. Alternatively, other SRP subunits may exist, but have escaped detection because they are more loosely bound to this particle. The concept of an *E. coli* SRP was highly controversial as it emerged from the phylogenetic comparisons (Bassford et al 1991; Beckwith 1991) because neither the protein nor the RNA component of *E. coli* SRP were identified genetically as secretion mutants or in biochemical assays that monitor protein translocation across the bacterial plasma membrane. By now, however, a substantial body of experimental evidence has accumulated in support of this idea. In particular, *E. coli* SRP54 can be specifically cross-linked to signal sequences in crude extracts (Luirink et al 1992), and it can replace mammalian SRP54 functionally in signal sequence recognition when it is assembled with mammalian SRP proteins and RNA into a chimeric particle (Bernstein et al 1993). Most importantly, the in vivo depletion of SRP54 from *E. coli* cells leads to translocation defects of some periplasmic proteins (Phillips & Silhavy 1992).

SRP Receptor

Mammalian SRP receptor is a heterodimeric integral membrane protein composed of SRα and SRβ (Tajima et al 1986) that is found only in the ER membrane (Meyer et al 1982b). The existence of a proteinaceous receptor was first shown when protease treatment of microsomal membranes rendered them translocation inactive, and when activity was restored by adding a cytosolic fragment back to the proteolyzed membranes (Meyer & Dobberstein 1980b; Walter et al 1979). This allowed the purification of the 52-kd cytosolic fragment (Meyer & Dobberstein 1980a), which was derived from a 70-kd protein that had been purified independently as a factor that releases the elongation arrest of presecretory proteins induced by SRP (Gilmore et al 1982a,b; Meyer et al 1982a).

From a comparison of the sequence of SRα with that of the soluble fragment released by protease treatment, it is clear that SRα is anchored to the ER membrane through an N-terminal domain (Lauffer et al 1985), presumably via interactions with SRβ (Tajima et al 1986) (Figure 4C). This N-terminal portion contains two hydrophobic regions that may contact the hydrophobic core of the lipid bilayer. However, SRα can be extracted from membranes with chaotropic agents (J Miller et al, in preparation), which indicates that it is not a bona fide integral membrane protein, and SRα synthesized in vitro can post-translationally assemble into membranes (Andrews et al 1989). After it was discovered that protein targeting is a GTP-dependent process, a reinspection of the SRα sequence revealed a GTPase domain at the C-terminus of the protein, which is released as the soluble fragment upon proteolysis (Connolly & Gilmore 1989). Consistent with this finding, SRα binds GTP. The GTPase domain of SRα is closely related to the GTPase domain of SRP54 (Figure 4A), and the sequence similarity extends through the N domain of SRP54.

FtsY, an *E. coli* protein of unknown function, was identified as an SRα homologue based on sequence similarity that extends through the N and GTPase domains of SRα and SRP54 (Bernstein et al 1989; Römisch et al

1989). The N-terminal portion of FtsY, however, bears no sequence similarity to the N-terminal region of SRα that anchors it to the ER membrane, and no *E. coli* SRβ subunit has been identified.

Sequence analysis of SRβ shows that it contains a standard transmembrane segment with an uninterrupted stretch of 25 hydrophobic amino acids (J Miller et al, in preparation). Surprisingly, SRβ also contains a predicted GTPase domain and experimentally binds GTP, thus bringing the number of GTPases that interact during protein targeting to three. The GTPase domain of SRβ, however, is not closely related to those of SRP54 and SRα, but is instead in its own new subfamily as a distant relative of the small GTPases SAR and ARF that are involved in vesicular trafficking.

The SRP receptor is unlikely to be an integral part of the translocon because it is present in membranes in substoichiometric amounts with respect to membrane-bound ribosomes (Tajima et al 1986). Hence it is likely that SRP receptor, like SRP, functions catalytically to promote the formation of the ribosome-translocon junction.

MECHANISM OF SRP-DEPENDENT PROTEIN TARGETING

For the purpose of this discussion, we divide the functional cycle of SRP into three distinct steps: (*a*) signal sequence recognition, which results in the recruitment into the targeting complex of those ribosomes that synthesize proteins destined for translocation across or integration into the ER membrane; (*b*) elongation arrest, which modulates the translational activity of the ribosome in the targeting complex; and (*c*) targeting, which leads to the release of SRP from the targeting complex concomitant with the formation of the ribosome-translocon junction. Upon targeting, the signal sequence of the nascent polypeptide chain has been delivered into a sealed aqueous compartment comprised of translocon components.

Selection of Signal Sequences

Current evidence suggests that signal sequences are positively selected by their ability to bind to a signal sequence-binding site on the M domain of SRP54. Historically, SRP was first shown to recognize information contained in the nascent polypeptide chain (as opposed to the mRNA encoding the protein) by experiments in which the structure of the signal sequence of a nascent protein was selectively altered by the incorporation of amino acid analogues (Walter et al 1981). Photo cross-linking experiments then showed that the signal sequence in the targeting complex is in close proximity to—and presumably bound to—the M domain of SRP54 (High & Dobberstein 1991; Krieg et al 1986; Kurzchalia et al 1986; Zopf et al 1990). The notion that SRP54 directly and selectively binds signal sequences is further supported by experiments showing that synthetic functional signal peptides—but not mutant signal peptides that are inactive as signal sequences in vivo and differ from functional signal peptides only by single amino acid substitutions—inhibit GTP binding to SRP54 (Miller et al 1993). However, the results from these indirect experiments still remain to be confirmed by assays that monitor the direct binding of signal peptides to SRP54.

The characteristic feature of ER-directed signal sequences is a core comprising about 8–12 hydrophobic amino acids that presumably forms an α-helix (von Heijne 1985). Because their amino acid sequences are not conserved, such signal sequence cores must each have a different shape. The predicted structural characteristics of the M domain of SRP54 suggest a model of how signal sequences may bind to SRP despite this structural diversity. Most of the unusually abundant Met residues in SRP54M are predicted to reside on one face of a group of strongly amphipathic α helices (Bernstein et al 1989; Hann et al 1989) that have been proposed to form or contribute to a signal sequence binding groove (Bernstein et al 1989). A unique feature of Met side chains is their flexibility; the side chains of Leu and Ile, amino acids of comparable hydrophobicity, are branched and hence comparatively rigid. Thus the flexible hydrophobic Met side chains would project like bristles of a brush into such a groove and provide a hydrophobic environment with sufficient plasticity to allow signal sequence binding despite the heterogeneity in amino acid sequence. This hypothesis is supported by the remarkable phylogenetic conservation of both the unusual abundance of Met residues and their position on the predicted α helices in the M domain of SRP54 homologues from mammalian cells to bacteria. Met is typically a relatively rare amino acid and is often replaced in phylogenetic comparisions by other hydrophobic amino acids. Its conservation in SRP54 therefore indicates an importance of the Met side chains that is both structural and functional.

A paradigm for the involvement of Met side chains in the binding of heterogeneous hydrophobic surfaces is provided by calmodulin. This dumbbell-shaped molecule binds to a variety of different target proteins by clamping down on amphipathic helices exposed on the surface of the target proteins (O'Neil & DeGrado 1990). Biochemical studies, as well as NMR and crystallographic analyses, show that patches of exposed Met side chains on calmodulin provide the interaction surfaces that contact the hydrophobic parts of the amphipathic helices in the target proteins. Similar structural analyses are needed for SRP54 to validate or disprove the speculative hypothesis presented above.

SRP has a low affinity for ribosomes that are not engaged in translation, but its affinity is increased by three to four orders of magnitude when a signal sequence is expressed and exposed outside the ribosome as part of a nascent chain (Walter et al 1981). It is likely that SRP normally cycles between a ribosome-bound state and a free state, thereby scanning nascent polypeptide chains for signal sequences. Regardless of their secretory activity, cells contain about one SRP for every ten ribosomes, which suggests that SRP cannot remain bound to any given ribosome waiting for a signal sequence to emerge. Moreover, in both bacteria and yeast there is evidence that SRP interacts with ribosomes at a discrete step in the elongation cycle (most likely before the translocation step catalyzed by elongation factor EF-G or eEF2, respectively) (Brown 1989; S Ogg & P Walter, unpublished). Thus free SRP may transiently bind to ribosomes during any elongation cycle. SRP then may either remain bound to the ribosome if a signal sequence has associated with SRP54, or dissociate rapidly and move to another ribosome if no signal sequence has been detected.

As nascent chains grow longer, their affinity for SRP decreases (Siegel & Walter 1988a). This could be because the signal sequence is no longer favorably positioned with respect to the ribosome-bound SRP and/or because the signal sequence is rendered less accessible through aberrant folding of the nascent chain. It is attractive to speculate that the other SRP subunits and SRP RNA help position SRP on the ribosome such that the signal sequence binding site on SRP54M and the nascent chain exit site on the large ribosomal subunit become juxtaposed. How long a nascent chain that contains a signal sequence on its amino terminus can be extended and still be recognized by SRP varies greatly between different proteins. For most proteins the affinity for SRP drops drastically after they have been elongated beyond a certain point (Siegel & Walter 1988a). Other proteins, however, can still be recognized by SRP after they are synthesized to full length, provided that translation has not terminated, i.e. that the protein remains ribosome-bound and covalently attached to tRNA (Garcia & Walter 1988). SRP will not promote post-translational translocation of signal sequence-bearing proteins that have been released from the ribosome (Garcia & Walter 1988). A reported activity (Crooke et al 1988; Sanz & Meyer 1988) of mammalian SRP to promote post-translational translocation in yeast and *E. coli* in vitro systems may not reflect a physiological pathway, but rather result from nonspecific hydrophobic interactions that retard dead-end protein folding or aggregation.

Elongation Arrest

The notion that SRP in the targeting complex interacts intimately with the ribosome is best supported by direct SRP-dependent effects on translation. When SRP is included in in vitro translation systems in the absence of ER membrane vesicles, it blocks elongation after the signal sequence has become exposed outside the ribosome (Walter & Blobel 1981). In some cases a discrete-sized protein fragment that corresponds to the elongation-arrested secre-

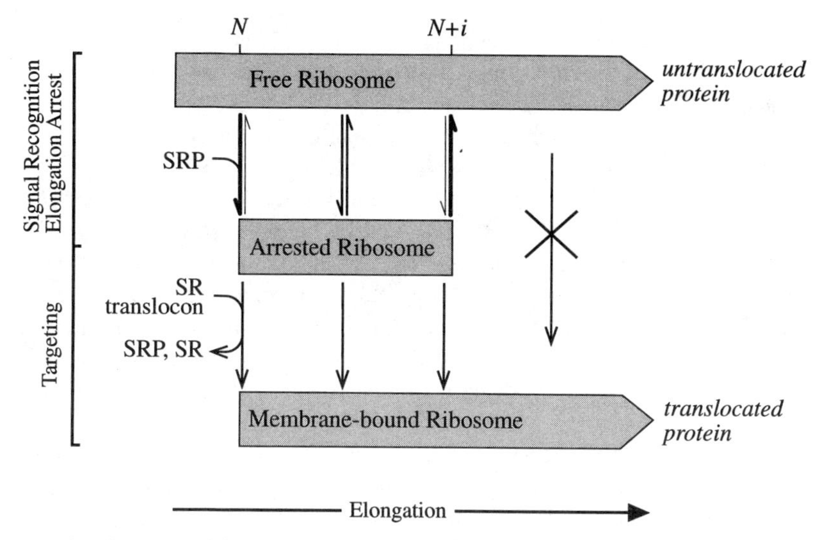

Figure 5 Kinetic model for elongation arrest and targeting (adapted from Rapoport et al 1987). SRP interacts with a ribosome and arrests translational elongation after N amino acids have been polymerized and the signal sequence is exposed outside the ribosome. The formation of the targeting complex will only occur within a window of nascent chain lengths (N to N+i). As nascent chains grow longer, their affinity for SRP decreases and drops drastically if elongation proceeds beyond this window of opportunity. In the absence of membranes, translation elongation is delayed by SRP because the ribosome spends time idling in the elongation-arrested targeting complex. In the presence of membranes on the other hand, the SRP in the targeting complex interacts with the SRP receptor, which leads to the formation of the ribosome-translocon junction, an efficient and energy-consuming reaction that leads to the synthesis of translocated protein.

tory protein can be observed by gel electrophoresis (Meyer et al 1982a; Walter & Blobel 1981); in other cases the arrested forms are more heterogeneous and difficult to detect (Anderson et al 1982; Lipp et al 1987). The positions of paused ribosomes along mRNAs confirm that SRP arrests translation just after the signal peptide emerges from the ribosome and becomes available for binding to SRP (Wolin & Walter 1988). Interestingly, the SRP enhances pausing of ribosomes at sites that are natural stutter points in the translation of the mRNA. The distribution of stutter points in the mRNA may therefore determine the spectrum of arrested nascent chain fragments. Stutter points that cause the nascent chain to pause just after a signal sequence has emerged from the ribosome may be advantageous to cells because they may increase the efficiency of signal sequence recognition by SRP.

To date, a strict block of elongation has only been observed in in vitro assays composed of heterologous components, e.g. mammalian SRP and a wheat germ translation system. In the yeast *Yarrowia lipolytica,* mutations in SRP RNA can cause an inhibition of the translation of a secretory protein (He et al 1992; Yaver et al 1992). This phenomenon may be indicative of an elongation block, although this still remains to be shown more directly. In assays composed exclusively of mammalian components, the SRP-induced elongation arrest is transient, thus causing a kinetic delay in protein elongation (Wolin & Walter 1989). Similarly, in yeast cells in which the SRP receptor has been genetically depleted, the synthesis of presecretory proteins is not detectably reduced (Ogg et al 1992), which indicates that the lack of SRP receptor in vivo does not lead to an irreversible elongation arrest. These observations are consistent with a kinetic model (Figure 5, adapted from Rapoport et al 1987) that treats the association of SRP with the ribosome/signal sequence as an equilibrium-binding reaction. According to this model, elongation in the targeting complex is completely blocked, but elongation resumes when SRP dissociates from the targeting complex.

Molecular dissection of SRP, either by fragmentation with nuclease (Siegel & Walter 1986) or by partial reconstitution (Siegel & Walter 1985), has mapped the elongation arrest function of SRP to a discrete domain comprising the Alu-portion of SRP RNA and SRP9/14 (indicated in Figure 3A). Thus SRP can also be visualized as having one end involved in elongation arrest, while the other end is involved with the signal sequence and ER membrane components. Partial SRPs that lack the elongation arrest domain (or that contain the Alu-domain of the RNA, but without SRP9/14 or with alkylated SRP9/14 bound) still promote signal recognition and protein targeting, which indicate that elongation arrest is not a prerequisite for protein translocation (Siegel & Walter 1985, 1988c). Because most signal sequence-bearing nascent proteins lose the ability to be translocated if elongation proceeds too far, however, elongation arrest helps to maintain the translocation competence of the nascent chain by delaying its elongation. Thus one physiologically important function of elongation arrest may be to increase the fidelity of protein translocation. If the function of SRP or the SRP receptor in cells could be regulated (which presently is not known), then it is also conceivable that elongation arrest may be used as a convenient on-off switch by which cells could adapt the synthesis of secretory proteins to the secretory needs of the cell.

Because elongation arrest is selective for signal sequence-bearing proteins, it requires recognition of a signal sequence by SRP54, and SRP subparticles that lack SRP54 are inactive. An SRP subparticle, termed SRP(-54G), that lacks only the N/G domains of SRP54 because it was reconstituted with purified M domain in place of SRP54 still elicits elongation arrest activity (Zopf et al 1993). This shows that the M domain of SRP54 is sufficient for signal sequence binding and, when reconstituted with the remaining SRP subunits, is also sufficient to transmit this information to the ribosome to elicit elongation arrest. The affinity of the isolated M domain or of SRP(-54G) for signal sequences is reduced, however, thus indicating that the presence of the N/G domains on SRP54 contributes to signal sequence binding (Zopf et al 1993). SRP(-54G) is completely inactive in promoting protein translocation across microsomal membranes presumably because it cannot interact normally with the SRP receptor. Thus the N/G domains have a dual function: they influence signal sequence recognition by promoting a tighter association between signal sequences and the M domain and, as discussed below, they play an essential role in targeting.

Targeting

Targeting to the ER membrane is mediated by multiple GTPases that comprise domains of SRP54, SRα, and SRβ. In numerous biological processes, GTPases function as molecular switches that provide unidirectionality and accuracy (Bourne et al 1990). Through GTP binding and hydrolysis, GTPases can exist in at least three discrete conformations: a nucleotide-free, a GTP-bound, and a GDP-bound conformation. Interconversion between these states in a defined sequence causes the GTPase to interact in temporal succession with its effectors, thereby regulating the biological process. In most cases, the conversion of one conformer to another is controlled by other molecules: GTP hydrolysis is often facilitated by the action of specific GTPase activating proteins, GDP release by the action of specific guanine nucleotide release factors, and GTP binding by the action of specific guanine nucleotide loading factors.

The GTPase domains of SRP54 and SRα are required to promote the progression from targeting complex to the formation of the ribosome-translocon junction. As mentioned above, SRP(-54G) does not promote targeting and, likewise, microsomal membranes that have been proteolytically depleted of the GTPase domain of SRα are inactive. The severed soluble domain of SRα can be added back to proteolyzed membranes to restore activity, thus confirming that SRα is the only membrane protein essential for this process that was destroyed by the protease treatment.

Interestingly, non-hydrolyzable GTP analogues can substitute for GTP in targeting (Connolly & Gilmore 1989; High et al 1991a). After targeting in the presence of non-hydrolyzable GTP analogues, however, SRP and SRP receptor remain locked together as a stable complex (Connolly et al 1991). Therefore, it is thought as a minimum that GTP binding to SRP54 and SRα is required for targeting, and GTP hydrolysis is required to allow the regeneration of free SRP and SRP receptor that can then engage in another round of targeting. GTP bound to SRP is hydrolyzed upon interaction with the SRP receptor, which thus functions as a GTPase-activating protein for SRP54 (Miller et al 1993).

The unprecedented direct interaction between three GTPases involved in targeting has made attempts to decipher the contributions of individual GTPases to the overall reaction challenging. Biochemical characterization of

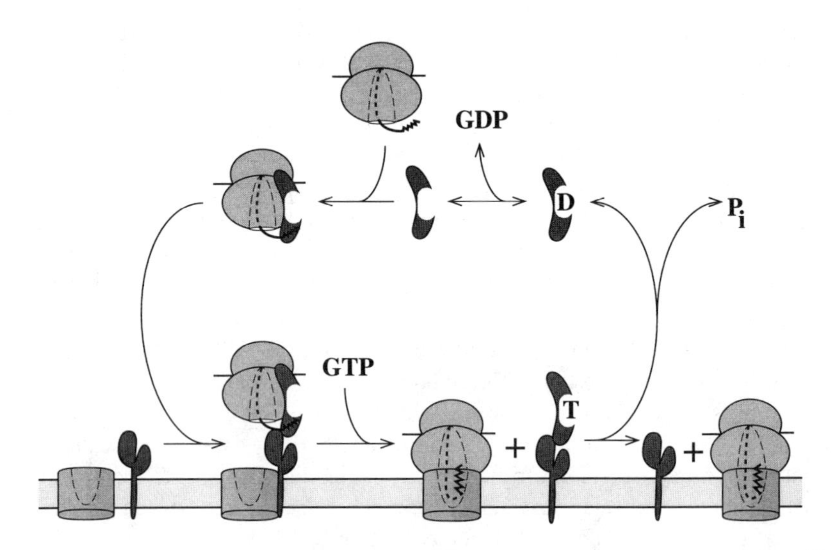

Figure 6 Model for GTP utilization by SRP54 (adapted from Miller et al 1993). As discussed in detail in the text, SRP54 is proposed to undergo a series of sequential conformational changes that drive SRP unidirectionally through cycles of protein targeting. In the experiments that led to the proposal of this model, guanine nucleotide occupancy of SRP54 was assessed by UV cross-linking (Miller et al 1993). It is possible, however, that the state depicted here as nucleotide-free does, in fact, contain a nucleotide that is bound in such a way that it fails to cross-link to SRP54 upon UV irradiation. GDP release from SRP54 may occur spontaneously or may be facilitated by binding of signal sequences and/or ribosome. T = SRP54-bound GTP; D = SRP54-bound GDP.

the interaction of purified SRP and SRP subparticles with purified SRP receptor has yielded initial insights into the mechanism by which the GTPase domain of SRP54 may regulate targeting (Figure 6, adapted from Miller et al 1993). According to these studies, the guanine nucleotide-bound state of SRP54 is influenced by at least two ligands: a signal peptide stabilizes a nucleotide-free state in SRP54 (and perhaps even stimulates the release of GDP), and interaction with the SRP receptor significantly increases the affinity of SRP54 for GTP. If these results hold true for the complete targeting complex (which needs to be confirmed experimentally), then SRP arrives at the membrane held in a nucleotide-free state by the signal sequence. The SRP receptor may then function as a guanine nucleotide-loading protein that promotes GTP binding to SRP54 and concomitantly reduces the affinity of SRP54 for the bound signal sequence. Interestingly, however, purified SRP receptor fails to stimulate GTP loading of SRP54 in the presence of signal peptides. One likely explanation for this observation is that additional components, such as translocon subunits, are required to effect signal sequence release in the purified in vitro system. This is an appealing notion because the requirement for translocon subunits would introduce a check point: unless appropriate translocon subunits have been recruited and proper translocation is ensured, the signal sequence is not released from SRP and the reaction cannot proceed.

In this model of targeting, SRP54 emerges in a central role in which its GTPase domain is used to integrate information received from both the nascent chain and the ER membrane. Available evidence points to extensive allosteric communication between the structurally separate domains of SRP54: binding of a signal sequence to the M domain prevents nucleotide binding to the GTPase domain (Miller et al 1993) and, conversely, chemical modification of the GTPase domain prevents signal sequence binding to the M domain (Lütcke et al 1992). However, the prediction that GTP binding to the GTPase domain lowers the affinity for signal sequences has not been confirmed experimentally. SRP54 may resemble EF-Tu in which large conformational changes accompany the exchange of bound GDP for bound GTP (Berchtold et al 1993). Consistent with this observation, nucleotide occupancy of the GTPase domain could, for example, change the relative arrangement of the GTPase and M domains such that the N/G domains no longer promote tight association between signal sequences and the M domain. One can speculate that the N domain, which is likely to be closely juxtaposed to the M domain (Figure 4B), could cover the signal sequence-binding groove on the M domain and that the GTPase domain may provide a hinge module that determines whether this cover is in the open or closed position.

The GTPase domains in SRα and SRβ may also provide check points in the targeting pathway that are monitored by guanine nucleotide switches. While the importance of guanine nucleotide binding to SRβ for targeting has yet to be demonstrated, it has clearly been shown for SRα. Using proteolyzed microsomal membranes that were repopulated with a mutant form of SRα that has a lower affinity for GTP, it was shown that SRα needs to be in a GTP bound state for the targeting reaction to progress through the cycle shown in Figure 6 (Rapiejko & Gilmore 1992). Thus it is likely that only GTP-bound SRα will promote GTP binding to SRP54. As the GTPase domains of SRα and SRP54 define a unique subgroup in the superfamily of GTPases and hence may function similarly, SRα may progress through a cycle of GTP binding and hydrolysis similar to that of SRP54. It is possible that, just as SRP recruits ribosomes with nascent chains from the cytosol to the membrane, so the SRP receptor may recruit translocon components within the plane of the membrane. In this view, SRP and SRP receptor function as molecular "match makers" during the assembly of the ribosome-translocon junction.

SRP68/72 are also required for targeting in the mammalian system because SRP subparticles that lack these subunits are inactive (Siegel & Walter 1988c). Moreover, SRPs that were selectively alkylated with N-ethylmaleimide (NEM) on their SRP68/72 subunits were unable to promote targeting of the nascent chains to the ER membrane, apparently because the NEM modification interfered with an interaction between the SRP68/72-modified SRP and the SRP receptor (Siegel & Walter 1988c). The challenge now is to decipher what component(s), other than SRP RNA, SRP68/72 binds during this process and how these interactions facilitate targeting.

The interactions between the *E. coli* SRP and FtsY, the bacterial homologue of SRα, closely mimic those of their mammalian counterparts (Miller et al 1994). In particular, the *E. coli* SRP binds tightly to FtsY in a GTP-dependent manner. This interaction leads to a stimulation of GTP hydrolysis, which can be inhibited by synthetic signal peptides. These results provided the first experimental evidence that FtsY has SRP receptor-like properties.

Beyond Targeting

The interaction of the targeting complex with the SRP receptor initiates the events that result in the binding of the ribosome to the membrane surface, the resumption of protein synthesis, the release of the signal sequence from SRP54, and the release of both SRP and SRP receptor from ribosome and translocon. The signal sequence is released from the SRP54 on the cytoplasmic side of the membrane during this process, but its exact location at the membrane is not well defined. In particular, it is not clear whether the signal sequence is bound to a protein in the translocon, perhaps after a direct transfer from SRP54. Such binding would have to be transitory, however, because the signal peptide is ultimately cleaved from the nascent chain by the signal peptidase on the lumenal side of the ER membrane.

Fluorescent probes that are sensitive in their emission characteristics to the hydrophobicity of their environment have been incorporated into the signal sequences of nascent chains. Measurements using such nascent chains indicate that, after completion of targeting, the signal sequence is initially in an aqueous environment, sealed off from both the cytoplasm and the lumen of the ER (Crowley et al 1994, 1993). This aqueous compartment appears to be formed by integral membrane proteins that reside in the rough ER, since photo cross-linking studies have shown that the signal sequence is positioned adjacent to at least two ER membrane proteins (High et al 1993; Krieg et al 1989; Wiedmann et al 1987) termed Sec61α and TRAM (Görlich et al 1992a; Görlich et al 1992b). Sec61α is part of a complex of three integral membrane proteins, Sec61α, Sec61β, and Sec61γ (Hartmann et al 1994). Reconstitution studies of purified components into artificial proteoliposomes have shown that the only components required to catalyze protein translocation across a lipid bilayer are the SRP receptor, the Sec61 complex and—for some proteins—TRAM (Görlich & Rapoport 1993). Thus they provide a minimal translocon for SRP-dependent protein translocation.

The endproduct of an equivalent SRP-dependent targeting reaction in *E.*

coli is less well defined because the existence of membrane-bound ribosomes in *E. coli* is not as well established. It is plausible, however, that signal sequence recognition and targeting lead to a ribosome/membrane junction much like that observed in mammalian cells. The *E. coli* SecY/E proteins, two known components of an *E. coli* translocon, are similar in primary structure to the α and γ subunits of the Sec61 complex, and hence could play similar roles (Görlich et al 1992b; Hartmann et al 1994). Alternatively, the co-translational features might hold true only for signal sequence recognition by the *E. coli* SRP and for FtsY-mediated targeting; the nascent chain may then be handed over to other cytoplasmic components, such as chaperonins, for subsequent post-translational delivery to a translocon.

ALTERNATIVE TARGETING ROUTES

For the vast majority of proteins, translocation across mammalian ER membranes has a strict requirement for co-translational delivery of the nascent chain. In contrast, translocation of proteins across bacterial (reviewed in Bassford et al 1991; Randall & Hardy 1989) and yeast membranes (reviewed in Meyer 1988), and in a few cases across mammalian membranes (Schlenstedt et al 1990), can occur post-translationally, i.e. signal sequence–bearing proteins that have been released from ribosomes can be translocated from a soluble pool. SRP and SRP receptor are not involved in post-translational translocation, and it is therefore imperative to define and delineate their role (and the role of the pathway that they catalyze) in organisms in which post-translational translocation is prevalent.

SRP-independent Targeting

In yeast, protein targeting to the ER can occur by redundant pathways in vivo. One inevitably has to arrive at this conclusion because *S. cerevisiae* mutant cells lacking SRP or SRP receptor are viable, even though they grow poorly and the translocation of some proteins across the ER membrane is severely impaired (Hann & Walter 1991). Thus every protein that has to cross or become integrated into the ER membrane during its biogenesis and that is essential for cell viability must be targeted via alternate, SRP- and SRP receptor-independent pathways efficiently enough to sustain cell growth. Whereas *S. cerevisiae* cells remain viable, *S. pombe, Y. lipolytica,* and *E. coli* cells die when genes encoding SRP components are genetically disrupted, which indicates that SRP-dependent protein targeting is usually an essential pathway and that *S. cerevisiae* cells have evolved a particularly effective means of bypassing it.

The molecular details of alternative targeting pathways in the SRP-deficient *S. cerevisiae* mutant cells are presently unclear. The translocation of most soluble and membrane proteins into the lumen of the ER is impaired in

SRP/SRP receptor-deficient cells but, surprisingly, different proteins show translocation defects of varying severity (Hann & Walter 1991). There are two conceptually distinct explanations for this. First, SRP/SRP receptor-independent targeting could occur post-translationally: precursor proteins are released from ribosomes and are maintained in a soluble and translocation-competent state by interactions with cytosolic chaperonins (Chirico et al 1988; Deshaies et al 1988) and other putative targeting factors (Figure 7, pathway C). The folding characteristics of a particular preprotein may thus determine how efficiently it can be maintained in a translocation-competent form. This could explain why translocation defects observed in SRP and SRP receptor-deficient cells vary in magnitude for different proteins. In vitro studies corroborate this notion. The yeast pheromone prepro-α-factor, for example, which can be efficiently translocated post-translationally in vitro, shows only minor translocation defects in SRP- and SRP receptor-depleted cells in vivo.

According to a second hypothesis, SRP/SRP receptor-independent targeting could occur co-translationally: ribosomes synthesizing precursor proteins engage with the ER membrane independent of SRP and SRP receptor, but prior to termination of protein synthesis (Figure 7, pathway B). If co-translational targeting is obligate for a given precursor protein, then the kinetics of its elongation would affect the efficiency of its membrane translocation. If elongation is slow, for example, then a longer time frame would be available for the nascent precursor protein to engage with the ER membrane before it is elongated too far or is completed and released from the ribosome. Experimental results do not allow us to distinguish whether pathway B or pathway C or both operate in SRP-deficient yeast cells.

The fact that alternative pathways can be used in these mutant cells, however, does not imply that such pathways are major routes in wild-type cells. To the contrary, we consider it likely that all proteins that show translocation defects in the mutant cells are co-translationally targeted by SRP and SRP receptor to the ER membrane in wild-type cells (Figure 7, pathway A), and that they become re-routed into alternative targeting pathways only in the mutant cells lacking SRP and/or SRP receptor function or if the SRP/SRP receptor system becomes saturated. As SRP is thought to scan all nascent chains emerging from ribosomes, any protein that expresses a signal sequence that can bind to SRP with a reasonable affinity would be shunted into the co-translational pathway. Some rare proteins, however, that show no translocation defects in SRP-depleted cells may have evolved signal sequences that do not interact efficiently with SRP. Such proteins, e.g. preprocarboxypeptidase Y, may not use the SRP-dependent targeting pathway even in wild-type cells (Bird et al 1987; Hann & Walter 1991).

It is likely that a similar scenario of redundant targeting pathways also exists in *E. coli*. Because of the fast growth rates of bacterial cells, however, many

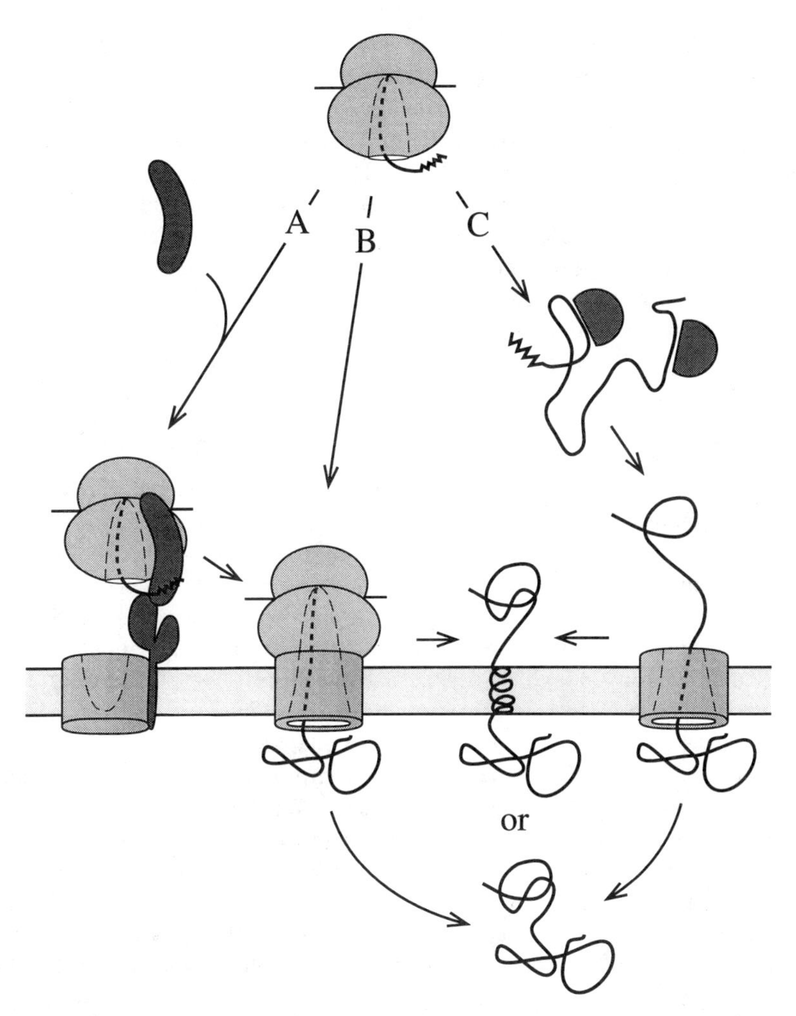

Figure 7 Three possible pathways for protein translocation in yeast (adapted from Hann & Walter 1991). SRP- and SRP receptor-dependent co-translational targeting, pathway A, can be efficiently bypassed in yeast. Pathway B shows SRP-independent co-translational targeting: SRP and SRP receptor are bypassed and the ribosome attaches to the ER membrane to yield an interaction indistinguishable from that achieved in pathway A. Attempts to identify such a pathway across mammalian ER were unsuccessful (Garcia & Walter 1988). Pathway C shows SRP-independent post-translational targeting. Prior to targeting, protein synthesis terminates and the precursor is released into a soluble pool. Translocation competence may be maintained by interactions with chaperonins, indicated by semi-circles. The preprotein interacts with a translocon, which is likely to share core components with the translocon used in pathways A and B, but which is not necessarily identical to it (see Figure 8).

abundant bacterial periplasmic proteins may have evolved such that they can use either SRP-dependent or SRP-independent pathways efficiently. *E. coli* has a particular chaperone, SecB, that seems to be dedicated to maintain preproteins competent for post-translational translocation (Kumamoto 1991). SecB-deficient cells die on rich medium, but are viable on minimal medium, which causes them to grow more slowly. The death of SecB-deficient cells on rich medium can be prevented if other chaperones are overproduced (Altman et al 1991; Wild et al 1992). This indicates that the SecB-mediated post-translational pathway is most important in fast growing cells, where an SRP-mediated pathway may be overwhelmed. The existence of at least partially redundant pathways may be the reason why genetic analyses have not identified an SRP or SRP receptor in bacteria.

Modular Translocons?

Proteins that are delivered to the membrane as short nascent chains emerging from the ribosome have different requirements for translocation than fully synthesized proteins that are delivered post-translationally. Conceptually, co-translational translocation is the easiest mode to envision (Figure 8A). The translocation pore that is formed by the translocon underneath the tightly attached ribosome can be thought of as an extension of the tunnel in the large ribosomal subunit through which the nascent chain exits (Simon & Blobel 1991; Crowley et al 1994, 1993). The translocon and the ribosome form a tight seal and the nascent chain therefore has no alternative but to move through the pore and into the ER lumen as translation proceeds and the polypeptide grows. Co-translational translocation is therefore dictated by the topography of the ribosome-membrane junction and probably driven by passive diffusion, although some method of active transport cannot be ruled out. In evolutionary terms this is an attractive mechanism because there are no contraints on the particular sequence of the nascent chain. SRP and SRP receptor mediate the immediate membrane attachment of short nascent chains and, as the protein is never exposed to the cytosol, there is no chance for it to fold, misfold, or aggregate into conformations that might then be difficult or impossible to translocate.

A different and mechanistically more complex model emerged from in vitro studies of post-translational translocation in *E. coli* (Figure 8B). Here the translocon uses a dedicated ATPase, the SecA protein, to insert the protein into the translocation channel and then pushes the protein, presumably by ratchet-like movements, through the membrane (Wickner et al 1991). As the preprotein is fully synthesized at this point, the translocon must recognize the signal sequence, unravel the chain, and release it from chaperones such as SecB. Moreover, somehow the permeability barrier of the membrane must be

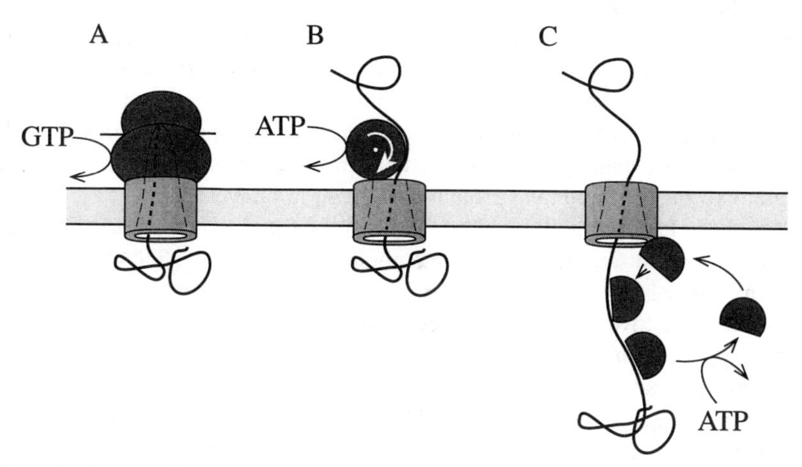

Figure 8 Three possible ways to drive protein translocation across membranes. These three modes, using chemical energy to drive protein translocation, are discussed in the text. (*A*) Translation elongation drives the nascent chain through the membrane pore (indirectly, the energy of GTP hydrolysis by elongation factors and that contributed by the charged tRNAs is utilized); (*B*) an ATP-consuming enzyme moves the polypeptide chain through the membrane pore; and (*C*) ATP-driven cycles of chaperonin binding and release pull the polypeptide chain through the membrane pore. It is possible that two, or maybe even three of these mechanisms collaborate to translocate or integrate certain proteins.

retained during translocation because in bacteria the membrane is also used to maintain a proton gradient.

A third mechanism for effecting translocation involves the sequential attachment of lumenal chaperones to pull a polypeptide across the membrane from the cytosolic side to the lumenal side (Figure 8C). There is good evidence in yeast, for example, that the ER lumenal hsp70 homologue BiP is required for protein translocation (Sanders et al 1992; Vogel et al 1990), and an analogous mechanism operates during the import of proteins into the mitochondrial matrix space (Hannavy et al 1993; Stuart et al 1994). A requirement for ER lumenal proteins was also shown for the mammalian system and may be particularly important for the completion of translocation, i.e. to pull the last section of a protein across the membrane after the ribosome has terminated its synthesis (Nicchitta & Blobel 1993).

It is likely that these three different modes of translocation all use a common "core translocon" comprising the evolutionarily conserved SecY/SecE (which contains a third uncharacterized protein called Band 1, or in eukaryotes, the Sec61α,β,γ protein complex) that provides the basic protein-conducting channel through which the protein chain crosses the membrane. To this core translocon, auxiliary components may be attached—perhaps transiently—to adapt the translocation process to particular needs dictated by whether the protein is delivered to the membrane co- or post-translationally or whether it is a soluble or membrane protein. For membrane protein synthesis, the core translocon has to provide not only a translocation pore, but also the means by which transmembrane segments in nascent membrane proteins are recognized. Then the translocation pore must be able to open on one side to allow the lateral exit of the membrane protein into the lipid bilayer.

Auxiliary factors might include, for example, ribosome receptors that may contribute to the seal between the membrane and ribosome: TRAM (already shown to be essential for the translocation of only some proteins; Görlich et al 1992a); the SRP receptor; the SecA protein; BiP; and the Sec62, 63, 71, and 72 proteins in yeast. The latter comprise a set of interacting membrane proteins important for protein translocation in yeast for which no homologues in mammalian cells or bacteria have been found. They may be dedicated to an SRP-independent translocation pathway and/or help the integration of membrane proteins. Because yeast cells contain much more Sec61 protein than they do Sec62 or Sec63 protein (Deshaies et al 1991), it is likely that translocons in a single cell are heterogeneous with regard to attached auxiliary components. The most challenging problem for future research is to decipher these complexities and to understand in mechanistic terms how protein translocation and membrane protein integration are catalyzed and modulated.

ACKNOWLEDGMENTS

Work in the authors' laboratories was supported by a grant from the Human Frontiers Science Program to PW and by grants from the National Institutes of Health to PW and AEJ.

Literature Cited

Althoff S, Selinger D, Wise JA. 1994. Molecular evolution of SRP cycle components: functional implications. *Nucleic Acids Res.* In press

Altman E, Kumamoto CA, Emr SD. 1991. Heat-shock proteins can substitute for SecB function during protein export in *E. coli*. *EMBO J.* 10:239–45

Amaya Y, Nakano A, Ito K, Mori M. 1990. Isolation of a yeast gene, SRH1, that encodes a homologue of the 54K subunit of mammalian signal recognition particle. *J. Biochem.* 107:457–63

Anderson DJ, Walter P, Blobel G. 1982. Signal recognition protein is required for the integration of acetylcholine receptor d subunit, a transmembrane glycoprotein, into the endoplasmic reticulum membrane. *J. Cell Biol.* 93:501–6

Andreazzoli M, Gerbi SA. 1991. Changes in 7SL RNA conformation during the signal recognition particle cycle. *EMBO J.* 10:767–77

Andrews DW, Lauffer L, Walter P, Lingappa VR. 1989. Evidence for a two-step mechanism involved in assembly of functional sig-

nal recognition particle receptor. *J. Cell Biol.* 108:797–810
Andrews DW, Walter P, Ottensmeyer FP. 1985. Structure of the signal recognition particle by electron microscopy. *Proc. Natl. Acad. Sci. USA* 82:785–89
Andrews DW, Walter P, Ottensmeyer FP. 1987. Evidence for an extended 7SL RNA structure in the signal recognition particle. *EMBO J.* 6:3471–77
Bassford P, Beckwith J, Ito K, Kumamoto C, Mizushima S, et al. 1991. The primary pathway of protein export in E. coli. *Cell* 65:367–68
Beckwith J. 1991. "Sequence gazing?" *Science* 251:1161
Berchtold H, Reshetnikova L, Reiser CO, Schirmer NK, Sprinzl M, Hilgenfeld R. 1993. Crystal structure of active elongation factor Tu reveals major domain rearrangements. *Nature* 365:126–32
Berneabeau C, Tobin EM, Fowler A, Zabin I, Lake JA. 1983. Nascent polypeptide chains exit the ribosome in the same relative position in both eucaryotes and procaryotes. *J. Cell Biol.* 96:1471–74
Bernstein HD, Poritz MA, Strub K, Hoben PJ, Brenner S, Walter P. 1989. Model for signal sequence recognition from amino-acid sequence of 54k subunit of signal recognition particle. *Nature* 340:482–86
Bernstein HD, Zopf D, Freymann DM, Walter P. 1993. Functional substitution of the signal recognition particle 54-kDa subunit by its *Escherichia coli* homolog. *Proc. Natl. Acad. Sci. USA* 90:5229–33
Bird P, Gething MJ, Sambrook J. 1987. Translocation in yeast and mammalian cells: not all signal sequences are functionally equivalent. *J. Cell Biol.* 105:2905–14
Blobel G, Dobberstein B. 1975. Transfer of proteins across the membrane. I. Presence of proteolytically processed and unprocessed nascent immunoglobulin light chains on membrane-bound ribosomes of murine myeloma. *J. Cell Biol.* 67:835–51
Bourne HR, Sanders DA, McCormick F. 1990. The GTPase superfamily: a conserved switch for diverse cell functions. *Nature* 348:125–32
Bourne HR, Sanders DA, McCormick F. 1991. The GTPase superfamily: conserved structure and molecular mechanism. *Nature* 349: 117–27
Brown S. 1989. Time of action of 4.5 S RNA in *Escherichia coli* translation. *J. Mol. Biol.* 209:79–90
Chirico WJ, Waters MG, Blobel G. 1988. 70K heat shock related proteins stimulate protein translocation into microsomes. *Nature* 332: 805–10
Connolly T, Gilmore R. 1989. The signal recognition particle receptor mediates the GTP-dependent displacement of SRP from the signal sequence of the nascent polypeptide. *Cell* 57:599–610
Connolly T, Rapiejko PJ, Gilmore R. 1991. Requirement of GTP hydrolysis for dissociation of the signal recognition particle from its receptor. *Science* 252:1171–73
Crooke E, Guthrie B, Lecker S, Lill R, Wickner W. 1988. ProOmpA is stabilized for membrane translocation by either purified E. coli trigger factor or canine signal recognition particle. *Cell* 54:1003–11
Crowley KS, Liao S, Worrell VE, Reinhart GD, Johnson AE. 1994. Secretory proteins move through the ER membrane via an aqueous, gated pore. *Cell.* Submitted
Crowley KS, Reinhart GD, Johnson AE. 1993. The signal sequence moves through a ribosomal tunnel into a noncytoplasmic aqueous environment at the ER membrane early in translocation. *Cell* 73:1101–15
Deshaies RJ, Koch BD, Werner WM, Craig EA, Schekman R. 1988. A subfamily of stress proteins facilitates translocation of secretory and mitochondrial precursor polypeptides. *Nature* 332:800–5
Deshaies RJ, Sanders SL, Feldheim DA, Schekman R. 1991. Assembly of yeast Sec proteins involved in translocation into the endoplasmic reticulum into a membrane-bound multisubunit complex. *Nature* 349:806–8
Dobberstein B. 1994. Protein transport: On the beaten pathway. *Nature* 367:599–600
Felici F, Cesareni G, Hughes JMX. 1989. The most abundant small cytoplasmic RNA of *Saccharomyces cerevisiae* has an important function required for normal cell growth. *Mol. Cell. Biol.* 9:3260–68
Franklin AE, Hoffman NE. 1993. Characterization of a chloroplast homologue of the 54-kDa subunit of the signal recognition particle. *J. Biol. Chem.* 268:22175–80
Garcia PD, Walter P. 1988. Full-length prepro-alpha-factor can be translocated across the mammalian microsomal membrane only if translation has not terminated. *J. Cell Biol.* 106:1043–48
Gilmore R. 1993. Protein translocation across the endoplasmic reticulum: a tunnel with toll booths at entry and exit. *Cell* 75:589–92
Gilmore R, Blobel G, Walter P. 1982a. Protein translocation across the endoplasmic reticulum. I. Detection in the microsomal membrane of a receptor for the signal recognition particle. *J. Cell Biol.* 95:463–69
Gilmore R, Walter P, Blobel G. 1982b. Protein translocation across the endoplasmic reticulum. II. Isolation and characterization of the signal recognition particle receptor. *J. Cell Biol.* 95:470–77
Görlich D, Hartmann E, Prehn S, Rapoport T. 1992a. A protein of the endoplasmic reticulum involved early in polypeptide translocation. *Nature* 357:47–52

Görlich D, Prehn S, Hartmann E, Kalies K-U, Rapoport TA. 1992b. A mammalian homolog of SEC61p and SECYp is associated with ribosomes and nascent polypeptides during translocation. *Cell* 71:489–503

Görlich D, Rapoport TA. 1993. Protein translocation into proteoliposomes reconstituted from purified components of the endoplasmic reticulum membrane. *Cell* 75:615–30

Gundelfinger ED, Carlo MD, Zopf D, Melli M. 1984. Structure and evolution of the 7SL RNA component of the signal recognition particle. *EMBO J.* 3:2325–32

Hann B, Stirling CJ, Walter P. 1992. SEC65 gene product is a subunit of the yeast signal recognition particle required for its integrity. *Nature* 356:532–33

Hann BC, Poritz MA, Walter P. 1989. *Saccharomyces cerevisiae* and *Schizosaccharomyces pombe* contain a homologue to the 54-kD subunit of the signal recognition particle that in *S. cerevisiae* is essential for growth. *J. Cell Biol.* 109:3223–30

Hann BC, Walter P. 1991. The signal recognition particle in S. cerevisiae. *Cell* 67:131–44

Hannavy K, Rospert S, Schatz G. 1993. Protein import into mitochondria: a paradigm for the translocation of polypeptides across membranes. *Curr. Opin. Cell Biol.* 5:694–700

Hartmann E, Sommer T, Prehn S, Görlich D, Jentsch S, Rapoport TA. 1994. Evolutionary conservation of components of the protein translocation complex. *Nature* 367:654–57

He F, Beckerich J-M, Gaillardin C. 1992. A mutant of 7SL RNA in *Yarrowia lipolytica* affecting the synthesis of a secreted protein. *J. Biol. Chem.* 267:1932–37

Herz J, Flint N, Stanley K, Frank R, Dobberstein B. 1990. The 68 kDa protein of signal recognition particle contains a glycine-rich region also found in certain RNA-binding proteins. *FEBS Lett.* 276:103–7

High S, Dobberstein B. 1991. The signal sequence interacts with the methionine-rich domain of the 54-kD protein of signal recognition particle. *J. Cell Biol.* 113:229–33

High S, Flint N, Dobberstein B. 1991a. Requirement for the membrane insertion of signal-anchor type proteins. *J. Cell Biol.* 113: 25–34

High S, Görlich D, Wiedmann M, Rapoport TA, Dobberstein B. 1991b. The identification of proteins in the proximity of signal-anchor sequences during their targeting to and insertion into the membrane of the ER. *J. Cell Biol.* 113:35–44

High S, Martoglio B, Görlich D, Andersen SSL, Ashford AJ, et al. 1993. Site-specific photocrosslinking reveals that Sec61p and Tram contact different regions of a membrane inserted signal sequence. *J. Biol. Chem.* 268: 26745–51

Janiak F, Walter P, Johnson AE. 1992. Fluorescence-detected assembly of the signal recognition particle: binding of the two SRP protein heterodimers to SRP RNA is noncooperative. *Biochemistry* 31:5830–40

Krieg UC, Johnson AE, Walter P. 1989. Protein translocation across the endoplasmic reticulum membrane: identification by photocrosslinking of a 39 kDa integral membrane glycoprotein as part of a putative translocation tunnel. *J. Cell Biol.* 109:2033–43

Krieg UC, Walter P, Johnson AE. 1986. Photocrosslinking of the signal sequence of nascent preprolactin to the 54-kilodalton polypeptide of the signal recognition particle. *Proc. Natl. Acad. Sci. USA* 83:8604–8

Kumamoto CA. 1991. Molecular chaperones and protein translocation across the *E. coli* inner membrane. *Mol. Microbiol.* 5:19–22

Kurzchalia TV, Wiedmann M, Girshovich AS, Bochkareva ES, Bielka H, Rapoport TA. 1986. The signal sequence of nascent preprolactin interacts with the 54K polypeptide of the signal recognition particle. *Nature* 320: 634–36

Larsen N, Zwieb C. 1991. SRP-RNA sequence alignment and secondary structure. *Nucleic Acids Res.* 19:209–15

Larsen N, Zwieb C. 1993. The signal recognition particle database (SRPDB). *Nucleic Acids Res.* 21:3019–20

Lauffer L, Garcia PD, Harkins RN, Coussens L, Ullrich A, Walter P. 1985. Topology of the SRP receptor in the endoplasmic reticulum membrane. *Nature* 318:334–38

Liao X, Selinger D, Althoff S, Chiang A, Hamilton D, Ma M, Wise JA. 1992. Random mutagenesis of *Schizosaccharomyces pombe* SRP RNA: lethal and conditional lesions cluster in presumptive protein binding sites. *Nucleic Acids Res.* 20:1607–15

Lingelbach K, Zwieb C, Webb JR, Marshallsay C, Hoben PJ, et al. 1988. Isolation and characterization of a cDNA clone encoding the 19 kDa protein of the signal recognition particle (SRP): expression and binding to 7SL RNA. *Nucleic Acids Res.* 16:9431–42

Lipp J, Dobberstein B, Haeuptle MT. 1987. Signal recognition particle arrests elongation of nascent secretory and membrane proteins at multiple sites in a transient manner. *J. Biol. Chem.* 262:1680–84

Luirink J, High S, Wood H, Giner A, Tollervey D, Dobberstein B. 1992. Signal sequence recognition by an *Escherichia coli* ribonucleoprotein complex. *Nature* 359:741–43

Lütcke H, High S, Römisch K, Ashford A, Dobberstein B. 1992. The methionine-rich domain of the 54 kDa subunit of signal recognition particle is sufficient for the interaction with signal sequences. *EMBO J.* 11: 1543–51

Lütcke H, Prehn S, Ashford AJ, Remus M, Frank R, Dobberstein B. 1993. Assembly of the 68- and 72-kD proteins of the signal recognition particle with 7S RNA. *J. Cell Biol.* 121:977–85

Meyer DI. 1988. Preprotein conformation: the year's major theme in translocation studies. *Trends Biochem. Sci.* 13:471–74

Meyer DI, Dobberstein B. 1980a. Identification and characterization of a membrane component essential for the translocation of nascent secretory proteins across the membrane of the endoplasmic reticulum. *J. Cell Biol.* 87: 503–8

Meyer DI, Dobberstein B. 1980b. A membrane component essential for vectorial translocation of nascent proteins across the endoplasmic reticulum: requirement for its extraction and reassociation with the membrane. *J. Cell Biol.* 87:498–502

Meyer DI, Krause E, Dobberstein B. 1982a. Secretory protein translocation across membranes—the role of the docking protein. *Nature* 297:647–50

Meyer DI, Louvard D, Dobberstein B. 1982b. Characterization of molecules involved in protein translocation using a specific antibody. *J. Cell Biol.* 92:579–83

Miller JD, Bernstein HD, Walter P. 1994. Interaction of *E. coli* Ffh/4.5S ribonucleoprotein and FtsY mimics that of mammalian signal recognition particle and its receptor. *Nature* 367:657–59

Miller JD, Wilhelm H, Gierasch L, Gilmore R, Walter P. 1993. GTP binding and hydrolysis by the signal recognition particle during initiation of protein translocation. *Nature* 366: 351–54

Nicchitta CV, Blobel G. 1993. Lumenal proteins of the mammalian endoplasmic reticulum are required to complete protein translocation. *Cell* 73:989–98

Nomura M, Traub P, Guthrie C, Nashimoto H. 1969. The assembly of ribosomes. *J. Cell Physiol.* 74:241–52

Nowotny V, Nierhaus KH. 1988. Assembly of the 30S subunit from *Escherichia coli* ribosomes occurs via two assembly domains which are initiated by S4 and S7. *Biochemistry* 27:7051–55

Nunnari J, Walter P. 1992. Protein targeting to and translocation across the membrane of the endoplasmic reticulum. *Curr. Opin. Cell Biol.* 4:573–80

O'Neil KT, DeGrado WF. 1990. How calmodulin binds its targets: sequence independent recognition of amphiphilic a-helices. *Trends Biochem. Sci.* 15:59–64

Ogg S, Poritz M, Walter P. 1992. The signal recognition particle receptor is important for growth and protein secretion in *Saccharomyces cerevisiae. Mol. Biol. Cell* 3:895–911

Palade G. 1975. Intracellular aspects of the process of protein secretion. *Science* 189:347–58

Phillips GJ, Silhavy TJ. 1992. The *E. coli ffh* gene is necessary for viability and efficient protein export. *Nature* 359:744–46

Poritz MA, Bernstein HD, Strub K, Zopf D, Wilhelm H, Walter P. 1990. An *E. coli* ribonucleoprotein containing 4.5S RNA resembles mammalian signal recognition particle. *Science* 250:111–17

Poritz MA, Strub K, Walter P. 1988. Human SRP RNA and E. coli 4.5S RNA contain a highly homologous structural domain. *Cell* 55:4–6

Randall LL, Hardy SJS. 1989. Unity in function in the absence of consensus in sequence: role of leader peptides in export. *Science* 243: 1156–59

Rapiejko PJ, Gilmore R. 1992. Protein translocation across the endoplasmic reticulum requires a functional GTP binding site in the α-subunit of the signal recognition particle receptor. *J. Cell Biol.* 117:493–503

Rapoport TA. 1992. Transport of proteins across the endoplasmic reticulum membrane. *Science* 258:931–36

Rapoport TA, Heinrich R, Walter P, Schulmeister T. 1987. Mathematical modeling of the effects of the signal recognition particle on translation and translocation of proteins across the endoplasmic reticulum membrane. *J. Mol. Biol.* 195:621–36

Römisch K, Webb J, Herz J, Prehn S, Frank R, et al. 1989. Homology of the 54K protein of signal recognition particle, docking protein, and two *E. coli* proteins with putative GTP-binding domains. *Nature* 340:478–82

Römisch K, Webb J, Lingelbach K, Gausepohl H, Dobberstein B. 1990. The 54-kD protein of signal recognition particle contains a methionine-rich RNA binding domain. *J. Cell Biol.* 111:1793–802

Samuelsson T. 1992. A mycoplasma protein homologous to mammalian SRP54 recognizes a highly conserved domain of SRP RNA. *Nucleic Acids Res.* 20:5763–70

Sanders SL, Schekman R. 1992. Polypeptide translocation across the endoplasmic reticulum membrane. *J. Biol. Chem.* 267:13791–94

Sanders SL, Whitfield KM, Vogel JP, Rose MD, W, Schekman RW. 1992. Sec61p and BiP directly facilitate polypeptide translocation into the ER. *Cell* 69:353–65

Sanz P, Meyer DI. 1988. Signal recognition particle (SRP) stabilizes the translocation-competent conformation of pre-secretory proteins. *EMBO J.* 7:3553–57

Schlenstedt G, Gudmundsson G, Bowman H, Zimmermann R. 1990. A large secretory protein translocates both cotranslationally, using signal recognition particle and ribosome, and post-translationally, without these ribonucleoparticles, when synthesized in the presence of mammalian microsomes. *J. Biol. Chem.* 265:13960–68

Selinger D, Brennwald P, Liao X, Wise JA. 1993a. Identification of RNA sequences and structural elements required for assembly of fission yeast SRP54 protein with signal recognition particle RNA. *Mol. Cell. Biol.* 13: 1353–62

Selinger D, Liao X, Wise JA. 1993b. Functional interchangeability of the structurally similar tetranucleotide loops GAAA and UUCG in fission yeast signal recognition particle RNA. *Proc. Natl Acad. Sci. USA* 90:5409–13

Siegel V, Walter P. 1985. Elongation arrest is not a prerequisite for secretory protein translocation across the microsomal membrane. *J. Cell Biol.* 100:1913–21

Siegel V, Walter P. 1986. Removal of the Alu structural domain from signal recognition particle leaves its protein translocation activity intact. *Nature* 320:81–84

Siegel V, Walter P. 1988a. The affinity of signal recognition particle for presecretory proteins is dependent on nascent chain length. *EMBO J.* 7:1769–75

Siegel V, Walter P. 1988b. Binding sites of the 19-kDa and 68/72-kDa signal recognition particle (SRP) proteins on SRP RNA as determined by protein-RNA "footprinting." *Proc. Natl. Acad. Sci. USA* 85:1801–5

Siegel V, Walter P. 1988c. Each of the activities of signal recognition particle (SRP) is contained within a distinct domain: analysis of biochemical mutants of SRP. *Cell* 52:39–49

Siegel V, Walter P. 1988d. Functional dissection of the signal recognition particle. *Trends Biochem. Sci.* 13:314–16

Simon SM, Blobel G. 1991. A protein-conducting channel in the endoplasmic reticulum. *Cell* 65:371–80

Stirling CJ, Hewitt EW. 1992. The *Saccharomyces cerevisiae SEC65* gene encodes a component of the yeast signal recognition particle with homology to human SRP19. *Nature* 356:534–37

Strub K, Moss J, Walter P. 1991. Binding sites of the 9- and 14-kilodalton heterodimeric protein subunit of the Signal Recognition Particle (SRP) are contained exclusively in the Alu domain of SRP RNA and contain a sequence motif that is conserved in evolution. *Mol. Cell. Biol.* 11:3949–59

Strub K, Walter P. 1989. Isolation of a cDNA clone of the 14-kDa subunit of the signal recognition particle by cross-hybridization of differently primed polymerase chain reactions. *Proc. Natl. Acad. Sci. USA* 86:9747–51

Strub K, Walter P. 1990. Assembly of the Alu domain of the signal recognition particle (SRP): dimerization of the two protein components is required for efficient binding of SRP RNA. *Mol. Cell. Biol.* 10:777–84

Struck J. CR, Toschka HY, Specht T, Erdmann VA. 1988. Common structural features between eukaryotic 7SL RNAs, eubacterial 4.5S RNA and scRNA and archaebacterial 7S RNA. *Nucleic Acids Res.* 16:7740

Stuart RA, Cyr DM, Craig EA, Neupert W. 1994. Mitochondrial molecular chaperones: their role in protein translocation. *Trends Biol. Sci.* 19:87–92

Tajima S, Lauffer L, Rath VL, Walter P. 1986. The signal recognition particle is a complex that contains two distinct polypeptide chains. *J. Cell Biol.* 103:1167–78

Ullu E, Murphy S, Melli M. 1982. Human 7SL RNA consists of a 140 nucleotide middle-repetitive sequence inserted in an Alu sequence. *Cell* 29:195–202

Ullu E, Tschudi C. 1984. Alu sequences are processed 7SL genes. *Nature* 312:171–72

Vogel JP, Misra LM, Rose MD. 1990. Loss of BiP/GRP78 function blocks translocation of secretory proteins in yeast. *J. Cell Biol.* 110: 1885–95

von Heijne G. 1985. Signal sequences. The limits of variation. *J. Mol. Biol.* 184:99–105

Walter P, Blobel G. 1980. Purification of membrane-associated protein complex required for protein translocation across the endoplasmic reticulum. *Proc. Natl. Acad. Sci. USA* 77:7112–16

Walter P, Blobel G. 1981. Translocation of proteins across the endoplasmic reticulum. III. Signal recognition protein (SRP) causes signal sequence and site specific arrest of chain elongation that is released by microsomal membranes. *J. Cell Biol.* 91:557–61

Walter P, Blobel G. 1982. Signal recognition particle contains a 7S RNA essential for protein translocation across the endoplasmic reticulum. *Nature* 299:691–98

Walter P, Blobel G. 1983. Disassembly and reconstitution of the signal recognition particle. *Cell* 34:525–33

Walter P, Gilmore R, Blobel G. 1984. Protein translocation across the endoplasmic reticulum. *Cell* 38:5–8

Walter P, Ibrahimi I, Blobel G. 1981. Translocation of proteins across the endoplasmic reticulum I. Signal Recognition Protein (SRP) binds to in vitro assembled polysomes synthesizing secretory protein. *J. Cell Biol.* 91: 545–50

Walter P, Jackson RC, Marcus MM, Lingappa VR, Blobel G. 1979. Tryptic dissection and reconstitution of translocation activity for nascent presecretory proteins across microsomal membranes. *Proc. Natl. Acad. Sci. USA* 76:1795–99

Walter P, Lingappa VR. 1986. Mechanism of protein translocation across the endoplasmic reticulum membrane. *Annu. Rev. Cell Biol.* 2:499–516

Wickner W, Driessen A. JM, Hartl F-U. 1991. The enzymology of protein translocation across the *Escherichia coli* plasma membrane. *Annu. Rev. Biochem.* 60:101–24

Wiedmann M, Kurzchalia TV, Hartmann E, Rapoport TA. 1987. A signal sequence receptor in the endoplasmic reticulum membrane. *Nature* 328:830–32

Wild J, Altman E, Yura T, Gross CA. 1992. DnaK and DnaJ heat shock proteins participate in protein export in *E. coli. Genes Dev.* 6:1165–72

Wolin SL, Walter P. 1988. Ribosome pausing and stacking during translation of a eukaryotic mRNA. *EMBO J.* 7:3559–69

Wolin SL, Walter P. 1989. Signal recognition particle mediates a transient elongation arrest of preprolactin in reticulocyte lysate. *J. Cell Biol.* 109:2617–22

Wood H, Luirink J, Tollervey D. 1992. Evolutionary conserved nucleotides within the *E. coli* 4.5S RNA are required for association with p48 in vitro and for optimal function in vivo. *Nucleic Acids Res.* 20:5919–25

Yaver DS, Matoba S, Ogrydziak DM. 1992. A mutation in the signal recognition particle 7S RNA of the yeast *Yarrowia lipolytica* preferentially affects synthesis of the alkaline extracellular protease: In vivo evidence for translational arrest. *J. Cell Biol.* 116:605–16

Zopf D, Bernstein HD, Johnson AE, Walter P. 1990. The methionine-rich domain of the 54 kd protein subunit of the signal recognition particle contains an RNA binding site and can be crosslinked to a signal sequence. *EMBO J.* 9:4511–17

Zopf D, Bernstein HD, Walter P. 1993. GTPase domain of the 54kD subunit of the mammalian signal recognition particle is required for protein translocation but not for signal sequence binding. *J. Cell Biol.* 120:1113–21

Zwieb C. 1991. Interaction of protein SRP19 with signal recognition particle RNA lacking individual RNA-helices. *Nucleic Acids Res.* 19:2955–60

Annu. Rev. Cell Biol. 1994. 10:121–52

VERTEBRATE LIMB DEVELOPMENT

Cheryll Tickle
Department of Anatomy and Developmental Biology, University College London, London, W1P 6DB England

Gregor Eichele
Department of Biochemistry, Baylor College of Medicine, Houston, Texas 77030

KEY WORDS: polarizing region, apical ectodermal ridge, homeobox genes, fibroblast growth factors, bone morphogenetic proteins, retinoids, *Sonic hedgehog*

CONTENTS

EMBRYOLOGY OF THE LIMB

Synopsis of Limb Morphogenesis

Limbs develop from small buds that protrude from the body wall and consist of mesenchymal cells encased in an ectodermal hull (Figure 1a). The initial formation of limb buds is not caused by an increase in cell proliferation at the sites where buds develop, but by a selective decrease in proliferation in tissue on either side of the future buds (Searls & Janners 1971). The regions of the

0743–4634/94/1115–0121$05.00

embryo that will give rise to limbs have been mapped (Chaube 1959). Early maps show that when the embryo has only a few somites, cells destined to form limbs lie posterior to Hensen's node (a knot-like structure situated at the midline of the early embryo, which acts as a signaling center similar to the Spemann organizer in amphibians). Later, as the node regresses, regions fated to form limbs are found anterior to the node and when transplanted, autonomously develop into limb-like structures (Stephens et al 1989).

Very little is known about the factors that specify the position where limbs will develop. It has been suggested that structures along the anteroposterior body axis are, in part, specified by combinatorial expression of *Hox* genes (*Hox* code; e.g. Hunt & Krumlauf 1992; McGinnis & Krumlauf 1992). Hence, one possibility is that the *Hox* code also determines the axial position at which limbs are formed. To the first approximation, the position of the forelimb correlates with anterior expression boundaries of *Hox* group 5, 6, and 7 genes (see Scott 1992 for nomenclature). For example, in the zebrafish, the pectoral fin develops just slightly cranial to the anterior boundary of *Hoxc-6* expression in mesoderm (Molven et al 1990). The situation in the mouse is similar in that the anterior expression boundary of *Hoxc-6* coincides with prevertebra 7 (e.g. Kessel & Gruss 1991). Furthermore, once buds appear this gene continues to be expressed in the anterior portion of the forelimb buds of mice, frogs, and chicks (Oliver et al 1988, 1990). Similarly, *Hoxb-5* expression extends up to the middle of somite 5 in the mouse, which is slightly cranial to the site where forelimbs form (Wall et al 1992). In

→

Figure 1 (*a*) Vertebrate limb development as illustrated with the chick wing. In all schemes anterior is on top and distal to the right. The wing bud, a small protrusion of the flank, grows distally and eventually develops into the characteristic shape of a wing. Shading represents chondrogenic regions and black cartilage. Ossification has not yet occurred at this stage. The limb skeleton is organized similarly in all vertebrate limbs. (*b*) Two signaling regions, the apical ectodermal ridge and the polarizing region, are thought to control the development of the limb pattern. Signals from the ridge (signal 1), possibly FGF-4 and/or FGF-2, maintain a region of proliferating and undifferentiated cells called the progress zone, which brings about the distal growth of the bud. In addition, such ridge signals (signal 2) maintain the polarizing region. In turn the polarizing region maintains the progress zone (signal 3). The polarizing region is thought to release a morphogen that specifies the pattern along the anteroposterior axis. Candidate morphogens are retinoic acid and the product encoded by the *Sonic hedgehog* gene. Whether signal 3 and the morphogen are related is not known. (*c*) Experimental manipulations that respecify the anteroposterior limb pattern are (from left to right) grafting of a polarizing region, local release of retinoic acid from an ion-exchange bead, and ectopic expression of *Sonic hedgehog*. In all instances a mirror-symmetrical *432234* pattern may be induced. In addition to skeletal elements muscles, nerves, and skin appendages are also duplicated. (*d*) When the ridge is removed from an early bud, a truncated limb with just a humerus develops. Simultaneous application of FGF-4 from a heparin bead to the posterior limb bud margin and to the apex of the bud results in a limb with humerus, radius, ulna, and digits. Although FGF-4 is highly effective in mimicking the chemical signal from the ridge, it fails to perform the mechanical role of the ridge. Instead of having the usual paddle-shaped geometry, buds become cylindrical and, as a result of this, digits are bunched together instead of lying side by side in the plane of the hand plate.

addition, this gene continues to be transcribed in an anteroproximal domain in forelimb buds of mouse and chick embryos (Wall et al 1992; Wedden et al 1989). That *Hoxb-5* and *Hoxc-6* genes are locally expressed in forelimbs, but not in hindlimbs, supports the idea that these genes are more likely to

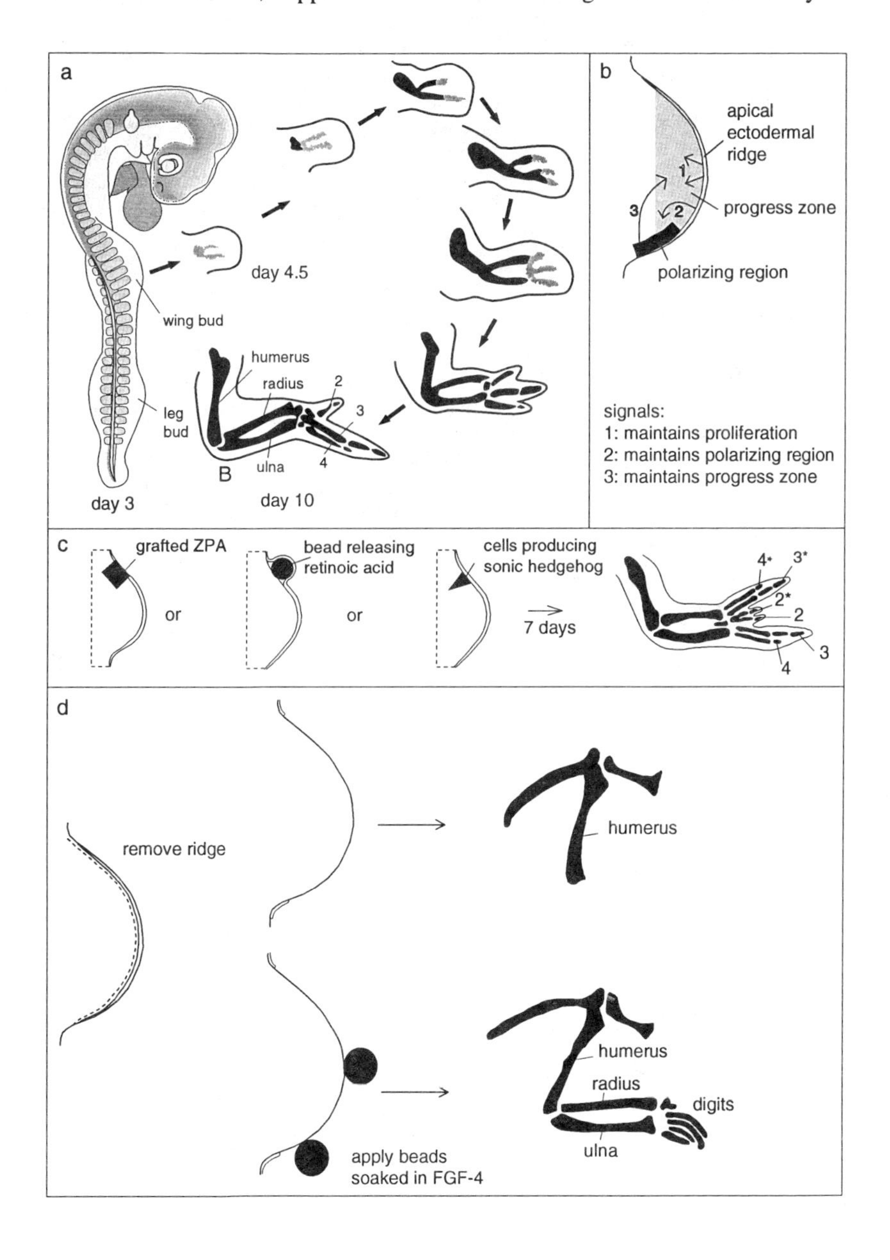

Table 1 Genes expressed in early vertebrate limb buds[a,b]

Gene	Expression
Genes expressed in limb mesenchyme	
Transcription factors encoded by Hox complexes	
Hoxa-10, Hoxa-11, Hoxa-13	progressively more 5' genes are restricted to more distal mesenchyme
Hoxd-9, Hoxd-10, Hoxd-11, Hoxd-12, Hoxd-13	progressively more 5' genes are restricted to more posterodistal mesenchyme
Hoxb-5	expressed anteriorly in forelimbs
Hoxc-6*	expressed anteriorly in forelimbs
Other transcription factors	
Evx-1	posterodistal mesenchyme
Msx-1	restricted to distal mesenchyme
Msx-2	anterior and distal mesenchyme
AP-2	distal mesenchyme
Retinoid receptors and retinoid binding proteins	
RARα	expressed at low levels throughout mesenchyme
RARβ	throughout mesenchyme but proximally enriched
RARγ	expressed at low levels throughout mesenchyme
RXRα, β	ubiquitous
CRABP I	distal mesenchyme
CRABP II	dorsally enriched, higher levels proximally
CRABP*	anterior to posterior protein concentration gradient with anterior high-point
Putative intercellular signaling molecules	
Wnt-5a	abundant in distal mesemchyme, low levels proximal mesenchyme
Sonic hedgehog	discrete posterior domain associated with polarizing region
BMP-2	posterior mesenchyme, approximately coex-tensive with polarizing region
BMP-4	anterior and posterior mesenchymal domains
FGF-2*	mesenchyme beneath ectoderm and apical ectodermal ridge
Genes expressed in apical ectodermal ridge	
Transcription factors	
dlx	
engrailed	
Msx-1	
Msx-2	
RARβ	
Putative intercellular signaling molecules	
FGF-2*	
Fgf-4	enriched posteriorly
BMP-2	
BMP-4	
Wnt-5a	

Genes expressed in limb ectoderm	
Transcription factors	
engrailed	enriched ventrally
Putative intercellular signaling molecules	
FGF-2*	dorsally enriched
BMP-2	
Wnt-5a	enriched ventrally
Wnt-7a	dorsal ectoderm only

[a] Only genes referenced in the text are listed; a comprehensive list would consist of several hundred entries.

[b] Expression data based on in situ hybridization analysis except for gene products marked by an asterisk, which indicates that expression pattern has been determined by immunohistochemistry.

be involved in axial positioning of the forelimb and not in limb morphogenesis per se. However, another possibility is that they may be involved in development of the shoulder girdle (Oliver et al 1990).

Once limb buds have formed, they rapidly grow at the tip (Hornbruch & Wolpert 1970), and while growth continues, mesenchyme cells at the base of the bud start to terminally differentiate into the various tissues that constitute a limb (Figure 1a). Eventually, the tip produces a flattened plate in which digit primordia can be discerned. Initially, digits are joined by thin webs of tissue, but these disappear and the digits eventually separate. One can describe the structure of limbs in terms of three axes: a proximodistal axis that extends between the shoulder and the tips of the digits; an anteroposterior axis that extends, e.g. in the hand, between thumb and little finger; and a dorsoventral axis that runs from the back of the hand to the palm (Figure 1a). Whether such a Cartesian reference frame is pertinent to the mechanism by which the pattern is formed has long been a subject of discussion.

Experimental analysis of limb development has employed mainly chick embryos because chicks are readily accessible to experimental manipulations in ovo. In the past, the effects of such manipulations on pattern have been monitored by whole mount staining of cartilage and by histology, but now many molecular markers are available that can provide a detailed picture of how specific experimental interferences influence molecular and cellular processes (Table 1). Experimental manipulations have led to the identification of two major signaling regions that are believed to mediate patterning and morphogenesis of developing limb buds (Figure 1b). One region is the apical ectodermal ridge (AER) that interacts with underlying mesenchyme, and the other is a small group of posterior mesenchyme cells, known as the polarizing region or zone of polarizing activity (ZPA), that signals to nearby mesenchyme cells.

Apical Ectodermal Ridge and Progress Zone

APICAL ECTODERMAL RIDGE The apical ectodermal ridge is the thickened epithelium that rims the tip of the bud (Figure 1b). It arises from surface ectoderm in the limb-forming region and consists of pseudostratified elongated cells that are closely packed and linked by extensive gap junctions (Fallon & Kelley 1977). The tight packing makes the AER rigid, and this is the likely basis for its ability to keep the bud flattened. The length of the ridge controls bud width and hence is related to the number of digits that form (see e.g. Lee & Tickle 1985).

Experiments in chick embryos show that ridge formation depends on signals from the underlying mesenchyme. When presumptive limb mesenchyme is transplanted to the flank, an AER is induced in flank ectoderm and an additional limb develops (Kieny 1968). Interactions between ridge and mesenchyme are reciprocal and continuous throughout bud development (reviewed in Tickle 1991). When as a result of treatment with retinoic acid the apical ridge is flattened, recombining it with limb mesenchyme from another embryo reestablishes the typical pseudostratified tissue architecture (Tickle et al 1989). The cells of the apical ridge appear to be a stable cell population. When a quail apical ridge is grafted to the dorsal surface of a chick wing bud, a secondary outgrowth is induced. The quail cells persist in the grafted AER throughout subsequent development and do not partition into adjacent ectoderm, nor do chick cells invade the ridge (Saunders et al 1976).

A major function of the ridge is to mediate bud outgrowth, and it maintains a zone of undifferentiated cells, the so-called progress zone, at the tip of the limb bud (see Figure 1b). When the ridge is cut away from chick limb bud, the bud stops growing out and a truncated limb will result (Saunders 1948; Summerbell 1974). Since skeletal structures along the proximodistal axis of the limb develop in sequence, with proximal structures forming first and distal structures last, removing the AER early results in a severely truncated limb. In contrast, with removal later on, proximal structures develop normally and only distal-most structures such as digits are missing. Although signals from the apical ridge are required for patterning along the proximodistal axis, these signals do not specify what type of structure the mesenchyme cells will form. This was shown in experiments in which the AER was exchanged between limb buds of different developmental stages. Irrespective of the age of the ridge, the limb pattern is governed by the mesenchyme (Rubin & Saunders 1972). Tissue recombination experiments in which mesoderm and ectoderm of hind- and forelimbs are interchanged also show that mesoderm and not ectoderm determines which type of limb is formed (Zwilling 1955).

PROGRESS ZONE The progress zone is defined as the distal region of the limb bud in which undifferentiated mesenchyme cells divide and are subject to a timing mechanism (Summerbell et al 1973). The progress zone behaves autonomously when transplanted. Thus grafting the tip of an early limb bud to a stump of an older bud results in a duplication of proximodistal skeletal elements, while the converse experiment produces a limb with missing structures. Based on such experiments, it has been suggested that the length of time cells spend in the progress zone specifies which structures they will form. Cells that spend a short time in the progress zone would form proximal structures, whereas cells (or their ancestors) that reside for a longer time would form more distal structures (Summerbell et al 1973). Time could possibly be measured by counting the number of cell divisions. Alternatively, the progressive activation of 5′ *HoxA* or *HoxD* cluster genes that are expressed along the proximodistal limb axis (see below) could provide a molecular clock (Dollé et al 1993).

Dorsoventral Polarity and Ectoderm

Ectodermal signals may be important in controlling dorsoventral pattern. When ectoderm from a right wing bud is recombined with mesoderm from a left wing bud so that the ectoderm is inverted dorsoventrally, limbs, which subsequently form from such recombinants, are also inverted dorsoventrally as judged, for example, by joint morphology and pattern of skin appendages (Pautou & Kieny 1973; MacCabe et al 1974). Reversal of the dorsoventral axis is restricted to more distal parts of the manipulated limb, while proximally, polarity appears to be governed by that of the mesoderm. The inversion of just distal structures suggests that only cells in the progress zone are responsive to changes in ectodermal signals. These experiments, however, do not show whether positional cues are produced by both ventral and dorsal ectoderm, or only by one of the two epithelia. However, there is also some evidence that when an outgrowth is induced from either the dorsal or ventral surface of a wing bud by grafting an additional apical ridge, the ectopic limb has either a double dorsal or a double ventral configuration, respectively (Saunders et al 1976; Shellswell & Wolpert 1977). This suggests that both dorsal and ventral ectoderm could be sources of dorsoventral patterning signals.

Polarizing Region

The polarizing region was discovered by Saunders & Gasseling (1968). In the course of exploring the control of programmed cell death that occurs at the posterior limb bud margin, they grafted tissue from this region of a chick wing bud to the anterior margin of a host wing bud. This operation had a dramatic effect on wing pattern: A polarizing activity originating in the graft induced an additional set of digits in anterior host cells. Thus instead of the normal *234* digit pattern (reading from anterior to posterior, Figure 1a), the manipulated wing displayed a *432234* pattern in which additional digits were arranged in mirror-image symmetry with respect to the normal digits (Figure 1c). Induced structures are derived from host tissue and hence the ZPA is a true signaling region (Saunders & Gasseling 1968).

Polarizing region cells cannot be identified histologically, but their distribution in limb buds has been mapped by cutting out small blocks of tissue and grafting those anteriorly in host wing buds to assay for digit-inducing activity (e.g. Honig & Summerbell 1985; Hinchliffe & Sansom 1985). Polarizing activity can be detected early in lateral body wall mesenchyme long before limb buds develop (Hornbruch & Wolpert 1991), and it persists at the posterior margin until the AER regresses at the time the hand plate is formed (Honig & Summerbell 1985). In the limb bud, the highest polarizing activity is always found near the tip of the bud just proximal to the progress zone. Hence the high point of activity is displaced distally as the limb bud grows out. How this is achieved is not known, but one possibility is that the ZPA is a transient cell population. Cells in the posterior part of the progress zone could serve as precursors and just after exiting the progress zone, as a result of the zone moving distally, these precursors could acquire polarizing activity that then enables them to signal to responsive cells in surrounding tissue. Later, polarizing region cells may die or contribute to the tissues that make up the limb (Bowen et al 1989). Another possibility is that the ZPA is a stable cell population that is gradually displaced distally and, as cells from the progress zone move around the ZPA, they are exposed to the digit-inducing signal.

Transplantation experiments have uncovered the following characteristics of the ZPA. (*a*) When ~ 30 polarizing region cells are grafted, a digit *2* is specified; ~ 70 cells evoke a digit *3*; and ≥ 100 cells are required to induce a digit *4* (Tickle 1981). An analogous progression is obtained by leaving ZPA grafts in situ for different lengths of time (Smith 1980). When a grafted ZPA is removed after 15 hr of contact, an additional digit *2* will form, while the induction of a digit *3* requires 17–24 hr of contact. Thus the extent of digit induction depends on the number of ZPA cells and also on the duration of contact between graft and host. (*b*) The digit next to the polarizing region is always more posterior in character than the digit(s) farther away (e.g. ZPA → *4* → *3* → 2) (Tickle et al 1975). Moreover, a *4* → *3* → *2* digit order is always maintained, i.e. digit *2* does not develop next to digit *4*. (*c*) The polarizing region can act on cells in the progress zone in both an anterior and a posterior direction. When placed at the apex of the wing bud, a polarizing region will evoke a sequence of digits *234* anterior to the graft (reading anterior to posterior) and a *434* pattern posteriorly. (*d*) Transplanted polarizing cells signal most efficiently in close contact with the apical ectodermal ridge (Tickle 1981). In the absence of the ridge, polarizing activity is reduced (Vogel & Tickle 1993). (*e*) The polarizing region is able to signal to responding mesenchyme cells at a distance because it induces wing digits through an intervening piece of leg mesoderm (Honig 1981). (*f*) Transplanted ZPA does not induce a new ZPA in the host (Smith 1979). (*g*) The polarizing activity is confined to the mesenchyme; neither the AER nor the rest of the ectoderm have activity. When an ectodermal jacket is separated from its mesenchymal core, rotated by 180°, and then recombined with the core so as to place posterior ectoderm over anterior mesoderm, the resulting limb appears normal (Zwilling 1956).

Two models attempt to explain these observations. One model assumes that the polarizing region produces a morphogen, i.e. a gradient-forming molecule that specifies in a concentration-dependent fashion the fate of a group of limb bud cells and hence the type of digit these cells will form. The model postulates distinct thresholds for each digit; low levels of morphogen specify a digit *2* and increasing concentrations specify digits *3* and *4*, respectively (Tickle et al 1975; see also Wolpert 1989).

A weakness of the morphogen model is that it singles out patterning along the anteroposterior axis from patterning along the proximodistal axis. But signals for distal growth and patterning along the anteroposterior axis are probably interdependent since the anteroposterior pattern can only form if the limb bud grows distally. The polar coordinate model attempts to take this into account by assuming that limb bud cells early in limb development already know where they are with respect to each other (they have coordinates or positional values; see Bryant & Gardiner 1992). By grafting polarizing cells that possess very posterior positional values to the anterior bud margin, a discontinuity in positional values is created. According to the model, cells respond by dividing and acquiring new positional values so as to fill in the discontinuity. An increase in cell proliferation has indeed been detected in anterior cells as early as 4 hr after a polarizing region graft (Cooke & Summerbell 1980). It is, however, difficult to explain the digit pattern when two polarizing regions are grafted to the limb bud on the basis of an intercalation model (Wolpert & Hornbruch 1981).

In summary, the most important distinction between a morphogen-based model and the polar coordinate model is that cell fate determination occurs either by long-range signaling encompassing several cell diameters or by short-range cell-cell interactions, respectively. Both models are premolecular and based solely on experimental manipulations. Furthermore, only in the first model does the ZPA have the special role of a signaling center. Once the mechanisms of action of molecules invoked in patterning (e.g. retinoic acid, fibroblast growth factors, *Sonic hedgehog*; see below) are understood, it should be possible to evaluate the merits of each model.

Conservation of Signaling in Vertebrate Limbs and Between Signaling Regions

Molecular mechanisms of epithelial-mesenchymal interactions and polarizing region signaling in the limb are probably common to all vertebrates. For example, when the mesenchymal core of a chicken wing bud is recombined with the ectodermal jacket of a rodent limb bud, outgrowth occurs and a normal wing pattern develops (Jorquera & Pugin 1971). It is also well-documented that cells from the posterior margin of a number of different vertebrates including humans can induce duplicated wing patterns when grafted to the anterior margin of the wing bud (Tickle et al 1976; MacCabe & Parker 1976; Fallon & Crosby 1977). Therefore proximodistal and anteroposterior signaling systems appear to be conserved, and it is the interpretation of the signal(s) that ultimately leads to species differences.

Other regions of the embryo also have polarizing activity and can induce duplicated wing patterns in the chick. Well-documented examples include Hensen's node and primitive streak of chick (Hornbruch & Wolpert 1986; Stoker & Carlson 1990) and mouse (Hogan et al 1992; Izpisúa-Belmonte et al 1992a), notochord, and floor plate (Wagner et al 1990). All these tissues are located in the midline of the embryo, and all have signaling activity in those regions from which they originate. A rather unusual tissue capable of acting as a polarizing region in the limb is the allantois (McLachlan & Pholonker 1988).

Polarizing activity in the flank appears before limb buds emerge but shortly after Hensen's node has lost its signaling activity (Hornbruch & Wolpert 1991). Signaling tissue is seen in early flank in lateral plate mesoderm at the level of the future wing buds. Subsequently, activity spreads out, and shortly before the buds emerge, polarizing activity increases and is found at posterior limb bud margins and the flank between wing and leg buds. It is well established that primitive streak, Hensen's node, notochord, and floor plate share a common origin (reviewed in Stern et al 1992), but whether such a lineage relationship exists between midline structures and cells with polarizing activity in the flank and limbs is not known. Fate mapping studies indicate that cells originating in the primitive streak are later found in the lateral regions of the embryo (Schoenwolf et al 1992), which raises the possibility that most if not all cells with polarizing activity share a common ancestry.

SIGNALING MOLECULES AND MOLECULAR RESPONSES

Fibroblast Growth Factors and Apical Ectodermal Ridge Signaling

A number of genes encoding regulatory molecules are expressed in the apical ectodermal ridge. These include *dlx* (Dollé et al 1992), *engrailed* (Davis et al 1991) (vertebrate homologues of *Drosophila distaless* and *engrailed*, respectively), bone morphogenetic proteins BMP-2 (Lyons et al 1990) and BMP-4

(Jones et al 1991), retinoic acid receptor-β (Schofield et al 1992), fibroblast growth factors FGF-2 and FGF-4 (Niswander & Martin 1992; Suzuki et al 1992), and FGF receptor 1 (Peters et al 1992; Noji et al 1993). At present no conclusive experiments about the function of most of these gene products in the ridge have been reported with the notable exception of the FGFs.

The fibroblast growth factor family comprises at least nine members (Miyamoto et al 1993), and studies in amphibians suggest a role of basic FGF in mesoderm formation (e.g. Amaya et al 1991). It now appears that FGFs have a central role in signaling between the apical ridge and underlying mesenchyme. Although *Fgf-3, Fgf-5,* and *Fgf-6* are apparently not expressed in developing limbs (see Niswander et al 1993), there is evidence that FGF-4 is important. First, transcripts of *Fgf-4* are found in the apical ridge of early mouse limb buds from the apex to almost the posterior boundary of the ridge (Niswander & Martin 1992; Suzuki et al 1992). How *Fgf-4* is expressed in the chick is not known. Second, in short-term organ cultures of mouse limb buds stripped of the ridge, FGF-4 maintains proliferation of cells in the bud that otherwise would have stopped growing (Niswander & Martin 1993a). Third, studies with developing chick embryos show that chick wing buds from which the AER was removed can be rescued by combined application of FGF-4 to apex and posterior limb bud margin (Figure 1d) (Niswander et al 1993). The significance of the requirement for apical and posterior co-application is that posteriorly provided FGF-4 maintains polarizing activity that is abolished if the ridge is removed. In turn, the polarizing activity preserves the progress zone and hence allows proper elaboration of the proximodistal pattern. Direct evidence for the ability of FGF-4 to maintain polarizing activity (Figure 1b, signal 2) comes from transplantation experiments in which posterior cells are taken from a bud in which an FGF-4-releasing bead had been placed posteriorly following ridge removal (Vogel & Tickle 1993). Such posterior cells are as effective in producing duplications as normal polarizing region cells when grafted to a host wing bud. In addition, FGF-4 can maintain polarizing activity in culture (Vogel & Tickle 1993). Evidence for the requirement of polarizing activity to maintain the progress zone (Figure 1b, signal 3) comes from the finding that simultaneous application of FGF-4 and retinoic acid to the apex of the bud also results in a wing with a complete proximodistal pattern. In this experiment, retinoic acid provides polarizing activity.

Although FGF-4 is highly effective in mimicking the chemical signal from the ridge, it fails to perform the mechanical role of the ridge. When the apical ridge is removed and FGF-4 is applied, the elongating bud does not acquire the usual paddle-shaped geometry, but becomes cylindrical and, as the result of this shape change, the digits are bunched together instead of lying side by side in the plane of the handplate (Figure 1d).

Antibodies specifically recognizing FGF-2 label the apical ridge, the dorsal surface ectoderm, and a thin layer of subectodermal mesenchyme in chick limb buds (Savage et al 1993). In contrast to FGF-4, FGF-2 shows no asymmetrical distribution in the apical ridge along the anteroposterior axis. Furthermore, these antibodies detect FGF-2 protein in the cell nuclei, cytoplasm, and also in extracellular space (Savage et al 1993). There are at least eight different transcripts of the *Fgf-2* gene in chicken embryos (Borja et al 1993).

Like FGF-4, FGF-2 can maintain the polarizing activity in culture (Anderson et al 1993). In addition, FGF-2 can mediate outgrowth in vivo after ridge removal (Fallon et al 1994). Nonetheless, FGF-2 also has striking effects when applied to intact limb buds. When limb bud cells are infected with a replication-defective retrovirus expressing either FGF-2 or FGF-2 fused to a immunoglobulin signal peptide, and these cells are implanted at the anterior limb bud margin of a host embryo, additional limb elements are formed (Riley et al 1993). The humerus and radius may be duplicated and extra digits can also form, most frequently a digit *2*. Digit duplications are reminiscent of those obtained by very low doses of retinoic acid (Tickle et al 1985), but FGF-2, unlike retinoic acid, also duplicates the humerus. It is proposed that ectopically expressed FGF-2 stimulates growth in anterior tissue (Riley et al 1993) and, as a result of this, primordia for elements that arise in the anterior wing bud half (humerus, radius and digit *2*) could grow disproportionally and split into two parts. In other words, FGF-2 evokes duplicate structures by stimulating proliferation of skeletal precursor cells and not by respecification of anterior limb bud cells. When FGF-2 is applied on a carrier bead to the anterior margin of the wing bud, only a small knob of extra cartilage forms. Hence an issue that needs further study is whether cells producing FGF-2 following retroviral infection are stimulated to produce other signaling molecules that could, at least in part, be responsible for the observed morphogenetic effects.

Putative Regulatory Genes Expressed in the Progress Zone

Mesenchyme cells in the progress zone (Figure 1b) specifically express several genes encoding regulatory proteins such as *Msx-1* and *Msx-2* (Hill et al 1989; Robert et al 1989), *Evx-1* (Niswander & Martin 1993b), *Wnt-5a* (Gavin et al 1990), and *AP-2* (Mitchell et al 1991). Transcripts of some genes such as *Msx-1* and *Msx-2* are expressed in both ridge and distal mesenchyme. If the products of these genes were concerned with the function of the progress zone, one would expect that gene expression would be regulated by the apical ridge (and FGF-4) and be coordinated with bud outgrowth. In addition, the protein would have the ability to maintain cells in an undifferentiated state. *Msx-1* appears to fulfill most of these criteria and is thus a good candidate for a molecule that maintains the progress zone. *Msx-2* and *Evx-1* show some of the expected properties, but their roles are less clear.

EXPRESSION OF MSX-1 AND MSX-2 The two related homeobox-containing genes, *Msx-1* and *Msx-2* (originally known as *Hox-7* and *Hox-8*), have been studied extensively. They have a divergent homeobox and are vertebrate homologues of the *msh* (muscle segment homeobox) gene of *Drosophila. Msx-1* and *Msx-2* are widely expressed in vertebrate embryos, particularly in regions where epithelial-mesenchymal interactions are known to occur, e.g. tooth, otocyst, and facial primordia (Hill et al 1989; Robert et al 1989). In early limb buds, transcripts of *Msx-1* and *Msx-2* soon become restricted to distal mesenchyme and apical ridge, with *Msx-1* being strongly expressed in mesenchyme and weakly in the ridge, and *Msx-2* expressed weakly in mesenchyme and more strongly in the ridge.

The evidence that signals from the ridge are required for *Msx-1* and *Msx-2* expression in distal mesenchyme of early buds comes from tissue transplantation studies within limb buds (Davidson et al 1991), ridge removal experiments (Robert et al 1991), and analysis of expression patterns in chick limb mutants (Coelho et al 1991). Manipulations reveal that changes in expression of the genes can occur quite rapidly. The rapidity of response is shown, for example, in the appearance of *Msx-1* transcripts at 5 hr in proximal cells placed distally (Davidson et al 1991), and the disappearance of *Msx-1* transcripts that can occur by 3 hr after ridge removal (Ros et al 1992). These rapid changes in gene expression are consistent with the idea that a continuous interplay between ridge and mesenchyme occurs during bud outgrowth. In the normal limb bud, there is a clear gradation in *Msx-1* mRNA abundance with the highest levels immediately adjacent to the ridge, and this pattern is reproduced in grafts of proximal tissue placed beneath the apical ridge (Brown et al 1993). This gradient in expression could reflect the response to a concentration gradient of a signaling substance produced by the ridge.

The effects of expressing *Msx-1* and *Msx-2* on cell differentiation have been tested. When cells of a myogenic cell line are transfected with *Msx-1,* they do not differentiate into muscle, and expression of muscle-specific genes is inhibited (Song et al 1992). This is consistent with the idea that *Msx-1* expression is instrumental in maintaining the undifferentiated state of mesenchyme cells in the progress zone. Expression of *Msx-2* in the cell line had no apparent effect on myogenic cell differentiation.

EVEN-SKIPPED HOMOLOGUES Two related vertebrate genes that are homologous to the pair rule gene *even-skipped* in *Drosophila, Evx-1* and *Evx-2,* lie just 5′ of the *HoxA* and *HoxD* clusters, respectively. In *Drosophila, even-skipped,* a pair rule gene, controls expression of segment polarity genes such as *wingless* and *hedgehog*. Homologues of some of these, including *Wnt* genes and *Sonic hedgehog,* are also expressed in vertebrate limbs (see below), but whether Evx-1 or 2 regulate their expression is not known. The pattern of

Evx-2 expression has not yet been reported, but *Evx-1* transcripts have been detected in mouse limb buds in posterior mesenchyme at the tip of the limb bud directly below the posterior part of the ridge that expresses FGF-4 (Niswander & Martin 1993b). In an organ culture system for mouse limb buds, *Evx-1* transcripts disappear in distal mesenchyme following removal of the apical ridge (Niswander & Martin 1993a). However, addition of FGF-4 to the culture medium maintains *Evx-1* expression in distal cells in the ridgeless buds. Experiments with cycloheximide show that regulation of *Evx-1* expression by FGF-4 requires protein synthesis, and hence the effects of FGF-4 are not direct. Since *Evx-1* expression can be regulated by an apical ridge signal, one possibility is that it is involved in maintenance of the progress zone. Alternatively, *Evx* genes, even though transcribed in the opposite direction of the *Hox* genes nearby, could function as additional gene members of the *HoxA* and *D* complexes (Bastian et al 1992). Activation of *Evx-1* expression appears to occur at the appropriate time (i.e. after all the *HoxA* complex genes have been turned on), but the spatial relationship of the expression domains has not been precisely compared.

Candidate Molecules that Establish Dorsoventral Limb Axis

At present, no molecules are known that affect the dorsoventral limb axis. Ectoderm rotation experiments (see above) suggest that the primary cues establishing this axis derive from the ectoderm covering the bud. Several putative regulatory molecules are expressed in the ectoderm in an asymmetrical manner. FGF-2 (Savage et al 1993) and transcripts of *Wnt-7a* (Parr et al 1993) are dorsally enriched with the *Wnt-7a* expression domain extending right up to the apical ridge (Parr et al 1993; Dealy et al 1993). In contrast, *engrailed-1* (Davis et al 1991) and transcripts of *Wnt-5a* (Gavin et al 1990; Parr et al 1993) are restricted primarily to the ventral ectoderm. In very early limb bud stages in mouse embryos, transcripts of bone morphogenetic protein *Bmp-2* (Lyons et al 1990) are also found in ventral ectoderm. As has been pointed out (Parr et al 1993), the overlap of *engrailed-1* and *Wnt-5a* in ventral ectoderm raises the possibility that these two signaling molecules interact along the lines proposed for *wnt-1* and *engrailed-1* in brain development (McMahon et al 1992). To test such ideas it will be important to misexpress and knock-out dorsal- and ventral-specific genes and determine whether this affects dorsoventral patterning.

The genes that are regulated in the limb by ectodermal signaling molecules are not known. Several regulatory molecules are expressed in subectodermal mesenchyme in a fashion reflecting dorsoventral asymmetry. For example, FGF-2, which is enriched in dorsal ectoderm, is also more abundant in dorsal mesenchyme (Savage et al 1993). Transcripts of the paralogous *Hoxa-13* and *Hoxd-13* genes are initially restricted to the dorsal half of the limb bud (Haack

& Gruss 1993; Francis et al 1994), which raises the possibility that they play a role in interpreting positional differences along this axis. In conclusion, much needs to be learned about the establishment of the dorsoventral limb axis, but several markers are now available to reexamine ectoderm rotations carried out by Saunders & Kieny and their associates (Pautou & Kieny 1973; MacCabe et al 1974).

Hox Genes Expressed in the Limb Bud

The genome of higher vertebrates contains 38 *Hox* genes that encode homeodomain proteins related to the *Drosophila* homeotic selector genes (reviewed in Hunt & Krumlauf 1992; McGinnis & Krumlauf 1992). *Hox* genes are organized into four clusters (*HoxA, B, C,* and *D*) whose organization is closely related to that of the *Drosophila Bithorax* and *Antennapedia* homeotic complexes. Genes at the 3′ end of each cluster are expressed prior to those at the 5′ end (temporal colinearity rule), and more 3′ genes are expressed in domains that extend more anteriorly than the domains of more 5′ genes (spatial colinearity rule). Overexpression and loss-of-function studies begin to reveal that *Hox* genes have important roles in establishment of cell identity along the anteroposterior body axis. Exposing embryos to retinoic acid during late gastrulation, for example, results in a marked rostral shift of the expression boundary of *Hoxb-1* (e.g. Conlon & Rossant 1992; Sundin & Eichele 1992), and this correlates with a specific change of the identity of vertebrae (Kessel & Gruss 1991) or segmental structures (rhombomeres) that make up the neural tube of the hindbrain (Marshall et al 1992). Knocking-out *Hoxa-2*, for example, results in homeotic transformation of structures of the second branchial arch into structures typical of the first arch (Gendron-Maguire et al 1993; Rijli et al 1993). The picture emerging from these studies is that *Hox* genes specify cell identity.

Expression of several *Hox* genes is initiated in the developing limb bud at the time when limb pattern is being specified. Expression patterns are often multifarious, but unlike the hindbrain or the developing vertebral column, the developing limb lacks an overt segmental organization, and it is therefore more difficult to correlate directly *Hox* expression patterns with specific morphological features. The expression of the three contiguous *Hoxa-10, 11,* and *13* genes initiates posterodistally and subsequently expands anteriad so that the final boundaries of *Hoxa-10, 11,* and *13* expression domains are perpendicular to the proximodistal limb axis (Yokouchi et al 1991; Haack & Gruss 1993). The proximal boundaries of expression of *Hoxa-10, 11,* and *13* are staggered so that the *Hoxa-10* boundary is most proximal and that of *Hoxa-13* most distal. Hence the locations of the proximal boundaries of these three genes reflect their relative positions on the chromosome. Haack & Gruss (1993) have hypothesized that the basis of the dynamic changes of *Hoxa* spatial expression domains could reflect a combination of growth and tissue movements. Cells in which expression of *Hoxa-10, 11,* and *13* is initiated are generated in a proliferative region located posterodistally, near the ZPA. These cells are then displaced anteriorly and proximally and eventually define the *Hoxa-10, 11,* and *13* expression domains as we see them in more developed limbs. Whether such cell displacement involves an actual translocation of cells within the limb bud or reflects more of an overall expansion of the limb rudiment that entails passive cell movements remains an open question.

The expression of *Hoxd-9, 10, 11, 12,* and *13* has been studied in mouse and chick (Oliver et al 1989; Dollé et al 1989, Izpisúa-Belmonte et al 1991; Nohno et al 1991; reviewed in Izpisúa-Belmonte & Duboule 1992). In the early chick wing bud, *Hoxd-9* and *10* expression are nearly uniform throughout the mesenchyme, while *Hoxd-11, 12,* and *13* domains are found more posterior. The expression domain of *Hoxd-13* is situated most posterior, while those of *Hoxd-10, 11,* and *12* extend into progressively more anterior limb bud mesenchyme. Thus *Hoxd* gene expression patterns in the limb comply with the spatial colinearity rule. The temporal colinearity rule is also obeyed: *Hoxd* gene transcripts appear sequentially with the more 3′ located *Hoxd-9* gene being transcribed first and *Hoxd-13* (the most 5′ gene) last. As the limb grows out, the expression domains of *Hoxd-9* to *12* appear to be pulled out with the distally extending bud. In contrast, *Hoxd-13* transcripts are confined to the cells at the tip of the bud (Izpisúa-Belmonte et al 1991) as are transcripts of *Hoxd-12* in the chick leg bud, but not in the wing bud (Mackem & Mahon 1991). At the time when the digits begin to differentiate, the domains of *Hoxd-12* and *13* transcripts have extended more anteriorly with the domain of *Hoxd-13* now lying almost at 90° to its original orientation (Yokouchi et al 1991).

Grafts of polarizing tissues (polarizing region, primitive streak, Hensen's node, genital tubercle, neural tube) to the anterior limb bud margin and also local release of retinoic acid result in recapitulation of *Hoxd-10* to *13* gene expression patterns within anterior limb bud mesenchyme (Izpisúa-Belmonte et al 1991, 1992a). Importantly, induction of *Hoxd* gene expression takes place in the same 3′ to 5′ sequence as during normal development with *Hoxd-11* being turned on before *Hoxd-13.* The ZPA in the intact limb bud can be shown to contain factors that efficiently activate *Hoxd* genes when anterior wing bud tissue or even primitive streak tissue is grafted to the posterior wing bud margin and hence brought in contact with the ZPA (Izpisúa-Belmonte et al 1992a). These grafts begin to express *Hoxd* genes, although under normal conditions these tissues would never do so. Furthermore, the kinetics of induction of *Hoxd* in anterior tissue grafted next to the ZPA is considerably more rapid than gene induction in anterior tissue that had received a ZPA graft. *Sonic hedgehog* expressed in the ZPA (Riddle et al 1993) and retinoic acid, which is present in the posterior half of the limb bud at a concentration of close to 100 nM (Thaller & Eichele 1987), may be signals capable of activating *Hoxd* genes in these grafts of anterior tissue and of primitive streak.

There seems to be a correlation between the strength of a polarizing signal, as determined by the extent of digit induction, and the extent of activation of the *HoxD* complex (Izpisúa-Belmonte et al 1992a). For instance, neural tube grafts induce patterns with only an additional digit *2,* and they may induce *Hoxd-10, 11,* and *12,* but not *Hoxd-13.* In contrast, polarizing region cells turn on all these genes and also result in a full duplication with additional digits *2, 3,* and *4.* Since ZPA expresses *Hoxd-10, 11, 12,* and *13,* it is possible, by virtue of expressing these genes, that the ZPA cells are capable of acting as inducers. That this is probably not the case is illustrated by the fact that primitive streak tissue does not express these genes, yet is an effective inducer of digits. Accordingly, *Hoxd* genes, and possibly all *Hox* complex genes, appear to be involved in the interpretation of positional signals, but not in their production.

Signals from the AER are required for proper expression of *Hoxd* genes. If a retinoic acid-releasing bead is implanted at the anterior margin of a chick wing bud, *Hoxd-10* to *13* genes are sequentially activated in a 3′ to 5′ direction. Activation through retinoic acid (and also through a ZPA graft) begins in a confined region distal to the bead underneath the AER (Izpisúa-Belmonte et al 1992b). If the AER is removed after the retinoic acid bead is implanted, *Hoxd* genes are not induced. The continuous presence of the AER is required for retinoic acid to induce *Hoxd* genes, and its presence only during the priming phase (see Retinoids below) is not sufficient to turn on these genes. The AER is also required for activation of *Hoxd* genes in normal limb development. Removal of the ridge from early limb buds greatly reduces or even abolishes activation of *Hoxd* gene expression. Once these domains are established, AER removal results in a decrease in levels of expression of *Hoxd* genes, and the normally occurring expansion of the expression domains is abolished. These findings can be understood in terms of the AER release of a signaling molecule that directly cooperates with retinoic acid in inducing *Hoxd* gene transcription. Alternatively, the ridge could play a more passive role and maintain cells in the progress zone in a proliferative state that permits *Hox* gene expression. AER signals could be fibroblast growth factor(s) that are capable of substituting for the AER (see Fibroblast Growth Factors and Apical Ectodermal Ridge Signaling).

Undoubtedly, *Hoxa* and *d* gene expression patterns serve as useful molecular markers that reveal developmental processes that are otherwise difficult to detect. However, a more provocative hypothesis emerging from a substantial body of data partly reviewed above is that *Hox* genes encode positional information. Yokouchi et al (1991), for example, have noticed that when the expression domains of *Hoxa* and *Hoxd* genes are superimposed, the cartilage elements in the chick limb are prefigured. These authors postulate that expression of *Hoxa* genes is principally responsible for subdividing the limb along the proximodistal axis, whereas *Hoxd* gene expression exerts influence on the branching of the cartilage primordia that eventually composes the pattern along the anteroposterior axis. If such a combinatorial model is correct, one would expect that gain-of-function and loss-of-function mutations would result in distinct changes in limb pattern. In a normal chick leg bud, the presumptive regions for digits I and II express *Hoxd-9* and *d-10,* and *Hoxd-9, 10,* and *11,* respectively (note that digits in hindlimbs are denoted by roman numerals). When *Hoxd-11* is expressed throughout the leg bud by the use of a replication-competent retroviral vector, presumptive leg digit I region now expresses *Hoxd-9, 10,* and *11,* which is the *Hoxd* combination typical for digit II (Morgan et al 1992). Indeed, in 30% of the *Hoxd-11*-infected leg buds, digit I is "homeotically" transformed into a digit II. That an anterior extension of the *Hoxd-11* expression domain posteriorizes anterior cells is consistent with the rule that gain-of-function results in posterior transformation. This finding also bolsters the notion that a particular combination of *Hoxd* genes could encode positional identity. However, alteration of *Hoxd-11* expression in wing buds leads to the development of an additional digit *2* rather than a change in digit identity.

Of the *Hox* genes expressed in the limb, so far only *Hoxd-13* has been disrupted (Dollé et al 1993). Unexpectedly, the phenotypic changes in the limbs of *Hoxd13*$^{-/-}$ mice appear to involve growth retardation of distal limb elements. For example, by day 13.25 post coitum, anlagen for metacarpals and phalanges of certain digits were poorly separated in the mutant limbs, while in control limbs this was not the case. In mutant limbs, fingers and toes were narrower and shorter due to loss of phalangeal anlagen. Later in development, ossification of several metacarpals and phalanges of *Hoxd-13*$^{-/-}$ limbs were considerably delayed. In addition to developmental retardations, *Hoxd-13*$^{-/-}$ mice frequently displayed an additional rudimentary digit posterior to digit V, the most posterior element in normal mouse limbs. Several conclusions are drawn from these findings. (*a*) Although the overall pattern of digit condensation is similar in wild-type and mutant mouse limbs, the rate at which the pattern is formed is different. (*b*) *Hoxd-13*$^{-/-}$ limbs are altered along proximodistal and anteroposterior axes (lack of certain phalanges and an additional posterior digit), which suggests that expression of this gene is important for patterning along both axes. (*c*) The phenotypic changes are difficult to interpret within the framework of an anteroposterior morphogen gradient specifying digit identity because retardation is seen with some but not all digits. Instead of *Hoxd-13* encoding positional information, it is proposed that *Hoxd* genes control limb patterning by regulating the timing and extent of local growth rates (Dollé et al 1993). It remains to be seen whether the idea of *Hoxd* genes acting as morphogenetic clocks rather than encoding cell position can be validated by other *Hoxd* knock-outs. It is also possible that *Hoxd-13*$^{-/-}$ mice might not fully reveal the function of the *Hoxd-13* gene. The expression domains of *Hoxd-13* and of the paralogous *Hoxa-13* gene are overlapping (Yokouchi et al 1991; Haack & Gruss 1993), and *Hoxa-13* could perhaps partly replace *Hoxd-13*.

Signaling Molecules Operating Along the Anteroposterior Limb Axis

RETINOIDS Retinoic acid is the first chemically defined, naturally occurring molecule to be identified that profoundly affects the pattern not only of the developing (Figure 1c) and regenerating limb (reviewed in Brockes 1991), but also of the anteroposterior body axis (reviewed in Linney 1992; Hofmann & Eichele 1994). Retinoids are small hydrophobic molecules that exert their action by binding and activating nuclear receptors. These receptors are ligand-dependent transcription factors that control the expression of target genes (e.g. Evans 1988). Two families of retinoid receptors have been identified, the retinoic acid receptors (RARα, β, and γ), which bind either all-*trans*-retinoic acid (referred to simply as retinoic acid in the subsequent text) or 9-*cis*-retinoic acid, and the retinoid-X-receptors (RXRα, β, and γ) whose ligand is 9-*cis*-retinoic acid (reviewed in Mangelsdorf et al 1994). Studies in chick and mouse limb have shown that retinoic acid is present endogenously in limb buds and that it is enriched in the posterior region that contains the zone of polarizing activity (Thaller & Eichele 1987; Satre & Kochhar 1989). The concentration of retinoic acid and of 3,4-didehydro-retinoic acid (another endogenous RAR ligand) is considerably higher in undifferentiated limb buds (Thaller & Eichele 1990) than in buds undergoing terminal differentiation, and the decrease in retinoid concentration slightly precedes that of the loss of polarizing activity (Honig & Summerbell 1985). Whether 9-*cis*-retinoic acid is endogenously present in the limb bud is not yet known. Limb buds are capable of converting the precursors retinol and retinal into retinoic acid (Thaller & Eichele 1988), and they can also degrade this compound to downstream metabolites. Although these findings show that retinoic acid metabolism occurs locally in the limb bud, it is not known at present whether the ZPA is a highpoint of retinoic acid production. However, Hensen's node, a potent polarizing tissue (Hornbruch & Wolpert 1986; Hogan et al 1992), can act as source of retinoic acid (Hogan et al 1992; Chen et al 1992).

Retinoid receptor expression patterns have been studied in considerable detail by in situ hybridization in mouse embryos (Ruberte et al 1992 and references therein; reviewed in Mendelsohn et al 1992) and to a lesser extent in the chick. In the mouse limb bud, *RARα* and *RARγ* are expressed throughout the mesenchyme but *RARβ* is restricted to proximal regions. For the chick limb bud, only *RARβ* expression has been reported, and in this case, two transcripts representing splice variants are differentially expressed (Smith & Eichele 1991; Schofield et al 1992). Less information is available about the expression of RXRs, but studies in the mouse show that RXRα transcripts are present in almost all tissues (Mangelsdorf et al 1992; Dollé et al 1994). *RXRγ* is expressed in the chick limb albeit at low levels (Rowe et al 1991; Thaller et al 1993; Hofmann et al 1994). How the receptor proteins are expressed in mouse or chick is not known, but Hill et al (1993) discovered that in newt regeneration blastemata, RARδ$_1$ receptor protein (a mammalian RARγ1 homologue) is expressed in only a fraction of cells, which raises the possibility that in developing limbs, receptor protein expression may also be more restricted than the in situ hybridization data would suggest.

Cellular retinol binding proteins (CRBP I and II) and cellular retinoic acid binding proteins (CRABP I and II) are also expressed in limb buds (see Mendelsohn et al 1992 for a review). CRABPs may regulate the concentration of free retinoic acid capable of binding to the RARs, but studies in cultured cells also demonstrate that CRABPs accelerate degradation of retinoic acid (Boylan & Gudas 1992). Although these binding proteins are strikingly enriched in the limb bud, their role remains unclear.

The presence of various components of the retinoid signaling pathways is consistent with a role of retinoids in limb development. Clues to the possible nature of this role came from experiments in which locally applied retinoic acid induced duplicate wing patterns virtually indistinguishable from those obtained by polarizing region grafts (Figure 1c; Tickle et al 1982, 1985; Summerbell 1983). A particularly striking demonstration of the morphogenetic effect of exogenously applied retinoic acid is produced when a retinoic acid-impregnated bead is implanted into a wing bud whose posterior half, including the ZPA, has been surgically removed (Eichele 1989; Tamura et al 1990). Such truncated wing buds would give rise to a wing with just a humerus and a radius, but when a retinoic acid-releasing bead (or ZPA; Saunders & Gasseling 1968) is implanted at the anterior margin, a complete set of wing structures will develop. The polarity of the digit pattern in such rescued wings is reversed. Instead of the sequence *234* (Figure 1a), a *432* pattern is generated with digit *4* at the anterior wing margin. Hence, retinoic acid alone, in the absence of a native polarizing region, can determine the polarity of the resulting pattern.

Both ZPA grafting and retinoic acid application display dose-dependence. Grafting progressively more ZPA cells or releasing increasing amounts of retinoic acid will evoke digits in a 2 → 3 → 4 sequence (Tickle 1981; Tickle et al 1985). An important point is that the dose of retinoic acid required to induce a full set of digits yields a concentration of (applied) retinoic acid in the limb bud tissue that is in the same range as the endogenous retinoic acid concentration (Thaller & Eichele 1990; Thaller et al 1993). This is consistent with the idea that the induction of digits by retinoic acid reflects a physiological process. Furthermore, the dose-differential between minimal (digit *2*-induced) and maximal response (all three digits induced) is merely a few-fold, which suggests that limb bud cells can discern and respond to small concentration differences of the kind provided by the shallow gradient of endogenous retinoic acid.

The existence of a gradient of endogenous or applied retinoic acid in combination with dose-response data is consistent with a mechanism in which a retinoic acid concentration gradient provides positional information. However, it is possible that it is not the gradient but the locally elevated concentration of retinoic acid at the bead that is the critical factor determining the response. Such a local high-point could, for example, induce a new ZPA, and increased doses of retinoic acid could give rise to an increasing number of polarizing region cells. Experiments in which the effects of retinoic acid and a synthetic retinoic acid analogue on digit pattern were compared suggested that these retinoids do not just act locally (Eichele et al 1985). However, these studies do not rule out the possibility of local and long-distance effects. Indeed, when tissue from a region immediately next to the retinoic acid-releasing bead is transplanted to another wing bud, it will induce digits in the host (Wanek et al 1991; Noji et al 1991; Tamura et al 1993). For grafts to become effective, an exposure time of ≤ 15 hr is required (Wanek et al 1991), which suggests that within that time span retinoic acid converts tissue next to the bead into a polarizing region. The new ZPA would then produce the morphogen proper that would be responsible for digit induction. However, the time course of digit induction is surprisingly similar between polarizing region grafting and retinoic acid application. A minimal contact time between ZPA graft and host tissue of approximately 15 hr is required to induce a digit *2* (Smith 1980). Interaction extending up to 17–24 hr will evoke additional digits *2* and *3*. Likewise, to induce digits, retinoic acid needs to be provided for at least 10 hr (Eichele et al 1985). Exposure to retinoic acid beyond the priming period yields wing patterns with digits of progressively more posterior character (2 → 3 → 4). A maximal response is reached after a total treatment time of 16–18 hr, thereafter the retinoic acid-releasing bead can be removed.

If retinoic acid and polarizing region grafts act via the same pathway, then genes normally expressed in the posterior half of the limb bud will display the same kinetics of induction when they are ectopically induced in anterior limb bud tissue. Since the induction of a polarizing region by retinoic acid takes ≥ 15 hr (Wanek et al 1991), ZPA grafts should induce posterior-specific genes significantly more rapidly than retinoic acid if retinoic acid first induced ZPA. With interspecies grafts between mouse and chick, expression of *Hoxd-*

10, 11, 12, and *13* genes in responding mesenchyme can be monitored precisely. Data for *Hoxd-11* can readily be compared, and the gene is particularly significant in this respect since ectopic expression of *Hoxd-11* results in changes in digit identity, which implicates it in the specification of cell position along the anteroposterior limb axis (Morgan et al 1992). *Hoxd-11* transcripts are detected in anterior cells after 20 hr with retinoic acid (Izpisúa-Belmonte et al 1991). A chick ZPA appears to induce *Hoxd-11* within 16 hr, but a mouse polarizing region takes more than 24 hr (Izpisúa-Belmonte et al 1992a). Thus the time of activation is, on the whole, similar for ZPA and retinoic acid, and the reasons for the slightly different response times to tissue and retinoic acid are not known. Because chick ZPA and retinoic acid induce *Hoxd-11* within ~16 and ~20 hr, respectively, yet retinoic acid evokes a ZPA in about 15 hr, there appears to be a problem with fitting a two-step process (retinoic acid signal followed by a second polarizing signal) into the available time. Collectively these data argue more in favor of the idea that retinoic acid acts directly on these genes and not by first making a ZPA (see Hofmann & Eichele 1994 for further discussion).

The chief observations supporting a role of retinoic acid in normal development are that it occurs in the limb bud (Thaller & Eichele 1987, 1990); it is enriched in the ZPA; the concentration of retinoic acid required for digit induction is in the same range as the endogenous retinoic acid concentration (Thaller & Eichele 1987, 1990; Thaller et al 1993); and limb buds express all the presently known macromolecular constituents of the retinoid signaling system [receptors, binding proteins, and retinoid metabolism enzymes (Thaller & Eichele 1988)] at the time when the pattern of digits is specified. Several posterior-specific genes such as *Hoxd-10, 11, 12,* and *13* are activated following retinoic acid treatment, which is also consistent with a physiological role of this molecule. However, *RAR*β expression is induced in anterior limb bud mesenchyme by retinoic acid but not by polarizing region grafts (Noji et al 1991). The *RAR*β gene contains a retinoic acid response element (RARE; de Thé et al 1990; Sucov et al 1990), and it has been argued that failure of induction of expression with polarizing grafts and the absence of *RAR*β transcripts at the tip of the limb bud are the result of endogenous retinoic acid being bound to CRABP and thus incapable of having any signaling function in the limb (Tabin 1991; Bryant & Gardiner 1992). The validity of this argument hinges on whether expression of the *RAR*β gene is an appropriate indicator for detecting retinoic acid in the limb bud. Other genes with RAREs such as *Hoxd-10* (Moroni et al 1993) and CRABP II (Durand et al 1992; Ruberte et al 1992) are expressed in the limb bud (see also Mendelsohn et al 1992, for a discussion of this issue).

To date much of the research on retinoids has been based on all-*trans*-retinoic acid and RARs, but this isomer binds only to RARs, not to RXRs.

Importantly, RARs by themselves do not bind to RAREs, but require RXRs as heterodimeric partners (reviewed in Mangelsdorf et al 1994). Thaller et al (1993) reported that 9-*cis*-retinoic acid (the isomer that binds to RXRs and RARs) is 20–30 times more potent in inducing digits than the all-*trans* isomer. If the 9-*cis* isomer were operating solely via RARs, this difference in potency between the two isomers would not be expected because both have similar affinities for RARs (Allenby et al 1993). Therefore, it appears as though RXRs, possibly in the form of RXR-RAR heterodimers, are involved in mediating the retinoid signal. An unresolved issue in invoking RXRs is that TTNPB, a synthetic retinoic acid analogue that cannot convert into a 9-*cis*-retinoic acid-like conformation and is incapable of binding to RXRs, also induces duplications (Eichele et al 1985). A possible explanation is that TTNPB binds to the RAR subunit in the RXR-RAR heterodimer and that the limb bud supplies endogenous 9-*cis*-RA for binding to the RXR subunit.

The role of retinoids in limb development is far from being fully understood. There is agreement that exogenously applied retinoic acid induces a signal cascade, but whether this cascade reproduces the one that controls antero-posterior patterning in normal development is controversial (see Tabin 1991; Bryant & Gardiner 1992). Much headway has been made in the past five years in understanding the astonishingly complex mechanism of action of retinoids in eukaryotic cells (reviewed in Mangelsdorf et al 1994). It is now important to assimilate appropriately these advances in the context of the limb bud system.

SONIC HEDGEHOG Riddle et al (1993) isolated *Sonic hedgehog* (*Shh*), a vertebrate homologue of the *Drosophila* segment polarity gene *hedgehog*. *Shh* encodes a protein consisting of 425 amino acids, with a hydrophobic leader sequence, which suggests that it is secreted. *Shh* transcripts colocalize with the ZPA in the chick wing bud, and *Shh* is also expressed in posterior mesenchyme of mouse limb buds and zebrafish pectoral fin buds (Echelard et al 1993; Krauss et al 1993). In the chick wing bud, the domain of *Shh* expression gradually displaces distally, reminiscent of the distal displacement of the polarizing region (Honig & Summerbell 1985). *Shh* is strongly expressed in several embryonic signaling tissues capable of inducing pattern duplications (see Conservation of Signaling in Vertebrate Limbs and Between Signaling Regions), but polarizing cells along the flank (Hornbruch & Wolpert 1991) do not express this gene. When cells transfected with a replication-competent retroviral vector containing the *Shh* coding region are implanted at the anterior wing bud margin, mirror-image duplications of the wing patterns can develop that are, on the whole, similar to those obtained by ZPA grafting or retinoic acid treatment (Figure 1c). Likewise, *Shh* also induces *Hoxd* genes with *Hoxd-11* and *Hoxd-13* being strongly expressed 24 and 36 hr after infecting the bud.

Local application of retinoic acid at the anterior wing bud activates *Shh* expression in mesenchyme of the treated bud. Twenty four hr after retinoic acid exposure was initiated, *Shh* mRNA was detected distal of the bead in a zone abutting the apical ridge, a region in which retinoic acid also induces *Bmp-2* (Francis et al 1994) and *Hoxd* genes (Izpisúa-Belmonte et al 1991, 1992a).

Based on these findings, *Sonic hedgehog* appears to pattern the anterioposterior limb axis (Riddle et al 1993). In the mechanism illustrated in Figure 1b, *Shh* could be the signal that maintains the progress zone and hence the ridge, and simultaneously could act as a positional signal along the anteroposterior axis. At the cellular level, several mechanisms of action can be envisaged (Riddle et al 1993). For example, *Shh* protein might be a diffusible morphogen forming a concentration gradient. Another possibility is that *Shh* only affects cells immediately bordering the ZPA. These cells would in turn produce a second signal affecting their immediate neighbors etc. The resulting cascade of signaling molecules could propagate in the mesenchyme cells or spread through the apical ridge (Riddle et al 1993). The second possibility seems unlikely because normal development occurs when the ridge is cut off and replaced by FGF-4-releasing beads (Figure 1c; Niswander et al 1993). In *Drosophila,* hedgehog protein appears to be made as a precursor protein whose aminoterminal part contains a membrane-spanning region and an extracellular domain (Tabata & Kornberg 1994 and references therein). The extracellular part may be cleaved off and released. In the *Drosophila* wing disc, for example, the hh protein is found in the posterior compartment and also detected in the anterior compartment in a stripe abutting the compartment boundary (Tabata & Kornberg 1994). In the eye disk, *hh* is expressed posterior to the morphogenetic furrow, but its range of action extends into the furrow where it activates *decapentaplegic* (*dpp*), which again indicates that hh protein diffuses away from the site of its synthesis (e.g. Heberlein et al 1993; Ma et al 1993). Compared with the dimensions of the chick limb bud, such distances are relatively small, and it will be critical to localize *Shh* in the limb bud using antibodies. *dpp* and *Bmp-2* are both members of the TGF-β family. It is intriguing that in early limb buds, transcripts of *Shh* and *Bmp-2* are found in nearly identical domains along the posterior limb bud margin (Riddle et al 1993; Francis et al 1994; see below). Thus it is possible that *Shh* controls the expression of *Bmp-2* in the vertebrate limb because in *Drosophila* there is an interaction between *hh* and *dpp.*

Riddle et al (1993) found that retinoic acid induces *Shh,* which then activates *Hoxd-11* expression as does retinoic acid. Based on this finding, it is proposed that retinoic acid first induces cells to express *Shh* and in turn *Shh* activates *Hoxd* genes. However, both retinoic acid and *Shh* take about the same length of time to induce *Hoxd-11* (20 compared with 24 hr). This, together with the finding that retinoic acid takes 24 hr to induce *Shh,* argues against the sequential operation of retinoic acid and *Shh* in activating *Hoxd* genes and suggests that different pathways may be taken.

BONE MORPHOGENETIC PROTEINS Bone morphogenetic proteins (BMPs) have well-recognized properties in inducing bone differentiation, but may also play an important role in early limb patterning (Lyons et al 1990; Jones et al 1991; Francis et al 1994). BMPs are members of the TGF-β superfamily and were isolated from extracts of bone that had the ability to cause ectopic bone formation (Rosen & Thies 1992; Wozney et al 1993). The BMP family may comprise at least eight members. BMP-2 to BMP-7 are proteins that stimulate bone differentiation and are produced in a latent form. *Bmp-2* and *Bmp-4* are closely related and are vertebrate homologues of the *Drosophila* gene *dpp*. Members of the TGF-β family have been shown to form homodimers and heterodimers (Sampath 1990).

Bmp-2 and *Bmp-4* transcripts are found in very early limb buds. In chick limb buds, *Bmp-2* transcripts are localized to posterior mesenchyme where the polarizing region is located. *Bmp-4* transcripts are also present in posterior mesenchyme, but in addition transcripts are found in the distal-most and anterior mesenchyme. *Bmp-2* and *Bmp-4* could act as heterodimers in posterior mesenchyme. Other *Bmp* genes may also be expressed in the mesenchyme of developing limbs (Wozney et al 1993).

Retinoic acid application and polarizing region grafting lead to an ectopic domain of *Bmp-2* transcripts in anterior mesenchyme, but when beads soaked in BMP-2 are implanted anteriorly, additional digits do not form (Francis et al 1994). Together these findings suggest that BMP-2 is not part of the signaling system that respecifies anterior cells to form posterior structures, but instead is expressed in response to the polarizing region signal. Consistent with this idea is the close correspondence, spatially and temporally, between *Bmp-2* transcripts and *Hoxd-13* transcripts in normal and manipulated limb buds. The role of BMP-2 may also involve the mediation of the interactions between mesenchyme and epithelium that are necessary to maintain the ridge, and thus BMP-2 would link pattern specification with morphogenesis.

Bmp-2 and *Bmp-4* transcripts are also present in the apical ridge (Lyons et al 1990; Jones et al 1991; Francis et al 1994). The expression in both epithelial and mesenchymal tissues near the tip of the limb is reminiscent of expression patterns in developing teeth. In teeth, BMP-4 can induce *Msx-1* expression (Vainio et al 1993), but it is not known whether the same pathway operates in the limb. Later in limb development, *Bmp-2* and *Bmp-4* transcripts are found in conjunction with developing cartilage. Therefore, like the related *dpp* gene in *Drosophila,* which has been shown to act at several different stages in development (see for example, Ferguson & Anderson 1992; Heberlein et al 1993), the vertebrate homologues may be necessary for initial patterning and also for later skeletal morphogenesis in the limbs and elsewhere. The mouse mutant "short ear," a mutation in the *Bmp-5* gene, displays localized cartilage defects implicating this gene and possibly other family members (Storm et al 1994) in skeletal development (reviewed in Kingsley 1994).

CONCLUSIONS

The developing limb is one of the foremost systems in which to study cell and tissue patterning in vertebrates. The study of the developing limb has also been influential in understanding innervation mechanisms (Lance-Jones 1988) and terminal differentiation of connective tissues and muscles (Emerson 1993). Additional problems that are illustrated in the developing limb include angiogenesis and programmed cell death (Abbadie et al 1993). Here we have focused on how the limb pattern is laid down in the embryo. The framework for analysis of developmental mechanisms involved is provided by extensive embryological data that addresses interactions in the limb bud. We have reviewed recent work that has begun to unravel the molecular basis of these interactions. Progress has been made in identifying some of the signals and early response genes that turn out to be regulatory genes themselves. It will be important to understand how these initial steps are interpreted, i.e. how cells respond to the regulatory signals. In other words, the challenge is to figure out the mechanisms by which, for example, the expression pattern of *Hox* genes translates into a particular morphology. Even in well-studied systems such as *Drosophila,* this has turned out to be a formidable problem (Botas 1993). Although the developing limb poses important questions in its own right, an understanding of the processes can be extended to other less amenable regions of the embryo. It is striking that the same or similar collections of molecules (growth factors and retinoids and their receptors, proteins encoded by *Hox* genes, Sonic hedgehog) seem to recur throughout vertebrate development, first in establishment of the body plan, and then in patterning of different regions such as the hindbrain, face, etc, in addition to limbs. Finally, an aim in studying vertebrate limb development is to shed light on the basis of congenital malformations. Recent work has identified a few genes that underlie specific limb abnormalities in humans or in mice (e.g. Schimmang et al 1993; Trumpp et al 1992; Jackson-Grusby et al 1992). It is particularly encouraging that clinical genetics and embryology are beginning to meet.

Literature Cited

Abbadie C, Kaburn N, Bouali F, Smardova J, Sthelin D, Vandenbunder B, Enrietto PJ. 1993. High levels of c-rel expression are associated with programed cell death in the developing avian embryo and in bone marrow cells in vitro. *Cell* 75:889–912

Allenby G, Bocquel M-T, Saunders M, Kazmer S, Speck J, Rosenberger M, et al. 1993. Retinoic acid receptors and retinoid X receptors: Interaction with endogenous retinoic acids. *Proc. Natl. Acad. Sci. USA* 90:30–34

Amaya Y, Musci, TJ, Kirschner, MW. 1991. Expression of a dominant negative mutant of the FGF receptor disrupts mesoderm formation in Xenopus embryos. *Cell* 66:257–70

Anderson R, Landry M, Muneoka K. 1993. Maintenance of ZPA signaling in cultured mouse limb bud cells. *Development* 117: 1421–33

Bastian H, Gruss P, Duboule D, Izpisúa-Belmonte J-C. 1992. The murine even-skipped-like gene Evx-2 is closely linked to the Hox-4 complex, but is transcribed in the opposite direction. *Mammal. Genet.* 3:241–43

Borja AZ, Meijers C, Zeller R. 1993. Expression of alternatively spliced bFGF first coding exons and antisense mRNAs during chick embryogenesis. *Dev. Biol.* 157:110–18

Botas J. 1993. Control of morphogenesis and differentiation by HOM/HOX genes. *Curr. Opin. Cell Biol.* 5:1015–22.

Bowen J, Hinchliffe JR, Horder T, Reeve AMF. 1989. The fate map of the chick forelimb-bud and its bearing on hypothesized developmental control mechanism. *Anat. Embryol.* 179: 269–83

Boylan JF, Gudas LG. 1992. The level of CRABP-I expression influences the amounts and types of all-*trans*-retinoic acid metabolites in F9 teratocarcinoma stem cells. *J. Biol. Chem.* 267:21486–91

Brockes JP. 1991. Some current problems in amphibian limb regeneration. *Philos. Trans. R. Soc. London Ser. B* 331:287–90

Brown JM, Wedden SE, Millburn GH, Robson LG, Hill RE, et al. 1993. Experimental analysis of the control of expression of the homeobox gene *Msx-1* in the developing limb and face. *Development* 119:41–48

Bryant SV, Gardiner DM. 1992. Retinoic acid, local cell-cell interactions, and pattern formation in vertebrate limbs. *Dev. Biol.* 152:1–25

Chaube S. 1959. On axiation and symmetry in the transplanted wing of the chick. *J. Exp. Zool.* 140:29–77

Chen Y, Huang L, Russo A F, Solursh M. 1992. Retinoic acid is enriched in Hensen's node and is developmentally regulated in the early chicken embryo. *Proc. Natl. Acad. Sci. USA* 89:10056–59

Coelho CND, Krabbenhoft KM, Upholt WB, Fallon JF, Kosher RA. 1991. Altered expression of the chicken homeobox-containing genes *GHox-7* and *GHox-8* in the limbbuds of limbless mutant embryos. *Development* 113:1487–93

Conlon RA, Rossant J. 1992. Exogenous retinoic acid rapidly induces anterior ectopic expression of murine *Hox-2* genes in vivo. *Development* 116:357–68

Cooke J, Summerbell D. 1980. Cell cycle and experimental pattern duplication in the chick wing during development. *Nature* 287:697–701

Davidson DR, Crawley A, Hill RE, Tickle C. 1991. Position-dependent expression of two related homeobox genes in developing vertebrate limbs. *Nature* 352:429–31

Davis CA, Holmyard, DP, Millen, KJ, Joyner, AL. 1991. Examining pattern formation in mouse, chicken and frog embryos with an En-specific antiserum. *Development* 111: 287–98

Dealy CN, Roth A, Ferrari D, Brown AMC, Kosher RA. 1993. *Wnt-5a* and *Wnt-7a* are expressed in the developing chick limb bud in a manner suggesting roles in pattern formation along the proximodistal and dorsoventral axes. *Mech. Dev.* 43:175–86

de Thé H, Vivanco-Ruiz MDM, Tiollais P, Stunnenberg H, Dejean A. 1990. Identification of a retinoic acid responsive element in the retinoic acid β receptor gene. *Nature* 343: 177–80

Dollé P, Dierich A, LeMeur M, Schimmang T, Schuhbaur B, et al. 1993. Disruption of the *Hoxd-13* gene induces localized heterochony leading to mice with neotenic limbs. *Cell* 75:431–41

Dollé P, Fraulob V, Kastner P, Chambon P. 1994. Developmental expression of murine retinoid X receptor (RXR) genes. *Mech. Dev.* 45:91–104

Dollé P, Izpisúa-Belmonte J-C, Falkenstein H, Renucci A, Duboule D. 1989. Coordinate expression of the murine HOX-5 complex homeobox-containing genes during limb pattern formation. *Nature* 342:767–72

Dollé P, Price M, Duboule D. 1992. Expression of the mouse Dlx-1 homeobox gene during facial, ocular and limb development. *Differentiation* 49:93–99

Durand B, Saunders M, Leroy P, Leid M, Chambon P. 1992. All-*trans* and 9-*cis*-retinoic acid induction of CRAPB II transcription is mediated by RAR-RXR heterodimers bound to DR1 and DR2 repeated motifs. *Cell* 71:73–85

Echelard Y, Epstein DJ, St-Jacques B, Shen L, Mohler J, et al. 1993. Sonic hedgehog, a member of a family of putative signaling molecules, is implicated in the regulation of CNS polarity. *Cell* 75:1417–30

Eichele G. 1989. Retinoic acid induces a pattern of digits in anterior half wing buds that lack the zone of polarizing activity. *Development* 107:863–68

Eichele G, Tickle C, Alberts B. 1985. Studies on the mechanism of retinoid-induced pattern duplications in the early chick limb bud: temporal and spatial aspects. *J. Cell Biol.* 101:1913–20

Emerson CP. 1993. Skeletal myogenesis: genetics and embryology to the fore. *Curr. Opin. Genet. Dev.* 3:265–74

Evans RM. 1988. The steroid and thyroid hormone receptor superfamily. *Science* 240: 889–95

Fallon JF, Crosby GM. 1977. Polarizing zone activity in limb buds of amniotes. In *Vertebrate Limb and Somite Morphogenesis,* ed. DA Ede, JR Hinchliffe, M Balls, pp 55–71. Cambridge: Cambridge Univ. Press

Fallon JF, Kelley RO. 1977. Ultrastructural analysis of the apical ectodermal ridge during vertebrate limb morphogenesis. II Gap junctions as distinctive ridge structures common to birds and mammals. *J. Embryol. Exp. Morph.* 41:223–32

Fallon JF, Lopez A, Ros MA, Savage MP, Olwin BB, Simandl K. 1994. FGF-2: Apical ectodermal ridge growth signal for chick limb development. *Science* 264:104–6

Ferguson EL, Anderson KV. 1992. Decapentaplegic acts as a morphogen to organize dorsal-ventral pattern in the Drosophila embryo. *Cell* 71:451–61

Francis PH, Richardson MK, Brickell PM, Tickle C. 1994. Bone morphogenetic proteins and a signalling pathway that controls patterning in the developing chick limb. *Development* 120:209–18

Gavin BJ, McMahon JA, McMahon AP. 1990. Expression of multiple novel *Wnt-1/int-1*-related genes during fetal and adult mouse development. *Genes Dev.* 4:2319–32

Gendron-Maguire M, Mallo M, Zhang M, Gridley T. 1993. *Hoxa-2* mutant mice exhibit homeotic transformation of skeletal elements derived from cranial neural crest. *Cell* 75: 1317–31

Haack H, Gruss P. 1993. The establishment of murine Hox-1 expression domains during patterning of the limb. *Dev. Biol.* 157:410–22

Heberlein U, Wolff T, Rubin GM. 1993. The TGF-β homolog *ddp* and the segment polarity gene *hedgehog* are required for propagation of a morphogenetic wave in the Drosophila retina. *Cell* 75:913–26

Hill DS, Ragsdale CW Jr, Brockes JP. 1993. Isoform-specific immunological detection of newt retinoic acid receptor delta 1 in normal and regenerating limbs. *Development* 117: 937–45

Hill RE, Jones PF, Rees AR, Sime CM, Justice MJ, et al. 1989. A new family of mouse homeobox-containing genes: molecular structure, chromosomal location and developmental expression of *Hox 7.1*. *Gen. Dev.* 3:26–37

Hinchliffe JR, Sansom A. 1985. The distribution of the polarizing zone (ZPA) in the legbud of the chick embryo. *J. Embryol. Exp. Morph.* 86:169–75

Hofmann C, Eichele G. 1994. Retinoids in development. In *The Retinoids, Biology, Chemistry, and Medicine,* ed. MB Sporn, AB Roberts, DS Goodman, pp. 387–441. New York: Raven

Hofmann C, Thaller C, Melia T, Eichele G. 1994. Retinoids and pattern formation. In *Retinoids from Basic Science to Clinical Application,* ed. MA Livrea, L Packer. In press

Hogan BLM, Thaller, C, Eichele, G. 1992. Evidence that Hensen's node is a site of retinoic acid synthesis. *Nature* 359:237–41

Honig L. 1981. Positional signal transmission in the developing chick limb. *Nature* 291:72–73

Honig LS, Summerbell D. 1985. Maps of strength of positional signalling activity in the developing chick wing. *J. Embryol. Exp. Morph.* 87:163–74

Hornbruch A, Wolpert L. 1970. Cell division in the early growth and morphogenesis of the chick limb bud. *Nature* 226:764–66

Hornbruch A, Wolpert L. 1986. Positional signaling by Hensen's node, when grafted to the chick limb bud. *J. Embryol. Exp. Morph.* 94:257–65

Hornbruch A, Wolpert L. 1991. The spatial and temporal distribution of polarizing activity in the flank of the pre-limb-bud stages in the chick embryo. *Development* 111:725–31

Hunt P, Krumlauf R. 1992. Hox codes and positional specification in vertebrate embryonic axes. *Annu. Rev. Cell Biol.* 8:227–56

Izpisúa-Belmonte J-C, Brown JM, Crawley A, Duboule D, Tickle C. 1992a. Hox-4 gene expression in mouse/chicken heterospecific grafts of signalling regions to limb buds reveals similarities in patterning mechanisms. *Development* 115:553–60

Izpisúa-Belmonte J-C, Brown JM, Duboule D, Tickle C. 1992b. Expression of Hox-4 genes in the chick wing links pattern formation to the epithelial-mesenchymal interactions that mediate growth. *EMBO J.* 11:1451–57

Izpisúa-Belmonte J-C, Duboule D. 1992. Homeobox genes and pattern formation in the vertebrate limb. *Dev. Biol.* 152:26–36

Izpisúa-Belmonte J-C, Tickle C, Dollé P,

Wolpert L, Duboule D. 1991. Expression of the homeobox Hox-4 genes and the specification of position in chick wing development. *Nature* 350:585–89

Jackson-Grusby L, Kuo A, Leder P. 1992. A variant *limb deformity* transcript expressed in the embryonic mouse limb defines a novel formin. *Genes Dev.* 6:29–37

Jones CM, Lyons KM, Hogan BLM. 1991. Involvement of *Bone Morphogenetic Protein-4* (BMP-4) and Vgr-1 in morphogenesis and neurogenesis in the mouse. *Development* 111:531–42

Jorquera B, Pugin E. 1971. Sur le compartement du mésoderme et de l'ectoderme du bourgeon de membre dans les échanges entre le poulet et le rat. *C. R. Acad. Sci. Paris Série D* 272:1522–25

Kessel M, Gruss P. 1991. Homeotic transformations of murine vertebrae and concomitant alteration of Hox codes induced by retinoic acid. *Cell* 67:89–104

Kieny M. 1968. Variation de la capacite inductrice du mésoderme et de la comptence de l'ectoderme au cours de l'induction primaire du bourgeon de membre chez l'embryon de poulet. *Arch. Anat. Microsc. Morphol. Exp.* 57:401–18

Kingsley D. 1994. The TGF-β superfamily: new members, new receptors, and new genetic tests of function in different organisms. *Genes Dev.* 8:133–46

Krauss S, Concordet J-P, Ingham PW. 1993. A functionally conserved homolog of the Drosophila segment polarity gene *hh* is expressed in tissues with polarizing activity in zebrafish embryos. *Cell* 75:1431–44

Lance-Jones C. 1988. Development of neuromuscular connections: guidance of motoneuron axons to muscles in the embryonic chick hindlimb. *Ciba Found. Symp.* 138:97–115

Lee J, Tickle C. 1985. Retinoic acid and pattern formation in the developing chick wing: SEM and quantitative studies of early effects on the apical ectodermal ridge and bud outgrowth. *J. Embryol. Exp. Morph.* 90:139–69

Linney E. 1992. Retinoic acid receptors: Transcription factors modulating gene regulation, development, and differentiation. *Curr. Top. Dev. Biol.* 27:309–50

Lyons KM, Pelton RW, Hogan BLM. 1990. Organogenesis and pattern formation in the mouse: RNA distribution patterns suggest a role for *Bone Morphogenetic Protein-2A* (BMP2A). *Development* 109:833–44

Ma C, Zouh Y, Beachy PA, Moses K. 1993. The segment polarity gene *hedgehog* is required for the propagation of the morphogenetic furrow in the Drosophila eye. *Cell* 75:927–38

MacCabe JA, Errick J, Saunders JW Jr. 1974. Ectodermal control of the dorsoventral axis in the leg bud of the chick embryo. *Dev. Biol.* 39:69–82

MacCabe JA, Parker BW. 1976. Polarizing activity in the developing limb of the Syrian hamster. *J. Exp. Zool.* 195:311–17

Mackem S, Mahon KA. 1991. *Ghox 4.7*: a chick homeobox gene expressed primarily in limb buds with limb-type differences in expression. *Development* 112:791–806

Mangelsdorf DJ, Borgmeyer U, Heyman RA, Zhou JY, Ong ES, et al. 1992. Characterization of three RXR genes that mediate the action of 9-*cis* retinoic acid. *Genes Dev.* 6: 329–44

Mangelsdorf DJ, Umesono K, Evans RM. 1994. The retinoid receptors. See Hofmann and Eichele 1994, pp. 319–49

Marshall H, Nonchev S, Sham MH, Muchamore I, Lumsden A, Krumlauf R. 1992. Retinoic acid alters hindbrain Hox code and induces transformation of rhombomeres 2/3 into a 4/5 identity. *Nature* 360: 737–41

McGinnis W, Krumlauf R. 1992. Homeobox genes and axial patterning. *Cell* 68:283–302

McLachlan JC, Phoplonker MH. 1988. Limb reduplicating effects of chorio-allantoic membrane and its components. *J. Anat.* 158: 147–55

McMahon AP, Joyner AL, Bradley A, McMahon JA. 1992. The midbrain-hindbrain phenotype of Wnt-1/Wint-1- mice results from stepwise deletion of engrailed-expressing cells by 9.5 days postcoitum. *Cell* 58: 1075–84

Mendelsohn C, Ruberte E, Chambon P. 1992. Retinoid receptors in vertebrate limb development. *Dev. Biol.* 152:50–61

Mitchell PJ, Timmons PM, Hébert JM, Rigby PWJ, Tjian R. 1991. Transcription factor AP-2 is expressed in neural crest cell lineages during mouse embryogenesis. *Genes Dev.* 5: 105–19

Miyamoto M, Naruo K, Seko C, Marsumoto S, Kondo T, Kurokawa T. 1993. Molecular cloning of a novel cytokine cDNA encoding the ninth member of the fibroblast growth factor family, which has a unique secretion property. *Mol. Cell Biol.* 13:4251–59

Molven A, Wright CVE, Bremiller R, De Robertis EM, Kimmel CB. 1990. Expression of a homeobox gene product in normal and mutant zebrafish embryos: evolution of the tetrapod body plan. *Development* 109:279–89

Morgan BA, Izpisúa-Belmonte J-C, Duboule D, Tabin CJ. 1992. Targeted misexpression of Hox-4.6 in the avian limb causes apparent homeotic transformations. *Nature* 358:236–39

Moroni MC, Vigano, MA, Mavilio, F. 1993. Regulation of the human HOXD4 gene by retinoids. *Mech. Dev.* 44:139–54

Niswander L, Martin GR. 1992. Fgf-4 expression during gastrulation, myogenesis, limb and tooth development in the mouse. *Development* 114:755–68

Niswander L, Martin GR. 1993a. FGF-4 and BMP-2 have opposite effects on limb growth. *Nature* 361:68–71

Niswander L, Martin GR. 1993b. FGF-4 regulates expression of *Evx-1* in the developing mouse limb. *Development* 119:287–94

Niswander L, Tickle C, Vogel A, Booth I, Martin GR. 1993. FGF-4 replaces the apical ectodermal ridge and directs outgrowth and patterning of the limb. *Cell* 75:579–87

Nohno T, Noji S, Koyama E, Ohyama K, Myokai F, et al. 1991. Involvement of the Chox-4 chicken homeobox genes in determination of antero-posterior axial polarity during limb development. *Cell* 64:1197–1205

Noji S, Koyama E, Myokai F, Nohno T, Ohuchi H, et al. 1993. Differential expression of the three chick FGF receptor genes, FGFR 1, FGFR 2 and FGFR 3, in limb and feather development. In *Limb Development and Regeneration,* ed. JF Fallon, PF Goetinck, RO Kelley, DL Stocum, pp. 645–54. New York: Wiley-Liss

Noji S, Nohno T, Koyama E, Muto K, Ohyama K, et al. 1991. Retinoic acid induces polarizing activity but is unlikely to be a morphogen in the chick limb bud. *Nature* 350:83–86

Oliver G, De Robertis EM, Wolpert L, Tickle C. 1990. Expression of a homeobox gene in the chick wing bud following application of retinoic acid and grafts of polarizing region tissue. *EMBO J.* 9:3093–99

Oliver G, Sidell N, Fiske W, Heizman C, Mohandas T, et al. 1989. Complementary homoeprotein gradients in developing limb buds. *Genes Dev.* 3:641–50

Oliver G, Wright CVE, Hardwicke J, De Robertis EM. 1988. A gradient of homeodomain protein in developing forelimbs of Xenopus and mouse embryos. *Cell* 55:1017–24

Parr BA, Shea MJ, Vassileva G, McMahon AP. 1993. Mouse *Wnt* genes exhibit discrete domains of expression in the early embryonic CNS and limb buds. *Development* 119:247–61

Pautou M-P, Kieny M. 1973. Interaction ecto-mésodermique dans l'établissement de la polarité dorso-ventrale du pied de l'embryon du poulet. *C. R. Acad. Sci. Paris Série D* 277:1225–28

Peters KG, Werner S, Chen G, Williams LT. 1992. Two FGF receptor genes are differentially expressed in epithelial and mesenchymal tissues during limb formation and organogenesis in the mouse. *Development* 114:233–43

Riddle RD, Johnson RL, Laufer, E, Tabin C. 1993. *Sonic hedgehog* mediates the polarizing activity of the ZPA. *Cell* 75:1401–16

Rijli FM, Mark M, Lakkaraju, S, Dierich A, Dollé P, Chambon P. 1993. A homeotic transformation is generated in the rostral branchial region of the head by disruption of *Hoxa-2,* which acts as a selector gene. *Cell* 75:1333–49

Riley BB, Savage MP, Simandl BK, Olwin BB, Fallon JF. 1993. Retroviral expression of FGF-2 (bFGF) affects patterning in chick limb bud. *Development* 118:95–104

Robert B, Lyons G, Simandl BK, Kuroiwa A, Buckingham M. 1991. The apical ectodermal ridge regulates *Hox-7* and *Hox-8* gene expression in developing chick limb buds. *Genes. Dev.* 5:2363–74

Robert B, Sassoon D, Jacq B, Gehring W, Buckingham M. 1989. *Hox-7,* a mouse homeobox gene with a novel pattern of expression during mouse embryogenesis. *EMBO J.* 8:91–100

Ros MA, Lyons G, Koshar RA, Upholt WB, Coelho CD, Fallon JF. 1992. Apical ridge dependent and independent mesodermal domains of *GHox-7* and *GHox-8* expression in chick limb buds. *Development* 116:811–18

Rosen V, Thies RS. 1992. The BMP proteins in bone formation and repair. *Trends Genet.* 8:97–102

Rowe A, Eager N, Brickell P .1991. A member of the RXR nuclear receptor family is expressed in neural-crest-derived cells of the developing chick peripheral nervous system. *Development* 111:771–78

Ruberte E, Friederich V, Morriss-Kay G, Chambon P. 1992. Differential distribution patterns of CRAPB I and CRABP II transcripts during mouse embryogenesis. *Development* 115:973–78

Rubin L, Saunders JW Jr. 1972. Ectodermal-mesodermal interactions in the growth of limb buds in the chick embryo: constancy and temporal limits of the ectodermal induction. *Dev. Biol.* 28:94–112

Sampath TK, Coughlin JE, Whetstone RM, Banach D, Corbett C, et al. 1990. Bone osteogenic protein is composed of OP-1 and BMP-2A, two members of the transforming growth factor-β superfamily. *J. Biol. Chem.* 265:13198–205

Satre MA, Kochhar DM. 1989. Elevations in the endogenous levels of the putative morphogen retinoic acid in embryonic mouse limb-buds associated with limb dysmorphogenesis. *Dev. Biol.* 133:529–36

Saunders JW Jr. 1948. The proximo-distal sequence of origin of limb parts of the chick wing and the role of the ectoderm. *J. Exp. Zool.* 108:363–404

Saunders JW Jr, Gasseling MT. 1968. Ectodermal-mesenchymal interactions in the origin of limb symmetry. In *Epithelial-mesenchy-*

mal Interactions, ed. R Fleischmajer, RE Billingham, pp. 78–97. Baltimore: Williams & Wilkins

Saunders JW Jr, Gasseling MT, Errick J. 1976. Inductive activity and enduring cellular constitution of a supernumerary apical ectodermal ridge grafted to the limb bud of the chick embryo. *Dev. Biol.* 50:16–25

Savage MP, Hart CE, Riley BB, Sasse J, Olwin BB, Fallon JF. 1993. Distribution of FGF-2 suggests it has a role in chick limb bud growth. *Dev. Dynamics* 198:159–70

Schimmang T, Lemaistre M, Vortkamp A, Rütter U. 1992. Expression of the zinc finger gene *Gli3* is affected in the morphogenetic mouse mutant *extra-toes (Xt). Development* 116:799–804

Schoenwolf GC, Garcia-Martinez V, Dias MS. 1992. Mesoderm movement and fate during avian gastrulation and neurulation. *Dev. Dynamics* 193:235–48

Schofield JN, Rowe A, Brickell PM. 1992. Position-dependence of retinoic acid receptor-β gene expression in the chick limb bud. *Dev. Biol.* 152:344–53

Scott MP. 1992. Vertebrate homeobox gene nomenclature. *Cell* 71:551–53

Searls RL, Janners MY. 1971. The initiation of limb bud outgrowth in the embryonic chick. *Dev. Biol.* 24:198–213

Shellswell GB, Wolpert L. 1977. The pattern of muscle and tendon development in the chick wing. See Fallon and Crosby 1977, pp. 71–86

Smith JC. 1979. Evidence for a positional memory in the development of the chick wing bud. *J. Embryol. Exp. Morph.* 52:105–13

Smith JC. 1980. The time required for positional signalling in the chick wing bud. *J. Embryol. Exp. Morph.* 60:321–28

Smith SM, Eichele, G. 1991. Temporal and regional differences in the expression pattern of distinct retinoic acid receptor-beta transcripts in the chick embryo. *Development* 111:245–52

Song K, Wang Y, Sassoon D. 1992. Expression of *Hox-7.1* in myoblasts inhibits terminal differentiation and induces cell transformation. *Nature* 360:477–81

Stephens TD, Beier RLW, Bringhurst DC, Hiatt SR, Prestridge M, et al. 1989. Limbness in the early chick embryo lateral plate. *Dev. Biol.* 133:1–7

Stern CD, Hatada Y, Selleck MAL, Storey KG. 1992. Relationship between mesoderm induction and the embryonic axes in chick and frog embryos. *Dev. Suppl.* pp. 151–56

Stoker KM, Carlson BM. 1990. Hensen's node, but not other biological signallers, can induce supernumerary digits in the developing chick limb bud. *Roux's Arch. Dev. Biol.* 198:371–81

Storm EE, Huynh TV, Copeland DM, Jenkins NA, Kingsley DM, Lee S. 1994. Limb alterations in *brachypodism* mice due to mutations in a new member of the TGFβ-superfamily. *Nature* 368:639–44

Sucov HM, Murakami KK, Evans RM. 1990. Characterization of an autoregulated response element in the mouse retinoic acid receptor typeβ gene. *Proc. Natl. Acad. Sci. USA* 87:5392–96

Summerbell D. 1974. A quantitative analysis of the effect of excision of the AER from the chick limb bud. *J. Embryol. Exp. Morph.* 32: 651–60

Summerbell D. 1983. The effect of local application of retinoic acid to the anterior margin of the developing chick limb. *J. Embryol. Exp. Morph.* 78:269–89

Summerbell D, Lewis J, Wolpert L. 1973. Positional information in chick limb morphogenesis. *Nature* 224:492–96

Sundin O, Eichele, G. 1992. An early marker of axial pattern in the chick embryo and its respecification by retinoic acid. *Development* 114:841–52

Suzuki HR, Sakamoto H, Yoshida T, Sugimura T, Terada, M, Solursh, M. 1992. Localization of Hst1 transcripts to the apical ectodermal ridge in the mouse embryo. *Dev. Biol.* 150: 219–22

Tabata T, Kornberg TB. 1994. Hedgehog is a signaling protein with a key role in patterning Drosophila imaginal discs. *Cell* 76:89–102

Tabin CJ. 1991. Retinoids, homeoboxes, and growth factors: toward molecular models for limb development. *Cell* 66:199–217

Tamura K, Aoki Y, Ide H. 1993. Induction of polarizing activity by retinoic acid occurs independently of duplicate formation in developing chick limb buds. *Dev. Biol.* 158: 341–49

Tamura K, Kagechika H, Hashimoto Y, Shudo K, Oshugi K, Ide K. 1990. Synthetic retinoids, retinobenzoic acids, Am80, Am580 and Ch55 regulate morphogenesis in the chick limb bud. *Cell Diff. Dev.* 32:17–26

Thaller C, Eichele G. 1987. Identification and spatial distribution of retinoids in the developing chick limb bud. *Nature* 327: 625–28

Thaller C, Eichele G. 1988. Characterization of retinoid metabolism in the developing chick limb bud. *Development* 103:473–83

Thaller C, Eichele G. 1990. Isolation of 3,4-didehydroretinoic acid, a novel morphogenetic signal in the chick wing bud. *Nature* 345:815–19

Thaller C, Hofmann C, Eichele G. 1993. 9-*cis*-retinoic acid, a potent inducer of digit pattern duplications in the chick wing bud. *Development* 118:957–65

Tickle C. 1981. The number of polarizing region cells required to specify additional digits in the developing chick wing. *Nature* 289: 295–98

Tickle C. 1991. Retinoic acid and limb patterning and morphogensis. In *Developmental Patterning of the Vertebrate Limb,* ed. JR Hinchliffe, J Hurle, D Summerbell, pp. 143–50. New York: Plenum

Tickle C, Alberts B, Wolpert L, Lee J. 1982. Local application of retinoic acid to the limb bud mimics the action of the polarizing region. *Nature* 296:564–66

Tickle C, Crawley A, Farrar J. 1989. Retinoic acid application to chick wing buds leads to a dose-dependent reorganization of the apical ectodermal ridge that is mediated by the mesenchyme. *Development* 106:691–705

Tickle C, Lee J, Eichele G. 1985. A quantitative analysis of the effect of all-*trans*-retinoic acid on the pattern of chick wing development. *Dev. Biol.* 109:82–95

Tickle C, Shellswell, G, Crawley A, Wolpert L. 1976. Positional signalling by mouse limb polarizing region in the chick limb bud. *Nature* 259: 396–97

Tickle C, Summerbell D, Wolpert L. 1975. Positional signalling and specification of digits in chick limb morphogenesis. *Nature* 254: 199–202

Trumpp A, Blundell PA, de la Pompa, J-L, Zeller R. 1992. The chicken limb deformity gene encodes nuclear proteins expressed in specific cell types during morphogenesis. *Genes Dev.* 6:14–28

Vainio S, Karavanova I, Jovett A, Thesleff I. 1993. Identification of BMP-4 as a signal mediating secondary induction between epithelial and mesenchymal tissues during tooth development. *Cell* 75:45–58

Vogel A, Tickle C 1993. FGF-4 maintains polarizing activity of posterior limb bud cells in vivo and in vitro. *Development* 119:199–206

Wagner M, Thaller C, Jessell T, Eichele G. 1990. Polarizing activity and retinoid synthesis in the floor plate of the neural tube. *Nature* 345:819–22

Wall NA, Jones CM, Hogan BLM, Wright CVE. 1992. Expression and modification of Hox 2.1 protein in mouse embryos. *Mech. Dev.* 37:111–20

Wanek N, Gardiner DM, Muneoka K, Bryant SV. 1991. Conversion by retinoic acid of anterior cells into ZPA cells in the chick wing bud. *Nature* 350:81–83

Wedden SE, Pang K, Eichele G. 1989. Expression pattern of homeobox-containing genes during chick embryogenesis. *Development* 105:639–50

Wolpert L. 1989. Positional information and prepattern in the development of pattern. In *Cell to Cell Signalling: From Experiments to Theoretical Models,* ed. A Goldbetter, pp. 133–43. New York: Academic

Wolpert L, Hornbruch A. 1981. Positional signalling along the antero-posterior axis of the chick wing. The effects of multiple polarizing region grafts. *J. Embryol. Exp. Morph.* 63:145–59

Wozney JM, Capparella J, Rosen V. 1993. The bone morphogenetic proteins in cartilage and bone development. In *Molecular Basis of Morphogenesis,* ed. M Bernfield, pp. 221–30. New York: Wiley-Liss

Yokouchi Y, Sasaki H, Kuriowa A. 1991. Homeobox gene-expression correlated with the bifurcation process of limb cartilage development. *Nature* 353:443–46

Zwilling E. 1955. Ectoderm-mesoderm relationship in the development of the chick embryo limb bud. *J. Exp. Zool.* 128:423–41

Zwilling E. 1956. Interaction between limb bud ectoderm and mesoderm in the chick embryo. I. Axis establishment. *J. Exp. Zool.* 132:157–72

Annu. Rev. Cell Biol. 1994. 10:153–80

MICROTUBULES IN PLANT MORPHOGENESIS: Role of the Cortical Array

Richard J. Cyr
Department of Biology, The Pennsylvania State University, University Park, Pennsylvania 16802

KEY WORDS: cytoskeleton, development, microtubule, plant

CONTENTS

INTRODUCTION

Like their protistan ancestors, plant and animal cells share many features common to eukaryotes. However, in the billion years since the divergence of

0743–4634/94/1115–0153$05.00

their phylogenetic lineages, notable differences, reflective of their different life styles, have evolved. For example, plant cells possess thick walls that support and physically bind cells together (Gifford & Foster 1989). Thus cellular locomotion is eliminated and, therefore, cellular behaviors available for morphogenetic processes are restricted (Sinnot 1960).

Despite these limitations in cellular motility, plants develop into complex multicellular organisms. They exhibit a multitude of gross morphologies, but certain basic morphological features are inherent to all species, and these can be discerned at the cellular level. Notably, organs extend because files of cells undergo orchestrated elongation. Organs increase in girth when cells swell after elongation, or as a result in a change in the elongative axis, which is usually preceded by a switch in cell division planes (Wardlaw 1952; Sinnot 1960; Esau 1977).

Microtubules (MTs) are involved in cellular elongation and cytokinesis. This review does not comprehensively cover all aspects of MT behavior in plants; several excellent treatments provide a broader overview (Hepler & Palevitz 1974; Gunning & Hardham 1982; Lloyd 1982, 1991a; Seagull 1989). This review does focus on those MT activities that affect the process of cellular elongation and considers the impact of the elongative process upon the ordering of the cytokinetic apparatus. A similar overview presented by Williamson in 1991 provided a springboard for this current work. The central thesis of this review states that the form and function of the cortical MT array is dependent upon the physical milieu created by the cortical cytoplasm, cell membrane, and cell wall.

CELLULAR PATTERNING IN GROWING PLANT ORGANS

Two cellular processes that affect morphology during development are elongation and cytokinesis. As depicted in Figure 1, the small, somewhat isodiametric, apical meristem cells in a developing organ (e.g. root) provide it with a constant supply of new cells which, in turn, can divide again or undergo cellular differentiation. Two fundamental features of cellular patterning are observed in virtually all elongated plant organs. First, vectorial cell expansion is either parallel or at right angles to the major axis of organ extension. The direction of cellular elongation is determined by the organized deposition of cellulose microfibrils (MFs) (Green 1962, 1980). The cellular elongative axis parallel to the major organ axis contributes to organ extension, while lateral elongation (or expansion) will either contribute to an increase in organ girth, or the initiation of a new lateral organ (e.g. a leaf from a stem; Green & Poethig 1982). Second, orientation of a new cell partition is either parallel or perpen-

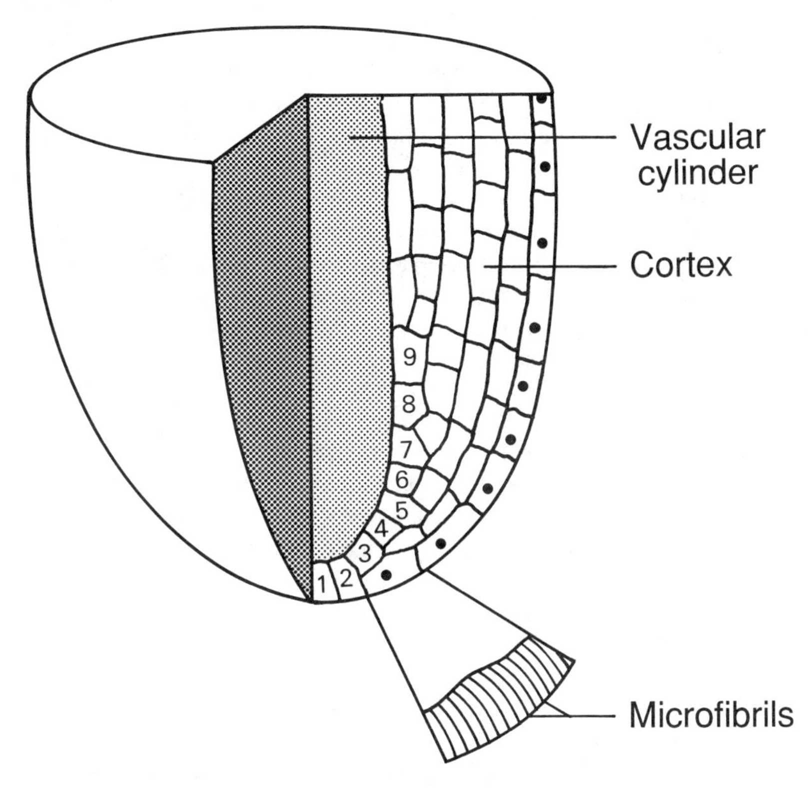

Figure 1 Cell division planes and the elongative cell axis contribute to basic morphology. This cartoon of root cortical cells (derived from data in Barlow & Adam 1989) illustrates how cell elongation and division planes contribute to the morphology of an elongated organ. The cells labeled 2, 5, 7, and 9 arose from divisions that were anticlinal (at right angles) to the vascular cylinder. The cells labeled 3, 4, 6, and 8 arose from divisions that were paraclinal (parallel) to the vascular cylinder. Each paraclinal division gave rise to a new file of cells and is comprised of cells resulting from anticlinal divisions only. One such file (all derivitives of the the sister cell of 3) is marked with dots. The net orientation of cellulose microfibrils in one elongated cell is shown.

dicular to the major axis of organ extension (Lindenmayer 1978; Lloyd & Barlow 1982; Barlow & Adam 1989). A new cell partition is created by the cell plate during cytokinesis. Figure 1 shows that these new partitions are oriented either parallel or perpendicular to the axis of cellular extension. This frequently observed relationship between division plane orientation and elongative axis led Hofmeister in 1863 to propose the two were causally related.

THE BIOPHYSICS OF PLANT CELL GROWTH AND THE ROLE OF THE CELL WALL

From a biophysical standpoint, plant cell growth involves at least three processes. First, osmotic pressure, generated by active ion accumulation in a central vacuole, provides the energy required for cell expansion (Steudle 1989; Cosgrove 1993c). Second, the surrounding wall must yield (Cosgrove 1993a) in a process that probably involves proton extrusion (Rayle & Cleland 1992), which leads to acidification and subsequent activation of the enzymes involved in bond rearrangements within the wall polymers (Cosgrove 1989). The wall becomes compliant (Cosgrove 1993b) once the wall strength falls below the threshold of osmotic pressure. Third, water flows into the cell, which results in an increase in cell volume (Cosgrove 1993c).

Plant cell walls are composite, polymeric structures comprised of proteins, polyphenolics, and a variety of carbohydrates including cellulose, xyloglucans, rhamnogalacturons, and pectins (Carpita & Gibeaut 1993). Cellulose MFs, the strongest wall polymers (their tensile strength and elastic modulus are similar to steel; Niklas 1992) are deposited at right angles to the elongative axis. This orderly scheme provides lateral reinforcement to the side walls, thereby restricting compliance to an axis that is normal to MF orientation (Green et al 1970; Gertel & Green 1977).

An extensive amount of observational and experimental data shows that most cells utilize cortical MTs to control the orientation of newly synthesized cellulose MFs (Eigsti & Dustin 1955; Newcomb 1969; Hepler & Fosket 1971; Giddings & Staehelin 1991; Morejohn 1991). Other mechanisms appear to be used by some tip-growing cells, such as root hair and pollen tubes (Emons 1982; Cimler et al 1985). In most cases, therefore, cortical MTs provide spatial information to affect the orderly synthesis of cellulose MFs; however, the details of this process are unknown. Three lines of evidence indicate that MT/plasma membrane (PM) associations are one component of this process. First, electron micrographs reveal opaque connections between cortical MTs and the PM (Hardham & Gunning 1978). Second, external application of the wall protein extensin, or charged polymers such as poly-L-lysine, to the outside of protoplasts results in the stabilization of cortical MTs (Akashi et al 1990). Third, when protoplasts are lysed on poly-L-lysine-coated slides, they leave behind "footprints" of cortical MTs that presumably are held to the glass surface by residual membrane components (Lloyd et al 1980; Doohan & Palevitz 1980). It is unclear how MTs interact with the PM, and therefore it is not known how PM-associated MTs affect cellulose orientation. Cellulose is synthesized by a cellulose synthase complex, which forms membrane rosettes that are believed to be capable of moving in the plane of the membrane (Staehelin & Giddings 1982). One hypothesis, which attempts to explain the ordering of MFs, is that the cortical MTs restrict this movement, thereby providing channels for the cellulose-synthesizing complex (Giddings & Staehelin 1988). As a result, the order of newly deposited cellulose fibrils reflects that of the underlying cortical MTs.

CYTOKINESIS

In plants, cytokinesis requires the deposition of a new cell plate (Gunning 1982; Lloyd 1991b). Plate formation can be considered a two-phase construction project. In the first phase, the cell plate is assembled via the fusion of vesicles that are donated and transported by the incipient daughter cells to the approximate area of final positioning (Hepler & Jackson 1968). These vesicles, along with the MTs responsible for their transport, in part, comprise the cytokinetic apparatus termed the phragmoplast (Figure 2d). Plate construction begins in the central region of this ensemble and proceeds outward in a centrifugal fashion (for example, see Palevitz 1986). As the cell plate forms, phragmoplast MTs appear to progress outward (Wick 1985), continuing to carry new vesicles to their fusion sites on the growing margin of the forming plate. As the growing margins of the new cell plate come close to the PM, the second or positioning phase of plate construction begins. The growing edges of the nascent plate move about in the vicinity of the PM before making contact, then fuse to specific areas of the PM (Palevitz 1986). Only unique areas of the PM are capable of participating in the fusion of the nascent cell plate, which indicates that a cortical mechanism exists to insure the proper placement of new plates. While the nature of this mechanism is unknown, it has been shown that the proper placement of the cell plate, at the conclusion of cytokinesis, is predicted by a transient band of cortical MTs known as the preprophase band (PPB) (Hepler & Palevitz 1974; Gunning 1982; Gunning & Wick 1985) that appears prior to division.

In a cycling cell, the interphase cortical array is dispersed along the length of the cell (Wick et al 1981; Williamson 1991; Figure 2). As the cell enters G_2 (Gunning & Sammut 1990), the cortical array appears to dissipate at both ends of the cell. As the cell enters prophase, only the narrow PPB of cortical MTs remains in the cortical region surrounding the nucleus (Figure 2). The PPB breaks down during prophase and is replaced by the spindle apparatus (Mineyuki et al 1991; Staiger & Lloyd 1991; Cleary et al 1992; Lloyd et al 1992; Baskin & Cande 1990). Although the PPB is a transient cortical array, it accurately predicts the future location of the cell plate. Therefore, the key to understanding cellular elongation, as well as cell plate orientation, resides in knowing what events influence the positioning of MTs in the cortical region during interphase (for cell elongation) and in G_2 (ultimately for cytokinesis).

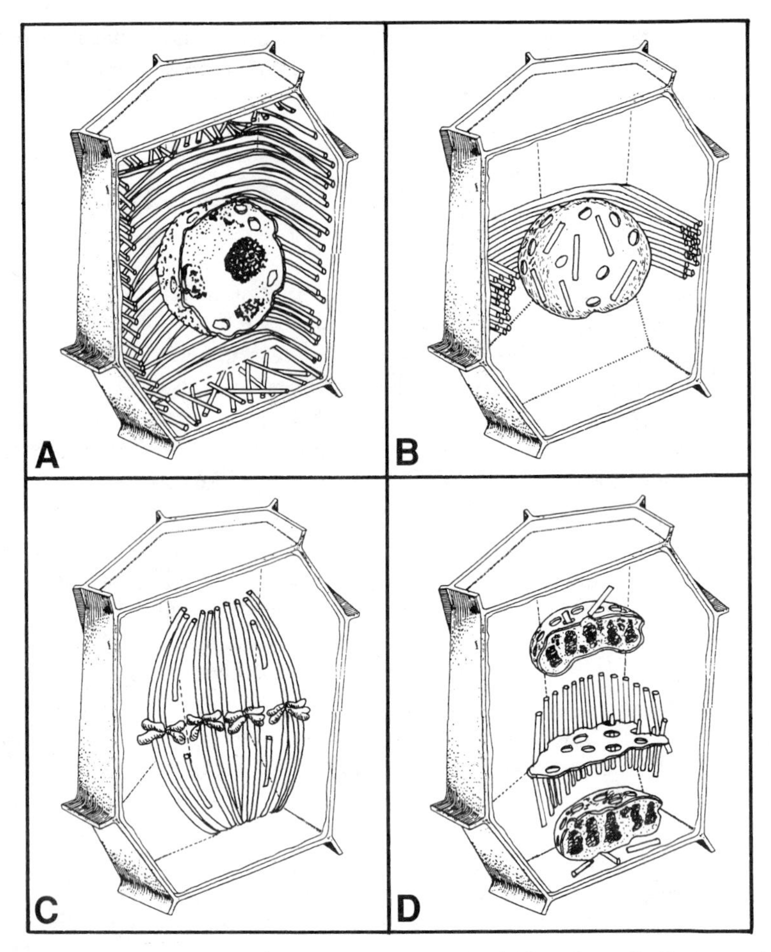

Figure 2 Microtubules are found in at least four arrays in dividing plant cells. During much of interphase, the cortical MT array is found transversely oriented to the major cellular axis (*A*). The preprophase band (PPB) appears prior to mitosis (*B*). The spindle apparatus is typically anastral (*C*). The phragmoplast appears during cytokinesis (*D*). Redrawn from Goddard et al (1994).

HOW DOES THE CELL ALIGN CORTICAL MTs?

The knowledge that cortical MTs are involved in ordering cellulose MFs and that the PPB predicts the future location of the cell plate is important to our understanding of cellular morphogenesis and leads to the question, how are cortical MTs positioned in their proper, nonrandom configurations? Mechanistically, cortical MTs might be configured via two basic routes. First, the MTs within the array can polymerize into the appropriate configuration or, second, the cell could move MTs into the correct orientation after assembly. It is possible that rearrangements occur via the concerted action of both processes (Palevitz 1991). Before reviewing the experimental data that deal with MT reorganization in the cortical array, it is appropriate to consider the conditions under which the array normally changes configurations.

Alignment of Cortical MTs in Unperturbed Cells

Plant cell shape, as well as the organization of the cortical MT array, changes dramatically during the life of a cell. For example, Marc & Hackett (1992) observed that the cortical arrays of individual cells in the extreme tip of *Hedera* (ivy) apices were not co-aligned with each other. However, the more basipetal mature tissue contained "packets" of cells with co-aligned arrays. Additionally, in cells of rapidly elongating tissue, the orientation of the cortical array tends to be transverse to the axis of elongation (Akashi & Shibaoka 1987; Wasteneys & Williamson 1987; Williamson 1991; Baluška et al 1992), and as the growth rate decreases, the MTs tend to become more oblique or, in some cases, even longitudinal (Wasteneys & Williamson 1987; Sakiyama & Shibaoka 1990; Baluška et al 1992). Reorientation of the cortical array also occurs before the deposition of secondary wall material in xylem tracheary elements (Gunning & Hardham 1982; Falconer & Seagull 1986) and differentiating guard cells (Marc et al 1989a). These studies provide ample data to indicate that MT reorganization is a normal component of cellular development.

Structure of the Cortical MT Array

Immunofluorescence microscopy is a valuable tool for analyzing the cortical array. This technique confirms the electron microscopic data by showing that the cortical array is typically organized into a net transverse orientation with respect to the elongative axis (Williamson 1991; Figure 2) and is valuable because the global arrangement of the cortical array can be readily discerned. However, as pointed out by Williamson (1991), its resolution is limited to about 400 nm. Hence, closely situated MTs will be imaged as a single entity. Hardham & Gunning (1978) serially reconstructed electron micrographs and revealed that the fine structure of the cortical array is a collection of short MTs that overlap in a staggered fashion. Thus, immunofluorescence micrographs probably represent the average arrangement of many short MTs that are laterally associated (bundled).

The lateral spacing between these bundled MTs is approximately 34 nm (Newcomb 1969; Doohan & Palevitz 1980; Lancelle et al 1987). Microtubule-associated proteins (MAPs) have been isolated from a number of plant sources, and they induce the bundling of taxol-stabilized MTs (Cyr & Palevitz 1989; Vantard et al 1991; Chang-Jie & Sonobe 1993; Schellenbaum et al 1993). The center-to-center spacing of MTs within bundles assembled in vitro is similar to that observed in vivo (Cyr & Palevitz 1989; Chang-Jie & Sonobe 1993). Therefore, it is reasonable to hypothesize that MAPs, which bundle MTs in vitro, serve a similar function within the cell, but this has not been proven.

While it is unclear how MTs associate with the PM, theoretically, two possibilities exist. First, tubulin has been identified in membrane preparations (Stephens 1986; Laporte et al 1993) and, therefore, MTs may associate directly with tubulin subunits in the PM. Second, the association may be via integral or periperal membrane proteins; however, such proteins have not been identified. These two possibilities are not necessarily mutually exclusive. Perhaps a peripheral bridging protein associates with membrane tubulin within the hydrophobic milieu of the PM, and with MTs in the hydrophilic environment of the cortical cytoplasm.

Actin microfilaments are also present in the cortex (Seagull et al 1987; Traas et al 1987; Sonobe & Shibaoka 1989; Ding et al 1991; Staiger & Lloyd 1991; Eleftheriou & Palevitz 1992; Hepler et al 1993; Zhang et al 1993; Staiger et al 1994), and sometimes in close proximity to MTs (Lancelle & Hepler 1991). MTs and microfilaments may form functional associations because treatment of cells with the microfilament disrupting agent cytochalasin D (CD) can interfere with the narrowing of the PPB during G_2 (Mineyuki & Palevitz 1990; Eleftheriou & Palevitz 1992). Additionally, CD treatment of developing cotton fibers and tracheids can affect the reorientation of MTs that occurs during differentiation (Fukuda & Kabayashi 1989; Seagull 1990). These studies indicate that actin microfilaments can, in some instances, alter the organizational status of cortical MTs; however, this finding is not universal. For example, CD does not affect the early reorganization of cortical MTs during protoplast regeneration (Kuss-Wymer 1994).

MT BEHAVIOR WITHIN THE CORTICAL ARRAY

Soluble tubulin can self-assemble to form a MT which, in turn, can depolymerize back into soluble tubulin. The dynamic equilibrium between soluble tubulin and insoluble MTs has been well studied in vitro and to a lesser extent in vivo. There are several excellent treatments of this topic (Soifer 1986; Cassimeris et al 1987; Mitchison 1988; Salmon 1989; Gelfand & Bershadsky 1991). This part of the review concentrates on the dynamic behavior of MTs within the cortical array of plant cells.

Two basic approaches are generally used to examine the dynamics of MTs in the cortical array. The first utilizes anti-MT drugs such as colchicine, ami-prophos methyl (APM) and oryzalin. Colchicine, oryzalin (and probably APM) act by binding to tubulin in a manner that prevents the growth of MTs (Morejohn 1991; Hugdahl & Morejohn 1993). Highly dynamic MTs are rapidly adding and subtracting dimers and are more susceptible to these drugs than are less dynamic MTs.

Drugs provide an indirect approach and certain caveats should be taken into account, including differences in the rate of drug penetration into the cells, as well as variations in the binding constants for various drugs. Furthermore, at substoichiometric concentrations to tubulin, colchicine stabilizes tubulin polymers by inhibiting the dynamicity of MTs (Vandecandelaere et al 1994).

A second and more direct approach involves microinjection of fluorescently derivitized tubulin into the cell. Once the tagged tubulin is incorporated into MTs, fluorescence recovery after photobleaching (FRAP) is used to directly measure the rate of subunit addition (pioneered in animal cells by Salmon et al 1984; Saxton et al 1984) into the MTs (a related technique utilizes a photoactivated form of derivitized tubulin that can be used similarly; Mitchison 1989). More recently the FRAP technique has been applied to MTs in plant cells (Hush et al 1994). Although direct, this method also has caveats. For example, MTs lying closer to each other than approximately 400 nM are imaged as one, and photolytic damage can alter the dynamics by severing MTs (Vigers et al 1988).

Cortical MTs are sensitive to anti-MT drugs to varying degrees. For example, young, rapidly multiplying tissue-cultured cells have relatively sensitive cortical MT arrays in comparison to the arrays in cells of older cultures, which are either not multiplying, or multiplying to a lesser extent (Falconer & Seagull 1987). Also, application of anti-MT drugs to growing organs indicates that cells in different stages of development have cortical arrays that possess MTs with different dynamic characteristics (Baluška & Barlow 1993). Treatment of *Nitella* internodal cells for 5 min with the anti-MT drug oryzalin results in a substantial depolymerization of the cortical array (Wasteneys et al 1993), which indicates the turnover time for tubulin in the cortical MTs can be less than 5 min. However, not all MTs were depolymerized in this time period, which indicates dynamic heterogeneity within the cortical array. One FRAP study performed on cortical MTs supports and extends the drug studies by showing that the cortical MTs of *Tradescantia* stamen hair cells are highly dynamic, with turnover times between 60–65 sec (Hush et al 1994).

The cortical MT array is composed of short overlapping MTs that are connected to the PM (and perhaps to one another). In spite of these connections, the MTs within this array are highly dynamic, but their dynamic status is variable.

How Might the Cell Control the Stability of Cortical MTs?

There are several possible mechanisms by which the cell might alter the stability of MTs within the cortical array. The most likely mechanisms are qualitative changes in tubulin gene expression (Joshi & Cleveland 1988; Sullivan 1988); detyrosination/tyrosination of alpha-tubulin (Gundersen & Bulinski 1986; Gundersen et al 1987; Webster et al 1987); acetylation of alpha-tubulin (Piperno et al 1987; Åström 1992; Bulinski et al 1988); changes in MAP interactions (Horio & Hotani 1986; Hotani & Horio 1988); phosphorylation of MT-protein (Sullivan 1988; Mizuno 1992; Gurland & Gunderson 1993); and fluctuations of calcium levels (Keith et al 1983; Cyr 1991).

Alpha-, beta-, and gamma-tubulin genes each comprise a separate gene family in plants (Goddard et al 1994) and multiple isoforms of alpha- and beta-tubulin have been described in a variety of plant tissues (Hussey et al 1991). The alpha and beta subunits are the most abundant forms of tubulin, existing in a 1:1 stoichiometry to one another and comprise the majority of tubulin found within MTs. The gamma-tubulin subunits are found in far lesser abundance and are thought to play a role in the nucleation of MTs (Goddard et al 1994). The expression of tubulin isoforms (as well as different gene family members) is heterogeneous within various tissues (Hussey et al 1991). The functional significance of this heterogeneity is unknown; however, it is intriguing to consider that various tubulin isoforms may differentially modulate the function of MTs. Luduena (1993) recently discussed how the various isoforms (in both plant and animal cells) might modulate the function of MTs in subtle, yet important, ways. In the context of this review it is worthwhile noting that the cortical array contains multiple isoforms of alpha- and beta-tubulin (Hussey et al 1987); it is intriguing to consider this may partially explain the heterogeneity in dynamic behavior within this array.

A number of studies have indicated a correlation between the stability of MTs and the detyrosination of alpha-tubulin. All alpha-tubulins (including those from plants) have a carboxyl-terminus composed of several glutamic acid residues followed by a terminal tyrosine (Fosket & Morejohn 1992). A preponderance of alpha-tubulins, lacking a terminal tyrosine, are found in some stable MTs in animal cells. The significance of detyrosinated tubulin is unclear because it does not appear to be a causative factor in the stabilization of MTs (Sullivan 1988). Glu-tubulin has not been reported in higher plants and attempts to label any MT structure in cultured plant cells with antibodies specific to glu-tubulin have failed (R Cyr, unpublished data).

MAPs have been isolated from a number of plant sources (Cyr & Palevitz 1989; Vantard et al 1991; Chang-Jie & Sonobe 1993; Schellenbaum et al 1993). One attribute of these proteins is the ability to stabilize MTs. MAPs are presumed to alter the stabilized state of MTs in a variety of nonplant cells (Chapin & Bulinski 1992; Lee 1993), and it is reasonable to propose that stabilizing MAPs might confer a degree of stability to cortical MTs. Some MAPs may be highly efficient at stabilizing MTs to extreme conditions. For example, crude MAP preparations from carrot cells confer cold stability to cold-labile MTs (Cyr & Palevitz 1989). This activity is physiologically relevant because the carrot cells from which these proteins were isolated have cold-stable cortical arrays (R Cyr, unpublished observation). Other MAPs may modulate stability states to a lesser degree. MAPs from maize and carrot are capable of stimulating the bulk assembly of low concentrations of tubulin (Cyr & Palevitz 1989a; Schellenbaum et al 1993), but no published studies have closely examined the dynamic characteristics of individual MTs assembled in the presence of plant MAPs. Comparative studies, using MAPs from cells that differ in the dynamic status of cortical MTs, may provide insight into the role these proteins play in affecting the dynamics of MTs within the cortical array. Ultimately, it will be necessary to demonstrate directly the significance of the in vitro findings with in vivo experimentation. Transgenic plants, which underexpress or overexpress specific MAPs, should prove beneficial to these studies.

Another mechanism that the cell may use to affect the dynamic status of cortical MTs is phosphorylation of MT proteins. Mizuno (1992) found that when cultured tobacco cells were treated with the kinase inhibitors, 6-DMAP and staurosporin, a subset of cortical MTs became cold stable. Furthermore, Sonobe (1990) reported that when lysed protoplasts are treated with ATP under specific conditions, cortical MTs disappear. Both findings support the idea that the phosphorylation of MT proteins may be necessary to maintain cortical MTs in a dynamic state. The target for phosphorylation could be tubulin (Koontz & Choi 1993) or a MAP (Lee 1993). An effect of phosphatase inhibition on cortical MTs in plant cells has not been demonstrated, but Gurland & Gunderson (1993) showed that okadaic acid (a phosphatase inhibitor) induced an increase in MT stability in cultured monkey cells. Future studies should reveal if plants possess a kinase/phosphatase system that serves to alter the dynamic status of cortical MTs.

Calcium is known to affect the MTs within the cortical array of plant cells (Cyr 1991). Although its inhibitory effect upon MTs in vitro is well documented (Bender & Rebhun 1986), its mode of action in vivo is less clear. Calmodulin (CaM) is required for Ca^{2+} to destabilize the cortical array of lysed carrot protoplasts. Furthermore, it appears that another CaM-binding protein is also required (Cyr 1991). Interestingly, a homologue to EF-1α has been identified that binds to CaM and MTs (Durso & Cyr 1994b). The binding of EF-1α to MTs results in the formation of bundles. Importantly, the MT bundling activity is abolished upon the addition of Ca^{2+}/CaM. Reports of interactions between EF-1α homologues and MTs are numerous, yet the phenomenon remains poorly understood (Durso & Cyr 1994a). Antibodies have been raised against an EF-1α homologue isolated from sea urchins. These antibodies not only demonstrate an association of this homologue to the spindles of sea urchins (Ohta et al 1988, 1990), but also with the cortical MT arrays of plant cells (Hasezawa & Nagata 1993). Data from biochemical and immunofluorescence studies indicate that an EF-1α homologue interacts in vivo with MTs, perhaps to affect their bundling. This bundling may be further modulated by a Ca^{2+}/CaM complex. If this hypothesis is correct, then the effect of Ca^{2+} on cortical MTs may involve a two-step process whereby Ca^{2+}/CaM unbundles MTs, which then depolymerize (also via a Ca^{2+}-mediated event).

MT-Organizing Centers in the Cortical Array

In animal cells, the assembly of a MT is typically limited to one (or a few) discrete MT organizing centers (MTOCs) (Mazia 1987). On the other hand, the MTOCs in plants are dispersed (see discussion in Palevitz 1991) as best evidenced by studies where MTs are depolymerized, then repolymerized, and the pattern of MT recovery examined closely. The majority of plant cells recovering from such treatment show dispersed MT foci in the cortex (Falconer et al 1988; Wasteneys & Williamson 1989). Additionally, when fluorescently labeled tubulin is microinjected into growing cells of *Nitella* and *Tradescantia* stamen hairs, dispersed fluorescent filaments appear in the cortex within minutes of injection (Wasteneys et al 1993). It could be argued that when tubulin is microinjected, a small region of unusually concentrated tubulin results, which could assemble without the benefit of a nucleating center; however, in this case one would expect assembly to be more prominent near the region of injection, but this was not observed.

One exception to dispersed cortical MTOCs was reported in differentiating guard cells by Marc and colleagues (1989b). If the cortical MTs of these cells are depolymerized with colchicine, then allowed to regrow (after inactivation of colchicne by photoisomerization), the cortical MTs appear to emanate only from the middle portion of the new ventral wall. Importantly, γ-tubulin appears to accumulate in the same area of the new ventral wall (McDonald et al 1993).

THE POSITIONING OF MTS IN THE CORTICAL ARRAY OF GROWING CELLS IS A CONSERVATIVE PROCESS

Cortical MTs in dividing, isodiametric cells of the apical meristem are randomly arranged (Baluška et al 1992). As the cell differentiates, the cortical MTs become more ordered into the characteristic transverse arrangement of an elongating cell (Laskowski 1990; Baluška et al 1992). Thus during normal cellular development, the cortical MT array progresses from a random to an ordered configuration.

Growing plant cells, recovering from either drug or cold-induced MT depolymerization, initially possess random cortical MT arrays that then become ordered (Hogetsu 1986b, 1987, Wasteneys & Williamson 1989, Wasteneys et al 1993). During recovery from experimental perturbation, the cortical MT array also progresses from a random to an ordered state.

Elongating plant cells have cortical MTs that are ordered in a net transverse orientation; upon enzymatic removal of the cell wall, a spherical protoplast is released, and the cortical array becomes relatively disordered. After culturing, the random cortical MTs become ordered (Lloyd et al 1980; Galway & Hardham 1986; Hasezawa et al 1988; Kuss-Wymer & Cyr 1992). Once again, there is a recapitulation of what is observed during normal development with a progression of cortical MT arrangements from a random to an ordered state.

If the ordering of the cortical array is considered to be a developmental process, then the organizational route should reveal key mechanistic steps in the ontogeny of the array. The consistent finding (under both normal and experimental conditions) that cortical MTs of growing cells are initially in a random state, which then proceeds into one that is ordered, provides evidence that there is only one fundamental mechanistic route that leads to the ordering of cortical MTs.

How are Cortical MTs Repositioned?

The cortical MTs of many elongating cells and growing protoplasts are dynamic. Microtubule dynamics are typically discussed in terms of dynamic instability and treadmilling. Dynamic instability refers to a steady state condition where MTs can exist in a minimum of two phase states, i.e. growing and shrinking, with the plus end of the MT dominating these phase transitions (Gelfand 1991; Cassimeris 1993). Phase transitions are thought to be governed by the state of GTP hyrolysis within the polymer, i.e. while a cap of GTP-tubulin remains at the MT end, it will continue growing; however, hydrolysis of the cap triggers the catastrophic shortening of the MT. Treadmilling refers to a steady state condition where there is a net growth of MTs at the plus ends, with a net shortening at the minus ends (Margolis & Wilson 1981). The FRAP data on *Tradescantia* stamen hairs (Hush et al 1994) are consistent with the cortical MTs behaving in a dynamically unstable fashion, and future experiments should clarify this point. Regardless of the nature of tubulin dynamics in the cortical array, it is clear that cortical MTs are in a constant state of flux, with turnover times as short as 60 sec (Hush et al 1994).

It is reasonable to propose that the spatial organization of the cortical array is dependent upon MTs being in a dynamic state. However, when protoplasts are treated with taxol (a MT-stabilizing drug), they elongate faster and to a greater extent than do untreated cells. Furthermore, the cortical arrays of taxol-treated protoplasts reorganize more quickly than do highly dynamic, untreated arrays (Kuss-Wymer & Cyr 1992; Kuss-Wymer 1994).

Two alternatives could explain how cortical MTs are oriented. One view is that MT dynamics are sufficient and necessary to explain MT positioning, i.e. orientation information is acquired at the time of nucleation. The other view is that MT positioning is not dependent on dynamics. Rather, MTs polymerize without the benefit of spatial information and then move into their correct position as intact units, i.e. spatial information would be acquired secondarily. The first alternative is not supported by the experimental depolymerization/repolymerization data, which show that MTs reappear disordered, then acquire order secondarily. Nonetheless, it is premature to discount a role for MT dynamics in the orientation of cortical MTs.

The two alternatives are not mutually exclusive (Palevitz 1991). Perhaps MTs move in toto, with the dynamic status of the array modulating the speed of this movement, the sensitivity to positional cues, or both. If the notion that both MT dynamics and movement play a role in the organization of the cortical array is accepted, then how is spatial fidelity achieved and maintained within this array?

HOW ARE CORTICAL MTS SPATIALLY ALIGNED?

The alignment of cortical MTs and cellulose MFs is analogous to a signal transduction event. In the broadest sense, any signal transduction pathway has three components: a sensor, which detects some environmental change; transduction machinery, which serves to interpret this information; and an effector event, which allows the cell to respond appropriately to the environmental stimulus. In the context of cortical cytoskeletal organization, the effector event is the alignment of cellulose MFs, whereas the transduction machinery is composed of MTs. The identity of the sensor and the nature of the signal to which it is sensitive remain unknown.

There are three candidates for the cue to which the cortical MT sensor responds; chemical, electrical, and biophysical. The evidence available to support each possibility is examined in the next section.

Evidence that a Chemical Signal is Involved in Providing Spatial Information to Orient Cortical MTs

Plant hormones are likely candidates for molecules that affect the positioning of MTs because there are several plant growth substances known to cause changes in the behavior and organization of cortical MTs.

Gibberellic acid (GA) is involved in cell and organ elongation and, owing to the involvement of MTs in cellular elongation, it is reasonable to hypothesize that this hormone may play a role in a signal transduction pathway that positions MTs. In many cases, treatment of plant organs with GA causes not only a rearrangement of cortical MTs (Shibaoka 1993), but also an increase in their stability (Sawhney & Srivastava 1974; Mita & Shibaoka 1984).

However, a relationship between GA treatment and MT stability is not always evident (Akashi & Shibaoka 1987; Sakiyama & Shibaoka 1990). Nonetheless, Shibaoka (1993) has proposed a relationship between MT stability and reorganization. That is, perhaps the two are causally related, although it is not clear if stable MTs reorient, or if reoriented MTs are stabilized. Shibaoka (1993) further suggested that perhaps GA functions to alter the association of MTs with the PM. In some cases, this also results in a change in the stability of the MTs. In order for MTs to convey spatial information to the wall, they must interact with the PM and, therefore, if GA enhances the association of cortical MTs to the PM, this would result in a more efficient conveyance of information.

Abscisic acid (ABA) is a plant hormone that can antagonize the effect of GA in stimulating organ elongation (Sakiyama-Sogo & Shibaoka 1993). The cortical MT arrays in cells obtained from ABA growth-suppressed organs are oriented obliquely, or longitudinally, to the former axis of elongation.

Ethylene is a hormone that redirects some organs to grow in girth, rather than length (Eisinger 1983; Baluška et al 1993). As with ABA treatment, the cortical MT array within the cells of treated organs show a change in orientation, from net transverse to longitudinal (Eisinger 1983; Roberts et al 1985; Baluška et al 1993).

While ample evidence shows that several hormones affect the cortical MT array, it is difficult to imagine how they could provide spatial information, with sufficient resolution, to accurately position the MTs within this array. It is more probable that they alter the characteristics of the array in a manner that predisposes the alignment of the MTs within the array. For example, regional changes in Ca^{2+} concentrations (serving as a second messenger) might alter the degree of MT-bundling and/or the degree of MT/PM associations. In addition, hormones might affect the cortical array by altering the expression of genes whose products modulate MT activity (a stabilizing or motor MAP might facilitate the physical movement of the MTs within the array, or a MT/PM linking protein, which alters the interaction between the MT and PM). It is also possible that hormones activate non-cytoskeletal genes whose products (or activities) then proceed to affect MT behavior in a secondary fashion, i.e. a kinase or phosphatase. Therefore, while a number of molecules affect the cortical array in a variety of ways, their role in providing a direct spatial cue to position MTs within the arrays is debatable. The main problem is one of spatial resolution. How could a diffusing molecule serve a role for the directional cuing of a MT? Even if we assume a concentration gradient, there is the problem of discrimination, i.e. how could a MT with a diameter of 25 nm detect a concentration difference over this small distance and with sufficient speed to alter its orientation?

Evidence that an Electrical Field is Involved in Providing Spatial Information to Orient Cortical MTs

An electrical field possesses both magnitude and directionality, hence it has the necessary vectorial characteristics to provide spatial information to the cortical array. Not only is this scenario theoretically attractive, but some experimental evidence supports its possibility. Isolated protoplasts of the alga *Mougeotia* elongate parallel to an imposed electric field (presumably with their cortical MTs oriented at right angles to the field vector; White et al 1990). Furthermore, when roots are placed in an electric field, their cortical MTs reorient (Hush & Overall 1991).

While these data clearly support the idea that MTs can spatially respond to an imposed electric field, the interpretation is hampered by two factors. First, there is a lack of knowledge concerning the nature of electrical fields within and around individual plant cells. The cytoplasmic electrical potential of plant cells can be quite high, with reported values for guard cells approaching − 250 mV (Assmann 1993). In the absence of channel activity, this potential will create a uniform field around the cell with global inward directions. The presence of ionic channels or pumps would predictably alter the magnitude of the field in regions of activity, without changing its direction. Thus with respect to the PM, the endogenous field vector is at right angles. The underlying MTs, in turn, would not be exposed to any differential force vector along their length and, therefore, it is difficult to envision how they could obtain meaningful spatial information.

The second interpretative problem relates to the observation that MTs have been observed to orient at right angles to imposed fields. Because MTs are predicted to behave as linear dipole rods, they would be expected to align parallel to an electric field. This discrepancy has led to the conclusion that MTs do not respond directly, but secondarily, to the electric field. Hush & Overall (1991) favor the idea that a PM/MT linking protein behaves as a dipole, thereby orienting the MTs. White and colleagues (1990) favor the interpretation that the imposed electric field induces a deformation of the PM which, in turn, imposes a strain on the underlying cortical MTs that alters their orientation. The notion that biophysically generated strains can influence MT orientation is intriguing and is covered in greater detail in the next section.

Evidence that a Biophysical Force is Involved in the Orientation of Cortical MTs

A biophysical cue for orienting MTs was originally suggested by Green and colleagues (1970). Although data for this hypothesis are largely correlative, it remains an appealing notion (Williamson 1990, 1991), primarily because plant growth is based on relatively large hydrostatic forces acting upon rigid cell walls. The rigid wall permits the transmission of stresses generated by one cell to be conveyed to adjoining cells. When cells act in concert, forces are exerted across tissues and organs. Although hydrostatic force is isotropic, the geometry of plant cells, together with their mechanically anisotropic walls, results in vectored transmission of stresses generated by turgor pressure. The fact that biophysical forces can be vectorially transmitted between cells, tissues, and organs theoretically allows the plant to convey information of structural significance between cells. The potential for transcellular propagation of a MT-alignment signal is attractive because cortical MT alignment within a given tissue often reveals remarkable co-alignment between adjacent cells (Hogetsu 1986a; Baskin et al 1992; especially see Figure 6a in Laskowski 1990). Although the idea of a biophysical MT-cuing mechanism is attractive, there is a paucity of direct evidence for its support.

Three studies have directly investigated biophysical MT-cuing mechanisms. Hush & Overall (1991) demonstrated that cells within roots, previously exposed to lateral non-injurious compression, possessed MTs that had reoriented perpendicular to the applied forces. Cleary & Hardham (1993) utilized *Lolium* leaves that were exposed to 50 MPa of pressure for 5–20 min, followed by a controlled decompression. Many cells within these treated tissues had cortical arrays that had reoriented. These studies provide direct evidence to support the idea that MTs are not only sensitive to mechanical forces but, additionally, that mechanical forces can influence the reorientation of the cortical arrays. It has also been shown that a brief centrifugation of freshly isolated protoplasts provides a cue for elongation at right angles to the applied centrifugal force (Kuss-Wymer 1994). Reversible application of the anti-MT herbicide APM revealed that an intact MT array was required for the perception of the cuing force. The finding that cortical MTs are necessary for cuing an elongative axis to a centrifugal force provides direct evidence that cortical MTs act as transducers, to convey spatial information to cellulose synthase, and as sensors, to receive physical force information. All three studies provide a framework for considering how cortical MTs might interact with wall components to influence cellulose deposition and how they may be oriented by forces that result from the anisotropic orientation of the MFs. In other words, MT orientation is determined by mechanical forces which, in turn, influence the deposition of cellulose MFs. The MFs affect the distribution of mechanical forces, which are experienced by the MTs, and so on. In this manner, a self-reinforcing mechanical feed-back loop is created. In considering how this system might work, it is necessary to regard the cytoplasm, the PM, and the wall as a functional continuum.

MTs, THE PM, AND THE CELL WALL AS A FUNCTIONAL CONTINUUM

Figure 3 summarizes the available data on the relationship between the two major wall polymers involved in elongation (cellulose and xyloglucans; Hayashi 1989; Carpita & Gibeaut 1993), the PM, and the underlying cortical MTs (Giddings & Staehlin 1991). The cellulose MFs do not undergo appreciable strain (stress-induced deformation) under commonly reported cell wall conditions. Xyloglucans[1]probably constitute the major strain-sensitive elements in the growing cell (Carpita & Gibeaut 1993). For growth to occur, the xyloglucan chains must yield, which results in wall relaxation in a direction at right angles to the cellulose MFs. The manner in which the exocellular matrix (ECM) of plant cells interacts with the PM is not well understood, but

[1]Xyloglucans are likely candidates for noncellulose carbohydrates that connect MFs to one another. However, there are a variety of other carbohydrates that may also be involved. For convenience of discussion (and illustrative clarity) I will refer only to this polymer. In the context of the model, it is not important if xyloglucans function alone or in concert with other polymers.

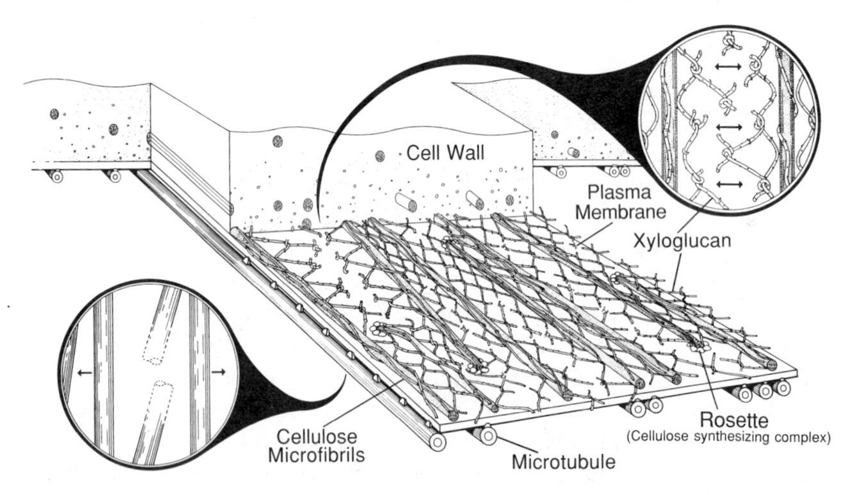

Figure 3 The MTs, PM, and wall form a functional continuum. This cartoon represents some events that occur in one portion of a cell with a compliant wall. In a growing primary wall there are a few layers of cellulose microfibrils (only the layer closest to the PM is drawn in detail). Other proteins and carbohyrates are present, but have been omitted for clarity. The insert on the upper right shows the predicted directions of strain in a compliant wall that has just undergone bond rearrangement in the xyloglucan chains. The insert on the lower left depicts the resulting strains predicted to affect the underlying MTs within the cell's cortex. Two MTs at right angles to the elongative axis have moved apart (being strain-protected by the overlying MFs), while one MT at an angle has been subjected to shear-induced breakage and is in the process of depolymerizing. Inspired by a drawing in Giddings & Staehlin (1991).

it probably involves integrin-type proteins that link the ECM to the PM (Wyatt & Carpita 1993). The intimate associations between the ECM and PM, and the PM with the cortical MTs, mean that biophysical forces resulting from alterations in the ECM will be transmitted to the underlying cytoskeleton. Thus the ECM, PM, and cortical cytoskeleton form a functional continuum that can affect the orientation of cortical MTs. In Figure 3, an instantaneous moment in the life of a growing cell is represented in which the xyloglucan chains have yielded, resulting in a longitudinal strain (*arrows*) that moves the cellulose MFs apart. The resulting effects upon cortical MTs are also depicted; those MTs at right angles to the strain are proposed to be protected by the overlying cellulose MFs, while one MT lying at an angle is shown subjected to shear-induced breakage followed by depolymerization.

In the model depicted in Figure 3, the anisotropic nature of the wall provides a protective milieu for some, but not all, MTs. The MFs bear the load (i.e do not undergo strain) and hence any "strain-protected" MT will continue to direct the synthesis of MFs in the favored direction. Perpetuation of cellulose reinforcement at right angles to the axis of elongation is insured. However, the situation is markedly different in the direction parallel to the elongative axis, where the wall is compliant. Any MT in this orientation will be subject to stress and hence be exposed to strain or shear-induced breakage (depending upon its angle to the major strain axis).

As described above, the MTs function as strain gauges. In order for MTs to function efficiently in such a capacity, they must depolymerize rapidly once they are broken (so they will not influence cellulose deposition). Thus a highly dynamic MT would be a more effective strain gauge than one that was less dynamic. Mechanistically, this rapid MT depolymerization could work in a simple way, i.e. if cortical MTs behave in a dynamically unstable fashion, then sheared MTs would be deprived of their GTP cap (see discussion in Gelfand & Bershadsky 1991) and would rapidly undergo catastrophic decay.

The idea of a strain-based biased-turnover mechanism for organizing cortical MTs was first proposed by Wasteneys & Williamson (1989). However, Williamson (1990, 1991) subsequently noted that one problem with such strain-based orientation models was that cells treated with an anti-MT agent (and consequently growing in girth) could reform a normal array after drug removal, then elongate. Normal growth, Williamson argued, should not resume if MTs respond directly to biophysical forces because the major growth direction should continue to be in girth (Williamson 1990, 1991). However, wall deposition prior to drug application may still affect the anisotropic distribution of forces experienced by the cortical MT array in a manner that allows them to cue into the proper position, i.e. the old wall could still have sufficient order to anisotropically distribute expansive forces. Therefore, it is premature to discount a biophysical model of biased turnover solely on this observation.

Williamson (1990) modified his strain-based model for cortical MT arrangement by invoking a hypothetical strain-sensitive MTOC. Although attractive, an entity with such character has not been identified. However, it is intriguing to consider the work of Liu and colleagues (1993), who described the presence of gamma-tubulin in the cortical MT array, which might serve as a nucleating center for MTs. Perhaps the short nucleating polymer formed by gamma-tubulin serves as a strain sensor, which could then affect the direction of further MT growth.

In the scheme depicted in Figure 3, the magnitude of strain predicted to affect the orientation of cortical MTs might be less than that discernible by a crude measurement of cell growth. Therefore, it is inappropriate to discount any biophysical model for the alignment of cortical MTs based on growth studies that lack sufficient resolution to detect strain at the subcellular level (Laskowski 1990; Sakiyama-Sogo & Shibaoka 1993; Shibaoka 1993). Conversely, a rigorous test of this model requires experimental approaches that possess sufficient resolution to detect minute strains in small areas of the cell. Ultimately, the validity of any model that attempts to describe the orientation of cortical MTs should be judged on two basic criteria: first, it must be testable; second, it must account for (if not mechanistically describe) the different MT arrangements in various developmental states.

In cells that are elongating slowly, or have ceased growing, cortical MT arrangements other than transverse are observed. As the growth of a cell diminishes, the orientation of both MFs and MTs change pitch relative to the axis of elongation. Can the model outlined in Figure 3 account for these alternative arrangements? It is the compliance of the cell wall that limits growth. Because xyloglucans are the major strain-sensitive polymer, it is expected that the number of bond breakages (or rearrangements) occurring in this polymer will decrease as growth slows. That is, irreversible linkages between MFs will increase. How the wall becomes rigidified is unclear; it may involve an increase in pectin cross-linking (by Ca^{2+} after de-esterification), an increase in cross-linking by wall proteins (extensins), or a combination of these events (Carpita & Gibeaut 1993). In plants grown under normal conditions, cellular growth rate reaches a maximum value (the time and magnitude are dependent upon the species, the organ, and the growth conditions) that is sustained for some period of time, followed by a decrease over a period of hours (Silk 1992). Wall rigidification is therefore not an instantaneous process. In a wall that is partially cross-linked (but still capable of yielding, albeit to a lesser extent), some movement of the MFs would be predicted. However, because the MFs are partially cross-linked, they no longer move independently away from one another. Rather, they move together—the cell increases in length because the partially cross-linked "net" of MFs rotate in unison towards the longitudinal axis. This is not a new idea; such a scheme is thought to contribute to the spiral or helical arrangement of MFs observed in cell walls, i.e. "multinet growth" (Roelofsen & Houwink 1953; Green 1960).

What then happens to the underlying MTs in a slowly growing cell? Although the MTs would experience some shear, the magnitude would be lower than in rapidly growing cells because the cell expansion rate is reduced. Some of the longer MTs might break and depolymerize, but shorter MTs would remain intact because they would be exposed to fewer shearing forces. An important physical feature of shorter cortical MTs is that if they are moved (instead of breaking), their angular movement would be greater; hence for any given amount of strain in the membrane to which they are associated, shorter MTs would reorient to a greater extent than would longer MTs. If we propose that only short MTs survive these shearing forces and that they proceed to direct cellulose synthesis (recall that the cortical array is composed of short and overlapping MTs), then the net pitch of the cortical MT array will be amplified until the net orientation of the array becomes longitudinal. An elongating cell undergoes these transitions of order. As the cell begins rapid expansion, the cortical array has a net transverse orientation; but as growth slows, the array becomes more helical. As growth ceases, the arrays may become longitudinal.

It has been suggested that the cortical MT array is a dynamic helix (Lloyd & Seagull 1985) that moves as a cohesive and integral unit during growth. This idea has not advanced much primarily due to the lack of a mechanistic explanation about how the entire array could move in toto. The scenario outlined above has all the mechanistic features to account for the occurrence of a helical array: partial rigidification of the wall, causing angular movement of cellulosic MFs towards the longitudinal axis, leading to a rotational movement of short MTs within the underlying cortical MTs. The change in pitch of the cortical MT array was originally suggested to come about via the resulting forces associated with wall extension; however, this notion was questioned on geometric grounds. That is, if one envisions a cell as a long rod, and the MTs as a continuous spiral structure, then even a large (50%) change in cell length would have a only a small effect upon the angular displacement (relative to the elongative axis) of the cortical MTs (Laskowski 1990). However, the cortical array is discontinuous, being composed of short overlapping segments. These shorter pieces would demonstrate a much greater angular movement per unit. If the cell expands uniformly along its length, then a myriad of short MTs could collectively reorient to the oblique or longitudinal position that is observed.

Can this model account for the random MT configuration seen in some cells? All cells are initially derived from apical meristematic areas containing varying numbers of cells. In shoot apical meristems, these cells are typically small and isodiametric, with random cortical MT arrays (Sakaguchi et al 1988,

1990; Baluška et al 1992). The forces that these cells experience are unknown. It is possible that for some period of their life, these cells are exposed to isotropic forces; hence, their cortical MTs would receive no net biophysical cue and therefore would be disorganized. Another intriguing possibility is that the functional coupling of cortical MTs to the PM is developmentally regulated, and the mechanism by which MTs form significant interactions with the PM is not expressed in rapidly dividing apical meristematic cells.

CONCLUSIONS AND PERSPECTIVES

Cortical MTs are morphogenetically important elements that provide spatial information for the orientation of cellulose MFs, and in the form of the PPB are implicated in cell plate positioning. Mechanistic details about how these MTs acquire spatial information remain unknown, but biophysical forces are likely candidates. Cortical MTs do not function in isolation, and future efforts should focus upon the behavior of MTs in the context of a functional continuum consisting of MTs, the PM, and the cell wall. As a starting point for discussions and experimentation, an hypothesis is presented herein to explain how individual components of this continuum are involved in orchestrating cellular elongation.

Future research efforts will confront several challenges. It will be necessary to identify the major molecular players regulating this continuum. Specifically, it will be essential to know how cortical MTs are linked to the PM as well as to learn more about how the wall associates with the PM. Future studies should focus upon understanding how plant MAPs affect the functioning of cortical MTs in vivo; transgenic approaches should prove especially useful in this area.

A better understanding about the role MT dynamics play in cortical MT functioning and how the cell regulates the dynamic status of cortical MTs needs further investigation. How MT dynamics are influenced by MAPs, intracellular signals, and perhaps even by biophysical forces, will be especially important areas to explore.

In the next decade, experimental approaches that examine the behavior of cortical MTs in the context of a living, growing cell will undoubtedly be the most enlightening. The ability to follow fluorescently tagged cytoskeletal proteins within living cells (in conjunction with experimental perturbations) should provide a clearer picture of the true repertoire of MT behaviors that enable these elements to contribute, in pivotal ways, to the process of plant morphogenesis.

ACKNOWLEDGMENTS

I thank several individuals who directly contributed to this review: Deborah Fisher for her comments, patience, and assistance during its preparation; Neil

Durso, Paul Green, Barry Palevitz, and Richard Williamson for helpful comments; Carol Wymer for both her comments on the review, as well as many discussions over the last several years. This work was supported by grants from the United States Department of Agriculture and the Department of Energy.

Literature Cited

Akashi T, Kawasaki S, Shibaoka H. 1990. Stabilization of cortical microtubules by the cell wall in cultured tobacco cells. *Planta* 182: 363–69

Akashi T, Shibaoka H. 1987. Effects of gibberellin on the arrangement and the cold-stability of cortical microtubules in epidermal cells of pea internodes. *Plant Cell. Physiol.* 28: 339–48

Assmann SM. 1993. Signal transduction in guard cells. *Annu. Rev. Cell Biol.* 9:345–75

Åström H. 1992. Acetylated α-tubulin in the pollen tube microtubules. *Cell Biol. Int. Rep.* 16:871–81

Baluška F, Barlow PW. 1993. The role of the microtubular cytoskeleton in determining nuclear chromatin structure and passage of maize root cells through the cell cycle. *Eur. J. Cell Biol.* 61:160–67

Baluška F, Brailsford RW, Hauskercht M, Jackson MB, Barlow PW. 1993. Cellular dimorphism in the maize root cortex: Involvement of microtubules, ethylene and gibberellin in the differentiation of cellular behaviour in postmitotic growth zones. *Bot. Acta* 106: 394–404

Baluška F, Parker JS, Barlow PW. 1992. Specific patterns of cortical and endoplasmic microtubules associated with cell growth and tissue differentiation in roots of maize (*Zea mays* L). *J. Cell Sci.* 103:191–200

Barlow PW, Adam JS. 1989. Experimental control of cellular patterns in the cortex of tomato roots. In *Structural and Functional Aspects of Transport in Roots,* ed. BC Loughman, O Gasparikova, J Kolek, pp. 21–24. Dordrecht: Kluwer

Baskin TI, Busby CH, Fowke LC, Sammut M, Gubler F. 1992. Improvements in immunostaining samples embedded in methacrylate: localization of microtubules and other antigens throughout developing organs in plants of diverse taxa. *Planta* 187:405–13

Baskin T, Cande W. 1990. The structure and function of the mitotic spindle in flowering plants. *Annu. Rev. Plant Physiol. Plant Mol. Biol.* 41:277–315

Bender PK, Rebhun LI. 1986. The calcium sensitivity of MAP-2 and tau microtubules in the presence of calmodulin. *Ann. NY Acad. Sci.* 466:392–409

Bulinski JC, Richards JE, Piperno G. 1988. Posttranslational modification of alpha-tubulin: Detyrosination and acetylation differentiate populations of interphase microtubules in interphase cells. *J. Cell Biol.* 106: 1213–20

Carpita NC, Gibeaut DM. 1993. Structural models of primary cell walls in flowering plants: consistency of molecular structure with the physical properties of the walls during growth. *Plant J.* 3:1–30

Cassimeris L. 1993. Regulation of microtubule dynamic instability. *Cell Motil. Cytoskeleton* 26:275–81

Cassimeris LU, Walker RA, Pryer NK, Salmon ED. 1987. Dynamic instability of microtubules. *BioEssays* 7:149–54

Chang-Jie J, Sonobe S. 1993. Identification and preliminary characterization of a 65-kDa higher-plant microtubule-associated protein. *J. Cell Sci.* 105:891–901

Chapin SJ, Bulinski JC. 1992. Microtubule stabilization by assembly-promoting microtubule-associated proteins: A repeat performance. *Cell Motil. Cytoskeleton* 23:236–43

Cimler BM, Andreasen TJ, Andreasen KI, Storm DR. 1985. P-57 is a neural specific calmodulin-binding protein. *J. Biol. Chem.* 260:10784–88

Cleary AL, Gunning BE, Wasteneys GO, Hepler PK. 1992. Microtubule and F-actin dynamics at the division site in living *Tradescantia* stamen hair cells. *J. Cell Sci.* 103: 977–88

Cleary AL, Hardham AR. 1993. Pressure induced reorientation of cortical microtubules in epidermal cells of *Lolium rigidum* leaves. *Plant Cell Physiol.* 34:1003–8

Cosgrove DJ. 1989. Characterization of long-term extension of isolated cell walls from growing cucumber hypocotyls. *Planta* 177: 121–30

Cosgrove DJ. 1993a. How do plant cell walls extend? *Plant Physiol.* 102:1–6

Cosgrove DJ. 1993b. Tansley review no. 46: Wall extensibility: Its nature, measurement and relationship to plant cell growth. *New Phytol.* 124:1–23

Cosgrove DJ. 1993c. Water uptake by growing cells: An assessment of the controlling roles of wall relaxation, solute uptake, and hydraulic conductance. *Int. J. Plant Sci.* 154:10–21

Cyr RJ. 1991. Calcium/calmodulin affects microtubule stability in lysed protoplasts. *J. Cell Sci.* 100:311–17

Cyr RJ, Palevitz BA. 1989a. Microtubule-binding proteins from carrot. I. Initial characterization and microtubule bundling. *Planta* 177:245–60

Ding B, Turgeon R, Parthasarathy MV. 1991. Microfilaments in the preprophase band of freeze substituted tobacco root cells. *Protoplasma* 165:209–11

Doohan ME, Palevitz B. 1980. Microtubules and coated vesicles in guard cell protoplasts of *Allium cepa* L. *Protoplasma* 149:389–401

Durso NA, Cyr RJ. 1994a. Beyond translation: Elongation factor 1-α (EF1-α) and the cytoskeleton. *Protoplasma.* In press

Durso NA, Cyr RJ. 1994b. A calmodulin-sensitive interaction between microtubules and a higher plant homolog of elongation factor 1-α. *Plant Cell.* In press

Eigsti OJ, Dustin P. 1955. *Colchicine in Agriculture, Medicine, Biology and Chemistry.* Ames: Iowa State College Press

Eisinger W. 1983. Regulation of pea internode expansion by ethylene. *Annu. Rev. Plant Physiol.* 34:225–40

Eleftheriou E, Palevitz B. 1992. The effect of cytochalasin D on preprophase band organization in root tip cells of *Allium. J. Cell Sci.* 103:989–98

Emons AC. 1982. Microtubules do not control microfibril orientation in a helicoidal cell wall. *Protoplasma* 113:85–87

Esau K. 1977. *Plant Anatomy.* New York: Wiley & Sons. 550 pp.

Falconer MM, Donaldson G, Seagull RW. 1988. MTOCs in higher plant cells: An immunofluorescent study of microtubule assembly sites following depolymerization by APM. *Protoplasma* 144:46–55

Falconer MM, Seagull RW. 1986. Xylogenesis in tissue culture II: Microtubules, cell shape and secondary wall patterns. *Protoplasma* 133:140–48

Falconer M, Seagull R. 1987. Amiprophosmethyl (APM): A rapid, reversible, anti-microtubule agent for plant cell cultures. *Protoplasma* 136:118–24

Fosket DE, Morejohn LC. 1992. Structural and functional organization of tubulin. *Annu. Rev. Plant Physiol.* 43:201–40

Fukuda H, Kabayashi H. 1989. Dynamic organization of the cytoskeleton during tracheary-element differentiation. *Dev. Growth Differ.* 31:9–16

Galway ME. 1986. Microtubule reorganization, cell wall synthesis and establishment of the axis of elongation in regenerating protoplasts of the alga *Mougeotia. Protoplasma* 135: 130–43

Gelfand VI, Bershadsky AD. 1991. Microtubule dynamics - mechanism, regulation, and function. *Annu. Rev. Cell Biol.* 7:93–116

Gertel ET, Green PB. 1977. Cell growth pattern and wall microfibrillar arrangment. *Plant Physiol.* 60:247–54

Giddings TH, Staehelin LA. 1988. Spatial relationship between microtubules and plasmamembrane rosettes during the deposition of primary wall microfibrils in *Closterium* sp. *Planta* 173:22–30

Giddings TH Jr, Staehelin LA. 1991. Microtubule-mediated control of microfibril deposition: A re-examination of the hypothesis. In *The Cytoskeletal Basis of Plant Growth and Form,* ed. CW Lloyd, pp 85–100. San Diego: Academic

Gifford EM, Foster AS. 1989. *Morphology and Evolution of Vascular Plants.* New York: Freeman. 626 pp.

Goddard RH, Wick SW, Silflow CD, Snustad DP. 1994. Microtubule components of the plant cell cytoskeleton. *Plant Physiol.* 104:1–6

Green PB. 1960. Multinet growth in the cell wall of *Nitella. J. Biophys. Biochem. Cytol.* 7:289–96

Green P. 1962. Mechanism for plant cellular morphogenesis. *Science* 138:1404–5

Green PB. 1980. Organogenesis- a biophysical view. *Annu. Rev. Plant Physiol.* 31:51–82

Green PB, Erickson RO, Richmond PA. 1970. On the physical basis of cell morphogenesis. *Ann. NY Acad. Sci.* 175:712–31

Green PB, Poethig RS. 1982. Biophysics of the extensin and initiation of plant organs. In *Developmental Order: Its Origin and Regulation,* ed. S. Subtelny, PB Green, pp. 485–509. New York: Liss

Gundersen GG, Bulinski J. 1986. Microtubule arrays in differentiated cells contain elevated levels of a post-translationally modified form of tubulin. *Eur. J. Cell Biol.* 42:288–94

Gundersen GG, Khawaja S, Bulinski JC. 1987. Postpolymerization detyrosination of α-tubulin: A mechanism for subcellular differentiation of microtubules. *J. Cell Biol.* 105: 251–64

Gunning BES. 1982. The cytokinetic apparatus: Its development and spatial regulation. In *The Cytoskeleton in Plant Growth and De-*

velopment, ed. C Lloyd, pp 230–88. New York: Academic

Gunning BES, Hardham AR. 1982. Microtubules. *Annu. Rev. Plant Physiol.* 33:651–98

Gunning BES, Sammut M. 1990. Rearrangements of microtubules involved in establishing cell division planes start immediately after DNA synthesis and are completed just before mitosis. *Plant Cell* 2:1273–82

Gunning BES, Wick SM. 1985. Preprophase bands, phragmoplasts and spatial control of cytokinesis. *J. Cell. Sci. Suppl.* 2:157–79

Gurland G, Gunderson G. 1993. Protein phosphatase inhibitors induce the selective breakdown of stable microtubules in fibroblasts and epithelial cells. *Proc. Natl. Acad. Sci. USA* 90:8827

Hardham AR, Gunning BES. 1978. Structure of cortical microtubule arrays in plant cells. *J. Cell Biol.* 77:14–34

Hasezawa S, Hogetsu T, Syono K. 1988. Rearrangement of cortical microtubules in elongating cells derived from tobacco protoplasts—A time-course observation by immunofluorescence microscopy. *J. Plant Physiol.* 133:46–51

Hasezawa S, Nagata T. 1993. Microtubule organizing centers in plant cells: Localization of a 49 kDa protein that is immunologically cross-reactive to a 51 kDa protein from sea urchin centrosomes in synchronized tobacco BY-2 cells. *Protoplasma* 176:64–74

Hayashi T. 1989. Xyloglucans in the primary cell wall. *Annu. Rev. Plant Physiol. Plant Mol. Biol.* 40:139–68

Hepler PK, Cleary AL, Gunning BES, Wadsworth P, Wasteneys GO, et al. 1993. Cytoskeletal dynamics in living plant cells. *Cell Biol. Int.* 17:127–42

Hepler PK, Fosket DE. 1971. The role of microtubules in vessel member differentiation in *Coleus. Protoplasma* 72:213–36

Hepler PK, Jackson WT. 1968. Microtubules and early stages of cell-plate formation in the endosperm of *Haemanthus katherinae* Baker. *J. Cell Biol.* 38:437–46

Hepler P, Palevitz B. 1974. Microtubules and microfilaments. *Annu. Rev. Plant Physiol.* 25:309–62

Hofmeister W. 1863. Zusatze und Berichtigungen zu den 1851 veröffentlichen Untersuchungen der Entwicklung höherer Kryptogamen. *Jahrb. Wiss. Bot.* 3:259–93

Hogetsu T. 1986a. Orientation of wall microfibril deposition in root cells of *Pisum sativum* L. var. Alaska. *Plant. Cell Physiol.* 27:947–51

Hogetsu T. 1986b. Re-formation of microtubules in *Closterium ehrenbergii* Meneghini after cold-induced depolymerization. *Planta* 167:437–43

Hogetsu T. 1987. Re-formation and ordering of wall microtubules in *Spirogyra* cells. *Plant Cell Physiol.* 28:875–83

Horio T, Hotani H. 1986. Visualization of the dynamic instability of individual microtubules by dark-field microscopy. *Nature* 321:605

Hotani H, Horio T. 1988. Dynamics of microtubules visualized by darkfield microscopy: treadmilling and dynamic instability. *Cell Motil. Cytoskeleton* 10:229–36

Hugdahl JD, Morejohn LC. 1993. Rapid and reversible high-affinity binding of the dinitroanaline herbicide oryzalin to tubulin from *Zea mays* L. *Plant Physiol.* 102:725–40

Hush J, Overall R. 1991. Electrical and mechanical fields orient cortical microtubules in higher plant tissues. *Cell Biol. Int. Rep.* 15:551–60

Hush JM, Wadsworth P, Callaham DA, Hepler PK. 1994. Quantification of microtubule dynamics in living plant cells using fluorescence redistribution after photobleaching. *J. Cell Sci.* 107:775–84

Hussey PJ, Snustad DP, Silflow CD. 1991. Tubulin gene experession in higher plants. See Lloyd 1991, pp. 15–28

Hussey PJ, Traas JA, Gull K, Lloyd CW. 1987. Isolation of cytoskeletons from synchronized plant cells: the interphase microtubule array utilizes multiple tubulin isotypes. *J. Cell Sci.* 88:225–30

Joshi HC, Cleveland DW. 1988. Differential utilization of the available beta tubulin isotypes in differentiating neurites. *J. Cell Biol.* 109:663–73

Keith C. 1983. Microinjection of calcium-calmodulin causes a localized depolymerization of microtubules. *J. Cell Biol.* 97: 1918–24

Koontz DA, Choi JH. 1993. Evidence for phosphorylation of tubulin in carrot suspension cells. *Physiol. Plant.* 87:576–83

Kuss-Wymer CL. 1994. *Microtubule cortical array reorientation and stability in plant cells.* PhD thesis. Penn. State Univ. College Park, PA. 165 pp.

Kuss-Wymer CL, Cyr RJ. 1992. Tobacco protoplasts differentiate into elongate cells without net microtubule depolymerization. *Protoplasma* 168:64–72

Lancelle SA, Cresti M, Hepler PK. 1987. Ultrastructure of the cytoskeleton in freeze-substituted pollen tubes of *Nicotiana alata. Protoplasma* 140:141–50

Lancelle SA, Hepler PK. 1991. Association of actin with cortical microtubules revealed by immunogold localization in *Nicotiana* pollen tubes. *Protoplasma* 165:167–72

Laporte K, Rossignol M, Traas JA. 1993. Interaction of tubulin with the plasma membrane-tubulin is present in purified plasmalemma and behaves as an integral membrane protein. *Planta 191:413–16*

Laskowski MJ. 1990. Microtubule orientation

in pea stem cells: a change in orientation follows the initiation of growth rate decline. *Planta* 181:44–52

Lee G. 1993. Non-motor microtubule-associated proteins. *Curr. Opin. Cell Biol.* 5:88–94

Lindenmayer A. 1978. Algorithms for plant morphogenesis. In *Theoretical Plant Morphology,* ed. R Sattler, pp. 37–81, Leiden: Leiden Univ. Press

Liu B, Marc J, Joshi HC, Palevitz BA. 1993. A γ- tubulin-related protein associated with the microtubule arrays of higher plants in a cell cycle-dependent manner. *J. Cell Sci.* 104: 1217–28

Lloyd CW, ed. 1982. *Cytoskeleton in Plant Growth and Development.* New York: Academic

Lloyd CW, ed. 1991a. *The Cytoskeletal Basis of Plant Growth and Form.* San Diego: Academic. 330 pp.

Lloyd CW. 1991b. Cytoskeletal elements of the phragmosome establish the division plane in vacuolated higher plant cells. See Lloyd 1991, pp. 245–58

Lloyd CW, Barlow PW. 1982. The co-ordination of cell division and elongation: The role of the cytoskeleton. See Lloyd 1982, pp. 203–28

Lloyd CW, Seagull RW. 1985. A new spring for plant cell biology: microtubules as dynamic helices. *Trends Biol. Sci.* 10:476–78

Lloyd CW, Slabas AR, Powell AJ, Lowe SB. 1980. Microtubules, protoplasts, and plant cell shape: An immunofluorescent study. *Planta* 147:500–6

Lloyd CW, Venverloo CJ, Goodbody KC, Shaw PJ. 1992. Confocal laser microscopy and three-dimensional reconstruction of nucleus-associated microtubules in the division plane of vacuolated plant cells. *J. Micros.* 166:99–109

Luduena RF. 1993. Are tubulin isotypes functionally significant. *Mol. Biol. Cell* 4:445–57

Marc J, Hackett W. 1992. Changes in the pattern of cell arrangement at the surface of the shoot apical meristem in *Hedera helix* L. following gibberellin treatment. *Planta* 186: 503–10

Marc J, Mineyuki Y, Palevitz B. 1989a. The generation and consolidation of a radial array of cortical microtubules in developing guard cells of *Allium cepa* L. *Planta* 179:516–29

Marc J, Mineyuki Y, Palevitz BA. 1989b. A planar microtubule-organizing zone in guard cells of *Allium:* experimental depolymerization and reassembly of microtubules. *Planta* 179:530–40

Margolis RL, Wilson L. 1981. Microtubule treadmills-possible molecular machinery. *Nature* 293:705

Mazia D. 1987. The chromosome cycle and the centrosome cycle in the mitotic cycle. *Int. Rev. Cytol.* 100:49–92

McDonald AR, Liu B, Joshi HC, Palevitz BA. 1993. γ-Tubulin is associated with a cortical-microtubule-organizing zone in the developing guard cells of *Allium cepa* L. *Planta* 191: 357–61

Mineyuki Y, Marc J, Palevitz B. 1991. Relationship between the preprophase band, nucleus and spindle in dividing *Allium* cotyledon cells. *J. Plant Physiol.* 138:640–49

Mineyuki Y, Palevitz BA. 1990. Relationship between preprophase band organization, F-actin and the division site in *Allium.* Fluorescence and morphometric studies on cytochalasin treated cells. *J. Cell Sci.* 97: 283–95

Mita T, Shibaoka H. 1984. Gibberellin stabilizes microtubules in onion leaf sheath cells. *Protoplasma* 119:100–9

Mitchison TJ. 1988. Microtubule dynamics and kinetochore function in mitosis. *Annu. Rev. Cell Biol.* 4:527–49

Mitchison T. 1989. Polewards microtubule flux in the mitotic spindle: Evidence from photoactivation of fluorescence. *J. Cell. Biol.* 109:637–52

Mizuno K. 1992. Induction of cold stability of microtubules in cultured tobacco cells. *Plant Physiol.* 100:740–48

Morejohn LC. 1991. The molecular pharmacology of plant tubulin and microtubules. See Lloyd 1991, pp. 29–44

Newcomb EH. 1969. Plant microtubules. *Annu. Rev. Plant Physiol.* 20:253–88

Niklas KJ. 1992. *Plant Biomechanics: An Engineering Approach to Plant Form and Function.* Chicago: Univ. Chicago Press

Ohta K, Toriyama M, Endo S, Sakai H. 1988. Localization of mitotic-apparatus-associated 51-kD protein in unfertilized and fertilized sea urchin eggs. *Cell Motil. Cytoskeleton* 10: 496–505

Ohta K, Toriyama M, Miyazaki M, Murofushi H, Hosoda S, et al. 1990. The mitotic apparatus-associated 51-kDa protein from sea urchin eggs is a GTP-binding protein and is immunologically related to yeast polypeptide elongation factor 1α. *J. Biol. Chem.* 265: 3240–47

Palevitz B. 1986. Division plane determination in guard mother cells of *Allium:* video time-lapse analysis of nuclear movements and phragmoplast rotation in the cortex. *Dev. Biol.* 117:644–54

Palevitz BA. 1991. Potential significance of microtubule rearrangement, translocation and reutilization in plant cells. See Lloyd 1991, pp. 45–56

Piperno G, LeDizet M, Chang X. 1987. Microtubules containing acetylated alpha-tubulin in mammalian cells in culture. *J. Cell Biol.* 104:289–302

Rayle DL, Cleland RE. 1992. The acid growth theory of auxin-induced cell elongation is alive and well. *Plant Physiol.* 99:1271–74

Roberts IN, Lloyd CW, Roberts K. 1985. Ethylene-induced microtubule reorientations: Mediation by helical arrays. *Planta* 164:439–47

Roelofsen PA, Houwink AL. 1953. Architecture and growth of the primary cell wall in some plant hairs and in the *Phycomyces* sporangiophore. *Acta Bot. Neerl.* 2:218–25

Sakaguchi S, Hogetsu T, Hara N. 1988. Arrangement of cortical microtubules in the shoot apex of *Vinca major* L. *Plant* 175:403–11

Sakaguchi S, Hogetsu T, Hara N. 1990. Specific arrangements of cortical microtubules are correlated with the architecture of meristems in shoot apices of angiosperms and gymnosperms. *Bot. Mag. Tokyo* 103:143–63

Sakiyama M, Shibaoka H. 1990. Effects of abscisic acid on the orientation and cold stability of cortical microtubules in epicotyl cells of the dwarf pea. *Protoplasma* 157:165–71

Sakiyama-Sogo M, Shibaoka H. 1993. Gibberellin A3 and abscisic acid cause the reorientation of cortical microtubules in epicotyl cells of the decapitated dwarf pea. *Plant Cell Physiol.* 3:431–37

Salmon ED. 1989. Microtubule dynamics and chromosome movement. In *Mitosis: Molecules and Mechanisms,* ed. JS Hyams, BR Brinkley, pp. 119–81. New York: Academic

Salmon ED, Leslie RJ, Saxton WM, Karow ML, McIntosh JR. 1984. Spindle microtubule dynamics in sea urchin embryos. *J. Cell Biol.* 99:2164–74

Sawhney VK, Srivastava LM. 1974. Gibberellic acid induced elongation of lettuce hypocotyls and its inhibition by colchicine. *Can. J. Bot.* 52:259–64

Saxton WM, Stemple DL, Leslie RJ, Salmon ED, Zavortink M, et al. 1984. Tubulin, dynamics in cultured mammalian cells. *J. Cell Biol.* 99:2175–86

Schellenbaum P, Vantard M, Peter C, Fellous A, Lambert A-M. 1993. Co-assembly properties of higher plant microtubule-associated proteins with purified brain and plant tubulins. *Plant J.* 3:253–60

Seagull R. 1989. The plant cytoskeleton. *Crit. Rev. Plant Sci.* 8:131–67

Seagull RW. 1990. The effects of microtubules and microfilament disrupting agents on cytoskeletal arrays and wall deposition in developing cotton fibers. *Protoplasma* 159:44–59

Seagull RW, Falconer MM, Weerdenburg CA. 1987. Microfilaments: Dynamic arrays in higher plant cells. *J. Cell Biol.* 104:995–1004

Shibaoka H. 1993. Regulation by gibberellins of the orientation of cortical microtubules in plant cells. *Aust. J. Plant Physiol.* 20:461–70

Silk WK. 1992. Steady form from changing cells. *Int. J. Plant Sci.* 153:S49–58

Sinnot EW. 1960. *Plant Morphogenesis.* New York: McGraw-Hill. 550 pp.

Soifer B, ed. 1986. *Dynamic Aspects of Microtubule Biology,* Vol. 466. New York: Ann. NY Acad. Sci. 978 pp.

Sonobe S. 1990. ATP-dependent depolymerization of cortical microtubules by an extract in tobacco BY-2 cells. *Plant. Cell. Physiol.* 31:1147–53

Sonobe S, Shibaoka H. 1989. Cortical fine actin filaments in higher plant cells visualized by rhodamine-phalloidin after pretreatment with m-maleimidobenzoyl N-hydroxysuccinimide ester. *Protoplasma* 148:80–86

Staehelin LA, Giddings TH. 1982. Membrane-mediated control of cell wall microfibrillar order. See Green & Poethig 1982, pp. 133–47

Staiger CJ, Lloyd CW. 1991. The plant cytoskeleton. *Curr. Opin. Cell Biol.* 3:33–42

Staiger CJ, Yuan M, Valenta R, Shaw PJ, Warn RM, et al. 1994. Microinjected profilin affects cytoplasmic streaming in plant cells by rapidly depolymerizing actin microfilaments. *Curr. Biol.* 4:215–19

Stephens RE. 1986. Membrane tubulin. *Biol. Cell* 57:95–110

Steudle E. 1989. Water flow in plants and its coupling to other processes: an overview. *Meth. Enzymol.* 174:183–225

Sullivan KF. 1988. Structure and utilization of tubulin isotypes. *Annu. Rev. Cell Biol.* 4:687–716

Traas JA, Doonan JH, Rawlins DJ, Shaw PJ, Watts J, et al. 1987. An actin network is present in the cytoplasm throughout the cell cycle of carrot cells and associates with the dividing nucleus. *J. Cell Biol.* 105:387–95

Vandecandelaere A, Martin SR, Schlilstra MJ, Bayley PM. 1994. Effects of the tubulin-colchicine complex on microtubule dynamic instability. *Biochemistry* 33:2792–801

Vantard M, Schellenbaum P, Fellous A, Lambert A-M. 1991. Characterization of maize microtubule-associated proteins, one which is immunologically related to tau. *Biochemistry* 30:9334–40

Vigers GP A, Coue M, McIntosh JR. 1988. Fluorescent microtubules break up under illumination. *J. Cell. Biol.* 107:1011–22

Wardlaw CW. 1952. *Phylogeny and Morphogenesis.* London: MacMillan. 536 pp.

Wasteneys GO, Gunning BES, Hepler PK. 1993. Microinjection of fluorescent brain tubulin reveals dynamic properties of cortical microtubules in living plant cells. *Cell Motil. Cytoskeleton* 24:205–13

Wasteneys GO, Williamson RE. 1987. Microtubule orientation in developing internodal cells of *Nitella:* a quantitative analysis. *Eur. J. Cell Biol.* 43:14–22

Wasteneys GO, Williamson RE. 1989. Reassembly of microtubules in *Nitella tasmanica:* quantitative analysis of assembly and orientation. *Eur. J. Cell Biol.* 50:76–83

Webster DR, Gundersen G, Bulinski JC, Borisy GG. 1987. Differential turnover of tyrosinated and detryrosinated microtubules. *Proc. Natl. Acad. Sci. USA* 84:9040–44

White R, Hyde G, Overall R. 1990. Microtubule arrays in regenerating *Mougeotia* protoplasts may be oriented by electric fields. *Protoplasma* 158:73–85

Wick S. 1985. Immunofluorescence microscopy of tubulin and microtubule arrays in plant cells. III. Transition between mitotic/cytokinetic and interphase microtubule arrays. *Cell Biol. Int. Rprts.* 9:357–71

Wick S, Seagull RW, Osborn M, Weber K, Gunning BES. 1981. Immunofluorescence microscopy of organized microtubule arrays in structurally stabilized meristematic cells. *J. Cell Biol.* 89:685–90

Williamson R. 1990. Alignment of cortical microtubules by anisotropic wall stresses. *Aust. J. Plant Physiol.* 17:601–13

Williamson RE. 1991. Orientation of cortical microtubules in interphase plant cells. *Int. Rev. Cytol.* 129:135–206

Wyatt SE, Carpita NC. 1993. The plant cytoskeleton-cell-wall continuum. *Trends Cell Biol.* 3:413–17

Zhang D, Wadsworth P, Hepler PK. 1993. Dynamics of microfilaments are similar, but distinct from microtubules during cytokinesis in living, dividing plant cells. *Cell Motil. Cytoskeleton* 24:151–55

Annu. Rev. Cell Biol. 1994. 10:181–205

ROLE OF PROTEIN MODIFICATION REACTIONS IN PROGRAMMING INTERACTIONS BETWEEN RAS-RELATED GTPASES AND CELL MEMBRANES

John A. Glomset and Christopher C. Farnsworth
Howard Hughes Medical Institute, Departments of Medicine and Biochemistry, and Regional Primate Research Center, University of Washington, SL-15, Seattle, Washington 98195

KEY WORDS: myristoylation, farnesylation, geranylgeranylation, carboxyl methylesterification, palmitoylation

---

CONTENTS

INTRODUCTION

Mammalian cells contain at least 50 related small GTPases that regulate processes as diverse as cell replication and differentiation, cytoskeletal organization, secretion, and endocytosis. The best known of these GTPases are the ras proteins. Mutant, constitutively active forms of ras have been identified in

0743–4634/94/1115–0181$05.00

human tumors and shown to play a role in cell transformation. Other small GTPases include ADP-ribosylation factor (ARF) proteins, rap proteins, ral proteins, members of the rho subgroup of proteins, and rab proteins. The molecular basis of the function of these GTPases largely remains to be clarified, but available evidence suggests that each GTPase undergoes a conformational change when it binds a molecule of GTP and that this exposes a binding site for specific target proteins (Stouten et al 1993). Exposure of the binding site appears to be controlled by factors that separately influence the ability of the GTPase to bind and hydrolyze GTP. In addition, the location of the GTPase within the cell may affect its interaction with regulatory and target proteins. The aim of this review is to discuss the accumulating evidence concerning factors that influence the translocation of small GTPases to and from cell membranes. These factors include enzymes that catalyze the covalent attachment of lipids to the N- or C-terminal regions of the GTPases, enzymes that catalyze the hydrolytic removal of some of the bound lipids, protein kinases, and guanine nucleotide exchange factors that form 1:1 complexes with the GTPases in the cytosol.

MODIFICATIONS OF THE N- OR C-TERMINI OF RAS-RELATED GTPASES

Only a few definitive structural studies of fully processed small GTPases have been done, but sequence analyses and in vitro incubation studies have identified motifs at the N- or C-termini of unprocessed GTPases that are recognized by specific enzymes, and the importance of these motifs generally has been confirmed by labeling studies done in vivo. The combined results of these studies suggest that most small GTPases are modified by lipids and that several types of lipid modification are involved. For example, ARF proteins have been shown to contain an N-terminal glycine residue that can be cotranslationally modified by an amide-linked, 14-carbon saturated fatty acid called myristic acid (Figure 1; Kahn et al 1988). The myristoyl co-enzyme A acyltransferase that catalyzes this reaction also catalyzes the myristoylation of other cell proteins including the src tyrosine kinase (Gordon 1991).

In contrast to the ARF proteins, most other small GTPases are posttranslationally modified at their C-termini (for a recent review, see Yamane & Fung 1993). For example, the amino acid sequences of unprocessed ras proteins end with a -CXXX motif (also referred to as a CAAX motif), where X is an aliphatic amino acid and the C-terminal X is usually methionine or serine. This motif is recognized by an enzyme that catalyzes the transfer of a 15-carbon isoprenoid group, called a farnesyl group, from farnesyl diphosphate to the sulfhydryl group of the cysteine residue. A membrane-bound endoprotease that preferentially reacts with prenylated substrates subsequently catalyzes the

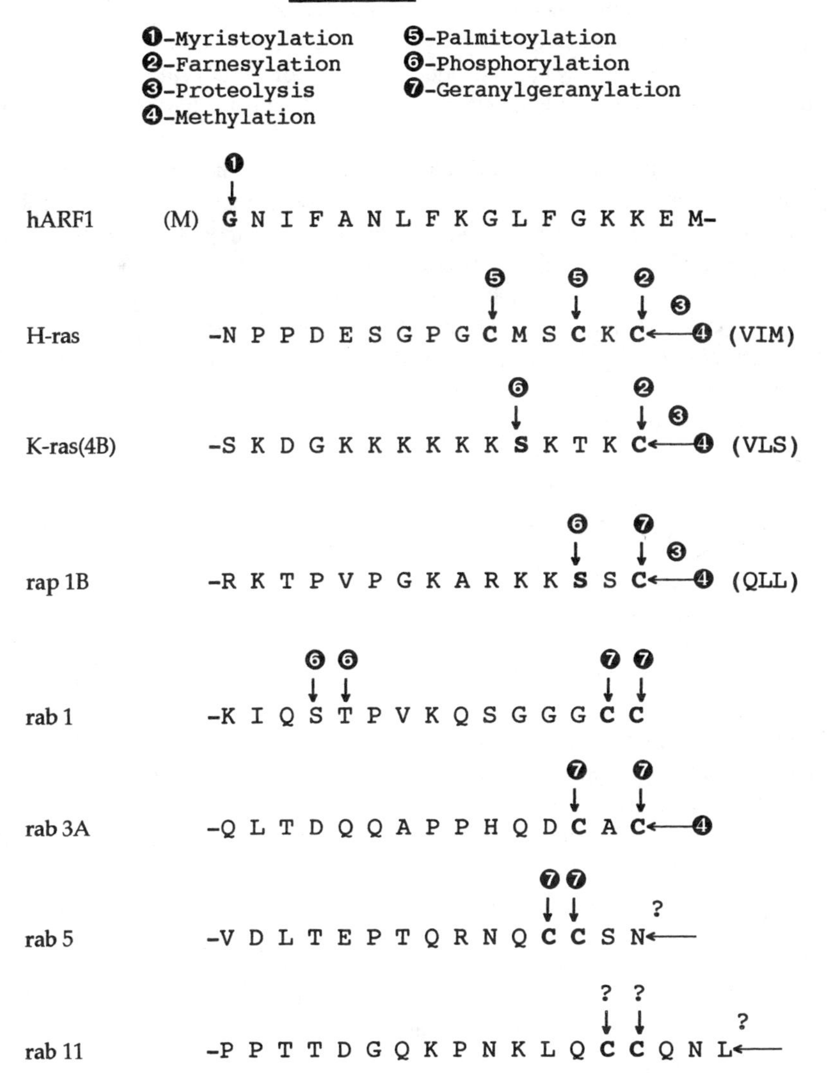

Figure 1 C- or N-terminal modification of selected, small GTPases. Amino acids in parenthesis are removed during processing.

removal of the C-terminal -XXX peptide from the farnesylated -CXXX sequence, and a membrane-bound methyltransferase that preferentially reacts with prenylated substrates catalyzes the methylesterification of the exposed carboxyl group of the cysteine residue. An intracellular enzyme that catalyzes the hydrolysis of farnesylated cysteinyl methyl esters also has been identified (Tan & Rando 1992).

Additional enzymes can catalyze the modification of upstream amino acids near the farnesylated, carboxyl methylesterified cysteine residue. Thus in K-ras(4B) (Figure 1), a serine that is located between the modified C-terminal cysteine residue and an upstream polybasic region can be phosphorylated by protein kinase A or protein kinase C (Ballester et al 1987). Furthermore, K-ras(4A), N-ras, and H-ras (Figure 1) each contain upstream cysteines that can be modified by palmitoyl thioesterification reactions (Hancock et al 1989). The palmitoyltransferase has not been isolated, but an activity that preferentially acts on the farnesylated form of N-ras has been identified in mouse fibroblast Golgi membranes (Gutierrez & Magee 1991). Furthermore, a palmitoyl-protein thioesterase that recognizes the guanine nucleotide-bound form of fully processed H-ras has been purified from bovine brain (Camp & Hoffman 1993).

The unprocessed forms of rap1A (not shown) and rap1B (Figure 1) each contain a C-terminal -CXXX motif that ends with leucine instead of methionine or serine. This motif is recognized by a soluble enzyme that transfers a 20-carbon isoprenoid group called a geranylgeranyl group (from geranylgeranyl diphosphate) to the sulfhydryl group of the cysteine residue. The geranylgeranyltransferase has been purified from rat brain and found to react with proteins or short peptides that end in -CXXL. After the geranylgeranylation reaction, the C-terminal -XXL peptide is removed by endoproteolysis and the exposed alpha carboxyl group of the geranylgeranylated cysteine residue is methylesterified as in the case of K-ras(4B). Moreover, a serine residue between the modified C-terminal cysteine residue and an upstream polybasic region can be phosphorylated by protein kinase A, Ca^{2+}/calmodulin-dependent protein kinase, or cyclic GMP-dependent protein kinase (Hata et al 1991; Altschuler & Lapetina 1993; Sahyoun et al 1991; Miura et al 1992).

The unprocessed forms of rap2A and rap2B (not shown in Figure 1) end with -CNIQ and -CVIL, respectively, and contain upstream cysteine residues (Ohmstede et al 1990). Rap2A is processed by farnesylation, whereas rap2B is processed by geranylgeranylation (Farrell et al 1992). In addition, at least one of the rap2 proteins is palmitoyl thioesterified (Béranger et al 1991). Presumably, both proteins are also processed by endoproteolysis and carboxyl methylesterification, but this has not been shown.

RalA and ralB have C-termini that end in -CCIL and -CCLL, respectively (Chardin & Tavitian 1989). The posttranslational modification of these proteins has not been studied directly, but a mutant form of ras that ended in -CCIL was shown to be geranylgeranylated (Moores et al 1991). Therefore, both ralA and ralB are probably modified by the same lipidation reactions that affect rap1A and rap1B. In addition, both ralA and ralB contain upstream polybasic regions and intervening serine residues that are predicted phosphorylation sites for one or more protein kinases including protein kinase A and protein kinase C.

Members of the rho subfamily of small GTPases also contain different posttranslational modification motifs at their C-termini. RhoA, rhoC, rhoG, rac1, rac2, human CDC42, and G25K contain -CXXL or -CXXF motifs that appear to be geranylgeranylated, endoproteolyzed, and carboxyl methylesterified (not shown in Figure 1). In addition, they contain upstream polybasic regions and in three cases (rhoA, rhoG, and human CDC42) also contain serine or threonine residues that may be targets for unidentified protein kinases. In contrast, the unprocessed C-terminal domains of rhoB and TC10 contain not only -CXXX motifs, but also upstream cysteine residues that are potential sites of palmitoyl thioesterification. The results of recent labeling studies suggest that rho B, whose unprocessed form ends with -CKVL, may be either geranylgeranylated or farnesylated and that it may also be palmitoyl thioesterified (Adamson et al 1992). The basis for the alternate prenylation of rhoB has not been determined, but the results can be questioned because the prenylation experiments were done in vitro or with transfected cells in vivo, and unknown amounts of expressed rhoB were present. Experiments with recombinant geranylgeranyltransferase have shown that the enzyme reacts preferentially with its peptide or protein substrates only at low substrate concentrations (Yokoyama & Gelb 1994). Therefore, analysis of native rhoB from animal tissues will be required to determine its normal prenylation status.

The unprocessed forms of many rab GTPases end in -CC, -CXC, -CCXX, or -CCXXX motifs, and the cysteine residues in these motifs appear to be geranylgeranylated (Figure 1). A rab geranylgeranyltransferase that catalyzes the modification of rab1A, rab3A, and rab5A proteins has been isolated and characterized (Seabra et al 1992a; Armstrong et al 1993; Andres et al 1993; Cremers et al 1994). The enzyme catalyzes the transfer of a geranylgeranyl group from geranylgeranyl diphosphate to the C-terminal regions of intact rab proteins, but does not react with corresponding cysteine-containing peptides. The enzyme evidently recognizes upstream regions of the rab GTPases (Seabra et al 1992b; Wilson & Maltese 1993; Sanford et al 1993; Béranger et al 1994a) even though it catalyzes the geranylgeranylation of cysteine residues at their C-termini. An incubation study with purified rab geranylgeranyltransferase demonstrated that it geranylgeranylates each of the two adjacent cysteine residues in rab1A, rab3A, and rab5A (Figure 1; Farnsworth et al 1994), but studies of rab proteins that end in -CCXXX still must be done. The result obtained with rab3A is consistent with a previous structural analysis of bovine brain rab3A, which demonstrated that each of the cysteine residues at the C-terminus of this protein is geranylgeranylated and that the C-terminus is also carboxyl methylesterified (Farnsworth et al 1991). Definitive structural analyses of other native rab proteins may be required to resolve a conflict in the literature. Molenaar et al (1988) overexpressed Ypt1p, a rab-related protein that ends in -GGCC, in yeast and found that a C-terminal domain in this protein could be labeled with radioactive palmitate; Newman et al (1992) obtained similar results in studies of Ypt3p, which ends in -SQCC. On the other hand, neither group found evidence for palmitoylation of these proteins when they were expressed at normal levels. Because the C-termini of Ypt1p and Ypt3p resemble that of rab1A, geranylgeranylation of each of the adjacent cysteine residues in these proteins would have been anticipated. Therefore, it is possible that the conflicting results obtained may have been the result of overexpression. This possibility warrants further study.

The alpha carboxyl methylesterification status of rab proteins remains to be fully clarified. Studies have shown that rab2, Ypt1p and Ypt3p, which end in -CC, and rab5A, which ends in -CCXX, are not methylesterified (Wei et al 1992; Newman et al 1992; Li & Stahl 1993). On the other hand, the native rab3A of bovine brain membranes is methylesterified (by an unknown enzyme), and rab proteins that end in -CCXXX have not been analyzed.

Lastly, upstream amino acids near the geranylgeranylated cysteine residues of rab proteins can be modified in some cases. Evidence has been obtained that the $p34^{cdc2}$ protein kinase can phosphorylate serine and threonine residues in the C-terminal region of rab1A (Figure 1) and a serine residue in the C-terminal region of rab4 (Bailly et al 1991). In addition, the insulin-activated extracellular-signal-regulated kinase ERK1 can also phosphorylate rab4, apparently on the same serine residue (Cormont et al 1994).

To summarize, available evidence indicates that ARFs are lipidated at their N-termini, whereas most other small GTPases are modified at their C-termini. Furthermore, several types of C-terminal modification have been observed. The C-termini of fully processed small GTPases may contain a single farnesylated, carboxyl methylesterified cysteine residue; a single geranylgeranylated, carboxyl methylesterified cysteine residue; or two adjacent geranylgeranylated cysteine residues that may or may not be carboxyl methylesterified. In addition, the C-termini of singly prenylated small GTPases may contain (*a*) an upstream polybasic region and an intervening serine or threonine residue that can be phosphorylated, or (*b*) one or more upstream cysteine residues that can be palmitoyl thioesterified. Finally, the C-termini of some doubly geranylgeranylated small GTPases contain upstream serine or threonine residues that can be phosphorylated.

FUNCTIONAL SIGNIFICANCE OF THE MODIFICATIONS

Several studies have demonstrated that lipids that are covalently attached to the N- or C-terminal domains of small GTPases promote hydrophobic interactions between the GTPases and cell membranes or cytosolic proteins. In addition, a growing body of evidence suggests that the modified N-termini and C-termini of these proteins may be of regulatory importance. Results related to these possibilities are discussed below.

ARF Proteins

These proteins were first identified as factors that promote the cholera toxin-dependent ADP ribosylation of the heterotrimeric GTP-binding protein, G_s (reviewed in Moss & Vaughn 1993). However, accumulating evidence indicates that they contribute to normal functions as well. (*a*) They appear to be required for trafficking of coated vesicles between the endoplasmic reticulum and the Golgi (Taylor et al 1994; Dascher & Balch 1994). (*b*) They promote recruitment of the adaptor protein AP-1 onto *trans*-Golgi membranes in preparation for the formation of clathrin-coated vesicles (Stamnes & Rothman 1993; Traub et al 1993). (*c*) They have been postulated to influence regulated exocytosis in adrenal chromaffin cells (Morgan & Burgoyne 1993), endosome-endosome fusion (Lenhard et al 1992), and nuclear vesicle fusion (Boman et al 1992). (*d*) They activate the hydrolysis of phosphatidylcholine by phospholipase D (Brown et al 1993).

The mechanisms that underlie these effects have yet to be characterized. However, it appears that a guanine nucleotide exchange factor promotes the translocation of cytosolic ARF to membranes by catalyzing the exchange of GTP for bound GDP (Randazzo et al 1993; Tsai et al 1994). Membrane-bound ARF_{GTP} evidently recruits a complex of "coatamer" proteins onto the membranes, which causes the formation of coated vesicles (Palmer et al 1993; Orci et al 1993a,b). In addition, subsequent hydrolysis of the bound GTP appears to be required for vesicle uncoating (Tanigawa et al 1993; Teal et al 1994).

The myristoylated N-terminal domains of ARF proteins clearly contribute to these events because soluble recombinant myristoylated human $ARF5_{GDP}$ exhibits GTP-dependent translocation to Golgi membranes, but recombinant non-myristoylated human $ARF5_{GDP}$ does not (Haun et al 1993). A role of the myristoyl group in anchoring ARF to membranes might be anticipated because myristoylated glycine and myristoylated glycine-containing peptides bind to phosphatidylcholine vesicles (Peitzsch & McLaughlin 1993). Nevertheless, more than simple myristoyl-dependent anchoring of ARF to membranes appears to be involved because unmyristoylated ARF1 binds to phospholipid vesicles in a GTPγS-dependent manner and promotes the GTPγS-dependent binding of cholera toxin to phospholipid-cholate micelles (Franco et al 1993).

It is not clear why myristoylated ARF_{GDP} binds to Golgi membranes, but unmyristoylated ARF_{GDP} does not. Binding may depend on an interaction between the myristoylated N-terminal domain of ARF_{GDP} and membrane-associated ARF-binding proteins rather than on an interaction between ARF_{GDP} and the Golgi membrane phospholipid bilayer (Franco et al 1993). Indeed, several observations appear to support this possibility. (*a*) Binding of myristoylated glycine to phospholipid vesicles occurs at 0°C (Peitzsch & McLaughlin 1993), but binding of myristoylated $ARF5_{GDP}$ to Golgi membranes occurs only at higher temperatures (Haun et al 1993). (*b*) The unmyristoylated N-terminal domain of ARF1 contains a repeat of hydrophobic residues, -IFXXLFXLF- (Figure 1), whose structure is compatible with that of an amphipathic helix. (*c*) A myristoylated, 17-amino acid peptide corresponding to this N-terminal domain formed an alpha helix in an aqueous environment (Kahn et al 1992). (*d*) A recombinant form of ARF1 from which this peptide had been deleted was devoid of ARF1 activity, but had an affinity for GTPγS that was more than 2000-fold greater than that of wild-type ARF1 (Kahn et al 1992). (*e*) High concentrations of the 17-amino acid peptide inhibited in vitro GTPγS-dependent accumulation of Golgi-coated buds and vesicles and in vitro intra-Golgi transport (Kahn et al 1992).

These observations raise the possibility that the myristoylated N-terminal domain of a cytosolic ARF protein may interact hydrophobically with a domain near its guanine nucleotide-binding region and thereby prevent exchange of bound GDP for GTP. If this is the case, translocation of the ARF protein to a membrane may require disruption of this intramolecular interaction followed by the temperature-dependent formation of intermolecular contacts between the myristoylated N-terminal region of the ARF protein and one or more membrane-associated ARF-binding proteins. Site-directed mutagenesis of specific amino acids in the N-terminal amphipathic domain of an ARF protein might be done to test these possibilities.

Ras Proteins

Ras proteins are being studied intensively in many laboratories because of their roles in cell signaling and carcinogenesis (Lowy & Willumsen 1993). Normally associated with the inner leaflet of the plasma membrane (Willingam et al 1980), Ras proteins are activated in response to growth factors and other agonists that promote cell proliferation and cell differentiation. Upon being activated, they initiate protein kinase cascades that lead to increased gene transcription. Although the molecular basis of their function is far from understood, recent studies have shown that epidermal growth factor activates ras proteins by a mechanism that depends on the induced tyrosine phosphorylation of its receptor (reviewed in Panayotou & Waterfield 1993). A soluble "adapter" protein, Grb2, binds to the tyrosine phosphorylated receptor and also binds the cytosolic, ras guanine nucleotide exchange protein Sos (Downward 1994; Buday & Downward 1993; Chardin et al 1993; Egan et al 1993). Sos then activates ras by catalyzing the exchange of GTP for bound GDP and exposing a binding site on ras for the protein serine/threonine kinase Raf-1 (Moodie et al 1993; Zhang et al 1993; Vojtek et al 1993; Van Aelst et al 1993; Warne et al 1993; Hallberg et al 1994). Upon binding to this site, Raf-1 initiates a protein kinase cascade involving mitogen-activated protein (MAP) kinase kinase and MAP kinase (Howe et al 1992; Dent et al 1992; Kyriakis et al 1992; Traverse et al 1993; MacDonald et al 1993). This cascade ultimately stimulates gene transcription (reviewed in Blenis 1993; Schlessinger 1993; Nishida & Gotoh 1993). Many variations of this mechanism appear to be possible because other receptor and nonreceptor protein kinases also activate ras proteins, because several different adapter proteins and ras guanine nucleotide exchange factors have been identified, and because different isoforms of ras are present in cells (Figure 1).

The lipidated C-terminal domains of ras proteins contribute to these signaling events by anchoring ras proteins to membranes. To define the determinants of this function, Hancock and colleagues generated mutant forms of H-ras by separately replacing each of the three upstream cysteine residues in the C-terminal region with serine residues or generated mutant forms of K-ras(4B), by substituting glutamines for lysines in the upstream polybasic region. When they expressed the mutant proteins in COS cells or NIH 3T3 cells and compared their behavior with that of normal ras proteins or mutant ras proteins that lacked a cysteine residue in the normal -CXXX sequence, they found that only the normal forms of H-ras and K-ras(4B) bound effectively to membranes and targeted to the plasma membrane (Hancock et al 1990). Further studies showed that alpha carboxyl methylesterification was required for optimal membrane binding of K-ras(4B) (Hancock et al 1991a) and that substitution of a geranylgeranylated cysteine residue for a farnesylated cysteine residue increased membrane binding (Hancock et al 1991b). Finally, experiments with chimeric proteins that had been generated by attachment of the C-terminal 10 amino acids of H-ras or the C-terminal 17 amino acids of K-ras(4B) to the C-terminus of a soluble protein (protein A) showed that these proteins became associated with membranes and targeted to the plasma membrane (Hancock et al 1991b). These results support two major conclusions: (*a*) The C-terminal, lipidated cysteine residues of ras proteins must act in concert with upstream palmitoylated cysteine residues or an upstream polybasic region in order to anchor ras proteins effectively to membranes. (*b*) The fully processed C-terminal regions of these proteins are all that is required to anchor otherwise soluble proteins to membranes.

The molecular basis for these requirements remains to be determined. The farnesyl group and palmitoyl groups of H-ras might anchor it to membranes by inserting into the lipid bilayer, whereas attachment of K-ras(4B) to the lipid bilayer might require both insertion of its farnesyl group into the lipid bilayer and interactions between its upstream polybasic region and the negatively charged head groups of acidic membrane phosphoglycerides. The results of recent computation studies (Black 1992) and measurements of the interaction of model-prenylated peptides with phosphoglyceride vesicles (Silvius & l'Heureux 1994) support this interpretation and so does the fact that acidic phospholipids are preferentially distributed in the cytoplasmic leaflet of the plasma membrane lipid bilayer (Devaux 1992). However, the possibility that the covalently attached lipid groups of ras proteins may interact directly with specific membrane proteins remains to be excluded. Indeed, precedent for this type of interaction exists. For example, bovine brain microsomes contain a high-affinity binding site for prenylated peptides (Thissen & Casey 1993). Furthermore, in the absence of membranes, the farnesylated form of Ras2 from *Saccharomyces cerevisiae* binds directly and with high affinity to its target protein, adenylyl cyclase, but the unprocessed form of RAS2 does not (Kuroda et al 1993).

The mechanisms that target ras proteins to the plasma membrane are likely to be complex. In the case of H-ras and other ras proteins that contain upstream cysteine residues in their C-terminal regions, these mechanisms probably involve the protein palmitoyltransferase that was mentioned earlier. In T15 cells, the palmitoyl groups of N-ras were shown to turn over with a $t_{1/2}$ of about 20 min, whereas the polypeptide chain of N-ras turned over much more slowly (Magee et al 1987). This suggests that the palmitoyltransferase may act in concert with a palmitoyl thioesterase to regulate the palmitoylation status of some ras proteins and promote their translocation to and from membranes. Experiments with quiescent cells in culture might test this possibility. Growth factors such as epidermal growth factor or platelet-derived growth factor could be added to the medium to initiate cell cycle traverse, and the cells could be examined for early synchronized changes in the palmitoylation and translocation of N-ras, H-ras, or K-ras(4A).

Soluble guanine nucleotide exchange factors also may regulate the translocation of ras proteins to and from the plasma membrane. Takai and colleagues identified a guanine nucleotide exchange factor, called smg guanine nucleotide exchange stimulator (GDS), that stimulates the exchange of GTP for bound GDP on K-ras(4B) and certain other members of the ras superfamily (Yamamoto et al 1990; Mizuno et al 1991). This exchange factor forms stable 1:1 molar complexes with the GTPases provided that they have posttranslationally modified C-termini. In addition, another soluble guanine nucleotide exchange factor called ras guanine nucleotide-releasing factor (GRF) acts preferentially on the processed forms of K-ras(4B) and H-ras (Orita et al 1993), and both exchange factors prevent K-ras(4B) from binding to membranes and promote the release of pre-bound K-ras(4B) from membranes (Kawamura et al 1993; Orita et al 1993). This suggests that the two guanine nucleotide exchange factors may promote translocation of K-ras(4B) from the plasma membrane to the cytosol in vivo and raises the possibility that ras GRF may do the same for other ras proteins. Whether phosphorylation of the C-terminal region of K-ras(4B) or turnover of the palmitoyl groups in the C-terminal regions of other ras proteins influences the translocating actions of smg GDS and ras GRF remains to be determined.

Another question concerns the relation between the contents of ras proteins in the plasma membrane and the functions of these proteins. Because activation of normal ras proteins by growth factors is mediated by guanine nucleotide exchange factors that translocate to the plasma membrane, ras proteins that are anchored there may be activated more efficiently than others. In addition, anchorage of ras proteins to the plasma membrane may facilitate downstream interactions between ras proteins and their target proteins. In support of the latter possibility, (*a*) a plasma membrane-enriched fraction from human embryonic retinal cells that had been transformed with oncogenic N-ras DNA contained increased amounts of both ras and its target protein, raf-1 (Traverse et al 1993); (*b*) serum treatment of COS cells activated raf-1 transiently and caused a concomitant translocation of raf-1 and two associated proteins to membranes (Wartmann & Davis 1994); and (*c*) in vitro studies of the interaction of activated ras proteins with raf-1 in the absence of membranes showed little or no change in raf-1 activity (Zhang et al 1993; Moodie et al 1993). This suggests that binding of both ras and raf-1 to the plasma membrane may be required for raf-1 activation (see note added in proof).

An alternate possibility is that ras proteins that have been activated on the plasma membrane may subsequently translocate to the cytosol and interact with target proteins. Itoh et al (1993a) showed that addition of the GTPγS-bound form of fully processed K-ras(4B) to a membrane-free, supernatant fraction from homogenates of immature *Xenopus* oocytes caused an approximately fivefold activation of both an endogenous MAP kinase and an exogenous recombinant MAP kinase fusion protein. In contrast, neither the GDP-bound form of fully processed K-ras(4B) nor the GTPγS-bound form of unprocessed K-ras(4B) showed much activity. Subsequent experiments showed that the ras-dependent activation of the MAP kinases required the presence of two endogenous oocyte proteins, a MAP kinase kinase and a protein called REKS, which stimulated the activity of the MAP kinase kinase (Itoh et al 1993b). These results suggest that the lipidated C-terminal region of K-ras(4B) may promote binding of K-ras(4B) to a cytosolic target protein in *Xenopus* oocytes and raise the possibility that this protein may be REKS.

How specific are the functional requirements for lipidated C-terminal domains in K-ras(4B) and other ras proteins? This question has yet to be addressed by experimentation with purified components. However, several investigators have altered the C- or N-terminal regions of ras proteins and assayed the ability of the modified ras proteins to transform cells. Hancock et al (1990) found that replacement of lysines with glutamines in the C-terminal polybasic region of oncogenically activated K-ras(4B) or replacement of upstream cysteines with serines near the C-terminus of oncogenically activated H-ras reduced, but did not abolish, the proteins' transforming activity. However, forms of oncogenic K-ras(4B) or H-ras that contained a serine instead of a cysteine in their -CXXX sequences were unable to transform cells.

A comparison of the activity of a geranylgeranylated form of non-oncogenic H-ras with that of normal H-ras yielded interesting results (Cox et al 1992). When the two proteins were separately overexpressed in NIH 3T3 cells, they had comparable cell-transforming activities. But when moderate levels of geranylgeranylated H-ras were expressed, growth was inhibited. This suggests that the normal functions of H-ras may differ from the oncogenic functions and that moderate expression of geranylgeranylated H-ras may interfere with binding of normal H-ras to its targets.

Other interesting results were obtained in experiments with chimeric proteins that contained an N-terminal myristoylated peptide attached to an oncogenic or a non-oncogenic form of H-ras that could not be farnesylated (Lacal et al 1988; Buss et al 1989). When the chimeric proteins were expressed in NIH 3T3 cells, they were myristoylated and became associated with membranes. Furthermore, chimeras that contained an oncogenic form of H-ras transformed cells. Indeed, even a myristoylated chimera that contained a non-oncogenic form of H-ras had transforming activity (Buss et al 1989). These results indicate that the requirement for membrane attachment of ras proteins in the cell transformation assay is relatively nonspecific. But further experimentation will be needed to determine whether myristoylated chimeras containing non-oncogenic H-ras interact effectively with purified guanine nucleotide exchange factors, target proteins, and negative regulators of normal ras proteins.

It would be of particular interest to examine the ability of a myristoylated chimera to interact with ras GTPase-activating proteins (GAPs) in experiments with reconstituted liposomes. Ras GAPs are soluble negative regulators of ras proteins that have also been implicated in downstream signaling events (reviewed in Boguski & McCormick 1993; Bollag & McCormick 1991). However, it is not clear whether ras GAPs normally inactivate ras proteins that are associated with membranes or ras proteins that are cytosolic. The N-terminal region of p120-GAP contains several potential membrane-binding domains including an SH2 domain that can bind to activated growth factor receptor tyrosine kinases in vitro and in vivo (Marengere & Pawson 1992; Soler et al 1993). But acidic lipids, such as those that are found on the cytoplasmic surface of the plasma membrane, inhibit GAP activity (Serth et al 1991). A comparison of the activity of p120-GAP toward a myristoylated, membrane-associated chimera of H-ras with that toward a depalmitoylated form of normal H-ras might help resolve this issue. To summarize: (*a*) Both farnesylation and palmitoylation of the C-terminal regions of H-ras, N-ras, and K-ras(4A) may be required for efficient binding to membranes, whereas binding of K-ras(4B) to membranes requires farnesylation, methylesterification, and the presence of an upstream polybasic region. (*b*) Special mechanisms may cause ras proteins to translocate to and from membranes. (*c*) These mechanisms may regulate interactions between ras proteins and their target proteins and negative effectors.

Rap Proteins

Rap proteins have been implicated in the control of a number of biological processes including cell replication, platelet activation, the generation of microbicidal oxygen radicals, and bud site selection in yeast (reviewed in Bokoch 1993). As in the case of ras proteins, the molecular basis of their effects has to be determined, but it appears that they are active when they contain bound GTP, and that their content of GTP is regulated by at least one guanine nucleotide exchange factor and at least two GAP activities. Furthermore, evidence has been accumulating that the lipidated C-terminal regions of rap proteins anchor them to membranes and regulate their function.

Much of this evidence relates to rap1B. The geranylgeranylated C-terminal region of this protein has been shown to be a membrane anchor (Hiroyoshi et al 1991). In addition, the geranylgeranylated form of rap1B, but not the unprocessed form, can form a 1:1 complex with smg GDS (Kawamura et al 1991). The formation of this complex can be regulated by a protein kinase A-dependent mechanism. As mentioned above, protein kinase A catalyzes the phosphorylation of a serine residue that is located between the geranylgeranylated C-terminus of rap1B and the nearby upstream polybasic region (Figure 1). This enhances the ability of smg GDS to bind rap1B, stimulate the exchange of GTP for bound GDP, and inhibit binding of rap1B to membranes (Hata et al 1991; Itoh et al 1991). Moreover, experiments with intact platelets have shown that activation of protein kinase A by prostaglandin E_1 or a prostaglandin analogue causes phosphorylation of rap1B and a marked increase in the content of rap1B in the cytosol (Lapetina et al 1989; Hata et al 1991).

The functional significance of these changes is far from understood, but prostaglandin E_1 is a negative regulator of platelets that increases the threshold dose of thrombin required for platelet activation. Thrombin activates platelets by stimulating the hydrolysis of phosphatidylinositol-4,5-bisphosphate and raising the concentration of Ca^{2+} in the cytosol. Therefore, it is of considerable interest that rap1B has recently been implicated in the regulation of plasma membrane Ca^{2+} transport (Corvazier et al 1992; Magnier et al 1994).

Experiments with human erythroleukemia cells have provided further evidence that the phosphorylation of rap1 may attenuate a thrombin-induced response (Lazarowski et al 1990). Thrombin activates the hydrolysis of phosphoinositides in these cells, but pretreatment of the cells with a prostaglandin analogue stimulated the phosphorylation of rap1 and markedly reduced the thrombin-dependent formation of inositol phosphates. Furthermore, when membranes from unstimulated cells were incubated with the catalytic subunit of protein kinase A in the presence of ATP, rap1 was phosphorylated and appeared in the supernatant fraction.

In the absence of prostaglandins or prostaglandin analogues, does the activation of platelets by thrombin affect the behavior of rap1B? A study by Fischer et al (1990) provides evidence that thrombin treatment causes the rap1B in platelets to associate quantitatively with the cytoskeleton. The molecular basis of this effect has to be determined. However, platelets change shape when they are activated; cytoskeletal material is formed that can be isolated by treating the platelets with Triton X-100 and centrifuging the extract at 13,000 to 15,600 × g; this material has been shown to contain actin, myosin, actin-binding protein, spectrin, vinculin, talin, the platelet integrin GP IIb/IIIb, pp60 src, pp60 yes, phosphoinositide-3-kinase, phosphoinositide(4)P 3-kinase, rap2B, and a rho protein (Fox et al 1993; Torti et al 1993; Zhang et al 1992, 1993). Therefore, it is possible that rap1B and other small GTPases may in some way contribute to the formation of this complex.

Rho-related Proteins

Rho-related proteins including rhoA, rhoB, rhoC, rac, and CDC42 have been implicated in the control of cytoskeletal organization, and rac1 and rac2 also regulate superoxide generation in neutrophils and macrophages (reviewed by A Hall, this volume). Although all these proteins appear to be posttranslationally modified at their C-termini, most are mainly present in the cytosol. They form 1:1 molar complexes with a cytosolic protein called rho guanine nucleotide dissociation inhibitor (rho GDI) and, to a lesser extent, they form 1:1 molar complexes with smg GDS (Ueda et al 1990; Kikuchi et al 1992). Like smg GDS, rho GDI forms complexes only with small GTPases that are posttranslationally processed (Hori et al 1991). In addition, rho GDI can prevent small GTPases from binding to membranes and can remove them from membranes in vitro and in vivo (Isomura et al 1991; Ando et al 1992; Miura et al 1993; Kishi et al 1993; Takaishi et al 1993). But unlike smg GDS, rho GDI apparently binds selectively to rho-related GTPases, interacts preferentially with GTPases that contain bound GDP (Hori et al 1991; Ando et al 1992), and inhibits rather than stimulates the exchange of GTP for bound GDP (Hiraoka et al 1992).

How then are GTPases that are bound to rho GDI activated? It seems likely that specific guanine nucleotide exchange factors disrupt complexes of rho GDI and rho-related GTPases and promote the exchange of GTP for bound GDP, but relatively little evidence for this exists at present (A Hall, this volume). It is known that acidic lipids such as phosphatidic acid and phosphoinositides can effectively disrupt a complex of rac and rho GDI (Chuang et al 1993) and that complexes of rac and rho GDI can be disrupted in vivo (Quinn et al 1993; Philips et al 1993). Furthermore, some of the small GTPases associated with rho GDI lack alpha carboxyl methylesters at their C-termini (Backlund 1992, 1993; Philips et al 1993) and activation of neutrophils by the chemoattractant N-formylmethionyl-leucyl-phenylalanine induces a transient carboxyl methylesterification of several small GTPases including rac (Philips et al 1993). Apparently, activation of the cells disrupts rho GDI complexes containing unmethylated GTPases, promotes exchange of GTP for bound GDP, and allows carboxyl methylesterification of the GTPases by a methyltransferase in the plasma membrane (Pillinger et al 1994).

Rab Proteins

Rab proteins play important roles in intracellular vesicle transport (reviewed in Ferro-Novick & Novick 1993; Simons & Zerial 1993). More than 30 different rab proteins have been identified, and it has been postulated that each vesicle transport step in the secretory and endocytotic pathways is associated with its own rab protein. Rab proteins bind to vesicles via their geranylgeranylated C-terminal domains (Khosravi-Far et al 1991; Rossi et al 1991; Kinsella & Maltese 1992; Musha et al 1992; Li & Stahl 1993; Giannakouros et al 1993; Lombardi et al 1993). In addition, the specific localization of rab proteins depends in part on their C-terminal domains.

Chavrier et al (1991) transiently expressed hybrid forms of rab2, rab5, and rab7 in baby hamster kidney cells and determined their intracellular locations. They found that exchanging the C-terminal 34- or 35-amino acids of these proteins was sufficient to change their distribution on endosomal and Golgi membranes. In contrast, exchanging shorter C-terminal regions containing the prenylation sites was not. Stenmark et al (1994) extended this work by examining the localization and function of hybrids of rab5 and rab6. When they replaced the C-terminal 32 amino acids of rab6, a protein that is normally localized to the Golgi, with the corresponding amino acids of rab5, a protein that is normally localized to the plasma membrane and early endosomes, the rab6-rab5 hybrid showed the same intracellular location as rab5. However, the hybrid protein did not appear to function like rab5 because overexpression of rab5 caused increased endocytosis, while overexpression of the hybrid did not unless further changes were made in a critical upstream sequence. Brennwald & Novick (1993) obtained similar results in experiments with Sec4-YPT1 hybrids in yeast.

Béranger and colleagues (1994b) also examined the role of structural domains in determining the localization and function of rab6, but used a somewhat different approach. They replaced the C-terminal 45 residues of rab6 with those of H-ras, or generated rab6 proteins that ended in -CC, -CCIL, -CKCCIL, or -CKCVLS. Their experiments with the rab6-H-ras hybrid provided evidence that sequences in the N-terminal 71 amino acids of rab6 also are required for localization to the Golgi, whereas their experiments with rab6 proteins that had other lipid modification motifs showed that most of these proteins targeted to Golgi membranes even though some were farnesylated and palmitoylated instead of geranylgeranylated. The only altered rab6 protein that was nonfunctional had a C-terminal -CVLS sequence and presumably was processed by single farnesylation, proteolysis, and carboxyl methylesterification.

A GDI for rab proteins has been isolated and cloned (Sasaki et al 1990; Matsui et al 1990). It forms 1:1 molar complexes with fully processed rab proteins, but not with unprocessed rab proteins (Araki et al 1991; Sasaki et al 1991; Soldati et al 1993; Ullrich et al 1993). A role for rab GDI in regulating the reversible binding of rab proteins to membranes has been proposed (Araki et al 1990), and critical evidence in support of this possibility has been obtained.

Ullrich et al (1993) permeabilized cells with streptolysin O, incubated the cells in the presence or absence of purified rab GDI, and showed that rab GDI completely removed endogenous rab2, rab5, rab7, rab8, rab9, and rab11 from membranes. In addition, they incubated permeabilized cells with a purified $rab5_{GDP}$-rab GDI complex and showed that this complex could deliver rab5, but not GDI, to a membrane compartment (Ullrich et al 1994). Furthermore, they observed that the binding of rab5 to membranes was saturable, that the appropriate membrane compartment was involved, and that the bound rab5 initially contained GDP. After binding, $rab5_{GDP}$ underwent nucleotide exchange in the presence of GTP, became constitutively active, and caused a time-dependent increase in the size of early endosomes. Soldati et al (1994) obtained similar results in a study of the interaction of $rab9_{GDP}$-rab GDI with semi-purified late endosomes. In addition, they observed that binding to the late endosomal membranes was rab specific, since incubation of the membranes with $rab3A_{GDP}$-rab GDI did not lead to uptake of the rab3A. Taken together, these results raise the possibility that (*a*) rab translocation involves specialized rab guanine nucleotide exchange proteins and (*b*) rab proteins remain bound to their target membranes when they are in an active GTP-bound state.

A membrane-associated guanine nucleotide exchange factor for Sec4 has recently been identified (Moya et al 1993). Furthermore, a soluble GRF that acts on rab3A has been partially purified (Burstein & Macara 1992). Whether this GRF acts on rab3A that is bound to membranes remains to be determined, but it shows a preference for native rab3A over unprocessed, recombinant rab3A, as does rab GDI, and thus may interact with rab3A primarily in the cytosol (Burstein et al 1993).

Once Rab proteins translocate to membranes and bind GTP, they may set in motion a series of protein-protein interactions that culminate in membrane fusion. Rab target proteins are presumably involved, but only one candidate target has been identified so far. Studies of rab3A, which is associated with neurotransmitter-containing synaptic vesicles, have shown that the GTP-bound form of this protein can bind to the N-terminal region of a soluble, cytosolic protein called rabphilin-3A (Shirataki et al 1992, 1993). On the other hand, the C-terminal half of rabphilin-3A contains two regions with homology to the C_2 domains of protein kinase C_{alpha}; these regions mediate Ca^{2+}-dependent binding of rabphilin to acidic phospholipids (Yamaguchi et al 1993). How rabphilin contributes to synaptic transmission remains to be determined, but rabphilin has been shown to colocalize with rab3A at the neurite tips of differentiated pheochromocytoma (PC12) cells (Wada et al 1994), and evidence for tight association of rab3A and proteins that have been implicated in membrane fusion has been obtained (Horikawa et al 1993).

CONCLUSIONS/FUTURE DIRECTIONS

The results discussed in this review provide evidence that the modified N- or C-terminal regions of small GTPases play a major role in programming interactions among the GTPases and cell membranes. The myristoylated, amphipathic N-terminal regions of ARF proteins help maintain these proteins in a GDP-bound state in the cytosol; promote interactions between these proteins and yet-to-be-isolated guanine nucleotide exchange stimulators; and serve as membrane anchors. Therefore, they critically affect the ability of ARF proteins to cycle between a cytosolic GDP-bound state and a membrane-associated GTP-bound state.

The prenylated C-terminal regions of other small GTPases play important roles as well. Results of studies of N-ras and H-ras have raised the possibility that GTPases that contain a single covalently bound prenyl group and one or more upstream palmitoyl groups may cycle between palmitoylated membrane-bound states and depalmitoylated cytosolic states under the control of palmitoyltransferase and palmitoyl thioesterase activities. On the other hand, results of studies of rap1B have raised the possibility that singly prenylated GTPases that contain an upstream polybasic region and an intervening serine or threonine residue may cycle on and off membranes under the control of protein kinase and protein phosphatase activities. Guanine nucleotide exchange factors that form soluble 1:1 molar complexes with singly prenylated GTPases also clearly affect the distribution of GTP-bound and GDP-bound GTPases between membranes and the cytosol, as do soluble guanine nucleotide exchange factors that form complexes with geranylgeranylated rab proteins. One of these guanine nucleotide exchange factors, rab GDI, has been shown to form complexes with many different rab proteins and to promote the translocation of rab proteins to and from membranes. Specificity in delivery appears to depend on upstream regions in rab proteins and on unidentified membrane-docking proteins.

Much more information is needed about the factors that control the distribution of small GTPases between membranes and the cytosol. These factors clearly include the membrane-associated guanine nucleotide exchange factors and GAPs that control the GTP content of the GTPases as well as the presumptive docking proteins that target specific GTPases to membranes. But some of the possibilities discussed in this review also warrant attention. To our knowledge, only one study has been done of the turnover of palmitoyl groups attached to small GTPases and that study used nonsynchronized cells (Magee et al 1987). In addition, only one study has been done of the relation between the palmitoylation of a small GTPase and its binding to membranes (Hancock et al 1990). Further studies of small GTPase palmitoylation are needed, and they should focus not only on ras proteins, but also on other palmitoylated small GTPases including rap and rho proteins. It will be important to determine whether the palmitoylation and depalmitoylation of these proteins are regulated and whether coordinate control is involved.

Additional studies of the phosphorylation of small GTPases also are warranted. Several studies of rap1B have provided evidence that the phosphorylation of a serine residue in the C-terminal region is of regulatory importance. But the possible regulatory roles of phosphorylation reactions involving other singly prenylated small GTPases such as K-ras(4B), rap1A, ralA and B, rhoG, and perhaps mammalian CDC42 remain to be examined. In addition, it is unclear how phosphorylation reactions involving rab proteins (Bailly et al 1991; Karniguian et al 1993) and rab GDI (Steel-Mortimer et al 1993) affect membrane programming.

Another question relates to the possible turnover of small GTPase alpha carboxyl methylesters and the relation of this turnover to membrane binding. Philips et al (1993) showed that small GTPases become carboxyl methylesterified when human neutrophils are stimulated with N-formylmethionyl-leucyl-phenylalanine and that the carboxyl methylesterified GTPases were present in membranes but not in the cytosol. Hancock et al (1991a) showed that alpha carboxyl methylesterification of K-ras(4B) is required for effective membrane binding. A study by Backlund (1992) provided evidence that GTP promotes the carboxyl methylesterification and translocation to membranes of G25K, an isoform of mammalian CDC42. Several studies have shown that alpha carboxyl methyltransferase inhibitors can block agonist-induced cell responses (reviewed in Clarke 1993). However, recent results provide evidence that the effects of these inhibitors may be unrelated to their action on cell carboxyl methyltransferases (Scheer & Gierschik 1993; Ma et al 1994). Thus further studies are needed to evaluate the roles of these enzymes in programming translocation of small GTPases between membranes and the cytosol.

Literature Cited

Adamson P, Marshall CJ, Hall A, Tilbrook PA. 1992. Post-translational modifications of $p21^{rho}$ proteins. *J. Biol. Chem.* 267:20033–38

Altschuler D, Lapetina EG. 1993. Mutational analysis of the cAMP-dependent protein kinase-mediated phosphorylation site of rap1b. *J. Biol. Chem.* 268:7527–31

Ando S, Kaibuchi K, Sasaki T, Hiraoka K, Nishiyama T, et al. 1992. Post-translational processing of *rac* p21s is important both for their interaction with the GDP/GTP exchange proteins and for their activation of NADPH oxidase. *J. Biol. Chem.* 267:25709–13

Andres DA, Seabra MC, Brown MS, Armstrong SA, Smeland TE, et al. 1993. cDNA cloning of component A of rab geranylgeranyl transferase and demonstration of its role as a Rab escort protein. *Cell* 73: 1091–99

Araki S, Kaibuchi K, Sasaki T, Hata Y, Takai Y. 1991. Role of the C-terminal region of *smg* p25A in its interaction with membranes and the GDP/GTP exchange protein. *Mol. Cell. Biol.* 11:1438–47

Araki S, Kikuchi A, Hata Y, Isomura M, Takai Y. 1990. Regulation of reversible binding of *smg* p25A, a *ras* p21-like GTP-binding protein, to synaptic plasma membranes and vesicles by its specific regulatory protein, GDP dissociation inhibitor. *J. Biol. Chem.* 265: 13007–15

Armstrong SA, Seabra MC, Südhof TC, Goldstein JL, Brown MS. 1993. cDNA cloning and expression of the α and β subunits of rat Rab geranylgeranyl transferase. *J. Biol. Chem.* 268:12221–29

Backlund PS Jr. 1992. GTP-stimulated carboxyl methylation of a soluble form of the GTP-binding protein G25K in brain. *J. Biol. Chem.* 267:18432–39

Backlund PS Jr. 1993. Carboxyl methylation of the low molecular weight GTP-binding protein G25K: regulation of carboxyl methylation of rhoGDI. *Biochem. Biophys. Res. Commun.* 196:534–42

Bailly E, McCaffrey M, Touchot N, Zahraoui A, Goud B, Bornens M. 1991. Phosphorylation of two small GTP-binding proteins of the Rab family by $p34^{cdc2}$. *Nature* 350:715–18

Ballester R, Furth ME, Rosen OM. 1987. Phorbol ester- and protein kinase C-mediated phosphorylation of the cellular kirsten ras gene product. *J. Biol. Chem.* 262:2688–95

Béranger F, Cadwallader K, Porfiri E, Powers S, Evans T, et al. 1994a. Determination of structural requirements for the interaction of rab6 and rabGDI and rab geranylgeranyltransferase. *J. Biol. Chem.* 269:13637–43

Béranger F, Paterson H, Powers S, de Gunzburg J, Hancock JF. 1994b. The effector domain of rab6, plus a highly hydrophobic C terminus, is required for Golgi apparatus localization. *Mol. Cell. Biol.* 14:744–58

Béranger F, Tavitian A, de Gunzburg J. 1991. Post-translational processing and subcellular localization of the Ras-related rap2 protein. *Oncogene* 6:1835–42

Black SD. 1992. Development of hydrophobicity parameters for prenylated proteins. *Biochem. Biophys. Res. Comm.* 186:1437–42

Blenis J. 1993. Signal transduction via the MAP kinases: proceed at your own RSK. *Proc. Natl. Acad. Sci. USA* 90:5889–92

Boguski MS, McCormick F. 1993. Proteins regulating Ras and its relatives. *Nature* 366: 643–54

Bokoch GM. 1993. Biology of the Rap proteins,

members of the ras superfamily of GTP-binding proteins. *Biochem. J.* 289:17–24

Bollag G, McCormick F. 1991. Regulators and effectors of *ras* proteins. *Annu. Rev. Cell Biol.* 7:601–32

Boman AL, Taylor TC, Melançon P, Wilson KL. 1992. A role for ADP-ribosylation factor in nuclear vesicle dynamics. *Nature* 358: 512–14

Brennwald P, Novick P. 1993. Interactions of three domains distinguishing the Ras-related GTP-binding proteins Ypt1 and Sec4. *Nature* 362:560–63

Brown HA, Gutowski S, Moomaw CR, Slaughter C, Sternweis PC. 1993. ADP-ribosylation factor, a small GTP-dependent regulatory protein stimulates phospholipase D activity. *Cell* 75:1137–44

Brown MS, Goldstein JL. 1993. Mad bet for rab. *Nature* 366:14–15

Buday L, Downward J. 1993. Epidermal growth factor regulates $p21^{ras}$ through the formation of a complex of receptor, Grb2 adapter protein, and Sos nucleotide exchange factor. *Cell* 73:611–20

Burstein ES, Brondyk WH, Macara IG, Kaibuchi K, Takai Y. 1993. Regulation of the GTPase cycle of the neuronally expressed ras-like GTP-binding protein rab3A. *J. Biol. Chem.* 268:22247–50

Burstein ES, Macara IG. 1992. Characterization of a guanine nucleotide-releasing factor and a GTPase-activating protein that are specific for the ras-related protein $p25^{rab3A}$. *Proc. Natl. Acad. Sci. USA* 89:1154–58

Buss JE, Solski PA, Schaeffer JP, MacDonald MJ, Der CJ. 1989. Activation of the cellular proto-oncogene product p21Ras by addition of a myristoylation signal. *Science* 243: 1600–3

Camp LA, Hofmann SL. 1993. Purification and properties of a palmitoyl-protein thioesterase that cleaves palmitate from H-Ras. *J. Biol. Chem.* 268:22566–74

Chardin P, Camonis JH, Gale NW, van Aelst L, Schlessinger J, et al. 1993. Human Sos1: a guanine nucleotide exchange factor for Ras that binds to GRB2. *Science* 260:1338–43

Chardin P, Tavitian A. 1989. Coding sequences of human ralA and ralB cDNAs. *Nucleic Acids Res.* 17:4380

Chavrier P, Gorvel J-P, Stelzer E, Simons K, Gruenberg J, et al. 1991. Hypervariable C-terminal domain of rab proteins acts as a targeting signal. *Nature* 353:769–72

Chuang TH, Bohl BP, Bokoch GM. 1993. Biologically active lipids are regulators of Rac GDI•complexation. *J. Biol. Chem.* 268: 26206–11

Clarke S. 1993. Protein methylation. *Curr. Biol.* 5:977–83

Cormont M, Tanti J-F, Zahraoui A, van Obberghen E, le Marchand-Brustel Y. 1994. rab4 is phosphorylated by the insulin-activated extracellular-signal-regulated kinase ERK1. *Eur. J. Biochem.* 219:1081–85

Corvazier E, Enouf J, Papp B, de Gunzburg J, Tavitian A, Levy-Toledano S. 1992. Evidence for a role of *rap*1 protein in the regulation of human platelet Ca^{2+} fluxes. *Biochem. J.* 281:325–31

Cox AD, Hisaka MM, Buss JE, Der CJ. 1992. Specific isoprenoid modification is required for function of normal, but not oncogenic, Ras protein. *Mol. Cel. Biol.* 12:2606–15

Cremers FPM, Armstrong SA, Seabra MC, Brown MS, Goldstein JL. 1994. REP-2, a rab escort protein encoded by the choroideremia-like gene. *J. Biol. Chem.* 269:2111–17

Dascher C, Balch WE. 1994. Dominant inhibitory mutants of ARF1 block endoplasmic reticulum to Golgi transport and trigger disassembly of the Golgi apparatus. *J. Biol. Chem.* 269:1437–48

Dent P, Haser W, Haystead TAJ, Vincent LA, Roberts TM, Sturgill TW. 1992. Activation of mitogen-activated protein kinase kinase by v-Raf in NIH 3T3 cells and in vitro. *Science* 257:1404–7

Devaux PF. 1992. Protein involvement in transmembrane lipid asymmetry. *Annu. Rev. Biophys. Biomol. Struct.* 21:417–39

Downward J. 1994. The GRB2/Sem-5 adaptor protein. *FEBS Lett.* 338:113–17

Egan SE, Giddings BW, Brooks MW, Buday L, Sizeland AM, Weinberg RA. 1993. Association of Sos Ras exchange protein with Grb2 is implicated in tyrosine kinase signal transduction and transformation. *Nature* 363: 45–51

Farnsworth CC, Kawata M, Yoshida Y, Takai Y, Gelb MH, Glomset JA. 1991. C terminus of the small GTP-binding protein *smg* p25A contains two geranylgeranylated cysteine residues and a methyl ester. *Proc. Natl. Acad. Sci. USA* 88:6196–200

Farnsworth CC, Seabra MC, Ericsson LH, Gelb MH, Glomset JA. 1994. Rab geranylgeranyl transferase catalyzes the prenylation of adjacent cysteines in the small GTPases, rab1A, rab3A, and rab5A. *Proc. Natl. Acad. Sci. USA.* Submitted

Farrell F, Torti M, Lapetina EG. 1992. rap proteins: investigating their role in cell function. *J. Lab. Clin. Med.* 120:533–37

Ferro-Novick S, Novick P. 1993. The role of GTP-binding proteins in transport along the exocytic pathway. *Annu. Rev. Cell Biol.* 9: 575–99

Fischer TH, Gatling MN, Lacal JC, White GC II. 1990. rap1B, a cAMP-dependent protein kinase substrate, associates with the platelet cytoskeleton. *J. Biol. Chem.* 265:19405–8

Fox JE, Lipfert L, Clark EA, Reynolds CC, Austin CD, Brugge JS. 1993. On the role of the platelet membrane skeleton in mediating signal transduction. Association of GP IIb-IIIa, $pp60^{c-src}$, pp62c-^{yes}, and the $p21^{ras}$ GTPase-activating protein with the membrane skeleton. *J. Biol. Chem.* 268:25973–84

Franco M, Chardin P, Chabre M, Paris S. 1993. Myristoylation is not required for GTP-dependent binding of ADP-ribosylation factor ARF1 to phospholipids. *J. Biol. Chem.* 268:24531–34

Giannakouros T, Newman CMH, Craighead MW, Armstrong J, Magee AI. 1993. Post-translational processing of *Schizosaccharomyces pombe* YPT5 protein. *J. Biol. Chem.* 268:24467–74.

Gordon JI, Duronio RJ, Rudnick DA, Adams SP, Gokel GW. 1991. Protein *N*-myristoylation. *J. Biol. Chem.* 266:8647–50

Gutierrez L, Magee AI. 1991. Characterization of an acyltransferase acting on $p21^{N-ras}$ protein in a cell-free system. *Biochim. Biophys. Acta* 1078:147–54

Hallberg B, Rayter SI, Downward J. 1994. Interaction of Ras and Raf in intact mammalian cells upon extracellular stimulation. *J. Biol. Chem.* 269:3913–16

Hancock JF, Cadwallader K, Marshall CJ. 1991a. Methylation and proteolysis are essential for efficient membrane binding of prenylated $p21^{K-ras(B)}$. *EMBO J.* 10:641–46

Hancock JF, Cadwallader K, Paterson H, Marshall CJ. 1991b. A CAAX or a CAAL motif and a second signal are sufficient for plasma membrane targeting of ras proteins. *EMBO J.* 10:4033–39

Hancock JF, Magee AI, Childs JE, Marshall CJ. 1989. All ras proteins are polyisoprenylated but only some are palmitoylated *Cell* 57: 1167–77

Hancock JF, Paterson H, Marshall CJ. 1990. A polybasic domain or palmitoylation is required in addition to the CAAX motif to localize $p21^{ras}$ to the plasma membrane. *Cell* 63:133–39

Hata Y, Kaibuchi K, Kawamura S, Hiroyoshi M, Shirataki H, Takai Y. 1991. Enhancement of the actions of *smg* p21 GDP/GTP exchange protein by the protein kinase A-catalyzed phosphorylation of *smg* p21. *J. Biol. Chem.* 266:6571–77

Haun RS, Tsai S-C, Adamik R, Moss J, Vaughn M. 1993. Effect of myristoylation on GTP-dependent binding of ADP-ribosylation factor to Golgi. *J. Biol. Chem.* 268:7064–68

Hiraoka K, Kaibuchi K, Ando S, Musha T, Takaishi K, et al. 1992. Both stimulatory and inhibitory GDP/GTP exchange proteins, smg GDS and rho GDI, are active on multiple small GTP-binding proteins. *Biochem. Biophys. Res. Commun.* 182:921–30

Hiroyoshi M, Kaibuchi K, Kawamura S, Hata Y, Takai Y. 1991. Role of the C-terminal region of *smg* p21, a *ras* p21-like small GTP-binding protein, in membrane and *smg* p21 GDP/GTP exchange protein interactions. *J. Biol. Chem.* 266:2962–69

Hori Y, Kikuchi A, Isomura M, Katayama M, Miura Y, et al. 1991. Post-translational modifications of the C-terminal of the rho protein are important for its interaction with membranes and the stimulatory and inhibitory GDP/GTP exchange proteins. *Oncogene* 6: 515–22

Horikawa HP, Saisu H, Ishizuka T, Sekine Y, Tsugita A, et al. 1993. A complex of rab3A, SNAP-25, VAMP/synaptobrevin-2 and syntaxins in brain presynaptic terminals. *FEBS Lett.* 330:236–40

Howe LR, Leevers SJ, Gómez N, Nakielny S, Cohen P, Marshall CJ. 1992. Activation of the MAP kinase pathway by the protein kinase raf. *Cell* 71:335–42

Isomura M, Kikuchi A, Ohga N, Takai Y. 1991. Regulation of binding of rhoB p20 to membranes by its specific regulatory protein, GDP dissociation inhibitor. *Oncogene* 6: 119–24

Itoh T, Kaibuchi K, Masuda T, Yamamoto T, Matsuura Y, et al. 1993a. The post-translational processing of *ras* p21 is critical for its stimulation of mitogen-activated protein kinase. *J. Biol. Chem.* 268:3025–28

Itoh T, Kaibuchi K, Masuda T, Yamamoto T, Matsuura Y, et al. 1993b. A protein factor for ras p21-dependent activation of mitogen-activated protein (MAP) kinase through MAP kinase kinase. *Proc. Natl. Acad. Sci. USA* 90:975–79

Itoh T, Kaibuchi K, Sasaki T, Takai Y. 1991. The *smg* GDS-induced activation of *smg*p21 is initiated by cyclic AMP-dependent kinase-catalyzed phosphorylation of *smg*p21. *Biochem. Biophy. Res. Comm.* 177:1319–24

Kahn RA, Goddard C, Newkirk M. 1988. Chemical and immunological characterization of the 21-kDa ADP-ribosylation factor of adenylate cyclase. *J. Biol. Chem.* 263: 8282–87

Kahn RA, Randazzo P, Serafini T, Weiss O, Rulka C, et al. 1992. The amino terminus of ADP-ribosylation factor (ARF) is a critical determinant of ARF activities and is a potent and specific inhibitor of protein transport. *J. Biol. Chem.* 267:13039–46

Karniguian A, Zahraoui A, Tavitian A. 1993. Identification of small GTP-binding rab proteins in human platelets: Thrombin-induced phosphorylation of rab3B, rab6 and rab8 proteins. *Proc. Natl. Acad. Sci. USA* 90:7647–51

Kawamura M, Kaibuchi K, Kishi K, Takai Y. 1993. Translocation of Ki-*ras* p21 between membrane and cytoplasm by *smg* GDS. *Biochem. Biophys. Res. Commun.* 190:832–41

Kawamura S, Kaibuchi K, Hiroyoshi M, Hata Y, Takai Y. 1991. Stoichiometric interaction of *smg* p21 with its GDP/GTP exchange protein and its novel action to regulate the translocation of *smg* p21 between membrane and cytoplasm. *Biochem. Biophys. Res. Commun.* 174:1095–102

Khosravi-Far R, Lutz RJ, Cox AD, Conroy L, Bourne JR, et al. 1991. Isoprenoid modification of rab proteins terminating in CC or CXC motifs. *Proc. Natl. Acad. Sci. USA* 88: 6264–68

Kikuchi A, Kuroda S, Sasaki T, Kotani K, Hirata KK, et al. 1992. Functional interactions of stimulatory and inhibitory GDP/GTP exchange proteins and their common substrate small GTP-binding protein. *J. Biol. Chem.* 267:14611–15

Kinsella BT, Maltese WA. 1992. *rab* GTP-binding proteins with three different carboxyl-terminal cysteine motifs are modified in vivo by 20-carbon isoprenoids. *J. Biol. Chem.* 267:3940–45

Kishi K, Sasaki T, Kuroda S, Itoh T, Takai Y. 1993. Regulation of cytoplasmic division of *Xenopus* embryo by *rho* p21 and its inhibitory GDP/GTP exchange protein (*rho* GDI). *J. Cell. Biol.* 120:1187–95

Kuroda Y, Suzuki N, Kataoka T. 1993. The effect of posttranslational modifications on the interaction of ras2 with adenylyl cyclase. *Science* 259:683–86

Kyriakis JM, App H, Zhang X-F, Banerjee P, Brautigan DL, et al. 1992. Raf-1 activates MAP kinase-kinase. *Nature* 358:417–21

Lacal PM, Pennington CY, Lacal JC. 1988. Transforming activity of *ras* proteins translocated to the plasma membrane by a myristoylation sequence from the *src* gene product. *Oncogene* 2:533–37

Lapetina EG, Lacal JC, Reep BR, Molina y Vedia L. 1989. A *ras*-related protein is phosphorylated and translocated by agonists that increase cAMP levels in human platelets. *Proc. Natl. Acad. Sci. USA* 86:3131–34

Lazarowski ER, Winegar DA, Nolan RD, Oberdisse E, Lapetina EG. 1990. Effect of protein kinase A on inositide metabolism and rap 1 G-protein in human erythroleukemia cells. *J. Biol. Chem.* 265:13118–23

Lenhard JM, Kahn RA, Stahl PD. 1992. Evidence for ADP-ribosylation factor (ARF) as a regulator of in vitro endosome-endosome fusion. *J. Biol. Chem.* 267:13047–52

Li G, Stahl PD. 1993. Post-translational processing and membrane association of the two early endosome-asociated rab GTP-binding proteins (rab4 and rab5). *Arch. Biochem. Biophys.* 304:471–78

Lombardi D, Soldati T, Riederer MA, Goda Y, Zerial M, Pfeffer, SR. 1993. Rab9 functions in transport between late endosomes and the *trans* Golgi network. *EMBO J.* 12:677–82

Lowy DR, Willumsen BM. 1993. Function and regulation of ras. *Annu. Rev. Biochem.* 62: 851–91

Ma Y, Shi Y, Lim YH, McGrail SH, Ware JA, et al. 1994. Mechanistic studies on human platelet isoprenylated protein methyltransferase: farnesylcysteine analogs block platelet aggregation without inhibiting the methyltranferase. *Biochemistry* 33:5414–20

MacDonald SG, Crews CM, Wu L, Driller J, Clark R, et al. 1993. Reconstitution of the Raf-1-MEK-ERK signal transduction pathway in vitro. *Mol. Cell. Biol.* 13:6615–20

Magee AI, Gutierrez L, McKay IA, Marshall CJ, Hall A. 1987. Dynamic fatty acylation of $p21^{N\text{-}ras}$. *EMBO J.* 6:3353–57

Magnier C, Bredoux R, Kovacs T, Quarck R, Papp B, et al. 1994. Correlated expression of the 97 kDa sarcoendoplasmic reticulum Ca^{2+}-ATPase and Rap1B in platelets and various cell lines. *Biochem. J.* 297:343–50

Marengere LE, Pawson T. 1992. Identification of residues in GTPase-activating protein Src homology 2 domains that control binding to tyrosine phosphorylated growth factor receptors and p62. *J. Biol. Chem.* 267:22779–86

Matsui Y, Kikuchi A, Araki S, Hata Y, Kondo J, et al. 1990. Molecular cloning and characterization of a novel type of regulatory protein (GDI) for *smg* p25A, a *ras* p21-like GTP-binding protein. *Mol. Cell. Biol.* 10: 4116–22

Miura Y, Kaibuchi K, Itoh T, Corbin JD, Francis SH, Takai Y. 1992. Phosphorylation of *smg* p21B/*rap* 1B p21 by cyclic GMP-dependent protein kinase. *FEBS Lett.* 297:171–74

Miura Y, Kikuchi A, Musha T, Kuroda S, Yaku H, et al. 1993. Regulation of morphology by *rho* p21 and its inhibitory GDP/GTP exchange protein (*rho* GDI) in Swiss 3T3 cells. *J. Biol. Chem.* 268:510–15

Mizuno T, Kaibuchi K, Yamamoto T, Kawamura M, Sakoda T, et al. 1991. A stimulatory GDP/GTP exchange protein for smg p21 is active on the post-translationally processed form of c-Ki-ras p21 and rhoA p21. *Proc. Natl. Acad. Sci. USA* 88:6442–46

Molenaar CMT, Prange, R, Gallwitz D. 1988. A carboxyl-terminal cysteine residue is required for palmitic acid binding and biological activity of the *ras*-related yeast YPT1 protein. *EMBO J.* 7:971–76

Moodie SA, Willumsen BM, Weber MJ, Wolfman A. 1993. Complexes of ras-GTP with raf-1 and mitogen-activated protein kinase kinase. *Science* 260:1658–61

Moores SL, Schaber MD, Mosser SD, Rands E, O'Hara MB, et al. 1991. Sequence dependence of protein isoprenylation. *J. Biol. Chem.* 266:14603–10

Morgan A, Burgoyne RD. 1993. A synthetic peptide of the N-terminus of ADP ribosylation factor (ARF) inhibits regulated exocytosis in adrenal chromaffin cells. *FEBS Lett.* 329:121–24

Moss J, Vaughn M. 1993. ADP-ribosylation

factors, 20,000 M_r guanine nucleotide-binding protein activators of cholera toxin and components of intracellular vesicular transport systems. *Cell. Signaling* 5:367–79

Moya M, Roberts D, Novick P. 1993. *DSS4–1* is a dominant suppressor of *sec4–8* that encodes a nucleotide exchange protein that aids Sec4p function. *Nature* 361:460–63

Musha T, Kawata M, Takai Y. 1992. The geranylgeranyl moiety but not the methyl moiety of the *smg*-25A/*rab*3A protein is essential for the interactions with membrane and its inhibitory GDP/GTP exchange protein. *J. Biol. Chem.* 267:9821–25

Newman CM, Giannakouros T, Hancock JF, Fawell EH, Armstrong J, Magee AI. 1992. Post-translational processing of *Schizosaccharomyces pombe* YPT proteins. *J. Biol. Chem.* 267:11329–36

Nishida E, Gotoh Y. 1993. The MAP kinase cascade is essential for diverse signal transduction pathways. *Trends Biol. Sci.* 18:128–31

Ohmstede C-A, Farrell FX, Reep BR, Clemetson KJ, Lapetina EG. 1990. rap2B: a ras-related GTP-binding protein from platelets. *Proc. Natl. Acad. Sci. USA* 87:6527–31

Orci L, Palmer DJ, Amherdt M, Rothman JE. 1993a. Coated vesicle assembly in the Golgi requires only coatomer and ARF proteins from the cytosol. *Nature* 364:732–34

Orci L, Palmer DJ, Ravazzola M, Perrelet A, Amherdt M, Rothman JE. 1993b. Budding from Golgi membranes requires the coatomer complex of non-clathrin coat proteins. *Nature* 362:648–51

Orita S, Kaibuchi K, Kuroda S, Shimizu K, Nakanishi H, Takai Y. 1993. Comparison of kinetic properties between two mammalian *ras* p21 GDP/GTP exchange proteins, *ras* guanine nucleotide-releasing factor and *smg* GDP dissociation stimulator. *J. Biol. Chem.* 268:25542–46

Palmer DJ, Helms JB, Beckers CJ, Orci L, Rothman JE. 1993. Binding of coatomer to Golgi membranes requires ADP-ribosylation factor. *J. Biol. Chem.* 268:12083–89

Panayotou G, Waterfield MD. 1993. The assembly of signalling complexes by receptor tyrosine kinases. *BioEssays* 15:171–77

Peitzsch RM, McLaughlin S. 1993. Binding of acylated peptides and fatty acids to phospholipid vesicles: pertinence to myristoylated proteins. *Biochemistry* 32:10436–43

Philips MR, Pillinger MH, Staud R, Volker C, Rosenfeld MG, et al. 1993. Carboxyl methylation of ras-related proteins during signal transduction in neutrophils. *Science* 259:977–80

Pillinger MH, Volker C, Stock JB, Weissmann G, Philips MR. 1994. Characterization of a plasma membrane-associated prenylcysteine-directed α carboxyl methyltransferase in human neutrophils. *J. Biol. Chem.* 269:1486–92

Quinn MT, Evans T, Loetterle LR, Jesaitis AJ, Bokoch GM. 1993. Translocation of rac correlates with NADPH oxidase activation. *J. Biol. Chem.* 268:20983–87

Randazzo PA, Yang YC, Rulka C, Kahn RA. 1993. Activation of ADP-ribosylation factor by Golgi membranes. *J. Biol. Chem.* 268:9555–63

Rossi G, Jiang Y, Newman AP, Ferro-Novick S. 1991. Dependence of Ypt1 and Sec4 membrane attachment on Bet2. *Nature* 351:158–61

Sahyoun N, McDonald OB, Farrell F, Lapetina EG. 1991. Phosphorylation of a Ras-related GTP-binding protein kinase, Rap-1b, by a neuronal Ca^{2+}/calmodulin-dependent protein kinase, CaM kinase Gr. *Proc. Natl. Acad. Sci. USA* 88:2643–47

Sanford JC, Pan Y, Wessling-Resnick M. 1993. Prenylation of rab5 is dependent on guanine nucleotide binding. *J. Biol. Chem.* 268:23773–76

Sasaki T, Kaibuchi K, Kabcenell AK, Novick PJ, Takai Y. 1991. A mammalian inhibitory GDP/GTP exchange protein (GDP dissociation inhibitor) for *smg* P25A is active on the yeast *SEC4* protein. *Mol. Cell. Biol.* 11:2909–12

Sasaki T, Kikuchi A, Araki S, Hata Y, Isomura M, et al. 1990. Purification and characterization from bovine brain cytosol of a protein that inhibits the dissociation of GDP from and the subsequent binding of GTP to *smg* p25A, a *ras* p21-like GTP-binding protein. *J. Biol. Chem.* 265:2333–37

Scheer A, Gierschik P. 1993. Farnesylcysteine analogues inhibit chemotactic peptide receptor-mediated G-protein activation in human HL-60 granulocyte membranes. *Fed. Eur. Biochem. Soc.* 319:110–14

Schlessinger J. 1993. How receptor tyrosine kinases activate Ras. *Trends Biol. Sci.* 18:273–75

Seabra MC, Brown MS, Slaughter CA, Südhof TC, Goldstein JL. 1992a. Purification of component A of rab geranylgeranyl transferase: possible identity with the choroideremia gene product. *Cell* 70:1049–57

Seabra MC, Goldstein JL, Südhof TC, Brown MS. 1992b. Rab geranylgeranyl transferase: a multisubunit enzyme that prenylates GTP-binding proteins terminating in Cys-X-Cys or Cys-Cys. *J. Biol. Chem.* 276:14497–503

Serth J, Lautwein A, Frech M, Wittinghofer A, Pingoud A. 1991. The inhibition of the GTPase activating protein— Ha-*ras* interaction by acidic lipids is due to physical association of the C-terminal domain of the GTPase activating protein with micellar structures. *EMBO J.* 10:1325–30

Shirataki H, Kaibuchi K, Sakoda T, Kishida S, Yamaguchi T, et al. 1993. Rabphilin-3A, a putative target protein for *smg* p25A/*rab*3A p25 small GTP-binding protein related to synaptotagmin. *Mol. Cell. Biol.* 13:2061–68

Shirataki H, Kaibuchi K, Yamaguchi T, Wada K, Horiuchi H, Takai Y. 1992. A possible target protein for *smg*-25A/*rab*3A small GTP-binding protein. *J. Biol. Chem.* 267: 10946–49

Silvius JR, l'Heureux F. 1994. Fluorimetric evaluation of the affinities of isoprenylated peptides for lipid bilayers. *Biochemistry* 33: 3014–22

Simons K, Zerial M. 1993. Rab proteins and the road maps for intracellular transport. *Neuron* 11:789–99

Soldati T, Riederer MA, Pfeffer SR. 1993. Rab GDI: A solubilizing and recycling factor for rab9 protein. *Mol. Biol. Cell* 4:425–34

Soldati T, Shapiro AD, Dirac Svejstrup AB, Pfeffer SR. 1994. Membrane targeting of the small GTP-binding protein rab9 is accompanied by nucleotide exchange. *Nature* 369:76–78

Soler C, Beguinot L, Sorkin A, Carpenter G. 1993. Tyrosine phosphorylation of ras GTPase-activating protein does not require association with the epidermal growth factor receptor. *J. Biol. Chem.* 268:22010–19

Stamnes MA, Rothman, JE. 1993. The binding of AP-1 clathrin adaptor particles to Golgi membranes requires ADP-ribosylation factor, a small GTP-binding protein. *Cell* 73: 999–1005

Steele-Mortimer O, Gruenberg J, Clague MJ. 1993. Phosphorylation of GDI and membrane cycling of rab proteins. *FEBS Lett.* 329:313–18

Stenmark H, Valencia A, Martinez O, Ullrich O, Goud B, Zerial M. 1994. Distinct structural elements of rab5 define its functional specificity. *EMBO J.* 13:575–83

Stouten PFW, Sander C, Wittinghofer A, Valencia A. 1993. How does the switch II region of G-domains work? *FEBS Lett.* 320: 1–6

Takaishi K, Kikuchi A, Kuroda S, Kotani K, Sasaki T, Takai Y. 1993. Involvement of *rho* p21 and its inhibitory GDP/GTP exchange protein (*rho* GD1) in cell motility. *Mol. Cell. Biol.* 13:72–79

Tan EW, Rando RR. 1992. Identification of an isoprenylated cysteine methyl ester hydrolase activity in bovine rod outer segment membranes. *Biochemistry* 31:5572–78

Tanigawa G, Orci L, Amherdt M, Ravazzola M, Helms JB, Rothman JE. 1993. Hydrolysis of bound GTP by ARF protein triggers uncoating of Golgi-derived COP-coated vesicles. *J. Cell Biol.* 123:1365–71

Taylor TC, Kanstein M, Weidman P, Melançon P. 1994. Cytosolic ARFs are required for vesicle formation but not for cell-free intra-Golgi transport: Evidence for coated vesicle-independent transport. *Mol. Biol. Cell* 5:237–52

Teal SB, Hsu VW, Peters PJ, Klausner RD, Donaldson JG. 1994. An activating mutation in ARF1 stabilizes coatomer binding to Golgi membranes. *J. Biol. Chem.* 269:3135–38

Thissen JA, Casey PJ. 1993. Microsomal membranes contain a high affinity binding site for prenylated peptides. *J. Biol. Chem.* 268: 13780–83

Torti M, Ramaschi G, Sinigaglia F, Lapetina EG, Balduini C. 1993. Association of the low molecular weight GTP-binding protein rap2B with the cytoskeleton during platelet aggregation. *Proc. Natl. Acad. Sci. USA* 90: 7553–57

Traub LM, Ostrom JA, Kornfeld S. 1993. Biochemical dissection of AP-1 recruitment onto Golgi membranes. *J. Cell Biol.* 123: 561–73

Traverse S, Cohen P, Paterson H, Marshall C, Rapp U, Grand RJA. 1993. Specific association of activated MAP kinase kinase kinase (Raf) with the plasma membranes of *ras*-transformed retinal cells. *Oncogene* 8:3175–81

Tsai S-C, Adamik R, Moss J, Vaughan M. 1994. Identification of a brefeldin A-insensitive guanine nucleotide-exchange protein for ADP-ribosylation factor in bovine brain. *Proc. Natl. Acad. Sci. USA* 91:3063–66

Ueda T, Kikuchi A, Ohga N, Yamamoto J, Takai Y. 1990. Purification and characterization from bovine brain cytosol of a novel regulatory protein inhibiting the dissociation of GDP from and the subsequent binding of GTP to *rho*B p20, a ras p21-like GTP-binding protein. *J. Biol. Chem.* 265:9373–80

Ullrich O, Horiuchi H, Bucci C, Zerial M. 1994. Membrane association of Rab5 mediated by GDP-dissociation inhibitor and accompanied by GDP/GTP exchange. *Nature* 368:157–60

Ullrich O, Stenmark H, Alexandrov K, Huber LA, Kaibuchi K, et al. 1993. Rab GDP dissociation inhibitor as a general regulator for the membrane association of rab proteins. *J. Biol. Chem.* 268:18143–50

Van Aelst L, Barr M, Marcus S, Polverino A, Wigler M. 1993. Complex formation between Ras and Raf and other protein kinases. *Proc. Natl. Acad Sci. USA* 90:6213–17

Vojtek AB, Hollenberg SM, Cooper JA. 1993. Mammalian ras interacts directly with the serine/threonine kinase raf. *Cell* 74:205–14

Wada K, Mizoguchi A, Kaibuchi K, Shirataki H, Ide C, et al. 1994. Localization of rabphilin-3A, a putative target protein for rab3A, at the sites of Ca^{2+}-dependent exocytosis in PC12 cells. *Biochem. Biophys. Res. Commun.* 198:158–65

Warne PH, Viciana PR, Downward J. 1993. Direct interaction of ras and the amino-terminal region of raf-1 in vitro. *Nature* 364: 352–55

Wartmann M, Davis RK. 1994. The native structure of the activated raf protein kinase is a membrane-bound multi-subunit complex. *J. Biol. Chem.* 269:6695–701

Wei C, Lutz R, Sinensky M, Macara IG. 1992. p23^{rab2}, a *ras*-like GTPase with a -GGGCC C-terminus, is isoprenylated but not detectably carboxymethylated in NIH3T3 cells. *Oncogene* 7:467–73

Willingham MC, Pastan I, Shih TY, Scolnick EM. 1980. Localization of the *src* gene product of the Harvey strain of MSV to plasma membrane of transformed cells by electron microscopic immunocytochemisty. *Cell* 19: 1005–14

Wilson AL, Maltese WA. 1993. Isoprenylation of rab1B is impaired by mutations in its effector domain. *J. Biol. Chem.* 268:14561–64

Yamaguchi T, Shirataki H, Kishida S, Miyazaki M, Nishikawa J, et al. 1993. Two functionally different domains of rabphilin-3A, *rab*3A p25/*smg* p25A-binding and phospholipid- and Ca^{2+}-binding domains. *J. Biol. Chem.* 268:27164–70

Yamamoto T, Kaibuchi K, Mizuno T, Hiroyoshi M, Shirataki H, Takai Y. 1990. Purification and characterization from bovine brain cytosol of proteins that regulate the GDP/GTP exchange reaction of *smg* p21s, *ras* p21-like GTP-binding proteins. *J. Biol. Chem.* 265:16626–34

Yamane HK, Fung BKK. 1993. Covalent modifications of G-proteins. *Annu. Rev. Pharmacol. Toxicol.* 32:201–41

Yokoyama K, Gelb MH. 1994. Mammalian protein geranylgeranyltransferase-I: Substrate specificity, kinetic mechanism, and metal requirements. *Biochemistry.* Submitted

Zhang J, Fry MJ, Waterfield MD, Jaken S, Liao L, Fox JEB, Rittenhouse SE. 1992. Activated phosphoinositide 3-kinase associates with membrane skeleton in thrombin-exposed platelets. *J. Biol. Chem.* 267:4686–92

Zhang X-F, Settleman J, Kyriakis JM, Takeuchi-Suzuki E, Elledge SJ, et al. 1993. Normal and oncogenic p21^{ras} proteins bind to the amino-terminal regulatory domain of c-Raf-1. *Nature* 364:308–13

ADDED IN PROOF

Conclusive evidence that plasma membrane association is required for raf activation comes from two recent studies (Leevers et al 1994; Stokoe et al 1994) in which the membrane localization signal of K-ras was fused to the C-terminus of raf. Upon expression, this chimera was targeted directly to the plasma membrane where it became constitutively active, independently of ras. Thus ras appears to function by recruiting raf to the membrane where it is apparently activated by additional factors.

Leevers SJ, Paterson HF, Marshall CJ. 1994. Requirements for Ras in Raf activation is overcome by targeting Raf to the plasma membrane. *Nature* 369:411–14

Stokoe D, Macdonald SG, Cadwallader K, Symons M, Hancock JF. 1994. Activation of Raf as a result of recruitment to the plasma membrane. *Science* 264:1463–67

Annu. Rev. Cell Biol. 1994. 10:207–49

STRUCTURE OF ACTIN BINDING PROTEINS: Insights about Function at Atomic Resolution

Thomas D. Pollard
Department of Cell Biology and Anatomy, Johns Hopkins Medical School, Baltimore, Maryland 21205

Steven Almo
Department of Biochemistry, Albert Einstein College of Medicine, New York, NY 10461

Stephen Quirk and Valda Vinson
Department of Cell Biology and Anatomy, and Department of Biophysics and Biophysical Chemistry, Johns Hopkins Medical School, Baltimore, Maryland 21205

Eaton E. Lattman
Department of Biophysics and Biophysical Chemistry, Johns Hopkins Medical School, Baltimore, Maryland 21205

KEY WORDS: actin, filament, annexin, DNaseI, gelsolin, villin, hisactophilin, myosin, profilin, severin, spectrin, tropomyosin

---

CONTENTS

0743–4634/94/1115–0207$05.00

OVERVIEW

During the past five years the investigation of actin and actin-binding proteins has taken a dramatic step forward. After years of searching for new proteins that bind to actin, now known to number more than 60 (Pollard 1993), the first atomic structures of these proteins have become available. Both multidimensional nuclear magnetic resonance (NMR) and X-ray crystallography have contributed to this burst of knowledge. Today we know the atomic structures of actin, actin filaments, the myosin head, and all or part of ten other proteins that bind actin monomers or actin filaments. This knowledge has elevated the field to a new level, providing insights about the mechanisms of action. Even more importantly for cell biologists, the availability of atomic structures makes it possible to design rational mutations for studies of mechanisms of action and to make informative derivatives for studies of function in live cells. This review provides a catalogue of the known structures available in March 1994. Our format consists of short sections on each protein with a figure of each structure, a summary of the methods used to determine the structure, and brief comments on the functional implications of the structure. For readers who want to learn more, we refer to other reviews of the rich literature on each protein.

ACTIN

Background

Actin is the main building block of one of the major cytoskeletal and motility systems in eukaryotic cells. Actin is the most abundant protein in many animal cells and one of the most abundant proteins on earth. It polymerizes into filaments that can transmit tension and resist deformation. The filaments provide tracks for ATP-driven myosin motor proteins. The interaction of actin and myosin generates the force for muscle contraction, cytokinesis, cytoplasmic streaming in plants, and many less well characterized movements of eukaryotic cells. The structure of the actin monomer, determined by Kabsch et al (1990), is now the foundation for all thinking about actin and its functions (Figure 1). The structure has been confirmed recently in studies of co-crystals of actin with gelsolin domain-1 (see below; McLaughlin et al 1993) and profilin (see below; Schutt et al 1993).

Structure Determination

The structure of actin was first determined at 2.4 Å resolution by X-ray crystallography of co-crystals of rabbit skeletal muscle actin and bovine pancreatic DNase I. An initial electron density map was determined at 3.5 Å resolution by multiple isomorphous replacement and solvent flattening. The structure of DNase I was solved independently (see below) and used for model building and phase combination to extend the resolution. The final Kabsch structure has an R-factor of 20.8% with root mean square (rms) deviations

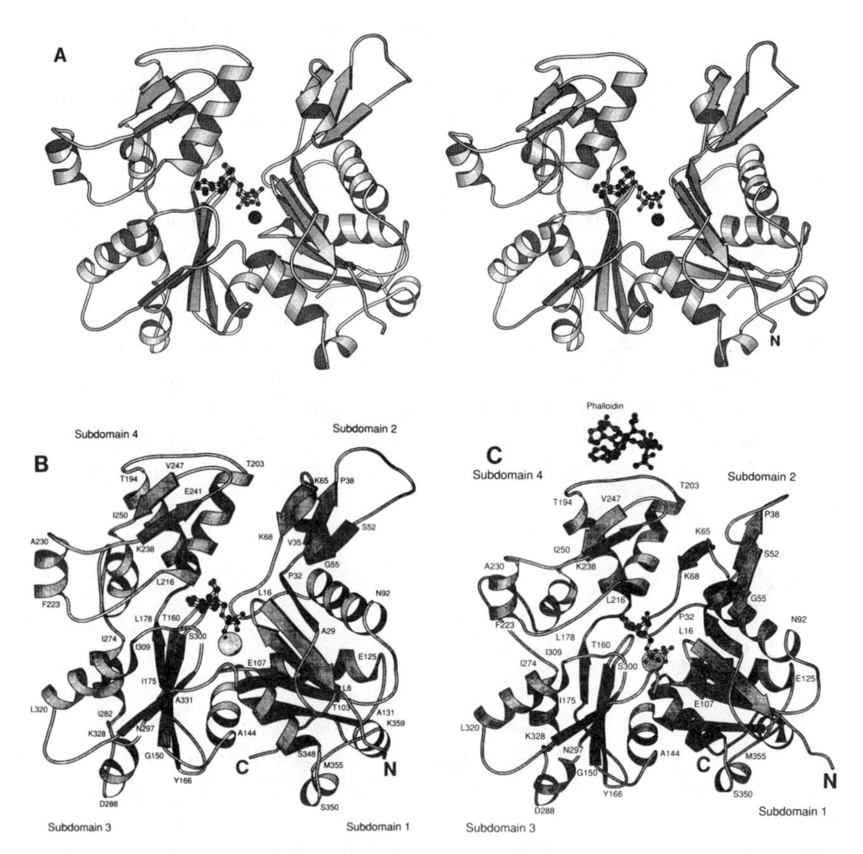

Figure 1 Actin monomer. (*A*) Stereo ribbon diagram of the elements of secondary structure of the actin molecule from the actin-DNase I cocrystal (Kabsch et al 1990). Drawings prepared with MOLSCRIPT (Kraulis 1991) using the X-ray coordinates of Kabsch et al (1990). (*B*) Ribbon diagram of the Kabsch model of actin including ATP, Ca^{2+}, and labels of key residues from Lorenz et al (1993). (*C*) Ribbon diagram of the actin subunit in the refined model of the phalloidin-stabilized actin filament (Lorenz et al 1993).

from ideal bond lengths of 0.017 Å and bond angles of 3.3°. Twenty-two backbone angles deviated from allowed regions of the Ramachandran plot. The confirmatory structures have similar resolution and crystallographic parameters.

Description of the Structure

The 375 residue polypeptide is folded into four domains that surround a cleft where an adenine nucleotide binds. The familiar orientation shown in Figure 1 illustrates the domain structure clearly but does not emphasize how flat the

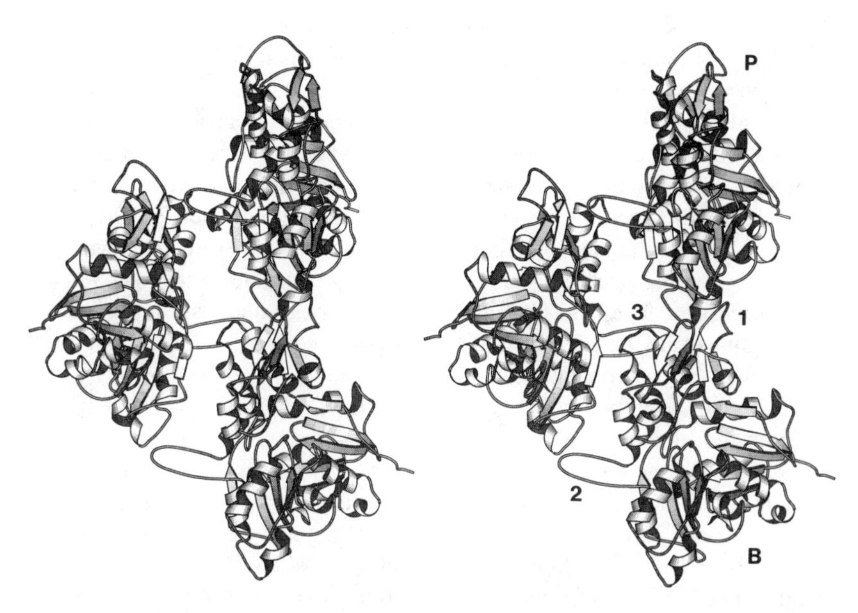

Figure 2 Actin filament. Three subunits from the Lorenz et al (1993) model of the actin filament stabilized with phalloidin. The axis of the filament is vertical. Stereo ribbon diagram of the elements of secondary structure prepared with MOLSCRIPT using coordinates provided by K Holmes of the Max Planck Institute, Heidelberg, Germany. B = barbed end of the filament; P = pointed end of the filament; 1 = DNase binding loop; 2 = hydrophobic loop; 3 = position of phalloidin.

protein is in the Z-dimension. This asymmetry can be appreciated by viewing reconstructions of the actin filament (Figure 2), which show side views of the subunits. The two halves of the molecule have similar folds, which suggest that the actin gene was formed by duplication. The fold of the whole molecule is remarkably similar to those of the glycolytic enzyme hexokinase and the ATP-binding domain of the heat shock cognate protein HSC-70, thus raising the possibility that all three have a common ancestor (Holmes et al 1993). The sequences of these three proteins are quite different except for the residues contacting the adenine nucleotide in the cleft.

The polypeptide begins and ends in subdomain 1 at the bottom right. In subdomain 1, it forms a beta sheet of mixed polarity before moving on to subdomain 2, which is composed of small antiparallel beta sheets with a short alpha helix on one side. The long DNase I binding loop, between proline-38 and serine-52, connects two of the beta strands. The chain then reenters subdomain 1 where it forms a helix-strand-helix-strand. These parallel strands complete the beta sheet in the core of the subdomain, while the two helices form the back side of the subdomain. From the last strand, the chain forms an alpha helix running from glutamine-137 to alanine-144, approximately parallel to the interface between the two halves of the molecule. Subdomain 3 is folded much like subdomain 1. Like subdomain 2, subdomain 4 has a central beta sheet crossed by an alpha helix, but the topology is different and the addition of three substantial alpha-helices makes subdomain 4 much larger than subdomain 2. After completing subdomain 4, the strand forms a helix-strand-helix-strand (like subdomain 1) to complete the beta sheet at the core of subdomain 3. The strand then crosses beneath the nucleotide-binding cleft to form four helices encircling the base of subdomain 1. The last of these helices was not seen in the original structure because the last three residues were removed by trypsin digestion. They are present in the co-crystal with gelsolin segment-1 (McLaughlin et al 1993). Reactive cysteine-374, used in many labeling studies, is located at the base of subdomain 1.

The nucleotide is bound in the central cleft of the protein with an apolar cavity for the adenine base, two hydrogen bonds for the ribose hydroxyls, and numerous hydrogen bonds to the di- or triphosphate groups. The structures of ATP- and ADP-actin are virtually identical except for the lack of a terminal phosphate. A Ca^{2+} is coordinated to oxygens of the alpha-, beta- and gamma-phosphates and to side chain oxygens from both sides of the cleft. This is thought to be the tightly bound divalent cation (K_d = 1 nM for Ca^{2+}).

Significance

The atomic structure of actin provides the framework for interpreting decades of biochemical and biophysical studies on the binding ligands (nucleotides, divalent cations, phalloidin, cytochalasins), the assembly of actin filaments, force production by interaction with myosin, and interactions of actin monomers and filaments with actin-binding proteins. We return to this structure in each of the following sections, just as investigators must do every time they design experiments with actin. The structure provided some surprises. One is the small difference between ATP-actin and ADP-actin, given the clear differences in their assembly properties. Another is the remarkable similarity to hexokinase and HSC70.

The availability of the atomic structure made possible theoretical analysis of the molecular dynamics of actin. Normal mode analysis suggests that the two halves of the molecule can move on a picosecond time scale about two hinges in the strands connecting subdomains 1 and 3 (Tirion & ben-Avraham 1993). The two halves flex apart from each other, thereby opening the cleft in a scissors-like motion. The two halves also rotate relative to each other in a propellor twist. The domains themselves are relatively rigid. The motion is confined to a few residues, especially the helix consisting of residues 137 to 144 running between subdomains 1 and 3, which acts as an elastic hinge. Rotation at alanine-331 is also required to open the cleft, and rotation at valine-339 provides for the twisting motion. These motions are expected to be important in the exchange of the bound nucleotide, opening a route for it to escape into the solvent. The beta-actin in the co-crystals with profilin has the two halves displaced by 5° (Schutt et al 1993), compared with the alpha-actin-DNase I co-crystal (Kabsch et al 1990), as allowed by these theoretical predictions.

ACTIN FILAMENT

Background

The basic geometry of the actin filament has been known for years from X-ray diffraction studies of muscle (for example, Huxley 1968) and oriented actin filament gels (Popp et al 1987), as well as from increasingly detailed reconstructions of electron micrographs of actin filaments prepared by negative staining (Bremer et al 1991; reviewed by Egelman 1994) and by rapid freezing in vitreous ice (Milligan et al 1990). The filament can be described in two ways: either as a right-handed double helix that repeats every 72 nm or as a left-handed single helix with 13 subunits that repeats after six turns. The subunit contacts are more extensive along the long pitch helix than between the two strands. The reconstructions from electron micrographs show enough detail to orient the actin molecule in the filament but do not reveal details of the atomic structure.

Structure Determination

The first atomic model of the actin filament was made using the atomic structure of the actin molecule and a rigid body search of subunit orientations that best fit X-ray diffraction patterns of oriented actin filament gels extending to 8 Å resolution (Holmes et al 1990a). A refined model including phalloidin, a cyclic peptide that stabilizes actin filaments (Figure 2), gives a calculated diffraction pattern that closely matches the experimental fiber diffraction patterns (Lorenz et al 1993). First, two surface loops were rebuilt by hand. Then the refinement consisted of a random search of individual subdomain orientations that optimized the fit of the calculated transform to the fiber diffraction data. Phalloidin was included after a search of locations near the known phalloidin cross-linking sites (Vandekerckhove et al 1985) that gave the best fit to the diffraction data. The phalloidin was manually rotated at this site to optimize the chemistry of the interactions with the adjacent subunits. Molecular dynamics refinement was used to optimize contacts between subunits. This minimized the energy of the model with reasonable stereochemistry but was not constrained by experimental data. The authors felt that the final model provides a unique fit to the experimental data. However, the fit was achieved by allowing significant changes in the orientation of the subdomains and some surface loops of the starting actin model (compare Figures 1A and 1B), distortions that have not been seen in the crystal structures of actin complexed with DNase I, gelsolin segment-1, or profilin. Although the main features of the model are probably valid, it is unlikely that the details are precisely correct. This model is supported by electron microscopic studies that have defined the molecular envelope of the subunits in the filament (Milligan et al 1990; Bremer et al 1991; Egelman 1994), localized the position of cysteine-374 (Milligan et al 1990), and mapped the N-terminus with antibody decoration (Orlova et al 1994).

Description of the Structure

The polymer model is stabilized by extensive hydrophobic contacts between the actin subunits, as well as by many potential hydrogen bonds. The longitudinal contacts along the long pitch helix are more robust than the lateral contacts. The main secondary structural elements of the actin crystal structure are maintained in the subunits of the filament model, but the α carbon positions underwent a rms displacement of 3.2 Å during refinement (Figure 1B). In particular, subdomain 2 moved closer to the filament axis and five other surface loops were repositioned to improve subunit contacts. The loop consisting of residues 262–274 extends across the polymer axis to form a hydrophobic plug between two adjacent subunits. It may act like the teeth of a zipper to stabilize the lateral contacts. In the filament model, the central cleft closes more tightly around the ADP and Ca^{2+} than in the monomer, in keeping with the slow rates of exchange of these ligands in the polymer. The packing of the subunits agrees with chemical cross-linking experiments (Elzinga & Phelan 1984; Hegyi et al 1992). The phalloidin contacts three subunits simultaneously, thus providing a structural basis for its ability to stabilize the filament. Further, its position in the refined model is consistent with mutational and cross-linking data.

Significance

This actin model provides a tentative structural basis for all of the experimental work on the assembly, dynamics, and interactions of actin filaments. For example, the structure of the actin-myosin complex (see below) is based on it. Similarly, models for the mechanisms of actin-binding proteins like gelsolin (below) start with this actin model. A completely different model for an actin polymer has been proposed by Schutt et al (1993) based on the packing of actin in co-crystals of actin and profilin. It is conceivable that some of the actin-actin interactions in these crystals are used in some actin assemblies, but the Lorenz et al (1993) model accounts for a much wider range of experimental data that have accumulated about conventional actin filaments.

ANNEXIN

Background

Annexins are a family of soluble proteins with calcium-dependent binding to phospholipids and membranes. Annexins comprise up to 2% of vertebrate cell proteins and are found in a wide variety of species and cell types. Twelve annexins (annexin-I-XII) have been identified and characterized (Swairjo & Seaton 1994). Independent discoveries of these proteins led to many different names for annexins including calpactins, lipocortins, chromobindins, endonexins, calcimedins, and synexin. Functions attributed to annexins include interaction with cytoskeletal proteins, inhibition of phospholipase A_2, inhibition of blood clotting, aggregation of membranes and vesicles, and formation of calcium-selective membrane channels. Some annexins [annexin-I, annexin-II, annexin-V, and an annexin-II heterotetramer (annexin-IIt), in which two annexin-II molecules combine with two smaller subunits similar to the S-100 protein and calmodulin], but not all, bind actin filaments in the presence of calcium (Gerke & Weber 1985; Glenney et al 1987; Khanna et al 1990). Some annexins may bundle actin filaments. More attention has been given to possible Ca^{2+}-dependent membrane channel activity of annexins (Swairjo & Seaton 1994), and actin binding needs to be reevaluated in light of the knowledge of the atomic structures.

Structure Determination

Huber and co-workers (1990) determined the first annexin structure by X-ray crystallography. Human annexin V formed both hexagonal and rhombahedral crystals in the presence of Ca^{2+}. The hexagonal crystals were analyzed by multiple isomorphous replacement and refined at 2.5 Å resolution. The rhombahedral crystals were analyzed by molecular replacement (Huber et al 1990). Further refinement gave R values of 18.4% at 2.3 Å resolution for the hexagonal crystals and 17.4% at 2.0 Å for the rhombahedral crystals (Huber et al 1992). Interestingly, calcium ions were observed only in the rhombahedral crystals. The crystal structures of human annexin I (Weng et al 1993) and chicken annexin V (Bewley et al 1993) were solved by molecular replacement and refined at 2.5 and 2.25 Å to R values of 17.7 and 19.7%, respectively. The structure of rat annexin V (Concha et al 1993) was solved by multiple isomorphous replacement and refined to an R value of 20.2% at 1.9 Å resolution.

Description of the Structure

Annexin sequences are characterized by a conserved core of four (or in the case of annexin-VI, eight) repeats of about 70 amino acids (Figure 3). The

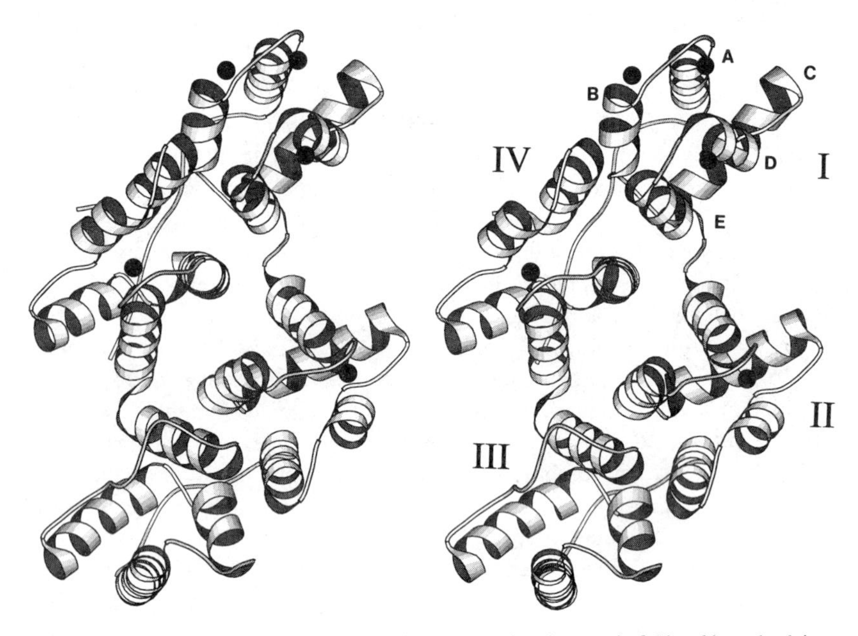

Figure 3 Human annexin V. Stereo ribbon diagram showing the protein fold and bound calcium ions (dark spheres) prepared with MOLSCRIPT using the X-ray coordinates of Huber et al (1992). The four domains are labeled I-IV. The helices in domain I are labeled A-E.

N-termini of the annexins vary in length and sequence. The core regions of annexins I and V have the same overall topology. Each of the four repeats is folded similarly into a domain consisting of five alpha helices (A-E). Two parallel helix-loop-helix structures comprising helices A and B, and helices D and E, are linked by helix C, which lies across the top, approximately perpendicular to the other four helices. Hydrophobic contacts between domains I and IV and II and III cause each of these pairs to form a tight module. The extended N-terminus links domains I and IV noncovalently. A second extended nonhelical connector links domains II and III. Interactions between the two modules are chiefly hydrophilic and contribute to a central hydrophilic channel enclosed by the four domains. Overall, the molecule resembles a slightly curved plate. The loops between the helices are located on the convex surface, while the four C helices and two extended connectors are on the concave surface. The available structures provide limited information on the variable region in annexins because annexin-V has a short variable region (15 amino acids), and the annexin-I that was crystallized was truncated at the N-terminus.

The affinity of most annexins for Ca^{2+} is rather low (K_d = 10–100 μM), both in the presence and absence of Mg^{2+} (Klee 1988). Only annexin VI has been found to have a single high affinity (K_d ~ 1 μM) Ca^{2+}-binding site. Addition of specific phospholipids increases the affinity of the annexins for Ca^{2+}. Annexin crystal structures show up to six calcium ions bound to two classes of sites. One class consists of residues from the AB loops, and the other class is formed by residues from DE loops. Neither site resembles the EF-hand calcium-binding sites found in calmodulin and related proteins. The AB sites are formed by three carbonyl oxygen ligands from the AB loop and bidentate carboxylate oxygens from an acidic residue in the following DE loop. These sites share structural features with the calcium-binding sites of phospholipase A_2. Ligands for the DE calcium-binding sites are two carbonyl oxygens from the DE loop and a bidentate carboxylate group from the E helix. AB sites appear to bind calcium with higher affinity than DE sites. The affinity of the AB sites in the four domains for calcium varies as follows: I > IV > II > III. Calcium binding to domain II appears to cause a conformational change that allows calcium to bind to the DE loop (Swairjo & Seaton 1994).

Significance

Considerable evidence shows that annexins bind phospholipid membranes via the convex surface so that the calcium-binding loops contact the membrane. Swairjo & Seaton (1994) review the extensive studies of annexins binding to membranes and forming ion channels. The mechanism of actin binding is less well understood. Initial investigations suggested that actin binding and filament bundling required close to millimolar Ca^{2+} concentrations; however, annexin-IIt can bundle actin filaments at near physiological (5–10 μM) Ca^{2+} concentrations (Ikebuchi & Waisman 1990). A nonapeptide from helix IVB of annexin-IIt, synthesized based on sequence similarity to an actin-binding site of myosin, inhibits actin bundling by annexin-IIt (Jones et al 1992). The regions of sequence similarity of annexins and gelsolin, noted by Burgoyne (1987), form the actin-binding helix of gelsolin (see below) and helix E in annexin. No evidence is available on the interaction of the E helix of annexin with actin. The possibility that some annexins bind actin under conditions of high intracellular calcium needs further investigation.

DEOXYRIBONUCLEASE I

Background

Pancreatic deoxyribonuclease I (DNase I, E.C. 3.1.21.1) degrades double-stranded DNA to 5′ oligonucleotides by hydrolysis of the P-O3′ phosphodiester bond (for a general review see Moore 1981). DNase I is considered to be a nonspecific hydrolytic enzyme because it does not recognize specific bases or sequence combinations. However, the frequency and pattern of single-stranded nicks in DNA produced by the enzyme depend on the sequence of the DNA. Most probably this is due to variation in the geometry of the DNA minor groove. DNase I is a useful probe for DNA conformation and for delineating areas of protein-DNA contact. Enzyme activity is slight for single-stranded DNA and depends on divalent cations. Bovine pancreatic DNase I is a 30.4-kd (260 amino acids) glycoprotein, but the carbohydrate moiety attached to asparagine-18 is not required for enzymatic activity.

DNase I also forms a 1:1 complex with monomeric actin. The complex prevents the formation of actin filaments (Pinder & Gratzer 1982) and inhibits enzymatic activity of DNase I (Lazarides & Lindberg 1974). The physiological significance of this interaction is unknown, but it has been exploited for actin purification and for the study of actin filament assembly.

Structure Determination

The initial 1:1 co-crystals of monomeric actin and DNase I (Mannherz et al 1977; Sugino et al 1979) diffracted poorly, so Suck et al (1981) used a new orthorhombic crystal to determine a model of the complex at 6 Å resolution by multiple isomorphous replacement. Oefner & Suck (1986) solved the structure of crystalline DNase I at 2 Å resolution by multiple isomorphous replacement. This high resolution structure not only allowed for a more detailed understanding of DNA hydrolysis, but it paved the way for a higher resolution model of the actin-DNase I complex (see above).

Description of the Structure

DNase I is comprised of a well-packed central core of two, six-stranded beta sheets with four alpha helices on each side (Figure 4). The molecule has a total of 16 beta strands. Three hydrophobic cores, two disulfide bonds (cysteine-98 to cysteine-101 and cysteine-170 to cysteine-206) and two structural calcium ions (which stabilize loop regions) contribute to the stability of the molecule.

The first hydrophobic core is formed in between the beta sheets, and the other two are formed on either side of this central sheet region. The disulfide bond formed between cysteine-170 and cysteine-206 is required for catalytic activity. The second disulfide bond (cysteine-98 to cysteine-101), contained in a small flexible loop, is not evident in the electron density. The first structural calcium is octahedrally coordinated in a loop formed by residues aspartate-198 through threonine-204. The ion is positioned by interactions with the backbone carbonyls of threonine-200, threonine-202, threonine-204, both β carboxylates of aspartate-198, the hydroxyl group of threonine-200, and two water molecules. The position of this loop prevents the reduction of the disulfide bond (cysteine-170 to cysteine-206) that is required for enzyme activity. The second calcium is tetrahedrally coordinated by the β carboxylates of aspartate-96,

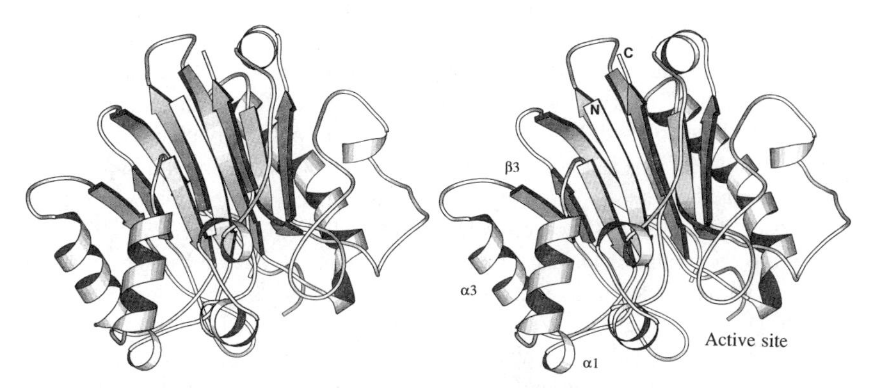

Figure 4 Bovine DNase I. Stereo ribbon diagrams of the elements of secondary structure prepared with MOLSCRIPT. The X-ray coordinates for the molecule were taken from the actin-DNase I co-crystal structure of Kabsch et al (1990). The active site for DNA hydrolysis is located in the groove between the β-sheets at the bottom of the figure. The loop between residues 101 and 104 is omitted just above the active site label. The β3/β4 loop from R70 to K74 just to the left of the active site label is thought to contact the minor groove of the DNA. DNase I binds to actin through three β-sheet backbone hydrogen bonds from the β3 strand to the actin DNase loop and through side chain interactions of residues on α1, α3, β3 and the β3/β4 loop.

aspartate-104, glutamate-109, the carbonyl of phenylalanine-106, and two water molecules. The site is in between two beta strands in the region of the loop glycine-97 to glycine-102. The stability of DNase I is also enhanced by many intramolecular hydrogen bonds and salt bridges. Interestingly, the longest helix (residues 136 to 155) has a 22° kink that separates the structure into two separate helices (residues 136–144 and 145–155) that are delineated by an intervening 310 helix. This kink allows the entire helical region to wrap around the central beta sheet core. This topology is stabilized by a salt bridge between aspartate-146 and arginine-182.

DNase I consists of two nearly identical regions of βαβαβββ topology that are related by a nearly perfect twofold axis of symmetry. The positions of the corresponding atoms in the independent domains (residues 1 to 120 and 121 to 257) have an rms deviation of 1.5 Å, when aligned on Cα positions, and are related to each other by a simple rotation of 177° (the near twofold axis) and a translation of 4.8 Å. This may indicate that DNase I was formed by replication of a primordial gene followed by sequence divergence. The topological duplication seen in the DNase I structure is not manifested in the primary sequence (Suck et al 1984).

The asparagine-18-linked carbohydrate is of the high mannose variety and extends from a large solvent channel on the surface of the enzyme. Two N-acetylglucosamine moieties are in contact with the protein and, along with the next mannose moiety, are clearly evident in the electron density. The remaining mannose molecules are disordered.

The active site of DNase I is centered around histidine-131. An exposed loop (arginine-70 to lysine-74) is proposed to contact the DNA minor groove, and phosphates of both DNA strands interact with positively charged residues on both sides of this loop. The DNA strand to be cleaved is bound in a groove between the beta sheets. A catalytic water associated with histidine-131 is the proposed nucleophile (Suck & Oefner 1986).

In the co-crystal of DNase I and monomeric actin, DNase I makes a major contact with actin subdomain 2. This interaction consists of hydrogen bonding, electrostatic, and van der Waals interactions between actin residues arginine-39, glutamine-41, valine-43, valine-45, lysine-61, and glycine-63, and DNase I residues aspartate-53, tyrosine-65, valine-67, glutamate-69, and histidine-44. Additionally, actin residues glycine-42, valine-43, and methionine-44 form a parallel beta strand in the DNase I beta sheet involving interactions with DNase I residues tyrosine-65, valine-66, and valine-67. Actin subdomain 4 residues glutamate-207 and threonine-203 and DNase I residues histidine-44 and glutamate-13 form a second minor contact . All contacts account for 1849 $Å^2$ of buried surface area (Kabsch & Vandekerckhove 1992). Contact between monomeric actin and DNase I alters the configuration of the main chain of DNase I residues tyrosine-65 to valine-67. This is the only significant structural deviation of the DNase I molecule upon binding actin.

Significance

The model of DNase I provided insight at an atomic level into the mechanism of DNA hydrolysis including a model for the interaction of DNase I with double-stranded DNA (Oefner & Suck 1986). The structure was also essential for the solution of the atomic structure of the actin bound to DNase I.

The binding of DNase I to actin explains the reciprocal effects of the two proteins on each other. Actin inhibits the DNase I by sterically blocking the active site around glutamate-13 (Suck et al 1988; Kabsch et al 1990). The binding site of the DNase I on actin provides at least two rationales for the inhibition of actin nucleotide exchange by DNase I (Hitchcock 1980). First, DNase I partially blocks the actin ATP-binding cleft and may hinder the escape of the nucleotide in the vertical direction as shown in Figure 1. Second, by bridging subdomains 2 and 4, it may suppress intramolecular motions (Tirion & ben-Avraham 1993) required for dissociation of the nucleotide. These two alternatives cannot be distinguished at this time, since we do not know the route of nucleotide dissociation from actin alone. The DNase I-binding site on actin is located at the pointed end of the actin filament (Figure 2). This explains why DNase I can bind to and cap the pointed end of actin filaments (Podolski & Steck 1988).

GELSOLIN AND VILLIN DOMAIN I

Background

Gelsolin and villin are structurally related actin-binding proteins that share some, but not all, biochemical activities. Under the control of Ca^{2+} and membrane polyphosphoinositides, they sever actin filaments, cap the barbed ends, and nucleate actin polymerization (reviewed by Matsudaira & Janmey 1988). These reactions may contribute to the rearrangement of actin filaments in the cortex during cellular motility and secretion. Villin also cross-links actin filaments.

The gelsolin family of capping proteins consists of three (fragmin, severin, gCap39) or six (gelsolin, villin) repeats of a 125–150 amino acid sequence motif. These proteins have been found in animals and protozoa. In the six domain proteins, the first three domains are more similar to the last three domains than they are to each other. Despite their sequence homology, the domains bind to different sites on actin monomers and filaments. In gelsolin, segment 1 binds with high affinity to actin monomers, segment 2 binds actin filaments, and segments 4–6 contain a second lower affinity monomer-binding site (reviewed by Weeds & Maciver 1993). Calcium promotes binding to actin monomers, but has no effect on binding to filaments. The first two domains are the minimum unit required for severing and capping. Gelsolin mutants lacking segment 1 decorate filaments but do not sever them, while gelsolin mutants without segment 2 do not bind actin filaments. Polyphosphoinositides (e.g. PIP_2) inhibit the severing activity of gelsolin and dissociate gelsolin capping the barbed end of filaments. Mutagenesis experiments have identified two binding sites for PIP_2, one near the N-terminus of segment 2 of gelsolin and villin and another near the C-terminus of segment 1 of gelsolin (see Weeds & Maciver 1993).

In addition to its severing and capping activities, villin cross-links actin filaments in microvilli. This cross-linking ability is due to an additional actin filament-binding site on a seventh domain, the headpiece, at the C-terminus of villin. The severing and capping activities of villin are only activated at calcium concentrations above 0.1 mM, whereas gelsolin is active at sub-micromolar calcium concentrations (Weeds & Maciver 1993). Since actin cross-linking by villin is calcium-independent, it can maintain actin bundles except under conditions of elevated calcium.

Structure Determination

Crystals of the complex of actin with human gelsolin segment 1 (residues 26–150), or a point mutant (N57C) of segment 1, were grown in polyethylene glycol and diffracted weakly to 2.5 Å resolution. The structure of the complex was determined by molecular replacement and model building that started with the Kabsch et al (1990) structure of actin. After phase-fitting the actin model, the calculated phases were adequate to identify the position of segment 1 of gelsolin. Solvent flattening revealed the first elements of secondary structure in segment 1, and further rounds of model building, refinement and phase calculations led to a full model of segment 1 that included two Ca^{2+} ions with good stereochemistry and an R-factor of 18.7%. The structure of recombinant villin segment 1 (residues 1 to 126, called fragment 14T) was solved by multidimensional NMR (Markus et al 1994). Using distance geometry, structures were calculated based on 720 NOE distance restraints, 113 dihedral angle restraints, 48 stereospecific assignments of methylene or methyl groups, and 43 hydrogen bonds. Superimposition of the backbone atoms of ten structures has an average rms deviation of 1.15 Å from a mean structure. These atomic structures provide a basis for understanding the severing, capping, and nucleation activities of these homologous proteins.

Description of the Structure

Gelsolin and villin have the same overall structures (Figure 5), but two secondary structure elements were either absent or not defined as secondary structure in gelsolin. The villin segment has a central five-stranded beta sheet with an alpha helix and small parallel beta sheet on one side and two alpha helices on the other side. An N-terminal helix, $\alpha 1$, is followed by strands $\beta 1$ and $\beta 2$, which are in the central sheet. Strand $\beta 3$ forms the first strand of the two-stranded parallel β sheet and is followed by strands $\beta 4$ and $\beta 5$, which are in the central sheet. A loop after strand $\beta 5$ leads into helix $\alpha 2$, which is on the same side of the sheet as the N-terminal helix. Helix $\alpha 2$ is followed by strand $\beta 6$, the final strand in the central sheet. Strand $\beta 6$ leads to helix $\alpha 3$ and the second strand of the parallel β sheet, $\beta 7$. The parallel beta sheet and helix $\alpha 3$ are on the opposite side of the central sheet from helices $\alpha 1$ and $\alpha 2$. Strands $\beta 2$ at the end of the central sheet and $\beta 3$ in the two-stranded sheet were identified as loops in gelsolin segment 1. Residues conserved in the gelsolin family repeats are in the hydrophobic core in both gelsolin and villin, which suggest that their conservation is required for correct folding and that the different segments in gelsolin and related proteins have similar folds.

In the gelsolin:actin complex (Figure 5c), gelsolin helix $\alpha 2$ binds in a cleft at the barbed end of the actin molecule between subdomains 1 and 3. The interface buries about 2000 $Å^2$ of surface area, including roughly equal numbers of nonpolar and polar atoms. Actin residues in the 137–144 helix, the 144–150 loop, the 166–170 loop, and the 339–348 helix provide many of the contacts with gelsolin helix 95–112 and adjacent loops. Two calcium ions were observed in the complex, one bound by segment 1 and actin and the other bound by segment 1 alone. The intermolecular calcium is bound by ligands from $\alpha 2$ and the loop

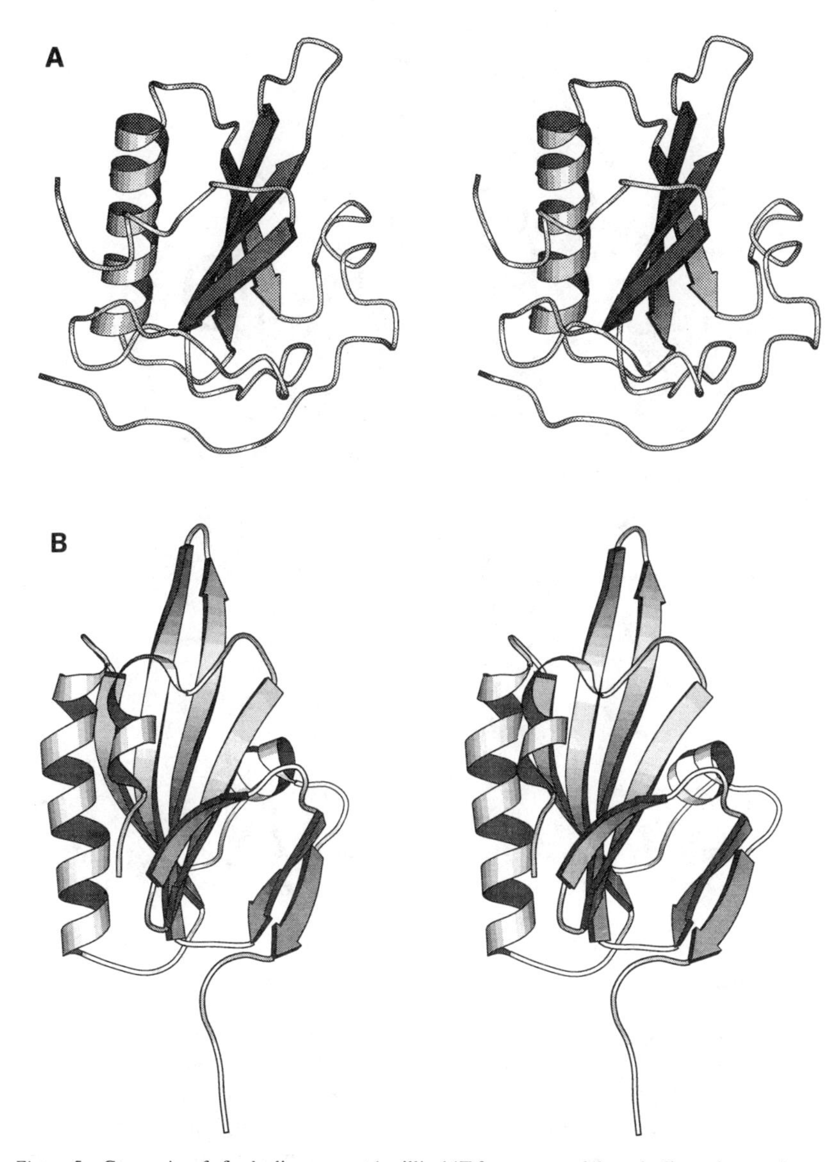

Figure 5 Comparisonf of gelsolin segment 1, villin 14T fragment, and the gelsolin-actin complex. Ribbon diagrams of the elements of secondary structure prepared with MOLSCRIPT. (*A*) Stereo ribbon diagram of the X-ray structure of human gelsolin segment 1 (McLaughlin et al 1993) provided by P McLaughlin, MRC Laboratory of Molecular Biology, Cambridge, England. (*B*) Stereo ribbon diagram of villin 14T fragment prepared from the NMR coordinates of (Markus et al 1994). (*C*) Ribbon diagram of the X-ray structure of gelsolin segment 1 bound to actin provided by P McLaughlin.

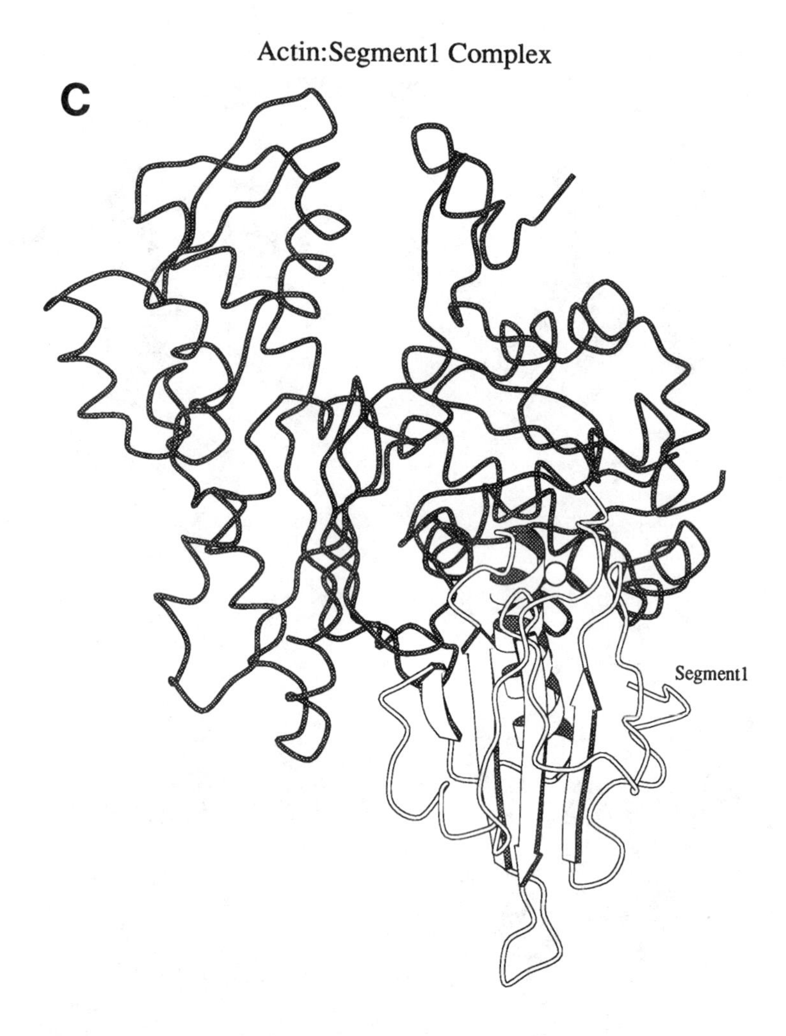

following it in segment 1 and by glutamate-167 in actin. It is presumed that this shared Ca^{2+} stabilizes the binding of the proteins and contributes to the 200-fold higher affinity of segment 1 for actin in the presence of Ca^{2+}. Ligands for the intramolecular calcium are provided by residues at the amino end of α2 and at the C-terminus of segment 1. This calcium only has six ligands, and it was suggested that PIP_2 might provide a seventh ligand. Similar Ca^{2+}-binding sites are present in the NMR structure of villin domain 1.

Significance

The structure of the gelsolin:actin complex provides valuable clues about the mechanism of actin filament severing and capping by segment 1. In the actin filament model of Holmes et al (1990a), the segment 1 binding site on the barbed end of each actin subunit (subdomains 1 and 3) is occupied by the interaction with subdomain 2 of the neighboring actin subunit. This is why segment 1 alone does not bind actin filaments. However, when targeted to the surface of an actin filament by segment 2 or another actin filament-binding protein, segment 1 rapidly severs the filament, presumably by intercalating into transient defects between actin subunits and destabilizing the polymer. Once bound with picomolar affinity to its site, segment 1 is unlikely to be displaced by the relatively weak (K_d 0.1 μM) forces that hold the filament together.

The segment 1 structure also provides some insights regarding the capping and nucleation of actin filaments by gelsolin. McLaughlin et al (1993) provide a model with segments 1 and 4 of one gelsolin molecule bound to adjacent subunits along the short pitch helix. Segments 2 and 3 are proposed to bridge these two binding sites, interacting with the exposed surface of the filament. Binding of two monomers in this orientation would form a nucleus for polymerization, while binding the barbed end of a filament would cap it (McLaughlin et al 1993). The roles of segments 5 and 6 are not specified.

HISACTOPHILIN

Background

Hisactophilin from *Dictyostelium discoidium* consists of 118 amino acids, of which 31 are histidine. To date the protein has only been found in *Dictyostelium.* It binds 1:1 to monomeric actin (Schleicher et al 1984). Binding to actin filaments depends on the pH, with a transition from tight (K_d ~ 0.1 μM) to negligible binding between pH 6.5 and 7.5 (Scheel et al 1989). Below pH 7 hisactophilin promotes the polymerization of actin under normally nonpolymerizing conditions (Scheel et al 1989). The stoichiometry of the reaction is ~ 1 hisactophilin per actin monomer, which suggests that hisactophilin coats the actin filament. Hisactophilin has no effect on the final viscosity of the actin filaments, thus it does not act to cap or sever the polymers. By fluorescent antibody staining, hisactophilin is concentrated in the cortex of the cell, usually coincident with actin filaments (Scheel et al 1989). Since chemoattractants stimulate actin polymerization, hisactophilin may act as a proton sensor, coupling extra-cellular events to hisactophilin-enhanced actin polymerization (Devreotes & Zigmond 1988). The three-dimensional structure in Figure 6 provides a rationale for this activity.

Structure Determination

Habazettl et al (1992a,b) determined the structure of ^{15}N-labeled, recombinant *Dictyostelium* hisactophilin by three-dimensional NMR. Twenty three-dimen-

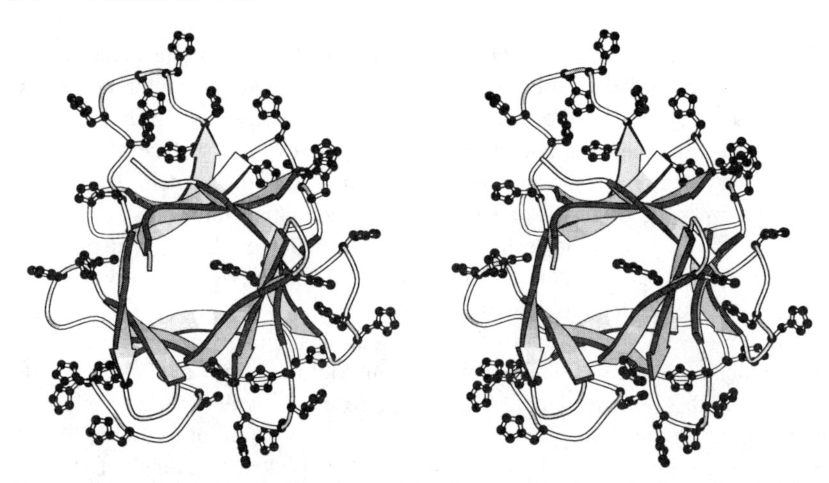

Figure 6 *Dictyostelium* hisactophilin. Stereo ribbon diagram of the elements of secondary structure and the histidine side chains prepared with MOLSCRIPT using the NMR coordinates of Habazettl et al (1992a). The N- and C-termini mark the beginning and end of β-strand 1 and β-strand 12, which are hydrogen-bonded to form part of the six-stranded barrel.

sional structures were calculated from 2541 intensity constraints and 27 backbone angle constraints. The rms deviation from the average structure was 0.84 Å for the backbone atoms and 1.75 Å for all atoms. The structure was the same at pH 6.7 and 5.7.

Description of the Structure

In spite of little primary sequence similarity, the architecture of hisactophilin is remarkably similar to interleukin-1β (Clore et al 1991) and fibroblast growth factor (Zhu et al 1991). The structure is highly symmetric, with 12 antiparallel beta strands arranged around a pseudo threefold axis. The structure can be naturally divided into three sectors S1, S2, and S3, each consisting of two pairs of antiparallel beta strands organized into a βββ-loop-β motif. Within each sector, strands 2 and 3 form a two-stranded hairpin. The 1 and 4 strands from S1, S2, and S3 interact cyclically to form a six-stranded barrel structure. This barrel is on the side of the molecule nearest to the viewer in Figure 6. The 2–3 hairpins from each sector cover one end of the barrel. This cover or flap is farthest from the viewer in Figure 6. The barrel cavity houses most of the molecule's hydrophobic side chains. Although the amino acid sequences of hisactophilin, of interleukin-1β, and of fibroblast growth factor do not appear to be related, some key hydrophobic residues in the barrel are identical or similar. Thus three phenylalanines critical in the stabilization of the interleukin-1β beta barrel are aligned with corresponding phenylalanines in hisactophilin when the structures are superimposed. Key hydrophobic residues in the three beta hairpins that act as a flap over the barrel end are similar in both proteins.

Significance

Twenty eight of the 31 histidine are disposed on the surface of the molecule, in the loops and turns joining the beta strands. NMR signals indicate that their pKs are nearly identical. Thus a small change in pH can transform an essentially neutral molecule into a polycation. The promotion of actin polymerization by coating the filament with hisactophilin is reminiscent of the impact of polylysine on polymerization and suggests a mechanism of action for hisactophilin. It will be important to learn whether other cells have such a pH-sensitive actin-binding protein and to establish whether hisactophilin is used as a pH-sensitive switch for actin assembly in the cell.

MYOSIN SUBFRAGMENT-1

Background

Myosin is the motor for all known actin filament-based cellular movements including muscle contraction, cytokinesis, folding of embryonic epithelia, cytoplasmic streaming and doubtless many others. A large family of myosin isoforms is now recognized, many or all of which are expressed in the wide range of eukaryotic cells (reviewed by Cheney et al 1993). This suggests that these isoforms existed in primitive eukaryotic cells more than 1 billion years ago. All myosins have a globular motor domain, called the myosin head or subfragment-1, with an active site for ATP hydrolysis and an actin-binding site. One or two head domains are attached to a variety of tails that act as adaptors to bind the myosin head to other myosins (to form filaments with many motor domains) or to other cellular structures including membranes (reviewed in Pollard et al 1991). The energy released by ATP hydrolysis is used to produce movement during transient interactions of the myosin head with actin filaments (Taylor 1992). The major unresolved question in the field is, how is this chemomechanical coupling accomplished? Since myosin heads can produce gliding movements of actin filaments in vitro (Toyoshima et al 1987), most of the attention has focused on the heads rather than the tails. Determination of the atomic structure of the myosin head (Rayment et al 1993b) was the major event in the field in the past decade because knowledge of the atomic structure greatly elevated our ability to investigate the energy transduction mechanism. The structure has not solved the problem, but it has provided the framework to move toward much deeper understanding.

Structure Determination

Growing crystals suitable for high resolution X-ray diffraction was the most challenging part of the work. Rayment et al (1993b) succeeded with chicken breast muscle myosin subfragment-1 produced by papain digestion of whole myosin. This fragment consists of 843 residues of the myosin heavy chain and two different light chains. After modification of all the lysine side chains by reductive methylation, S1 formed crystals that diffracted to 2.8 Å. Initial phases were obtained from multiple isomorphous heavy atom derivatives and improved by solvent flattening. Multiple cycles of model building, phase combination, and refinement resulted in a model with 1072 of the 1157 residues and an R-factor of 22.3% at 2.8 Å resolution with rms deviations of 0.018 Å from expected bond lengths and 2.5° from ideal bond angles. No solvent molecules are included.

Description of the Structure

The myosin head (Figure 7) is highly asymmetric with dimensions of 165 × 40 × 65 Å, divided into two large domains: the globular catalytic domain consisting of the first 783 residues of the heavy chain; and a light chain domain consisting of the two light chains wrapped around an 85 Å long alpha helix formed from the last 60 residues of the heavy chain.

Subfragment-1 is 48% alpha helix, but the central structural element that holds the catalytic domain together is a seven-strand beta sheet of mixed polarity located in the heart of the structure (Figure 7). The heavy chain weaves through this beta sheet five times at residues 116–126 (strands 1 and 2), 173–179 (strand 4), 247–268 (strands 6 and 7), 457–465 (strand 5), and 668–675 (strand 3) so the sheet anchors the other elements of secondary structure. Since the sheet also forms a major part of the nucleotide-binding site, it is likely that conformational changes associated with nucleotide binding and hydrolysis and the production of motion are centered in or around this sheet. The heavy chain has two sites particularly sensitive to proteolysis between residues 204 and 216 and between residues 626 and 647. Both are surface loops and neither is present in the model due to a lack of density in the map. The polypeptide leading up to the first two strands of the central beta

→

Figure 7 Chicken skeletal muscle myosin subfragment-1 from the X-ray structure of Rayment et al (1993). (*A*) Full view of the structure. (*B*) Detail of the beta-sheet in the core of the catalytic domain. The active site for binding ATP and the actin binding site are indicated. The essential (ELC) and regulatory (RLC) light chains are bound to the long C-terminal alpha helix. The residue numbers for the essential light chain are preceded by 3. Those for the regulatory light chain are preceded by 2. The residues at the edges of the breaks in the heavy chains and light chains are indicated. The side chains of C697 and C707 are included. Stereo ribbon diagrams of the elements of secondary structure were prepared with MOLSCRIPT and provided by I Rayment of the University of Wisconsin.

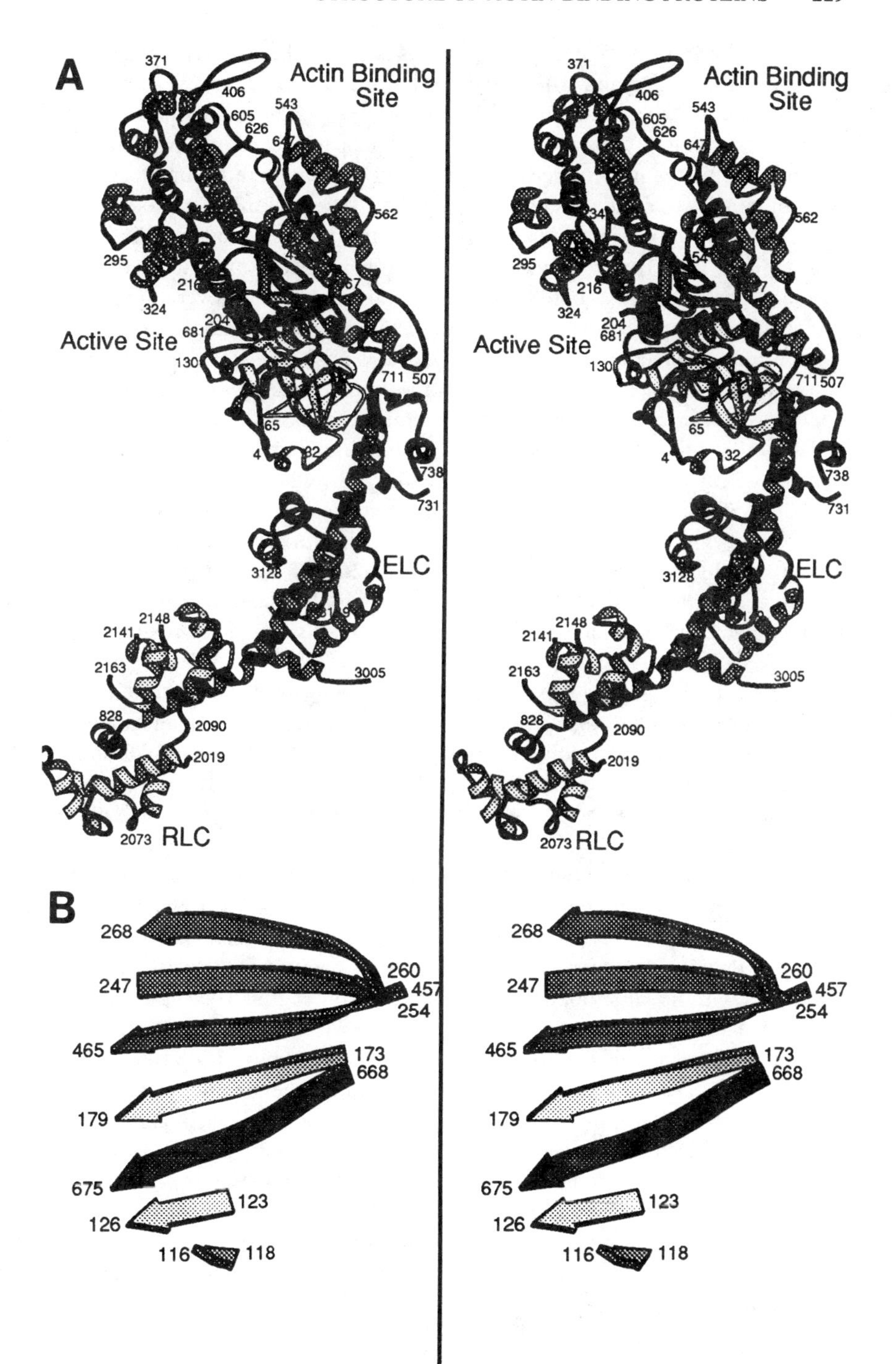
A
Actin Binding Site
Active Site
ELC
RLC
B

sheet consists of a six-strand antiparallel beta sheet similar in design to an SH3 (src homology 3) domain. After three short alpha-helices, the chain forms the fourth strand of the central sheet. The following chain forms the phosphate-binding loop with the classic GESGAGKT sequence and an alpha helix forming one side of the nucleotide-binding pocket. The structure is similar to the ATP-binding sites of adenylate kinase and Ras. After the missing proteolytic site (204–216), another alpha helix (218–233) forms another side of the ATP-binding site. The chain then passes twice through the central beta sheet and several alpha helices that border the ATP-binding site. Then several long (up to 45 Å) alpha helices form the top end of the globular domain. One of the surface loops connecting these helices includes arginine-403, a highly conserved residue mutated to glutamine in some human patients with the congenital heart disease called hypertrophic cardiomyopathy (Geisterfer-Lowrance et al 1990). (A mild defect in contractility results in compensatory cardiac hypertrophy and eventually causes fatal cardiac arrhythmias.) After forming strand 5 of the central beta sheet, the chain forms four long alpha helices and an antiparallel beta sheet and moves across a cleft to form the rest of the upper part of the globular domain. Missing residues 626–647 are particularly interesting because they are protected from proteolysis when bound to actin. After a long alpha helix beginning at 647, the chain passes through the central beta sheet the last time. After a loop on one side of the ATP site, the chain forms two short alpha helices. The second of these helices (700–707), located between cysteine residues at 697 and 707, may hold the key to understanding the force-producing conformational change (see below). After a small antiparallel beta sheet of three strands, the heavy chain forms the 85 Å alpha helix broken only by a 90° bend at a tryptophan-proline-tryptophan sequence near its end.

The light chains are members of the calmodulin family and consist of two globular domains connected by an alpha helix, although most of the divalent cation-binding sites are defective. The "essential" light chain next to the globular domain is wrapped around the heavy chain much like calmodulin associates with its target alpha helical peptides (Ikura et al 1992; Meador et al 1993). The density in this part of the map was weak, and 45 residues at the C-terminus were not visible. Remarkably, when subfragment-1 binds to actin, this invisible part of the essential light chain can be chemically cross-linked to actin near residue 362, 80 Å away. The distal light chain is called the regulatory light chain because it is phosphorylated on threonine-18 and serine-19 by the kinase that activates smooth muscle contraction. Neither these residues nor the rest of the N-terminus are visible in the map. A Mg^{2+} is present in the divalent cation-binding site at the N-terminus of the regulatory light chain. The light chain wraps around both of the right angle segments of heavy chain at the end of subfragment-1.

The light chains of scallop myosin bind to the heavy chain in a similar fashion (Xie et al 1994). A novel Ca^{2+} binding site on the essential light chain is stabilized by interactions of all three polypeptides. Ca^{2+} binding to this site controls the activity of scallop muscle myosin.

The atomic structure of subfragment-1 (Figure 7), the model of the actin filament (Figure 2), and reconstructions of actin filaments decorated with subfragment-1 were used to build models of the complex of the two proteins (Rayment et al 1993a; Schroeder et al 1993). The shape of subfragment-1 allows its unambiguous orientation in the EM model. Binding appears to be stabilized by electrostatic interactions between the myosin 626–647 loop and the N-terminus of actin, hydrophobic interactions between myosin residues phenylalanine-529, methionine-530, isoleucine-535, methionine-541, phenylalanine-542, and phenylalanine-543, and actin residues alanine-144, isoleucine-341, leucine-349 and phenylalanine-352, and the myosin 403–415 loop and actin 332–334.

Significance

The atomic structure of subfragment-1 has provided the structural basis to interpret more than 40 years of detailed biochemical and biophysical studies. The field has just begun to appreciate all of the implications, and no one has attempted to synthesize all of the details in print. Yount (1994) and Rayment & Holden (1994) have highlighted some of the key features linking the structure to the chemistry of nucleotide binding and hydrolysis. Some of the insights include the following: (*a*) The ATP site is well defined by the structure and earlier chemical cross-linking studies, although the crystal contained only a sulfate ion, presumed to correspond to the beta-phosphate of a nucleotide. The catalytic mechanism remains to be established. (*b*) The existence of an alpha helix between cysteines 697 and 707 was a major surprise, since these residues can be cross-linked when a nucleotide is bound to myosin. In the nucleotide-free myosin, these residues are 18 Å apart and exposed to solvent on opposite sides of the molecule! Nucleotide binding must cause a major conformational change in this highly conserved part of the myosin. Given its proximity to the light chain domain, the conformational change in this helix may affect the orientation of the light chain domain relative to the catalytic domain. (*c*) The C-terminal alpha helix, reinforced by its associated light chains, is an attractive candidate for a lever arm to amplify and transmit conformational changes in the catalytic domain associated with ATP hydrolysis and product release to the myosin thick filament. This provides support for a class of mechanisms for motility that depend upon bending within the myosin head rather than tilting of the myosin head on the surface of the actin filament (Huxley & Kress 1985; Rayment et al 1993a). (*d*) The actin-binding site is not precisely defined in terms of specific atomic interactions, but the contact regions on the surfaces of both proteins are established. The actin-binding site of myosin is located on the opposite side of the catalytic domain from the nucleotide site, so reduction in affinity of myosin for actin caused by nucleotide binding must (as predicted by biochemical studies) be due to a conformational change rather than a direct effect.

PROFILIN

Background

The profilins are a family of actin monomer-binding proteins found in vertebrates, invertebrates, plants, and fungi. Profilin is essential for viability of *Drosophila melanogaster* and mice and required for a normal spatial and temporal distribution of various actin-based structures in yeast and higher eukaryotic cells (reviewed by Machesky & Pollard 1993). Profilin was initially isolated based on its ability to form 1:1 stoichiometric complexes with actin monomers. It was thought that profilin could regulate the soluble and filamentous actin pools by a simple mass action mechanism. However, it is now apparent that there is insufficient profilin to account for all of the unpolymerized actin in vivo, and the likely candidate for monomer sequestration, at least in vertebrate white blood cells and platelets, is thymosin-β4. Profilin is thought to modulate the assembly of actin-based structures by catalyzing the exchange of the actin bound adenine nucleotides (i.e. ATP and ADP) and by shuttling actin subunits from thymosin-β4 to the barbed end of actin filaments (reviewed by Theriot & Mitchison 1993). The profilin-actin complex can be dissociated by phosphatidylinositol phosphates including PIP_2, and profilin may regulate phosphoinositide turnover by inhibiting some isoforms of phospholipase C. Profilins also bind poly-L-proline, although the naturally occurring proline-rich ligand(s) has not been identified. Some organisms utilize multiple isoforms of profilin with differing biochemical properties. *Physarum* and *Dictyostelium* express different isoforms during development, and plants express different isoforms in various tissues. Plant profilins are major human allergens.

Structure Determination

A large structural database for the profilins has recently accrued. The structures of *Acanthamoeba* profilin I (Vinson et al 1993) and human profilin (Metzler et al 1993) were determined by multidimensional NMR. *Acanthamoeba* profilin I structures were calculated using 915 NOE distance constraints, 55 backbone ϕ angle constraints, 23 side chain χ angle constraints, and 40 hydrogen bonds. Twelve structures with no significant constraint violations (NOE < 0.5 Å, angles $< 5°$) had average rms deviations from a mean structure of 1.2 Å for backbone atoms and 1.7 Å for all atoms. The human profilin structure was determined from 1005 NOE distance restraints, 85 backbone ϕ angle constraints, and 48 hydrogen bonds. Ten calculated structures clearly defined the three-dimensional fold. The structure of the complex of bovine β-actin and profilin was solved (Schutt et al 1993) by a combination of multiple isomorphous replacement and molecular replacement using the actin coordinates from the actin:DNase I complex (Kabsch et al 1990). The structure of the profilin-actin complex was refined at 2.55 Å resolution to an R-factor of 20.05% with good stereochemistry. The structure of *Acanthamoeba* profilin-I was solved using single isomorphous replacement phases from a single-site derivative in the space group C2 (Fedorov et al 1994) and was refined at 2.0 Å resolution to an R-factor of 17.9% with good geometry. The structure of *Acanthamoeba* profilin-II was solved by molecular replacement to 2.8 Å resolution using the profilin-I structure as the search model and has been refined to an R-factor of 18.6% with good geometry (Fedorov et al 1994). The structure of recombinant human profilin has recently been solved by multiple isomorphous replacement and refined at 2.3 Å resolution with an R-factor of 15.9% with good geometry (AA Fedorov & SC Almo, unpublished results).

Description of the Structures

All of the profilins have a similar polypeptide fold (Figure 8A). The *Acanthamoeba* and mammalian profilins are built around a central six-stranded antiparallel beta sheet (β1-1-β1-6). Two alpha helices, which correspond to the N- (α1) and C-termini (α4), are on one side of the large sheet and run approximately parallel to it. Positioned on the other face of the central sheet are two helices (α2) and (α3) and a small two-stranded beta sheet (β3) and (β4), which is orthogonal to the six-stranded sheet. Most of the extra 14 residues in vertebrate profilins are accommodated in surface-exposed loops or at the termini of secondary structural elements (Figure 8B). These loops are the most mobile parts of the backbone structure (Constantine et al 1993).

The crystal structure of the mammalian β-actin-profilin complex directly identified the residues in profilin responsible for binding actin. Residues from α3, α4, β5, β6, and β7 in the profilin contribute to the binding surface and specifically interact with subdomains 1 and 3 of actin. Independent biochemical and genetic evidence is consistent with the observed binding interface and suggests that despite the great sequence divergence seen in the profilins, all profilin-actin complexes share a similar binding mode.

Two NMR studies (Archer et al 1994; Metzler et al 1994) identified a cluster of aromatic and hydrophobic residues exposed to the solvent between the N- and C-terminal helices as the binding site for poly-L-proline. Similar residues participate in the binding of proline-rich peptides by src homology-3 (SH3) domains (Yu et al 1994), but the underlying polypeptide scaffolding shares no similarity with profilin.

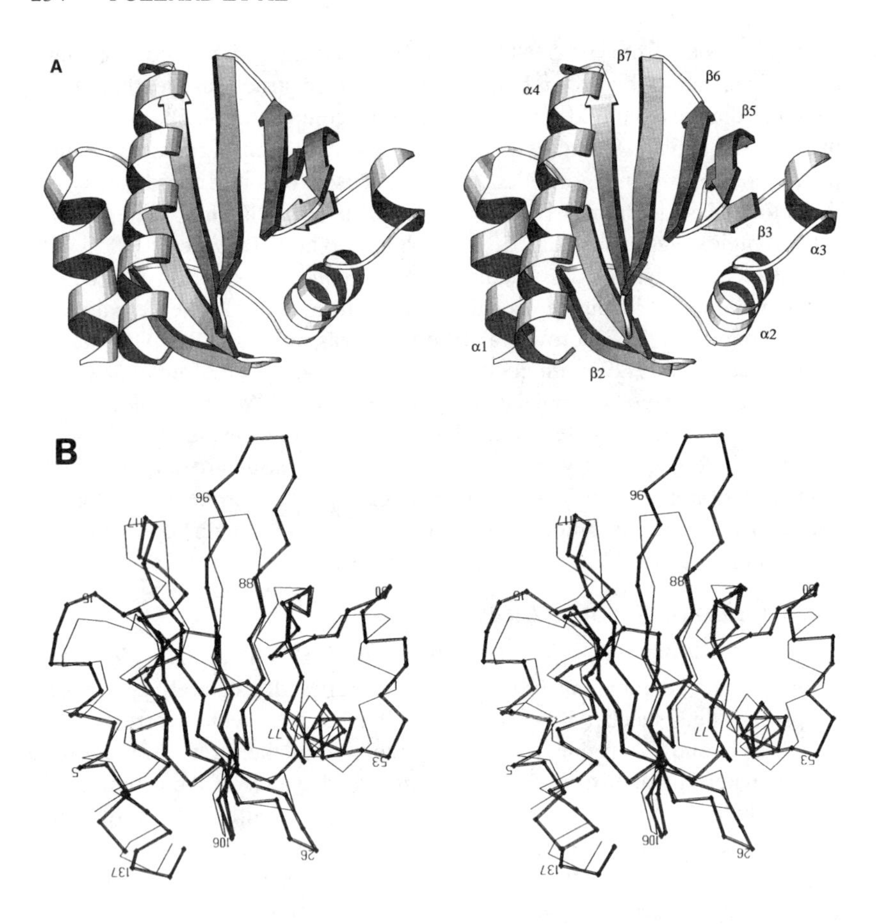

Significance

The structures demonstrate directly how profilin can bind simultaneously to actin and poly-L-proline. By binding to the face of the actin subunit, which corresponds to the barbed end of the actin filament, profilin sterically interferes with the binding of the actin-profilin complex to the pointed end of actin filaments. For the same reason, profilins have a weak effect on elongation at the barbed end of filaments, because the profilin in the complex does not interfere with binding of the complex to the barbed end of the filament. Given its rapid dissociation from actin, any profilin associated with the barbed end will dissociate before interfering with the binding of additional actin subunits. Binding to actin just beneath the nucleotide-binding cleft suggests that profilin might force open the cleft, which could account for the rapid dissociation of nucleotide from the complex. However, such a distortion is not evident when

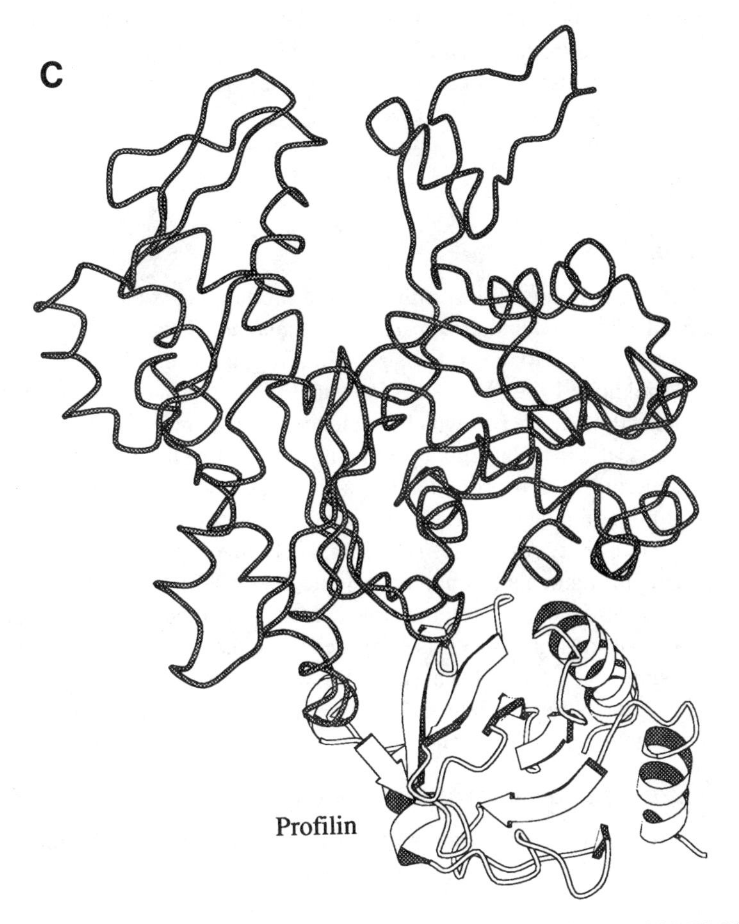

Figure 8 Profilin. (*A*) Stereo ribbon diagram of *Acanthamoeba* profilin-I drawn with MOLSCRIPT using the X-ray coordinates of Fedorov et al (1994). Most of the elements of secondary structure are labeled. (*B*) Stereo line drawing comparing the backbone structures of *Acanthamoeba* profilin (*single line*) and human profilin (*double line*). The X-ray coordinates for the human profilin come from unpublished work of A Fedorov & S Almo. (C) Model of the beta-actin-bovine profilin complex from the X-ray structure of Schutt et al (1993). Drawing provided by C Schutt of Princeton University.

the complex is packed in the crystal (Schutt et al 1993) so another mechanism may be involved. The profilin structures also provide a plausible model for the competing effects of actin and PIP_2. One of two surface patches on amoeba profilin-II, with a highly positive electrostatic surface potential, overlaps with the observed actin-binding site. Fedorov et al (1994) have postulated that these positive patches contribute to binding the negatively charged polyphosphoinositides. If this is correct, simple steric overlap might account for the observed disruption of the profilin-actin complex by PIP_2. The structure does not explain how profilin can inhibit phospholipase Cγ1, but not phospholipase Cβ (see Machesky & Pollard for review).

SEVERIN DOMAIN 2

Background

Fragmin and severin are 40-kd actin filament modulating proteins originally isolated from *Physarum* and *Dictyostelium discoideum.* They are functionally and structurally related to the proteins gelsolin, fragmin, and villin (for reviews see Vandekerckhove 1989; Weeds & Maciver 1993). Severin has three separate actin-related activities, each of which is calcium-dependent. Severin fragments actin filaments, it caps filaments by remaining bound to the barbed filament end after severing, and it promotes filament assembly by producing nuclei. Analysis of the amino acid sequence shows that severin is composed of three subdomains of roughly equal size. The second domain of severin is responsible for binding actin filaments (Eichinger & Schleicher 1992). Evolutionary analysis of the severin family of proteins indicates that the higher eukaryotic forms (gelsolin and villin) may have arisen by gene duplication of a three domain progenitor similar to severin because the three domains of severin are similar in sequence to the N-terminal three domains and the C-terminal three domains of villin and gelsolin (Schleicher et al 1988).

Determination and Description of the Structure

The structure of the second domain of severin (Figure 9) was determined by solution NMR (T Holak, personal communication) and will be described fully in a forthcoming paper. The domain is comprised of a central four-stranded beta sheet. This sheet contains beta strands 1, 2, 3, and 4 in an antiparallel conformation. A parallel orientation is seen between strands 4 and 5 to finish the sheet. Beta strand 2 points away from the central sheet. A large 16-residue alpha helix is associated with the two parallel beta strands in a βαβ topology, which is connected by two short turn regions. The two cysteine residues (41 and 87) form a disulfide bond that bridges beta strands 3 and 5. The severin domain 2 has three alpha helices. Helices 2 and 3 lie on either side of the central sheet region in a topology reminiscent of the villin or gelsolin segment 1 structures.

Significance

The severin domain 2 structure has several topological features in common with the gelsolin and villin domain 1, but the similarity is less than expected

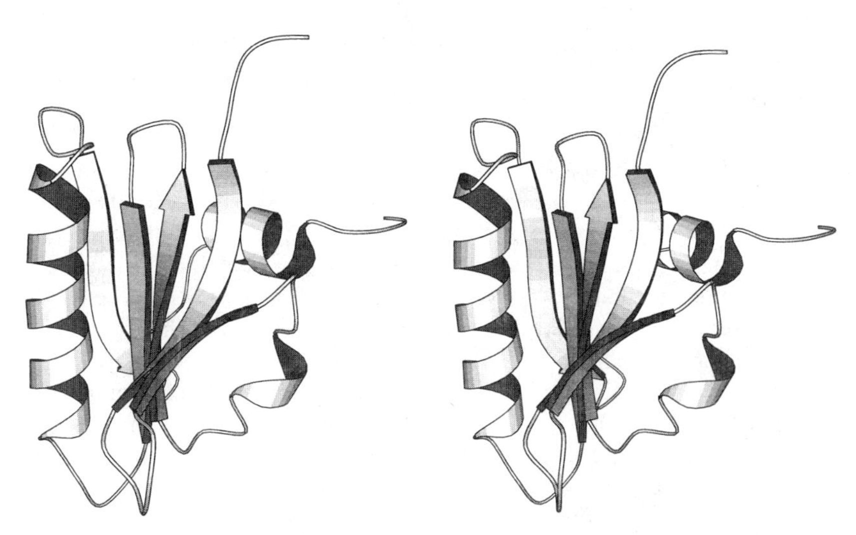

Figure 9 *Dictyostelium* severin domain-2. Stereo ribbon diagram of the elements of secondary structure prepared with MOLSCRIPT prepared from unpublished NMR coordinates provided by T Holak of the Max Planck Institute, Martensreid, Germany.

from previous comparisons of the amino acid sequences of the domains within this family of proteins. In both severin domain 2 and gelsolin segment 1, the central beta sheet is flanked by two alpha helices. The arrangement of three antiparallel strands, a helix, a parallel strand, and another helix on the opposite side is the same, but the details and the other elements of secondary structure differ. Profilin is also built from a central sheet flanked by helices, although the topology differs from the domains of the capping proteins. Others have speculated that many different actin-binding proteins arose from a progenitor molecule (Way & Weeds 1988). Although possible, this theory is not strongly supported by the available atomic structures.

SPECTRIN TRIPLE HELICAL DOMAIN

Background

The actin monomer binding and capping proteins discussed herein are thought to regulate the dynamics of individual actin filaments, while a large family of cross-linking proteins organizes actin filaments into supermolecular structures including networks and bundles. The cross-linking proteins are often dimeric, composed of an actin filament-binding domain and a dimerization element. This combination of modules results in a bivalent actin filament cross-linker (reviewed by Matsudaira 1991). Many actin filament cross-linking proteins, including α-actinin, dystrophin, spectrin, filamin, and ABP-120, share a homologous N-terminal domain that is responsible for binding to actin filaments (Matsudaira 1991). The structure of this domain remains to be determined.

The mechanisms of dimerization are diverse. The dimerization modules of filamin and ABP-120 consist of repetitive motifs of approximately 100 amino acids, predicted by sequence analysis to be largely beta sheet (Gorlin et al 1990). The tail of vertebrate filamin consists of 24 beta modules. Two subunits form a parallel homodimer joined by interactions of the most distal C-terminal beta modules. *Dictyostelium* ABP-120 has six beta modules that overlap to form antiparallel dimers.

The dimerization of spectrin, α-actinin, and dystrophin depends on a variable number (4 to 25) of domains of ~ 100 residues with a common sequence motif and a high content of alpha helix. The X-ray structure of the 14th repeat of α-spectrin from *Drosophila* provides a plausible model for the general structure of these repeating elements.

Structure Determination

The structure of the 14th repeat of α-spectrin from *Drosophilia* was solved by multiple isomorphous replacement and refined at 1.8 Å resolution to an R-factor of 20.3% with good geometry (Yan et al 1993). Based on this structure, an informative model has been put forth for the general structure of these repeating elements.

Description of the Structure

Two polypeptides associated to form a dimer in the crystal (Figure 10A). Each polypeptide is folded into three alpha helices, A, B, and C. Helices A and B are paired in an antiparallel fashion. The C helix from the second polypeptide associates with these two helices to form a three-helix bundle. The naturally occurring repeat is thought to be arranged somewhat differently, forming a left-handed three-helix bundle, approximately 50 Å long and 20 Å in diameter (Figure 10B). This bundle contains 4400 $Å^2$ of buried surface area, which corresponds to 38% of the total surface area. Helices A and B pack against each other with their side chains directly abutting, in a ridge-to-ridge fashion typical of coiled-coil proteins; however, the helices do not coil around each other like parallel coiled-coils such as tropomyosin (see below). Helix C is axially displaced relative to the A and B helices such that its side chains interdigitate between the side chains of A and B, which gives rise to the type

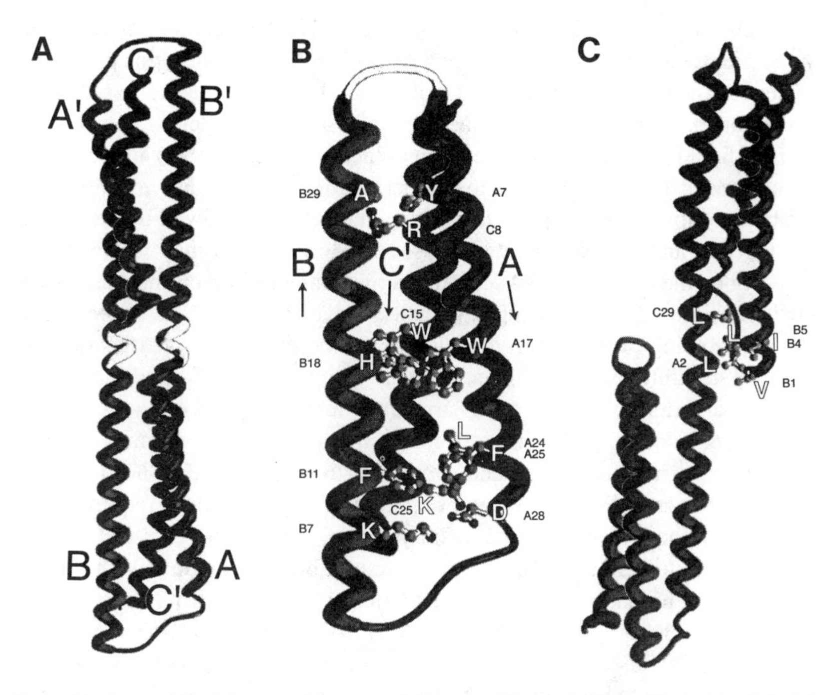

Figure 10 *Drosophila* alpha-spectrin segment 14 based on Yan et al (1993). (*A*) Antiparallel triple helical dimer found in the crystal. The three helices of one polypeptide are labeled A, B and C; those of the other A′, B′, and C′. The B-C connections are shaded. (*B*) Model for one triple helical domain made by connecting the C-terminus of helix B to the N-terminus of helix C′ with the white loop shown at the top of the model. Side chains of some of the key residues that maintain the separation of the helices are illustrated and labeled with the single letter code. Positions along the helices are indicated by letters and numbers. (*C*) Model for two successive triple helical segments. The conformations of the residues connecting helix C and A are hypothetical. Illustrations were prepared from transparencies provided by Y Yan of Harvard University.

of helical packing commonly seen in globular proteins. Both hydrophobic and ionic interactions stabilize the interactions of the helices.

Significance

The X-ray structure of the 14th repeat of α-spectrin naturally leads to a detailed model of the entire spectrin-repetitive segment (Figure 10B). Helix C is proposed to form a continuous helix with helix A of the subsequent repeating unit. The continuity of these helices places severe constraints on the relative orientation of subsequent repeats and indicates that each unit will be axially rotated by 60° in a right-handed manner relative to the preceding unit. The detailed structure of the repetitive segments found in different proteins depends on the precise length of the contiguous A and C helices because this parameter influences the distance and angular orientation between repeats.

The mechanical properties of these actin assemblies depend upon the rate of cross-linker binding and dissociation (Wachsstock et al 1994) and may also be influenced by properties of the dimerization domains. For example, the number of triple helical repeats, the segmental motions allowed by the particular repeats, and the constraints imposed by dimerization may influence the elasticity of the cross-link between the actin filaments, which themselves are relatively rigid when cross-linked (Wachsstock et al 1994). The properties of the cross-linking proteins will also determine the physical properties of actin filament structures such as filament polarity and density of bundling, which lead to the variety of higher order structures observed in vivo.

To cross-link actin filaments, spectrin must form a heterotetramer with an actin-binding module on each end. Both side-to-side and tail-to-head interactions between two different types of spectrin polypeptides are required. The structure of the triple helical domain provides an attractive model for the head-to-tail interaction. β-spectrin has an actin-binding module at the N-terminus and an unmatched C helix at its C-terminus. α-spectrin has an incomplete repeat unit composed of only the A and B helices at its N-terminus. Yan et al (1993) proposed that the two subunits are bound together, tail-to-head, when these unmated elements combine to form a discontinuous triple helix bundle. The second type of association is side by side with an antiparallel orientation that pairs each alpha chain with a beta chain. This produces a long floppy molecule with an actin-binding module on each end (Elgaester et al 1986). Analysis of two-dimensional crystals of α-actinin has provided a clear model for the arrangement of the triple helical repeats in these antiparallel, side by side associations (Taylor & Taylor 1993).

THYMOSIN-β4 AND ACTOBINDIN

Background

Thymosin β4 is small peptide of 43 amino acids that was first isolated from calf thymus. Thymosin β4 was thought to be a thymic hormone until it was realized that the peptide is widely distributed and abundant in many vertebrate cells and tissues (for reviews see Voelter et al 1987; Nachmias 1993). Then Safer et al (1990) discovered that thymosin-β4 is a potent inhibitor of actin polymerization by binding actin monomers. Given a concentration of approximately 200 μM thymosin-β4 in human platelets and a K_d of 0.7 μM, all or most of the unpolymerized actin in platelets should be complexed with thymosin-β4. Thus thymosin-β4 is the major known actin monomer sequestrating protein in the cell. Preliminary experiments show that DNase I competes with thymosin-β4 for actin monomer binding, which indicates that binding sites on the actin monomer overlap. Thymosin β4 has no known actin filament severing or capping activity. Thymosin β4 is rich in polar and negatively charged amino acids. Primary sequence alignments show that thymosin-β4 shares a short patch of sequence similarity with other actin-binding proteins (actobindin, tropomyosin, and α-actinin), although the region of sequence similarity is small (Safer et al 1991).

Structure Determination

Circular dichroism (CD) and ^{1}H-NMR of thymosin-β4, performed in aqueous solution, show no evidence for any ordered structure. This is consistent with the fact that thymosin-β4 consists of only 43 amino acids with no disulfide bonds. Such peptides are unlikely to contain significant ordered secondary structures. CD studies showed that the addition of 60% (v/v) trifluoroethanol-d_3 and 50% (v/v) hexafluoroisopropyl-d_2 alcohol stabilized some alpha-helical secondary structure in thymosin-β4 (Zarbock et al 1990). The solution structure of thymosin-β4 was evaluated by two-dimensional NMR experiments in the alcohol/water mixture to assign sequence-specific resonances. Simulated annealing and energy minimization were employed to calculate an ensemble of final structures. Later NMR experiments (Czisch et al 1993) at low temperature in water were undertaken to analyze the aqueous conformation of thymosin-β4.

Description of the Structure

In an alcohol/water environment, thymosin-β4 contains two alpha-helical regions that encompass residues proline-4 to lysine-16 and serine-30 to alanine-40, which account for 55% of the total residues. The remaining regions of the molecule appear to be highly mobile, and a unique structure cannot be defined. Because the disordered regions include the segment connecting the two helices, and there are no measurable NOEs between residues in the two helices, the relative orientations of the helices cannot be determined. In water at 1°C, NMR studies indicate that thmyosin-β4 does not adopt a unique structure, but there is discernable secondary structure in the regions of residue 5–16 and 31–37. These are the same segments that form the alpha helices in alcohol. However, in pure water these residues showed interactions characteristic of both alpha helix and beta strands. At 14°C, these regions are more disordered. Even when bound to actin, thymosin-β4 does not adopt a helical conformation (M Czisch & T Holak, personal communication).

Significance

The structure of thymosin-β4 in solution is not clearly determined, and it is probably best described as a large ensemble of interconverting conformations.

While the lack of a well-defined structure in solution makes it impossible to predict specific interactions with actin, it is likely that thymosin-β4 does adopt a unique conformation upon binding actin. This disorder-order transition has important implications for the energetics of complex formation. The association reaction must be accompanied by a large, unfavorable loss in configurational entropy associated with a decrease in the number of accessible states (i.e. going from a large number of conformations in the unbound state to a single conformation in the bound state). There are also entropic effects associated with side chain conformations and solvation. While the magnitudes of the entropic contributions are difficult to estimate, entropy can be a major determinant in the energetics of complex formation and may have substantial physiological importance. If free thymosin-β4 had the same conformation adopted in the complex, the association would be much tighter than actually observed ($K_d \sim 0.1$ μM) because there would no longer be the unfavorable loss of configurational entropy. The utilization of entropy in this manner can prevent the formation of irreversible complexes and may ensure that complexes dissociate on a physiologically relevant time scale. The dissociation rate constant of the complex is estimated to be about $1 s^{-1}$ (Goldschmidt-Clermont et al 1992).

Actobindin, an 88 amino acid polypeptide from *Acanthamoeba* (Vandekerckhove et al 1990), may behave similarly to thymosin-β4. Actobindin binds actin monomers and inhibits the nucleation of actin filaments. By CD spectroscopy, actobindin is a 15 and 22% beta structure, with a majority of the time-averaged structure not adopting periodic secondary structure (Vancompernolle et al 1991). As actobindin tightly binds two actin molecules, it is likely that substantial ordered structure is gained upon complex formation, and the entropy change associated with this ordering will affect the affinity. The entropic changes associated with disorder-order transitions may be common among actin-binding peptides and of general importance in controlling assembly reactions.

TROPOMYOSIN

Background

Tropomyosin is a parallel, two-stranded alpha-helical coiled-coil. Muscle tropomyosins are 41 nm long and some nonmuscle tropomyosins are about 35 nm long. Tropomyosin molecules overlap by about 3 nm, head-to-tail, to form long polymers that bind in the long-pitch groove of the actin filament helix (reviewed by Squire et al 1990; Holmes et al 1990b; Phillips et al 1986). Each tropomyosin associates with seven actin monomers, each actin binding to one of the seven so-called alpha-repeats found in the tropomyosin amino acid sequence (McLachlan & Stewart 1975, 1976). Tropomyosin, together with the trimeric Ca^{2+}-binding protein, troponin, mediates the Ca^{2+} control of striated muscle contraction. Troponin molecules bind to a unique site on each tropomyosin molecule, so they are spaced 38 nm apart along the actin filament in muscle. Each troponin is believed to control one tropomyosin molecule, spanning seven actin subunits. Ca^{2+} binding to troponin determines the position of the tropomyosin in the groove of the actin filament (Lehman et al 1994). In the inactive or "off" state, tropomyosin interferes sterically with the interaction of myosin with actin filaments. In the active state, Ca^{2+} causes tropomyosin to move deeper into the long pitch groove of the actin filament, which allows myosin to interact productively with the actin filament.

Structure Determination

Tropomyosin crystallizes in the form of endless filaments similar in structure to the tropomyosin bound to the thin filaments in muscle. The large volume fraction of solvent present in these crystals and the consequent molecular motions have limited the resolution to 9 Å. The structure was solved by refining and extending a uniform wire model based on electron microscopy (Phillips et al 1986; Whitby et al 1992). The path of the chain was represented using refinable space curve parameterization, and the use of anisotropic temperature factors normal to the chain axis allowed the long and short transverse dimensions to be distinguished.

Description of the Structure

In the crystal the axis of the tropomyosin coiled-coil forms a super coil with an elliptical cross-section with major and minor axes of 46 and 20 Å (Figure 11). The supercoil has a pitch of about 400 Å, which is smaller than the ~ 750 Å pitch of tropomyosin in the thin filament. Both in the crystal and in the thin filament, the pitch of the coiled-coil is close to the canonical value, about 140 Å. Ample evidence for the type of flexibility needed by tropomyosin to carry out its functions exists in the diffuse background of the X-ray diffraction pattern. Modeling studies of the motions of the coiled-coil within the lattice show local motions with rms amplitudes close to 8 Å and demonstrate a preference for deformation across the narrow face of the structure. The two strands were determined to be parallel and unstaggered by visualization of mercury atoms bound to cysteine residues in the two chains. Differences in the contour lengths measured for tropomyosin in the crystal lattice and calculated from the known parameters of the coiled-coil suggest that eight or nine residues overlap at the head-to-tail junctions between tropomyosin molecules. The electron density suggests that this region is globular, comprising a complex, four-stranded cluster. The actin-binding sites coincide with the α-subsites of the weak 14-fold pattern of apolar and acidic residues detected by Fourier transform analysis of the sequence (McLachlan & Stewart 1975, 1976).

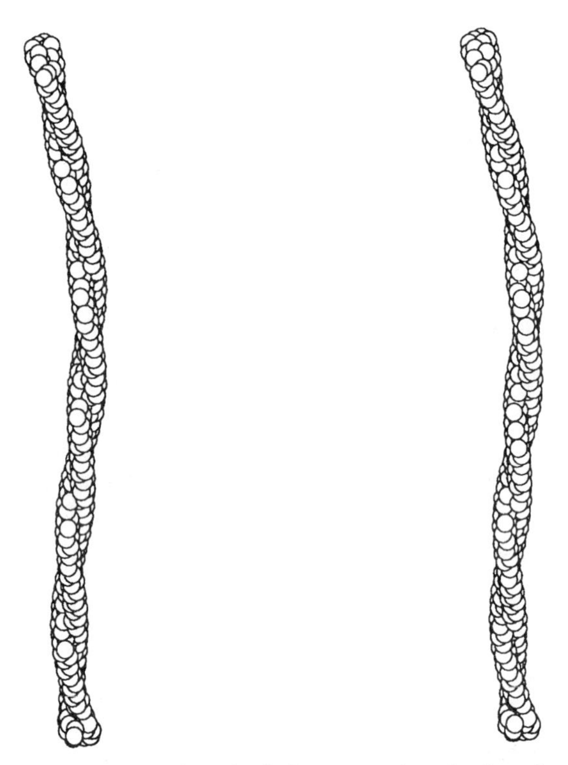

Figure 11 Tropomyosin. Stereo view of a single tropomyosin molecule made with the program MacImdad© from the coordinates of Whitby et al (1992). Spheres are centered at the positions of the alpha carbon atoms, and have radius 0.3 nm, which gives them a volume equal to that of an average amino acid. Structure determined by a combination of solvent flattening and refinement applied to an initial model derived from electron microscopy.

Significance

Many of the observed structural features of tropomyosin shed light on existing theories of muscle regulation. Most prominently, the tropomyosin supercoil observed in the crystal is reminiscent of the supercoil taken up by tropomyosin on the actin filament, thus confirming that this interaction is energetically reasonable. Also the flexibility shown by tropomyosin in the crystal is sufficient to allow for the independent control of tropomyosin segments by a single troponin. As expected, regularities in the amino acid sequence have clear structural correlates. For example, the unbroken string of heptad repeats (*a b c d e f g*) in which *a* and *d* are apolar residues corresponds to a hydrophobic seam between the two strands of the coiled-coil.

CONCLUDING REMARKS

The structural studies reviewed here provide a new starting point for studying the molecular basis of cell motility and the assembly of the actin cytoskeleton. In many ways the biological processes dependent on actin and its associated proteins are simply complex self-assembly reactions driven by mass action. Now we know the shapes and sites of interaction of many of the molecular building blocks in this complicated, dynamic jigsaw puzzle. In several cases we know the rate constants for the interactions. The unmet challenge is to learn how the molecular associations are controlled inside the cell. In the simplest cases, such as the all-or-none contraction of striated muscles, we have some ideas about how Ca^{2+} acts on the molecular switch that turns the myosin motors on and off. However, without the atomic structure of the whole troponin complex, the mechanism is not clear. More complex systems, even smooth muscle, represent more formidable challenges. In nonmuscle cells, the assembly of the actin system is much more dynamic than in muscle, and assembly and contractility must be controlled temporally and spatially. Rapidly expanding knowledge about the molecular structures of the components of the actin system should provide the insights required to design and interpret the experiments that will reveal the underlying processes that regulate motility and cytoplasmic structure.

ACKNOWLEDGMENTS

The authors wish to thank Dr. Tad Holak for providing coordinate and MOLSCRIPT files for hisactophilin and severin and for permission to mention unpublished thymosin-β4 experiments. We are also grateful to Drs. Ken Holmes, Paul McLaughlin, Ivan Rayment, C. Schutt, and Y. Yan for help with the figures. Our original research on actin-binding proteins is supported by the National Institutes of Health research grants GM-26338 and GM-35171

Literature Cited

Archer SJ, Vinson VK, Pollard TD, Torchia DA. 1994. Elucidation of the poly-L-proline binding site in *Acanthamoeba* profilin-I by NMR spectroscopy. *FEBS Lett.* 337:125–51

Bewley M, Boustead C, Walker J, Waller D. 1993. Structure of chicken annexin V at 2.25Å resolution. *Biochemistry* 32:3923–29

Bremer A, Millonig RC, Sutterlin R, Engel A, Pollard TD, Aebi U. 1991. The structural basis for the intrinsic disorder of the actin filaments: the "lateral slipping" model. *J. Cell Biol.* 115:689–703

Burgoyne RD. 1987. Gelsolin and P36 share a similar domain. *Trends Biochem. Sci.* 12:85–86

Cheney RE, Riley MA, Mooseker MS. 1993. Phylogenetic analysis of the myosin superfamily. *Cell Motil. Cytoskeleton* 24:215–23

Clore GM, Wingfield PT, Gronenborn A. 1991. High resolution three-dimensional structure of interleukin 1β in solution by 3- and 4-dimensional NMR spectroscopy. *Biochemistry* 30:2315–23

Concha NO, Head JF, Kaetzel MA, Dedman JR, Seaton BA, 1993. Rat annexin V crystal structure: Ca^{2+} induced conformational changes. *Science* 261:1321–24

Constantine KL, Friedrichs MS, Bell AJ, Lavoie TB, Mueller L, Metzler WJ. 1993. Relaxation study of the backbone dynamics of human profilin by 2-dimensional ^{1}H-^{15}N NMR. *FEBS Lett.* 336:457–61

Czisch M, Schleicher M, Horger S, Voelter W, Holak TA. 1993. Conformation of thymosin-β4 in water determined by NMR spectroscopy. *Eur. J. Biochem.* 218:335–44

Devreotes PN, Zigmond SH. 1988. Chemotaxis in eukaryotic cells: a focus on leukocytes and *Dictyostelium. Annu. Rev. Cell Biol.* 4:649–86

Egelman E. 1994. The ghost of ribbons past. *Curr. Biol.* 4:79–81

Eichinger L, Schleicher M. 1992. Characterization of actin- and lipid-binding domains in severin a Ca^{2+}-dependent F-actin fragmenting protein. *Biochemistry* 31:4779–87

Elgaester A, Stokke BT, Mikkelsen A, Branton D. 1986. The molecular basis of erythrocyte shape. *Science* 234:1217–23

Elzinga M, Phelan JJ. 1984. F-actin is intermolecularly crosslinked by N,N′-p-phenylenedimaleimide through lysine-191 and cytstein-374. *Proc. Nat. Acad. Sci. USA* 81:6599–602

Federov AA, Magnus KA, Graupe H, Lattman EE, Pollard TD, Almo S. 1994. X-ray structures of isoforms of the actin binding protein profilin that differ in their affinity for polyphosphoinositides. *Proc. Natl. Acad. Sci. USA.* In press

Geisterfer-Lowrance AAT, Kass S, Tanigawa G, Vosberg H-P, McKenna W, et al. 1990. A molecular basis for familial hypertrophic cardiomyopathy: a beta-cardiac myosin heavy chain gene missense mutation. *Cell* 62:999–1006

Gerke V, Weber K. 1985. Calcium-dependent conformational changes in the 36 kDa subunit of intestinal protein I related to the cellular 36 kDa target of Rous sarcoma virus tyrosine kinase. *J. Biol. Chem.* 260:1688–95

Glenney JR, Tack B, Powell MA. 1987. Calpactins: two distinct Ca^{2+} regulated phospholipid and actin binding proteins isolated from lung and placenta. *J. Cell Biol.* 104:503–11

Goldschmidt-Clermont PJ, Furman MI, Wachsstock DH, Safer D, Nachmias VT, Pollard TD. 1992. The control of actin nucleotide exchange by thymosin-β4 and profilin. A potential regulatory mechanism for actin polymerization in cell. *Mol. Biol. Cell* 3:1015–24

Gorlin JB, Yamin R, Egan S, Stewart M, Stossel TP, et al. 1990. Human endothelial actin binding protein (ABP280, human nonmuscle filamin): a molecular leaf spring. *J. Cell Biol.* 111:1089–105

Habazettl J, Gondol D, Wiltscheck R, Otlewski J, Schleicher M, Holak TA. 1992a. Structure of hisactophilin is similar to interleukin-1β and fibroblast growth factor. *Nature* 359:855–58

Hazabettl J, Schleicher M, Otlewski J, Holak TA. 1992b. Homonuclear 3-dimensional NOE-NOE NMR/spectra for structure determination of proteins in solution. *J. Mol. Biol.* 228:156–69

Hegyi G, Michel H, Shabanowitz J, Hunt DF, Chattterje N, et al. 1992. Gln-41 is intermolecularly cross-linked to Lys-113 in F-actin by N-(4-azidobenzoyl)-putrescine. *Protein Sci.* 1:132–44

Hitchcock SE. 1980. Actin-DNase I interaction. Depolymerization and nucleotide exchange. *J. Biol. Chem.* 255:5668–73

Holmes KC, Popp, D, Gebhard W, Kabsch W. 1990a. Atomic model of the actin filament. *Nature* 347:44–49

Holmes KC, Popp D, Gebhard W, Kabsch W. 1990b. The structure of F-actin calculated from X-ray fibre diagrams and the 0.6 nm crystal structure. In *Molecular Mechanisms in Muscular Contraction,* ed JM Squire, pp. 49–64. Boca Ratan: CRC Press

Holmes KC, Sander C, Valencia A. 1993. A new ATP-binding fold in actin, hexokinase and Hsc70. *Trends Cell Biol.* 3:53–59

Huber R, Beredes R, Burger A, Schneider M, Karshikov A, Luecke H. 1992. Crystal and molecular structure of annexin V after refinement. Implications for structure, membrane binding and ion channel formation of the annexin family of proteins. *J. Mol. Biol.* 223:683–704

Huber R, Romisch J, Paques E. 1990. The crystal and molecular structure of human annexin V, an anticoagulant protein that binds to calcium and membranes. *EMBO J.* 9:3867–74

Huxley HE, Brown W. 1967. The low-angle X-ray diagram of vertebrate striated muscle and its behavior during contraction and rigor. *J. Mol. Biol.* 30:383–434

Huxley HE, Kress M. 1985. Crossbridge behavior during muscle contraction. *J. Muscle Res. Cell Motil.* 6:153–62

Ikebuchi NW, Waisman DM. 1990. Calcium-dependent regulation of actin filament bundling by lipocortin-85. *J. Biol. Chem.* 265:3392–400

Ikura M, Clore GM, Gronenborn AM, Zhu G, Klee CB, Bax A. 1992. Solution structure of a calmodulin-target peptide complex by multidimensional NMR. *Science* 256:632–38

Jones PG, Moore GJ, Waisman DM. 1992. A nonapeptide to the putative F-acting binding site of annexin-II tetramer inhibits its calcium-dependent activation of actin filament bundling. *J. Biol. Chem.* 267:13993–97

Kabsch W, Mannherz HG, Suck D, Pai EF, Holmes KC. 1990. Atomic structure of the actin:DNase I complex. *Nature* 347:37–49

Kabsch W, Vandekerckhove J. 1992. Structure and function of actin. *Annu. Rev. Biophys. Biomol. Struct.* 21:49–76

Khanna NC, Helwig ED, Ikebuchi W, Fitzpatrick S, Bajiwa R, Waisman DM. 1990. Purification and characterization of annexin proteins from bovine lung. *Biochemistry* 29: 4852–62

Klee CB. 1988. Ca^{2+}-dependent phospholipid- (and membrane-) binding proteins. *Biochemistry* 27:6645–53

Kraulis PJ. 1991. MOLSCRIPT: a program to produce both detailed and schematic plots of protein structure. *J. Appl. Cryst.* 24:946–50

Lazarides E, Lindberg U. 1974. Actin is the naturally occurring inhibitor of deoxyribonuclease-I. *Proc. Nat. Acad. Sci. USA* 71:4742–46

Lehman W, Craig R, Vibert P. 1994. Ca^{2+} induced tropomyosin movement in *Limulus* thin filaments revealed by 3-dimensional reconstruction. *Nature* 368:65–67

Lorenz M, Popp D, Holmes KC. 1993. Refinement of the F-actin model against x-ray fiber diffraction data by the use of a directed mutation model algorithm. *J. Mol. Biol.* 234: 826–36

Machesky LM, Pollard TD. 1993. Profilin as a potential mediator of membrane-cytoskeletal communication. *Trends Cell Biol.* 3:381–85

Mannherz HG, Kabsch W, Leberman R. 1977. Crystallization of skeletal muscle actin: pancreatic DNase I complex. *FEBS Lett.* 73: 141–43

Markus MA, Nakayama T, Matsudaira P, Wagner G. 1994. Solution structure of villin 14T, a domain conserved among actin severing proteins. *Protein Sci.* 3:70–81

Matsudaira P. 1991. Modular organization of actin cross-linking proteins. *Trends Biochem. Sci.* 16:87–92

Matsudaira P, Janmey P. 1988. Pieces in the actin severing puzzle. *Cell* 54:139–40

McLachlan AD, Stewart M. 1975. Tropomyosin coiled-coil interactions: Evidence for an unstaggered structure. *J. Mol. Biol.* 98: 293–304

McLachlan AD, Stewart M. 1976. The 14-fold periodicity in alpha-tropomyosin and the interaction with actin. *J. Mol. Biol.* 103:271–98

McLaughlin PJ, Gooch JT, Mannherz HG, Weeds AG. 1993. Structure of gelsolin segment 1-actin complex and the mechanism of filament severing. *Nature* 364:685–92

Meador WE, Means AR, Quiocho FA. 1993. Target recognition by calmodulin: 2.4Å structure of a calmodulin-peptide complex. *Science* 257:1251–55

Metzler WJ, Bell AJ, Ernst E, Lavoie TB, Mueller L. 1994. Identification of the poly-L-proline binding site on human profilin. *J. Biol. Chem.* 269:4620–25

Metzler WJ, Constantine KL, Friedrichs MS, Bell AJ, Ernst EG, et al. 1993. Characterization of the three-dimensional structure of human profilin: ^{1}H, ^{13}C and ^{15}N NMR assignments and global folding pattern. *Biochemistry* 32:13818–29

Milligan RA, Whittaker M, Safer D. 1990. Molecular structure of F-actin and location of surface binding proteins. *Nature* 348: 217–21

Moore S. 1981. Deoxyribonuclease I. In *The Enzymes,* 18:281–96. New York: Academic

Nachmias VT. 1993. Small actin-binding proteins: the β-thymosin family. *Curr. Opin. Cell Biol.* 5:56–62

Oefner C, Suck D. 1986. Crystallographic refinement and structure of DNase I at 2Å resolution. *J. Mol. Biol.* 192:605–32

Orlova A, Yu X, Egelman EH. 1994. Three-dimensional reconstruction of a co-complex of F-actin with antibody Fab fragments to actin's amino terminus. *Biophys. J.* 66:276–85

Phillips Jr GN, Fillers JP, Cohen C. 1986. Tropomyosin crystal structure and muscle regulation. *J. Mol. Biol.* 192:111–31

Pinder JC, Gratzer WB. 1982. Deoxyribonuclease I interaction using a pyrene-conjugated actin derivative. *Biochemistry* 21:4886–90

Podolski JL, Steck T. 1988. Association of DNase I with the pointed ends of actin filaments in human red blood cells. *J. Biol. Chem.* 263:638–45

Pollard TD. 1993. Actin and actin binding proteins. In *Guidebook to the Cytoskeletal and Motor Proteins,* ed. T Kreis, R Vale, pp. 3–11. Oxford/New York/Tokyo: Oxford Univ. Press. 276 pp.

Pollard TD, Doberstein SK, Zot HG. 1991. Myosin-I. *Annu. Rev. Physiol.* 53:653–81

Popp D, Lednev VV, Jahn W. 1987. Methods of preparing well-oriented sols of F-actin containing filaments suitable for X-ray diffraction. *J. Mol. Biol.* 197:679–84

Rayment I, Holden HM. 1994. The three-dimensional structure of a molecular motor. *Trends Biochem. Sci.* 19:129–34

Rayment I, Holden HM, Whittaker M, Yohn CB, Lorenz M, et al. 1993a. Structure of the actin-myosin complex and its implications for muscle contraction. *Science* 261:58–65

Rayment I, Rypniewski WR, Schmidt-Base K, Smith R, Tomchick DR, et al. 1993b. Three dimensional structure of myosin subfragment-1: a molecular motor. *Science* 261:50–58

Safer D, Elzinga M, Nachmias VT. 1991. Thymosin β4 and Fx an actin sequestering peptide are indistinguishable. *J. Biol. Chem.* 266: 4029–32

Safer D, Golla R, Nachmias VT. 1990. Isolation of a 5 kDa actin sequestering peptide from human blood platelets. *Proc. Natl. Acad. Sci. USA* 87:2536–40

Scheel J, Ziegelbauer K, Kupke T, Humbel BM, Noegel AA, et al. 1989. Hisactophilin, a histidine-rich actin-binding protein from *Dictyostelium discoideum. J. Biol. Chem.* 264:2832–39

Schleicher M, Andre E, Hartmann H, Noegel AA. 1988. Actin-binding proteins are conserved from slime molds to man. *Dev. Genet.* 9:521–30

Schleicher M, Gerisch G, Isenberg G. 1984. New actin binding proteins from *Dictyostelium discoideum. EMBO J.* 3:2095–2100

Schroeder RR, Manstein DJ, Jahn W, Holden HM, Rayment I, et al. 1993. Three-dimensional atomic model of F-actin decorated with *Dictyostelium* myosin S1. *Nature* 364: 171–74

Schutt CE, Myslik JC, Rozycki MD, Goonesekere NCW, Lindberg U. 1993. The structure of crystalline profilin-β-actin. *Nature* 365:810–16

Squire JM, Luther PK, Morris EP. 1990. Organization and properties of the striated muscle sarcomere. In *Molecular Mechanisms in Muscular Contraction,* ed JM Squire, pp. 1–48. Boca Ratan: CRC Press

Suck D, Kabsch W, Mannherz HG. 1981. Three-dimensional structure of the complex of skeletal muscle actin and bovine pancreatic DNase I at 6Å resolution. *Proc. Natl. Acad. Sci. USA* 78:4319–23

Suck D, Lahm A, Oefner C. 1988. Structure refined to 2Å of a nicked DNA octanucleotide complex with DNase I. *Nature* 332:464–68

Suck D, Oefner C. 1986. Structure of DNase I at 2.0Å resolution suggests a mechanism for binding and cutting DNA. *Nature* 321:620–25

Suck D, Oefner C, Kabsch W. 1984. Three dimensional structure of bovine pancreatic DNase I at 2.5Å resolution. *EMBO J.* 3: 2423–30

Sugino H, Sakabe N, Sakabe K, Hatano S, Oosawa F, et al. 1979. Crystallization and preliminary crystallographic data of chicken gizzard G-actin-DNase I complex and *Physarum* G-actin-DNase I complex. *J. Biochem.* 86:257–60

Swairjo MA, Seaton BA. 1994. Annexin structure and membrane interactions: a molecular perspective. *Annu. Rev. Biophys. Biomol. Struct.* 23:193–213

Taylor EW. 1992. Mechanism and energetics of actomyosin ATPase. In *The Heart and Cardiovascular System,* ed HA Fozzard, pp. 1281–93. New York: Raven. 2nd. ed.

Taylor KA, Taylor DW. 1993. Projection image of smooth muscle alpha-actinin form two-dimensional crystals formed on positively charged lipid layers. *J. Mol. Biol.* 230:196–205

Theriot JA, Mitchison TJ. 1993. The three faces of profilin. *Cell* 75:835–38

Tirion MM, ben-Avraham D. 1993. A normal mode analysis of G-actin. *J. Mol. Biol.* 230: 186–95

Toyoshima YY, Kron SJ, McNally EM, Niebling KR, Toyoshima C, Spudich JA. 1987. Myosin subfragment-1 is sufficient to move actin filaments in vitro. *Nature* 328:536–39

Vancompernolle K, Vanderkerckhove J, Bubb MR, Korn ED. 1991. The interfaces of actin and *Acanthamoeba* actobindin. Identification of a new actin-binding motif. *J. Biol. Chem.* 266:15427–31

Vandekerckhove J. 1989. Actin-binding proteins. *Curr. Opin. Cell Biol.* 1:15–22

Vandekerckhove J, Deboben A, Nassal M, Wieland T. 1985. The phalloidin binding site of F-actin. *EMBO J.* 4:2815–18

Vanderkerckhove J, VanDamme J, Vancompernolle K, Bubb MR, Lambooy PK, Korn ED. 1990. The covalent structure of *Acanthamoeba* actobindin. *J. Biol. Chem.* 265:12801–05

Vinson VK, Archer SJ, Lattman EE, Pollard TD, Torchia DA. 1993. Three-dimensional structure of *Acanthamoeba* profilin-I. *J. Cell Biol.* 122:1277–83

Voelter W, Echner E, Kalbacher H, Dinh TQ, Kapurniotu A, et al. 1987. In *Peptides 1986,* ed. D Theodoropoulos, pp. 581–84. Berlin/New York: de Gruyter

Wachsstock DH, Schwarz WH, Pollard TD. 1994. Crosslinker dynamics determine the mechanical properties of actin gels. *Biophys. J.* 66:801–9

Way M, Weeds A. 1988. Nucleotide sequence of pig plasma gelsolin. Comparison of protein sequence with human gelsolin and other actin-severing proteins shows strong homologies and evidence for large internal repeats. *J. Mol. Biol.* 203:1127–33

Weeds A, Maciver S. 1993. F-actin capping proteins. *Curr. Opin. Cell Biol.* 5:63–69

Weng X, Luecke H, Song I, Kang D, Kim S, Huber R. 1993. Crystal structure of human annexin I at 2.5Å resolution. *Protein Sci.* 2: 448–58

Whitby FG, Kent H, Stewart F, Stewart M, Xie X, et al. 1992. Structure of tropomyosin at 9Å resolution. *J. Mol. Biol.* 227:441–52

Yan Y, Winograd E, Viel A, Cronin T, Harrison SC, Branton D. 1993. Crystal structure of the repetitive segments of spectrin. *Science* 262: 2027–30

Yount R. 1994. Subfragment-1. The first crystalline motor. *J. Muscle Res. Cell Motil.* 14: 547–51

Yu H, Chen JK, Feng S, Dalgarno DC, Brauer AW, Schreiber SL. 1994. Structural basis for the binding of proline-rich peptides to SH3 domains. *Cell* 76:933–45

Zarbock J, Oschkinat H, Hannappel E, Kalbacher H, Voelter W, Holak TA. 1990. Solution conformation of thymosin-β4: a nuclear magnetic resonance and simulated annealing study. *Biochemistry.* 20:7814–21

Zhu X, Komiya H, Chirino A, Faham S, Fox GM, et al. 1991. Three-dimensional structures of acidic and basic fibroblast growth factors. *Science* 251:90–93

Xie X, Harrison DH, Schlichting I, Sweet RM, Kalabokis VN, et al. 1994. Structure of the regulatory domain of scallop myosin at 2.8Å resolution. *Nature* 368:306–12

Annu. Rev. Cell Biol. 1994. 10:251–337

RECEPTOR PROTEIN-TYROSINE KINASES AND THEIR SIGNAL TRANSDUCTION PATHWAYS

Peter van der Geer and Tony Hunter

Molecular Biology and Virology Laboratory, The Salk Institute, P.O. Box 85800, San Diego, California 92186

Richard A. Lindberg

Department of Immunology, AMGEN, 1840 Dehavilland Drive, Thousand Oaks, California 91320

KEY WORDS: receptors, protein-tyrosine kinases, SH2 domains, signal transduction, protein-tyrosine kinase substrates

---

CONTENTS

0743–4634/94/1115–0251$05.00

INTRODUCTION

Cells modulate their activities in response to signals from their surrounding environment. Single cell organisms respond mainly to nutrient cues, many of which are able to cross the cell membrane. In multicellular organisms, however, a much more constant milieu bathes the individual cells and reduces the need for cells to respond to nutritional signals. Instead, the need to coordinate the activities of one cell with those of its neighbors has resulted in the evolution of complex intercellular signaling pathways.

Many of the activities of vertebrate cells are controlled by extracellular signaling molecules. The signals are often transduced across the cellular membrane by transmembrane receptors. One type of receptor has intrinsic protein-tyrosine kinase (PTK) activity and belongs to the receptor protein-tyrosine kinase (RPTK) family. RPTKs have been found in all multicellular eukaryotic organisms. These receptors are involved in the regulation of many cellular programs; the most notable are the control of cell growth and differentiation. Because of these properties, this class of molecules is also important in the genesis of many neoplasias.

RPTKs are activated by polypeptide ligands commonly known as growth factors, but more properly called cytokines. Most known ligands for RPTKs are secreted, soluble proteins, but membrane-bound proteins as well as extracellular matrix proteins may also prove to activate RPTKs. RPTKs are closely related in their catalytic domains, and this has allowed homology cloning of many so-called orphan RPTKs, which by definition lack an identified ligand. Much recent progress in elucidating the function of RPTKs has come from the identification of cognate ligands for orphan RPTKs. The study of intracellular pathways that are modulated by ligand activation of RPTKs is contributing greatly to our understanding of how cells can be regulated by extracellular signals. Signaling by RPTKs involves ligand-mediated receptor dimerization, which results in transphosphorylation of the receptor subunits and activation of the catalytic domains for the phosphorylation of cytoplasmic substrates. The identities of many RPTK substrates have been uncovered recently. Progress in understanding how the ligand-mediated RPTK signal is propagated via these substrates into known cellular responses has been very rapid in the past few years. In this review we discuss the structure and classification of the members of the expanding RPTK family and cover recent advances in the understanding of RPTK-activated signaling pathways (for earlier reviews see Cantley et al 1991; Chao 1992; Fantl et al 1993; Schlessinger & Ullrich 1992; Ullrich & Schlessinger 1990; Williams 1989; Yarden & Ullrich 1988).

STRUCTURE

Our RPTK classification is based on structure rather than on function. However, closely related RPTKs generally prove to have similar functions.

Overall Basic Structure

RPTKs are type I transmembrane proteins, with their N-termini outside the cell and single membrane-spanning regions. The following structural features are common to all RPTKs. Starting at the N-terminus, the generic RPTK has a signal peptide that targets the protein to the secretory pathway. This is followed by an extracellular domain of several hundred amino acids that contains a distinctive pattern of Cys residues and often a characteristic array of structural motifs. Most RPTK extracellular domains are modified by N-linked glycosylation and some may also have O-linked sugars. The transmembrane domain consists of a stretch of hydrophobic residues that are followed by several basic residues that function as a stop-transfer signal. On the cytoplasmic side of the membrane there is a juxtamembrane region, which precedes the catalytic domain. The catalytic domain is related to that of cytoplasmic PTKs and the protein-serine/threonine kinases (see below) and is about 250 residues in length, excluding inserts. The phosphotransfer reaction is catalyzed by this autonomously functioning domain. There is a region C-terminal to the catalytic domain that varies from a few residues up to 200 residues. The functions of this C-terminal tail vary among members of the RPTKs. The molecular topologies of one member from each of the 14 subfamilies (see below) are shown in Figure 1.

The Ligand-Binding Domain

The ligand-binding domain is the most distinctive feature in the RPTKs and is composed of combinations of various recognizable sequence motifs. The EGF receptor has two Cys-rich regions that are related in amino acid sequences. Related Cys-rich regions are also found in the insulin receptor. The exact function of these domains is not known, but they may maintain structural integrity as opposed to being directly involved in ligand contact. The PDGF receptor subfamily members have either five (e.g. PDGF β receptor) or seven (e.g. Flk1) immunoglobulin-like (Ig) domains. The FGF receptor subfamily has three Ig-like domains, but there are also forms generated by alternative

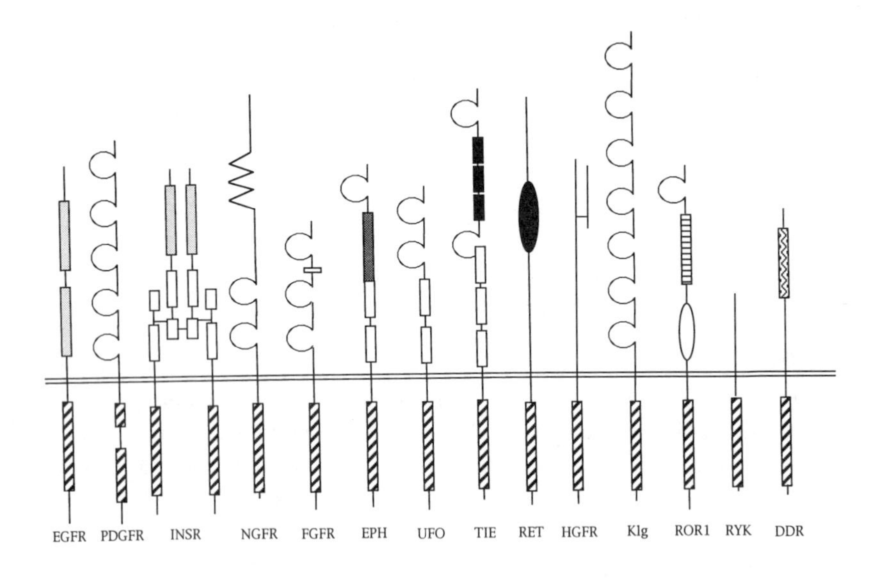

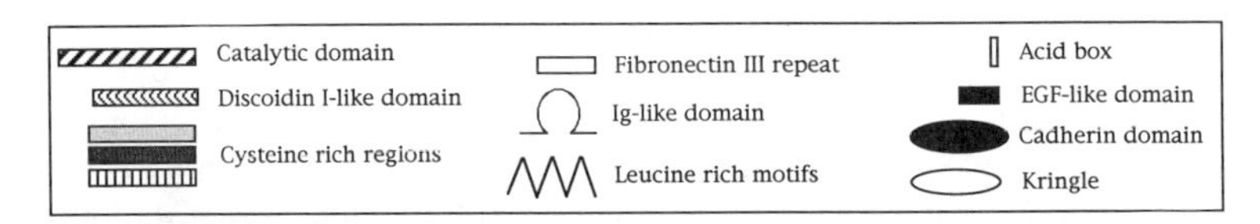

Figure 1 Structural topology of RPTKs. Abbreviations: EGFR, epidermal growth factor receptor; PDGFR, platelet-derived growth factor receptor; INSR, insulin receptor; NGFR, nerve growth factor receptor; FGFR, fibroblast growth factor receptor; HGFR, hepatocyte growth factor receptor (Johnson et al 1993; Masiakowski & Carroll 1992; O'Bryan et al 1991; Partanen et al 1992; Schneider 1992; Schneider & Schweiger 1991; Ullrich & Schlessinger 1990).

splicing that have two Ig-like domains. The insulin receptor, in addition to the Cys-rich domain, has three fibronectin type III (FNIII) repeats (FNIII and Ig repeats have an overall similar structure and may be evolutionarily related). The insulin receptor is derived from a precursor that is proteolytically cleaved to yield a mature receptor in which the α and β subunits are covalently linked by disulfide bonds and function as a heterotetrameric molecule. The NGF receptor contains two Ig-like domains and a recently defined Leu-rich motif, which is thought to function in cell adhesion. The EPH subfamily has an Ig-like domain, a Cys-rich region that is unrelated to those in the EGF and insulin receptors, and two FNIII repeats. AXL has two FNIII repeats and two Ig-like domains. TIE has Ig-like and FNIII domains and, uniquely, three EGF repeats.

RET has a domain related to those found in cadherins, which are cell surface adhesion receptors that interact homotypically via Ca^{2+} bridges. The HGF receptor is a heterodimer that, like the insulin receptor, is formed by proteolytic cleavage of a precursor, which results in a receptor with two subunits held together by a disulfide bond. Klg has seven Ig-like domains; in this respect it is similar to Flk1. The ROR1 RPTK has a unique Cys-rich region, an Ig-like domain, and a kringle motif in its ligand-binding domain. DDR has a discoidin I-like domain, also known as a factor V/VIII-like phospholipid-binding domain.

The functions of the structural motifs found in these ligand-binding domains are not clear. Some of them presumably specify direct interaction with the cognate ligand. The limited mapping of RPTK ligand-binding sites that has been carried out thus far indicates that they lie in relatively short regions of the extracellular domain, which suggests that RPTK extracellular domains have other functions. Dimerization of two receptor molecules is critical for RPTK signaling, and some motifs in the extracellular domain may serve this function. However, other functions such as interactions with other cell surface proteins facilitating adhesion or direct communication between cells, or interactions with the extracellular matrix, represent candidate functions for RPTK extracellular domains.

The Conserved Catalytic Domain

Most protein kinases share a conserved catalytic domain. Sequences on either side of this domain and some large inserts within the domain itself are not required for activity and are likely to have functions that are not involved in phosphotransfer. The catalytic domain is composed of ~ 250 amino acids and has been delineated by sequence similarities (Hanks et al 1988). Experimental determination of the boundaries of the catalytic domain by deleting in from the ends and testing for activity has shown that the region required for activity is similar to that predicted by similarity in primary amino acid sequence homology.

Within the catalytic domains of distinct RPTKs, sequence identities range from 32 to 95%. Alignment of PTK catalytic domain sequences revealed that there are 13 residues conserved in all of the 54 sequences presented (Hanks 1991). Several other residues would be invariant if not for the ERBB3 RPTK, which may not be fully active as a protein kinase, and the divergent sequences of some of the PTKs from lower organisms, such as *Dictyostelium* and *Drosophila.*

The determination of the crystal structures of the cAMP-dependent protein kinase (PKA) catalytic subunit, CDK2, ERK2, and casein kinase I has revealed the conserved architecture of the protein kinase catalytic domain (Bossemeyer et al 1993; De Bondt et al 1993; Knighton et al 1991a,b; Zhang et al 1994).

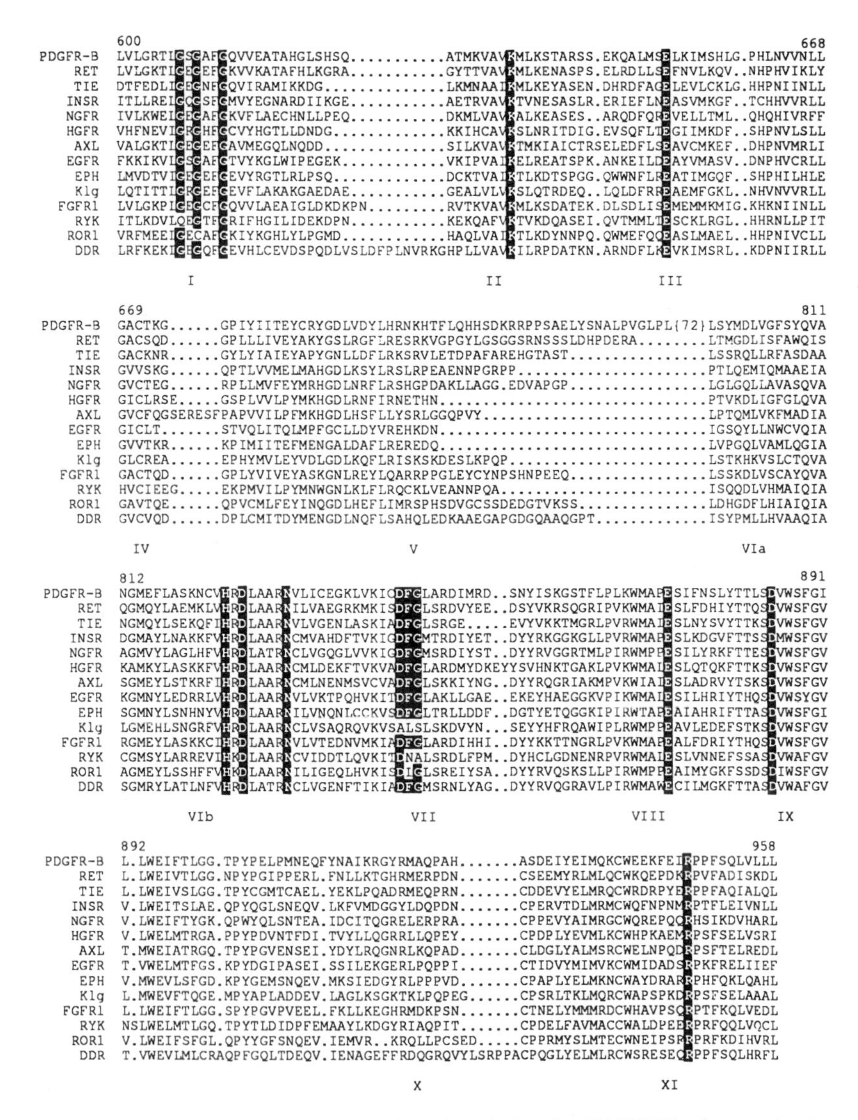

Figure 2 Amino acid sequence alignment of the catalytic domains of 14 RPTKs. One member from each of the 14 subfamilies described in Table 1 is shown. Highly conserved residues among all protein kinases are indicated by black boxes with white lettering. The numbering is for human PDGFR β.

The structure of a PTK has not been reported, but given that many residues are conserved in both protein-serine/threonine kinases and PTKs, their catalytic domain structures will probably be similar. The relatedness in primary amino acid sequence between RPTKs and PKA has enabled modeling of the structure of the catalytic domain of the EGF receptor (Knighton et al 1993). Using the information from alignments and the three-dimensional structure solved for PKA, we can predict the functions of some of the conserved residues in PTKs. Fourteen RPTKs are aligned in Figure 2, and these sequences represent one each from the subfamilies defined below. In the following discussion, the residue number refers to the human PDGF β receptor with the corresponding number for PKA given in parentheses.

The catalytic domain structure is composed of two lobes. Mg^{2+}/ATP and the protein substrate are brought together in the cleft, which allows phosphotransfer to be catalyzed. The N-terminal lobe is responsible for binding Mg^{2+}/ATP. The motif responsible is the GXGXXG (21 amino acids) K635 (K72). The glycine fold holds the phosphate moieties of the nucleotide and the lysine residue, which is covalently modified by the ATP analogue p-fluorosulfonylbenzoyl adenosine and is required for ATP binding by interacting with the α and β phosphates. Moving towards the C-terminus, the next residue conserved in all protein kinases is E651 (E91) (also believed to be involved in coordinating Mg^{2+}/ATP), which forms a salt bridge to K635 (K72). In the other lobe, HRDLAARN (residues 824–831 for the PDGF receptor, 164–171 for PKA) forms the catalytic loop. The Asp is believed to be the catalytic base. This region is highly conserved in both protein kinase families, yet discriminates between PTKs and protein-serine/threonine kinases (e.g. this sequence is HRDLKPEN in PKA). The aspartate of DFG, residues 844–846 (184–186), functions in the chelation of Mg^{2+}. E873 (E208) and R948 (R280) are thought to form ion bridges that stabilize the two lobes, and D885 (D220) stabilizes the catalytic loop. When the structure of a PTK catalytic domain is solved, it will probably confirm these similarities, but more interestingly, it should highlight the differences between the two families and provide insight into how the different phosphoamino acid acceptor specificities are achieved.

The sequences of the catalytic domains of 52 vertebrate RPTKs have been analyzed as described by Hanks & Quinn (1991), and the results are represented as a molecular phylogenetic tree (Figure 3). The tree is an unrooted relatedness tree that clusters similar sequences and indicates relative relatedness by branch lengths. The 52 vertebrate RPTKs chosen for this analysis represent what are putatively distinct gene products (see below). This tree is useful in classifying the RPTKs and makes predictions about which receptors will have similar functions and bind related ligands.

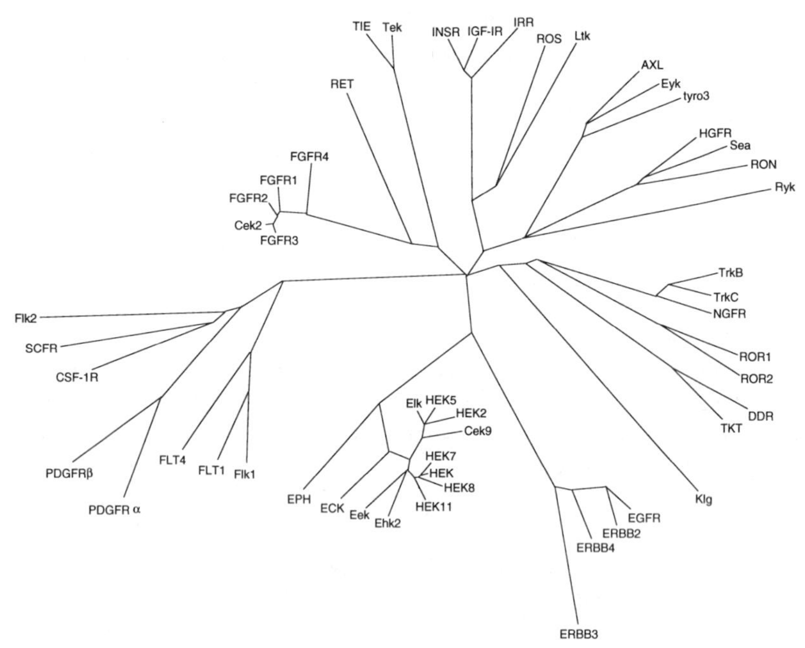

Figure 3 Molecular phylogeny of 52 RPTKs. The catalytic domain sequences for the receptors were analyzed as described previously (Hanks & Quinn 1991). The RPTKs shown are distinct vertebrate gene products (i.e. none are orthologues). The receptors are named as the receptor for a known ligand or following previous conventions (Hunter 1991). Abbreviations are IGF-1R, insulin-like growth factor-I receptor; CSF-1R, colony stimulating factor-1 receptor; and SCFR, stem cell factor receptor. Other abbreviations are defined in Figure 1.

CLASSIFICATION

The classification proposed here makes use of the types of structural analyses shown in Figures 1 and 3. There is a good concordance between the two sets of information, and the phylogenetic tree shown in Figure 3 clusters the RPTKs that have similar catalytic domain sequences. The similarities among the sequences are reflected graphically in the tree. Subfamily groupings are readily apparent as bifurcations of the long branches of the tree, which separate a group of RPTKs (or single RPTKs). It is striking that these groupings would be similar if one instead used the overall molecular topology shown in Figure 1 to define the subfamilies.

The nomenclature used for this review considers all protein kinases (including protein-serine/threonine kinases) as a superfamily and the PTKs as a family. Groupings of PTKs that have similar overall structure and cluster in phylogenetic trees are referred to as subfamilies. We have grouped the RPTKs into subfamilies as shown in Table 1. Many of the receptors listed in Table 1 have been isolated as cDNAs multiple times. When this occurs in the same species, the identity is obvious, but when closely related genes are isolated from different species, it is difficult to determine whether they are orthologues, especially if little is known about the gene or the gene product. Chromosomal mapping or cross-species genomic Southern hybridizations with probes from both species can indicate whether the genes are orthologous. However, in most cases only sequence data are available. In some cases, the alternative names obviously represent the same RPTK, e.g. EGF receptor, HER, and c-ErbB. In other cases, genes from different species have been grouped together under the assumption that they are from orthologous genes, e.g. the FGFR2 group. cDNAs for this gene have been isolated from three species: human, TK14 and K-SAM; mouse, Bek; and chicken, Cek3. The catalytic domains for these are about 99% identical and are considered to represent products of orthologous genes. With other genes, it is not so clear. Cek2 is 95% identical to human FGFR3 in the catalytic domain and could be the chicken orthologue, but with only these data prediction is difficult. Sequences outside the catalytic domain are commonly more divergent between subfamily members than the catalytic domains and prove useful in determining whether two RPTKs from different species with closely related catalytic domains are orthologues or not. The predictions in Table 1 as to which RPTKs are true orthologues from different vertebrate species should be considered tentative. Because of the difficulty in defining orthologues, in Table 1 we have not included the few known examples of RPTKs from amphibians or fish, or RPTKs from invertebrates, but these RPTKs are discussed below.

FUNCTION

RPTKs are known to act as receptors for growth factors, for differentiation factors, and for factors that stimulate metabolic responses. The functions of the RPTKs are governed by three principles. First, expression of almost all RPTKs is restricted to specific cell types in the organism. The pattern of expression depends upon the nature of the regulatory elements in the RPTK gene promoter and enhancer. Second, in the cells where a particular RPTK is expressed, its function is dictated by the activating ligands that bind to its extracellular domain, and this in turn depends on the availability and distribution of the cognate ligands in the organism. Third, the response to ligand-dependent RPTK activation in a given cell depends upon the intracellular proteins that are targets for phosphorylation and/or regulation by the activated RPTK. The second and third properties are intrinsic to the structure of the RPTK, and thus these functions would be expected to parallel the structural relationships within RPTK subfamilies, whose members probably have similar functions.

Table 1 RPTKs divided into subfamilies[a]

PDGFR subfamily	PDGF receptor α (5)
	PDGF receptor β (5)
	SCF receptor, c-Kit (5)
	CSF-1 receptor, c-Fms (5)
	Flk2, Flt3 (5)
	FLT1, Quek1 (7)
	Flk1, KDR (7)
	FLT4, Quek2 (7)
FGFR subfamily	FGF receptor 1, FLG, Cek1
	FGF receptor 2, Bek, K-SAM, Cek3, TK14
	FGF receptor 3, FLG2
	FGF receptor 4, TKF
	Cek2
INSF subfamily	Insulin receptor
	IGF-1 receptor
	IRR
	ROS
	Ltk
EGFR subfamily	EGF receptor, HER, c-ErbB
	Neu, ERBB2, HER2
	ERBB3, HER3
	ERBB4, Tyro2
NGFR subfamily	NGF receptor, TrkA
	TrkB
	TrkC
HGFR subfamily	HGF receptor, MET
	c-Sea
	RON
EPH subfamily	EPH
	ECK
	Elk, Cek6
	Eek
	Cek4, Mek4, HEK
	Cek5, HEK5, Nuk
	Cek7, HEK7, Ehk1
	Cek8, Sek, HEK8
	Cek9
	Cek10, HEK2
	HEK11
	Ehk2
AXL subfamily	AXL, UFO, Ark
	Eyk, c-Ryk
	Tyro3, SKY, Brt, TIF
TIE subfamily	TIE
	Tek, TIE2
DDR subfamily	Tyro10, TKT
	DDR, Nep, TrkE, CAK
ROR subfamily	ROR1
	ROR2
	RET
	Ryk, Nyk-r, Vik, Nbtk-1, Mrk
	Klg

[a] The receptors are grouped according to structural characters and sequence homology in their catalytic domains. The number of Ig-like repeats in members of the PDGFR subfamily is indicated in parentheses. References not in Hunter (1991) are: Flk2 (Matthews et al 1991b); Flt3 (Rosnet et al 1991), Flk1 (Matthews et al 1991a); KDR (Terman et al 1991); FLT4 (Aprelikova et al 1992); FGFR3 (Keegan et al 1991); FGFR4 (Partanen et al 1991); TK14 (Houssaint et al 1990); FLG2 (Avivi et al 1991); TrkC (Lamballe et al 1991); ECK, (Lindberg & Hunter 1990); Eek, (Chan & Watt 1991); Cek4/Mek4 (Sajjadi et al 1991); HEK (Wicks et al 1992); Cek5 (Pasquale 1991); Nuk (Henkemeyer et al 1994); Sek (Gilardi-Hebenstreit et al 1992); UFO (Janssen et al 1991); AXL (O'Bryan et al (1991); Ark (Rescigno et al 1991); Eyk (Jia & Hanafusa 1994); Tie (Partanen et al 1992); Tek (Dumont et al 1992); Klg (Chou & Hayman 1991); Nyk-r (Paul et al 1992); Ryk (Hovens et al 1992); Vik (Kelman et al 1993); Nbtk-1 (Maminta et al 1992); Mrk (Yee et al 1993); ROR1 and 2 (Masiakowski & Carroll 1992); Tyro2 and Tyro3 (C Lai, personal communication); Tyro10 (Lai & Lemke 1994) Ceks 6, 7, 8, 9, and 10 (Sajjadi & Pasquale 1993); ERBB4 (Plowman et al 1993a); RON (Ronsin et al 1993); HEKs 5, 7, 8, and 11 (M Fox, personal communication); Ehks 1 and 2 (Maisonpierre et al 1993); HEK2 (Bohme et al 1993); Queks 1 and 2 (Eichmann et al 1993); TKT (Karn et al 1993); DDR (Johnson et al 1993); Nep (Zerlin et al 1993); TrkE (Di Marco et al 1993); CAK (Perez et al 1994); SKY (Ohashi et al 1994); Brt (Fujimoto & Yamamoto 1994); and TIF (Dai et al 1994).

This is evidently true in most cases where ligands are known, but it is not universal. For example, the closely-related insulin and IGF-1 receptors have different functions; activation of the insulin receptor induces a metabolic response, whereas activation of the IGF-1 receptor elicits a growth response.

The functions of RPTKs are most readily apparent where a ligand is known. However, because of the rapid pace of identification of new RPTKs, the cognate ligands are lacking for about half of the known RPTKs. With regard to ligand binding, even though the conserved structural motifs are apparent, extracellular domain sequences are more divergent within a subfamily than the catalytic domain sequences. Nevertheless, for most of the RPTK subfamilies there is evidence that they bind ligands that are themselves members of families (see below). For some RPTK subfamilies, no ligands have been identified, and therefore we cannot be certain that these orphan RPTKs will interact with ligand families, even though this is a strong prediction.

The identities of substrates for different RPTKs are still being uncovered. It is becoming apparent that many of the physiologically relevant substrates are proteins that bind with high affinity to RPTK autophosphorylation sites via a specific domain called the Src-homology 2 (SH2) domain (see below). This binding is sequence specific with the residues immediately to the C-terminal side of the P.Tyr being most important. Therefore, the extent to which autophosphorylation sites and the precise sequences around them are conserved among members of a subfamily will dictate whether individual subfamily members phosphorylate the same spectrum of substrates. Since autophosphorylation sites are commonly located outside the catalytic domain proper in regions that are less highly conserved, i.e. the juxtamembrane region, the kinase insert, and the C-terminal tail, even closely related RPTKs can bind and phosphorylate distinct but usually overlapping sets of substrates. For instance, this is the case for the PDGF α and β receptors.

It is clear that the cell type in which an RPTK is expressed affects the cellular response to receptor activation. For instance, in neuronal cells, NGF elicits a differentiation response upon binding to the NGF receptor (TrkA), whereas when the NGF receptor is expressed ectopically in fibroblasts, NGF induces a growth response. This may in part be due to the different repertoire of substrates available in different cell types, the strength and persistence of the signal generated, as well as to differences in the programmed responses of cells to activation of the same signal pathway. In other cases, activation of an ectopically expressed RPTK may result in the same response as in its normal host cell. For instance, CSF-1 treatment of fibroblasts expressing the CSF-1 receptor elicits a growth response, as it does in myeloid precursor cells, which normally express the CSF-1 receptor. This implies that many of the RPTK-activated signaling pathways are conserved between different cell types (see below). Conversely, different RPTKs expressed in the same cell can elicit

Table 2 Ligands for RPTKs

PDGF family	EGF family
PDGF A	EGF
PDGF B	TGFα
VEGF	amphiregulin
PLGF	HB-EGF
CSF-1	SDGF
SCF, steel factor, Kit ligand	betacellulin
Flt3/Flk2 ligand	cripto
	NDF, heregulin, GGF, ARIA
FGF family	
aFGF	NGF family
bFGF	NGF
int-2	BDNF
K-FGF	NT-3
FGF-5	NT-4, NT-5, NT-4/5
FGF-6	
KGF	HGF family
FGF-8	HGF
FGF-9	HGFL
Insulin family	B61 family
insulin	B61
IGF-1	
IGF-2	

different responses through phosphorylation of different substrates, which in turn trigger different signal pathways. Before considering RPTK signaling, we discuss each vertebrate RPTK subfamily (Table 1) and its corresponding family of ligands (Table 2).

The Platelet-Derived Growth Factor Receptor Subfamily

The receptors in the PDGF receptor subfamily all contain a long insert (kinase insert) in their catalytic domain, and they all possess various numbers of Ig-like domains in their ligand-binding domains. The prototypical member, PDGF receptor, has five Ig repeats, as do the CSF-1 and SCF receptors and Flt3 (also known as Flk2). Flt3 is expressed most highly in the brain and hematopoietic cells (Rosnet et al 1991), and its ligand has recently been identified (Lyman et al 1993). The genes corresponding to the CSF-1 and SCF receptors were originally identified as the retroviral transforming genes, v-*fms* and v-*kit,* respectively. Flk1, FLT1, and FLT4, the other RPTKs in this subfamily, have seven Ig repeats and, since they cluster on a secondary branch and all bind VEGF or VEGF-related ligands, one might consider them a separate subfamily. One interesting feature of the PDGFR subfamily is that the genes are organized as three chromosomal clusters—SCF receptor, PDGF α receptor, and Flk1 on chromosome 4q1; CSF-1 receptor, PDGF β receptor, and FLT4 on chromosome 5q3 (the CSF-1 and PDGF β receptor genes are only 500 bp apart); and FLT1 and Flk2 on chromosome 13q1. It seems likely that all these genes arose from a single progenitor RPTK gene by duplication and then reduplication of one member to form a cluster of three RPTK genes, which was followed by duplication and transfer of the whole cluster to two other chromosomes.

The ligands of the PDGF family receptors can be divided into two groups: PDGF A, PDGF B, VEGF, and PLGF; and CSF-1, SCF, and the Flt3 ligand. Even within these groups, the ligands are rather distantly related (e.g. there is only 18% identity between VEGF and PDGF B, but the cysteines involved in disulfide bridges are conserved) (Keck et al 1989; Leung et al 1989). However, all these ligands have a dimeric structure and are homodimers, except for PDGF, which can exist as an active AB heterodimer. CSF-1, SCF, and the Flt3 ligand are made as transmembrane proteins whose external domains are in a four-helix bundle motif (Bazan 1991). VEGF binds both Flt1 (de Vries et al 1992) and Flk1 (Millauer et al 1993; Terman et al 1992). The exact receptors that VEGF uses in vivo remain to be determined. PLGF is related to VEGF and can be expected to bind one of these RPTKs, or a yet to be discovered subfamily member (Maglione et al 1991). Of the known classes of RPTK ligands, the members of the PDGF family are the only ones that are covalently dimeric, and this may play a role in receptor dimerization. Ligand binding to the CSF-1 receptor and SCF receptor (c-Kit) requires Ig repeats 1, 2, and 3 (Lev et al 1993; Wang et al 1993).

A unique feature of the PDGF receptor subfamily is the presence of a split kinase domain, in which there is an insert of 65 to 97 amino acids (compared to the SRC nonreceptor PTK) between the ATP-binding portion of the catalytic domain and the substrate-binding portion (between subdomains V and VIA). The insert occurs at a position where there is a surface loop in PKA, and it is believed that the kinase insert will form a separate structure protruding from the surface of the globular catalytic domain. The kinase insert is quite widely diverged between subfamily members. Because it contains several autophosphorylation sites, the kinase insert is important for substrate recognition (see below). However, it is dispensable for catalytic activity, and deletion of the kinase insert does not abolish all RPTK-mediated responses (Reedijk et al 1990).

The Fibroblast Growth Factor Receptor Subfamily

Members of the FGF receptor subfamily, which also have Ig repeats in their extracellular domain, represent a highly related group, as indicated in Figure

3. The use of different promoters and alternative splicing gives rise to many different forms of these receptors, and there are species containing two or three Ig repeats. These receptors are depicted as having a kinase insert like the PDGF receptor subfamily but, although the group has 14 more residues in this region of the catalytic domain than in SRC, this length is not significantly different from that of many other RPTKs (see Figure 2). In addition, no significant function has been attributed to this region in the FGF receptor subfamily. Therefore, we show it in Figure 1 without a kinase insert. To date four human FGF receptors have been reported, but there may be additional members of this subfamily.

Nine related fibroblast growth factors that give rise to a myriad of biological responses are known. Although not all of the FGFs have been tested against every FGFR, the picture that emerges for the FGFR and FGF families is that each receptor binds a subset of the growth factors and vice versa. Ig repeats 2 and 3 have been implicated in ligand binding, with both repeats being capable of binding independently to individual FGFs (Cheon et al 1994). Alternative splicing of an exon encoding the C-terminal half of repeat 3 results in FGFR1 and FGFR2 receptors with different ligand-binding specificities (Crumley et al 1991; Werner et al 1992).

The Insulin Receptor Subfamily

This subfamily includes the insulin receptor, the closely-related IGF-I receptor and IRR, and the next two closest relatives, ROS and Ltk. ROS and Ltk may be found to bind ligands not related to insulin and constitute a different subfamily. The mature insulin receptor is a disulfide-bonded heterotetramer containing two copies each of the α and β subunits, which are derived by proteolytic cleavage from a precursor molecule. The insulin and IGF-1 receptors bind insulin and IGF-1, which are related factors. However, the IGF-1 receptor has a much higher affinity for IGF-1 than insulin, and in consequence the majority of the metabolic effects of insulin are mediated through the insulin receptor. Conversely, the growth-stimulatory effects of these factors are elicited via the IGF-1 receptor.

The ROS RPTK is the closest vertebrate relative to the *D. melanogaster sevenless* RPTK, which is involved in the differentiation of the R7 photoreceptor cell in each ommatidion in the eye. Although the ligand for ROS is unknown, the ligand for *sevenless* is the *boss* protein, which is present as a surface protein containing seven transmembrane domains on the R8 photoreceptor cell that lies adjacent to the R7 cell expressing the *sevenless* RPTK (Hart et al 1993). Thus far this is the only example of an RPTK ligand that is obligatorily a cell surface protein, although several soluble ligands that activate RPTKs are derived from precursors with transmembrane domains and are active when anchored in this fashion (e.g. an anchored form of TGFα can activate the EGF receptor) (Brachmann et al 1989; Wong et al 1989). Multiple forms of the Ltk RPTK exist that differ in their extracellular domains (Snijders et al 1993; Toyoshima et al 1993). One form of Ltk with a short extracellular domain of ~ 100 residues is located predominantly in the endoplasmic reticulum and is activated by disulfide-bond cross-linking under conditions of oxidative stress, apparently in the absence of a ligand (Bauskin et al 1991).

The Epidermal Growth Factor Receptor Subfamily

The EGF receptor subfamily currently comprises four members, although evidence for a fifth member has been published (Zhang et al 1992b). These receptors have similar overall topology and significant relatedness in amino acid sequence. An interesting feature of ERBB3 is the fact that several residues conserved in all PTKs are absent, including the aspartate that has been proposed to be the catalytic base in the PKA catalytic subunit, which in ERBB3 is an asparagine. There is evidence that ERBB3 can autophosphorylate on tyrosine (Kraus et al 1993; Fedi 1994), but one should consider the possibility that it has a function that does not require PTK activity. ERBB4 is the most recently described member of this subfamily and is most similar to the ERBB2 and EGF receptors, based on catalytic domain similarity.

Several of the cytokines listed in Table 2, including EGF and TGFα, are capable of binding the EGF receptor itself. The EGF-binding site in the EGF receptor extracellular domain has been mapped by expressing chicken/human EGF receptor chimeras (Lax et al 1989) and by chemical cross-linking studies of EGF to the region between the two Cys-rich domains, with some contribution from the N-terminal Cys-poor region of the receptor (Lax et al 1988; Woltjer et al 1992). NDF (Wen et al 1992) [also called heregulin (Holmes et al 1992), GGF (Marchionni et al 1993), and ARIA (Falls et al 1993)] were originally reported to bind to the ERBB2 protein. Recent evidence, however, suggests that NDF cannot bind ERBB2 directly, but instead binds another family member and may only involve ERBB2 in heterodimers (Culouscou et al 1993; Plowman et al 1993b). The NDF gene gives rise to multiple differentially spliced forms that may have different biological activities (Marchionni et al 1993). All the members of the EGF family are derived from precursors that have a single transmembrane domain. The sequences for the mature factor, which contains an EGF repeat, lie close to the transmembrane domain in the precursor, and the soluble factor is derived by processing of the precursor at specific sites. In some cases these precursors have very long N-terminal extensions (e.g. the EGF precursor is ~ 100 kd) and conserved cytoplasmic domains, and it is possible that these membrane-anchored precursors have a physiological signaling function.

The Nerve Growth Factor Receptor Subfamily

The NGF receptor subfamily has three members that share structure and sequence similarities and also bind ligands of similar nature. The original receptor in this family was called proto-Trk after the human oncogene product, TRK, which is a cytosolic nonmuscle tropomyosin-receptor kinase fusion protein. Subsequently, the Trk-related RPTKs, TrkB and TrkC, were identified and proto-Trk was renamed TrkA. There is significant similarity in the external domains of this family, and the discovery that TrkA binds NGF quickly led to the demonstration that TrkB and TrkC bind NGF-related neurotrophins (see Barbacid 1993; Chao 1992; Glass & Yancopoulos 1993 and references therein). All three *Trk* genes are expressed in neurons and generate a variety of alternatively spliced products. *trkB* and *trkC* encode truncated receptor molecules retaining the transmembrane domains but lacking the entire cytoplasmic domain, whose function is unknown. *trkC* also encodes a family of full-length proteins with one or two inserts in the catalytic domain, which appear to have different signaling properties (Lamballe et al 1993; Tsoulfas et al 1993; Valenzuela et al 1993).

The ligands for the Trk subfamily RPTKs are neurotrophins which, as their name implies, are cytokines that play a role in directional neurite extension and also promote the survival of neurons in culture. NGF stimulates the PTK activity of TrkA, which is now referred to as the high affinity NGF receptor. Subsequently, it has been shown that overlapping sets of these factors bind the three receptors (for review see Barbacid 1993; Chao 1992; Glass & Yancopoulos 1993). Although exactly which neurotrophins act as physiological regulators of the different Trk subfamily RPTKs is still being determined, it is currently believed that TrkA acts as a receptor for NGF, TrkB as a receptor for BDNF and NT4/5, and TrkC as a receptor for NT3. In neuronally derived cells, activation of Trk subfamily RPTKs by their cognate ligands elicits a differentiation response that includes neurite outgrowth. When expressed in fibroblasts, however, Trk subfamily RPTKs elicit a mitogenic response. NGF functions as a noncovalently associated dimer, and under some circumstances the binding of NGF and other neurotrophins to Trk RPTKs may be facilitated by the low affinity NGF receptor, p75.

The Hepatocyte Growth Factor Receptor Subfamily

The HGF receptor has been shown to be identical to the *MET* gene product (Bottaro et al 1991; Naldini et al 1991). This receptor was discovered as a transforming gene in the NIH3T3 cell transfection assay. MET is synthesized as a precursor that is proteolytically cleaved into two subunits that are held together by a disulfide bond in a fashion analogous to the insulin receptor (Giordano et al 1988) (see Figure 1). The receptor mediates the effects of a cytokine known as HGF or scatter factor. HGF was originally characterized as a growth factor for hepatocytes and other epithelially-derived cells, whereas scatter factor was identified as a motogenic factor able to stimulate the movement or scattering of epithelial cells (see Gherardi & Stoker 1991). The same epithelial cells can respond mitogenically or motogenically depending on the conditions, but it is not clear what governs these two different responses. Presumably, at least some of the substrates phosphorylated by the activated HGFR RPTK are different in the two situations, but what these substrates are remains to be determined. Little is known about Sea, which was discovered as a transduced retroviral oncogene (Smith et al 1989), but the similarity between Met and c-Sea (Huff et al 1993), and another recently discovered subfamily member RON (Ronsin et al 1993), indicates that they probably bind related ligands.

The structure of HGF is unique among the RPTK ligands (Nakamura et al 1989). It has four kringle domains and a domain that has homology with the serine protease catalytic subunit. However, because residues known to be part of the catalytic triad are altered in HGF, it is thought that HGF does not have protease activity. HGF has a signal peptide, and the mature form is found as two disulfide-linked peptides that are derived by proteolysis. A related molecule, HGFL, has been identified, but it has not been shown to bind any receptor thus far (Han et al 1991).

The EPH Subfamily

The EPH subfamily currently consists of twelve members and is the largest RPTK subfamily. Several of the receptors in this subfamily are expressed most highly in adult brain, i.e. Elk (Letwin et al 1988); Eek (Chan & Watt 1991); Cek4 (Sajjadi et al 1991); Cek5 (Pasquale 1991); Sek (Gilardi-Hebenstreit et al 1992); and Ehk1/2 (Maisonpierre et al 1993). This implies a function other than in mitogenesis or differentiation. However, these receptors might be involved in such processes earlier in development, and they could play a role in differentiated function in the adult. Some receptors such as EPH (Hirai et al 1987) and ECK (Lindberg & Hunter 1990) have a broader tissue distribution and are not highly expressed in neural tissues. Overexpression of EPH in NIH3T3 cells appears to confer tumorigenic properties, but the cells are not morphologically transformed (Maru et al 1990). Studies from mouse, chicken, and fish indicate that this subfamily is important during development (Gilardi-Hebenstreit et al 1992; Henkemeyer et al 1994; Pasquale 1991; Ruiz & Robertson 1994; Xu et al 1994). A potential ligand, B61, has been found for the ECK receptor (Bartley et al 1994). This molecule is a secreted protein, but may function as a cell-bound ligand because it is attached to the cell surface via a GPI anchor (Holzman et al 1990). This may be the first of a family of molecules that bind the EPH subfamily of receptors.

The AXL Subfamily

AXL was isolated by the NIH3T3 cell transformation assay from two chronic myelogenous leukemia patients (O'Bryan et al 1991). UFO, identical to AXL, was also isolated as a transforming gene by the same assay from the leukemic cells of a patient with a chronic myeloproliferative disorder (Janssen et al 1991). Another member of this subfamily, Eyk, was identified as the cellular counterpart of a retroviral oncogene, v-*ryk* (Jia & Hanafusa 1994). Tyro3 (Lai & Lemke 1991), the third member of the subfamily, has also been called SKY (Ohashi et al 1994), Brt (Fujimoto & Yamamoto 1994), and TIF (Dai et al 1994). This gene is most highly expressed in the brain and gonads. Exogenous expression of Tyro3 transforms rodent fibroblasts (C Lai, personal communication). These data indicate that the RPTKs in this subfamily have an oncogenic propensity. Ligands for this subfamily have not been identified.

Other Receptor Protein-Tyrosine Kinases

The remaining RPTKs are subfamilies that consist of one or two members. Ligands have not been identified for any of these RPTKs. TIE was identified in a PCR-based screen for novel PTKs (Partanen et al 1992), and it is expressed most highly in endothelial cells and some hematopoetic cell lines of early lineage. A close relative, which is also expressed in cells of the endothelial lineage, has recently been isolated in several laboratories. It was first reported as Tek (Dumont et al 1992), but has also been called TIE2. Several putative receptors in another subfamily have recently been reported that apparently correspond to two distinct genes: DDR (Johnson et al 1993) or Nep (Zerlin et al 1993) or TrkE (Di Marco et al 1993) or CAK (Perez et al 1994), and TKT (Karn et al 1993) or Tyro10 (Lai & Lemke 1994). This subfamily is characterized by a discoidin I-like domain, which has homology to a phospholipid-binding domain in several known proteins such as the coagulation factors V and VIII. The two RPTKs, ROR1, and ROR2, were discovered in a screen designed to find receptors in the Trk subfamily (Masiakowski & Carroll 1992). Their closest relatives are the Trks, but because of a unique extracellular region, it seems that the RORs represent a new subfamily. The RORs are expressed much more highly in the embryo than the adult. RET was originally identified by transfection of NIH3T3 cells as an oncogene that had been activated by a transfection-induced rearrangement (Takahashi & Cooper 1987). RET has been implicated in several forms of human cancer (Grieco et al 1990; Hofstra et al 1994; Mulligan et al 1993). Ryk, Nyk-r, Vik, Nbtk-1 and Mrk were all found in PCR-based screens and appear to be products of the same gene (Hovens et al 1992; Kelman et al 1993; Maminta et al 1992; Paul et al 1992; Yee et al 1993). This RPTK has a rather short extracellular domain (~ 200 amino acids) and some unusual substitutions at highly conserved residues in the catalytic domain. The cDNA for Klg was isolated from a chicken embryonic library with a v-*sea* probe (Chou & Hayman 1991). Klg stands for kinase-like gene, so-named because it was noted that the highly conserved DFG motif was absent, which led to the speculation that Klg lacks intrinsic PTK activity.

RPTKs from Other Organisms

In addition to the RPTKs discussed above, several other RPTKs have been identified in lower vertebrate and invertebrate species. In most cases, it is hard to be certain whether these are the true homologues of RPTKs of higher organisms, although some clearly fall into particular RPTK subfamilies. RPTK genes have been found in fish (Wittbrodt et al 1989; Xu et al 1994), the electric ray (Jennings et al 1993), and amphibians (Friesel & Dawid 1991). Many have been discovered in *Drosophila*: DER (Livneh et al 1985) (EGF receptor-related); DILR (Nishida et al 1986) (insulin receptor-related); *sevenless* (Hafen et al 1987); *torso* (Sprenger et al 1989); D*trk* (Pulido et al 1992); *DFGFR1* (Glazer & Shilo 1991) (also called *DFR2*; Shishido et al 1993); DFR1 (Shishido et al 1993) (FGF receptor-related); and D*ror* (Wilson et al 1993). In *C. elegans, let-23* (Aroian et al 1990) and two related RPTKs, *kin-15* and *kin-16* (Morgan & Greenwald 1993), have been reported. Putative RPTK genes have been identified in the coelenterate Hydra, but no RPTK genes have been found in simpler eukaryotes such as the slime molds, or in budding or fission yeasts. The functional insights that have been gained from the genetic analyses of these RPTKs are discussed below.

RPTK Ligands

The ligands that bind individual RPTKs have been discussed above and are listed in Table 2. Some RPTK ligands bind more than one RPTK, and vice versa, whereas other ligands bind only one RPTK, and the cognate receptor binds only one known ligand. Because of the complex patterns of ligand:RPTK interaction, we have not indicated which ligand binds which receptor in Table 2. In general, however, cross-talk occurs only within the matching subfamilies listed in Tables 1 and 2. The existence of families of related ligands that interact with subfamilies of RPTKs raises the important question of which ligands regulate which receptors in vivo. Although many ligands within a family may have the ability to bind and activate a specific RPTK in vitro, the temporal and spatial aspects of expression and processing of the ligands will dictate which pairs are relevant in vivo. The use of homologous recombination to inactivate genes in the mouse, including RPTK and ligand genes, coupled with the use of transgenes expressing dominant-negative forms of some ligands and their receptors, is beginning to provide answers to some of these questions.

The availability of RPTK ligands synthesized as membrane-anchored proteins can be regulated by processing to generate soluble forms. In some cases, the ligand has to be in an anchored form to provide an appropriately localized signal (e.g. SCF has to be membrane associated to allow hematopoietic stem cell development) (Brannan et al 1991). The availability of soluble ligands can also be regulated by binding of cytokines to extracellular matrix molecules, serum-binding proteins, co-ligands, and co-receptors. For instance, the FGFs bind very tightly to heparin sulfate proteoglycans (HSPG). This is not only a means of concentrating FGF locally in the extracellular matrix, but HSPG also proves to be necessary as a co-ligand for high affinity binding of FGF to the FGF receptor (Kan et al 1993). In other cases, there is a second surface protein that can bind the ligand, which may be involved in presentation of ligand to an RPTK. The low affinity NGF receptor is thought to have this function. In addition, several RPTK genes express alternatively spliced forms that lack the catalytic domain, which might modulate ligand availability positively or negatively. Finally, there is a report that estrogens can bind to and activate ERBB2 (Matsuda et al 1993), which raises the possibility that there may be other nonpeptide ligands or modulators of RPTKs.

Genetics

The gain of function mutations that gave rise to RPTK-derived oncogenes in vertebrates have been useful in pinpointing RPTKs that have strong mitogenic activity and in defining regions involved in receptor regulation, dimerization, and activation (Rodrigues & Park 1994). Likewise, the phenotypes of loss of function mutations in RPTK genes and their ligands in the mouse, *Drosophila,* and *C. elegans* have proved enlightening in terms of RPTK function and site of action. In the mouse, the *w* (white spotting) locus was originally identified by the appearance of a white spot on the belly of heterozygotes. Homozygotes have a more severe phenotype that includes lack of pigmentation, anemia, sterility, and death. The implications are that the *w* locus gene product is required for the normal development of melanocytes and of hematopoetic and germ cells. The *w* locus has been found to be the gene that encodes c-Kit, which is now called the SCF receptor (Geissler et al 1988). Several *w* alleles have been sequenced, and mutations that diminish PTK activity cause the severest phenotype (reviewed in Pawson & Bernstein 1990). Mutants in the *steel* locus have a similar phenotype to SCF receptor mutations, and it has been demonstrated that this locus encodes the ligand for SCF receptor (reviewed in Witte 1990). The originally unexpected dominant nature of the *w* mutant SCF receptors is explained by the ability of the kinase-inactive SCF receptor to form nonfunctional heterodimers with the normal SCF receptor in the same cell (discussed below), thus attenuating the SCF signal.

The mouse *patch* mutation, which also gives a coat color phenotype, deletes the PDGF α receptor gene, and the coat color defect could result from a loss of PDGF α receptor function, although the *patch* deletion is large and may involve other genes (Stephenson et al 1991). The mouse *op* mutant, which develops osteopetrosis in the homozygous state, shows a severe deficiency in macrophages and osteoclasts. *op* encodes CSF-1, and it was expected that the development of macrophages would be affected by the absence of CSF-1 because CSF-1 is required for growth and differentiation of myeloid precursors (Wiktor-Jedrezejczak et al 1990; Yoshida et al 1990). The failure to develop osteoclasts is more surprising, but osteoclast precursors are derived from the hematopoietic lineage and must require CSF-1.

Targeted gene inactivation in the mouse has provided a powerful tool for analyzing RPTK function in the mouse (Imamoto et al 1994). For instance, ablation of genes in the NGF receptor subfamily has the anticipated effects on neural function and development. $TrkB^-$ mice die shortly after birth, apparently because they are unable to feed (Klein et al 1993). This is probably due to the fact that several ganglia that normally express TrkB, including the facial, trigeminal, and dorsal root ganglia, are lacking neurons, which implies that TrkB function is required for neuronal development. $TrkA^-$ (Smeyne et al 1994) and $TrkC^-$ (Klein et al 1994) mice survive for a few weeks, but they have severe defects in the sympathetic nervous system and in muscle afferents, respectively. NGF^-, $BDNF^-$, and NT-3^- mice have also been obtained, and the phenotypes of these animals largely resemble those observed for the corresponding receptor knock outs, although the $BDNF^-$ mice do not show such a strong phenotype, probably because NT-4/5 also acts as ligand for TrkB (Crowley et al 1994; Ernfors et al 1994; Jones et al 1994). Ret^- mice survive until birth, then die shortly afterward (Schuchardt et al 1994). These mice lack kidneys and enteric neurons, which suggests that Ret is required for organogenesis and neurogenesis.

A knock out of the IGF-1 receptor results in a severe growth deficiency and IGF-$1R^-$ mice invariably die at birth (Liu et al 1993a). The corresponding IGF-1^- mice also exhibit a severe growth defect, but some animals survive and may be rescued by the presence of IGF-2, which can also bind to the IGF-1 receptor (Liu et al 1993a). *waved1* and *waved2* are mouse mutants that exhibit a similar wavy hair phenotype. *waved1* proves to be a mutation in the TGFα gene, and a TGFα knock out gives exactly the same wavy hair phenotype (Imamoto et al 1994). With the clue provided by the nature of the *waved1* mutation, it was shown that *waved2* is a mutation in the EGF receptor gene. The *waved2* EGF receptor has a Val 743 to Gly substitution in the catalytic domain. However, this does not result in a complete loss of function, and the activity of the EGF receptor seems to be severely affected only in the epidermis (Luetteke et al 1994). A complete EGF receptor knock out in the mouse causes embryonic lethality.

In *C. elegans* the *let-23* gene encodes a RPTK that resembles the EGF receptor subfamily (Aroian et al 1990). Let-23 is required for vulval induction. The putative ligand for Let-23 is an EGF-related protein encoded by the *lin-3* gene. The nature of the mutations in different *let-23* alleles has allowed the definition of three distinct C-terminal regions lying beyond the catalytic domain needed for Let-23 signaling in different cell types (Aroian et al 1994). Several *Drosophila* RPTK mutants have been found to affect development: (*a*) *torso* results in the lack of proper development of terminal structures in the embryo (Sprenger et al 1989); (*b*) *sevenless* mutations have only one observed effect on the developing fly, namely that photoreceptor cell R7 does not develop (Hafen et al 1987); (*c*) *faint little ball* (an allele of the DER, an EGF receptor homologue) exhibits severe developmental effects when homozygous, and the embryo dies very early with many cell types and structures affected (Schejter & Shilo 1989). Other mutant alleles of the DER locus are *torpedo* and *ellipse* (Baker & Rubin 1989; Price et al 1989). *ellipse* is of interest because it is a gain of function mutation that constitutively activates the DER. (*d*) *breathless* (an allele of *DFGFR1,* a FGFR-related gene) exhibits a deficiency in tracheal development (Glazer & Shilo 1991).

The expression of dominant-negative mutant RPTKs during embryogenesis has also proved to be a fruitful approach to understanding RPTK function in vertebrates. For instance, expression of a truncated dominant-negative form of FGFR1 under a keratin 10 promoter, which is utilized predominantly in the differentiating suprabasal cells of the epidermis, in transgenic mice disrupts the organization of keratinocytes and results in epidermal thickening (Werner et al 1993), which implies a role for FGF and FGFR1 in keratinocyte differentiation. The expression of a dominant-negative form of the *Xenopus* FGFR in *Xenopus* embryos blocks mesoderm induction (Amaya et al 1991), consistent with the proposed role of FGF in mesoderm induction. Finally, the expression of a dominant-negative mutant of Rtk1, a member of the EPH RPTK subfamily, in zebrafish blocks development at the stage when Rtk1 is normally expressed (Xu et al 1994). In addition, the growth of a glioblastoma in nude mice is inhibited by expression of a dominant-negative form of Flk-1, apparently because it suppresses the VEGF-induced endothelial cell proliferation needed for tumor angiogenesis (Millauer et al 1994). However, one limitation to the use of dominant-negative mutant RPTKs is that they may be able to dimerize with other subfamily members, and thus their effects are not as specific as those obtained by RPTK gene inactivation.

The conclusions from these genetic studies are that some RPTKs have very specific functions, while others act more globally. Although the interpretation of genetic data may be obscured because RPTKs exist as large families and probably have partially overlapping functions, and the phenotypes seen in null mutations may only reveal a small subset of the functions of the particular gene product, it is clear that genetic approaches offer major benefits in studying

RPTK function. New data from genetic methodologies should answer many questions about the function of RPTKs that cannot be addressed in other ways.

SIGNALING

RPTK Activation

Signal transduction is initiated when the ligand binds the receptor. The generalized series of events is as follows:

1. After ligand binding, the receptor dimerizes. This is presumed to occur by a ligand-induced conformational change in the external domain that results in dimerization of receptor. Emerging evidence suggests that the binding of a single ligand molecule can result in formation of an RPTK dimer, as is the case for the growth hormone receptor (Cunningham et al 1991). In the case of ligands that are dimers, such as PDGF, dimerization may in part be driven by each subunit of the ligand binding to a separate receptor molecule, and the binding of dimeric ligands generates a truly symmetrical receptor dimer. However, in the case of monomeric ligands such as EGF, there must be two distinct binding sites on the ligand and the interaction sites on the two receptor molecules must also be at least partly different.
2. Receptor dimerization leads to intermolecular autophosphorylation. The ligand-induced dimerization of the extracellular domains necessarily results in juxtaposition of the cytoplasmic domains. It is thought that contact between the two cytoplasmic domains induces a conformational change that stimulates catalytic activity, which leads to mutual transphosphorylation between the two receptor molecules in a dimer. Autophosphorylation within the dimer occurs at a distinct set of sites, most of which lie outside the catalytic domain.
3. Transphosphorylation results in activation of the dimeric RPTK for phosphorylation of cytoplasmic substrates. Based largely on the use of dominant-negative kinase-inactive mutant RPTKs, it has been concluded that transphosphorylation is essential for substrate phosphorylation. When dominant-negative kinase-inactive mutant RPTKs are coexpressed with a wild-type RPTK, signaling by the wild-type RPTK is suppressed in proportion to the relative levels of the wild-type and mutant RPTKs. The suppression is thought to occur because the kinase-inactive RPTK molecules form dimers with the wild-type molecules and become phosphorylated, but cannot in turn phosphorylate their wild-type partner. This result also implies that a secondary conformational change occurs upon transphosphorylation, which allows the catalytic domain to phosphorylate substrate molecules.
4. The autophosphorylated dimer recruits substrates that have an increased affinity for the receptor due to its autophosphorylation (see below). The

use of kinase-inactive RPTK mutants demonstrates that RPTK signaling is absolutely dependent on the phosphorylation of specific cytoplasmic substrates. This model has been discussed in recent reviews (Schlessinger & Ullrich 1992; Ullrich & Schlessinger 1990; Williams 1989; Yarden & Ullrich 1988).

A large amount of evidence supports the dimerization model of RPTK activation. For most RPTKs, it is possible to detect dimer formation following ligand binding by chemical cross-linking experiments. In the PDGF receptor system, the fact that PDGF heterodimers bind two different PDGF receptors has made it possible to observe the ligand-driven dimerization of receptors into heterodimers (Hammacher et al 1989). In this system, the PDGF heterodimer, PDGF-AB, and the PDGF homodimers, PDGF-AA and PDGF-BB, can be assayed for activities on PDGF α and β receptors independently, and on mixed populations. Because the A chain of PDGF binds only the α receptor, whereas the B chain binds both types of receptors, it was shown that dimerization occurs during receptor activation and that receptor homodimers and heterodimers are utilized by PDGF.

The mechanism whereby monomeric ligands induce receptor dimerization is intuitively less obvious than for dimeric ligands, but evidence that monomeric ligands like EGF can cause receptor dimerization is strong. In the growth hormone receptor system, where a single growth hormone molecule is present in a receptor dimer, the three-dimensional structure of the receptor/growth hormone dimer shows that growth hormone has two distinct receptor binding sites, which can bind receptor molecules independently (Cunningham et al 1991). Although there is no experimental evidence that any monomeric RPTK ligand has two binding sites, this seems to be a reasonable possibility. If this is true, then it should be possible to create dominant-negative mutant ligands in which one of the two receptor-binding sites has been mutated. Although there are three-dimensional structures of several RPTK ligands, to date there is no three-dimensional structure of an RPTK ligand-binding domain nor of an RPTK/ligand complex, which will be needed to elucidate the precise mechanism of ligand-mediated receptor dimerization.

In support of the dimerization model, the constitutively activated Neu oncoprotein, which has a single change in its transmembrane domain converting an uncharged valine to a charged glutamate, formed multimers in the absence of ligand (Weiner et al 1989). This could result in constitutive PTK activity in the absence of the ligand. Even more convincing are recent reports that PDGF, EGF, and SCF receptor mutants, which lack catalytic activity, act in a dominant-negative fashion to attenuate signal transduction by wild-type receptors (Kashles et al 1991; Nocka et al 1990; Reith et al 1993; Ueno et al 1991). The interpretation of this phenotype is that inactive heterodimers form upon ligand binding in which the wild-type receptor cannot be transphosphorylated by the inactive subunit, and therefore is not activated for substrate phosphorylation. There are also several examples of RPTK-derived oncoproteins, where constitutive PTK activation is achieved by the fusion of the RPTK cytoplasmic domain to a fragment of another protein that can dimerize. For instance, the Tpr-Met chimera is activated by dimerization occurring through a leucine zipper dimerization motif present in the Tpr sequence (Rodrigues & Park 1993).

In general, signal transduction by RPTKs is dependent entirely on PTK activity. Mutations of many varieties, from truncations to single point mutations that block ATP binding and abolish PTK activity, also eliminate most known cellular responses. However, there are scattered reports where a kinase-inactive mutant RPTK has been found to elicit a specific cellular response. For example, a kinase-inactive EGF receptor has been shown to activate MAP kinase following EGF treatment even though it is incapable of stimulating a mitogenic response (Campos-Gonzalez & Glenney 1992; Hack et al 1993). The mechanism for this is unclear, but it may occur by EGF-dependent activation of an EGF receptor-associated PTK (Selva et al 1993). In addition, there are some reports that translocation of a RPTK ligand into the nucleus of a treated cell via a surface RPTK can elicit specific cellular responses. For example, nuclear responses have been demonstrated for SDGF and aFGF (Kimura 1993; Wiedlocha et al 1994).

For many PTKs, phosphorylation of a conserved Tyr residue(s), present within the catalytic domain, appears to be important for activation of kinase activity. Absence of this residue (Tyr 416) reduces the protein kinase and transforming activities of an activated c-Src mutant, in which a negative regulatory phosphorylation site (Tyr 527) has been mutated (Kmiecik & Shalloway 1987; Piwnica-Worms et al 1987). Mutation of a residue homologous to Tyr 416 in Src, within the PDGF, CSF-1, insulin, and HGF receptors, reduces their kinase activity and their ability to initiate signal transduction (Ellis et al 1986; Fantl et al 1989; Kazlauskas et al 1991; Longati et al 1994; van der Geer & Hunter 1991). The EGF receptor is one of the few RPTKs for which there is no evidence that this residue is phosphorylated during activation. In most RPTKs, additional tyrosines are phosphorylated in response to ligand binding. These phosphorylated tyrosines act as binding sites on the activated receptor for proteins that contain SH2 domains.

Interaction of Activated RPTKs with SH2 Domain-Containing Proteins

SH2 domains, which are about 100 residues in length, were first identified as regions of homology between Src and Fps that lie outside the catalytic domain and that are not required for kinase activity (Pawson 1988; Sadowski et al

1986). Work with v-Crk, the transforming protein of CT10, a chicken sarcoma virus, led to the finding that SH2 domains bind specifically to P.Tyr residues (Mayer et al 1991). Subsequently, it was shown that each SH2 domain binds to P.Tyr in a sequence-specific fashion. Many SH2 domains have been found to bind P.Tyr-containing proteins in vivo and in vitro (Anderson et al 1990; Koch et al 1991; Moran et al 1990). Three-dimensional structures of several SH2 domains have now been obtained by X-ray crystallography and NMR analysis (Booker et al 1992; Overduin et al 1992; Waksman et al 1992). An important conclusion is that the N- and the C-terminal ends of the SH2 domain are close together in a globular structure. This allows the SH2 domain to protrude from rest of the protein and function independently to interact with available binding sites. The bound P.Tyr is contained in a pocket within the SH2 domain, which is lined by conserved amino acids that coordinate the tyrosine ring and the phosphate group. In the known SH2 domain structures, binding specificity is achieved by the interaction of amino acids C-terminal to the P.Tyr with the surface of the SH2 domain.

Based on structural and biochemical studies, it appears that only 3–5 residues C-terminal to the P.Tyr interact with the surface of the SH2 domain. SH2-binding specificities have been determined by comparing the binding sites of specific SH2 domains in different proteins and by incubation of SH2 domains with degenerate P.Tyr-containing peptide libraries, followed by sequencing of the bound peptides (Songyang et al 1993, 1994). Correlations between amino acids preferred for binding by individual SH2 domains at positions +1, +2, and +3 and the locations of contact residues in the SH2 domain structure for the amino acids in these positions have allowed predictions about the specificities of novel SH2 domains (Songyang et al 1993), and a binding sequence code is emerging. More recently, it has become apparent that residues on the N-terminal side of the P.Tyr may also play a role in the sequence-specific binding of some SH2 domains (Batzer et al 1994; Nishimura et al 1993).

Many substrates or targets for RPTKs contain SH2 domains (Table 3). The ability of activated RPTKs to bind SH2 domain-containing proteins provides a mechanism for recruiting substrates and targets to activated RPTK dimers. The binding of SH2 domain-containing proteins to autophosphorylated RPTKs can affect their activity in at least three ways, which are by no means mutually exclusive. First, the binding to the activated RPTK can facilitate efficient Tyr phosphorylation of the SH2 domain-containing protein, which in turn results in activation (or inhibition). The measured affinities for SH2 domain binding to a target P.Tyr-containing peptide are in the nM range (Felder et al 1993; Panayotou et al 1993), whereas the the K_ms for non-SH2 domain substrates are generally in the μM range. Thus the affinity of the RPTK for a substrate containing a SH2 domain can be increased by nearly three orders of magnitude. PLCγ is an example of a substrate where phosphorylation increases enzymatic

Table 3 Signaling PTK substrates and targets

Enzymes	
PLCγ1 and PLCγ2	SH2 (2), SH3, split PHD
RasGAP	SH2 (2), SH3, PHD
Src family kinases	SH2, SH3
SH-PTP1/HC-PTP/PTP1C	SH2 (2)
SH-PTP2/SYP/PTP1D	SH2[a] (2)
PI3 kinase p85/p110	SH2[a] (2), SH3
VAV	SH2, SH3 (2), PHD
Adaptors	
IRS-1 (and 4PS)	
NCK	SH2, SH3 (3)
CRK	SH2, SH3 (2)
GRB2 (not phosphorylated)	SH2, SH3 (2)
SHC	SH2
Structural proteins	
Annexins I (p35) and II (p36)	
Ezrin (p81) (and ezrin-related proteins)	
Clathrin H chain	
Vinculin (focal adhesion)	
Talin (focal adhesion)	
Tensin (focal adhesion)	SH2
Paxillin	
Zixin	
p80/85 (cortactin)	SH3
AFAP-110 (F-actin-binding protein)	
p120 (armadillo/β catenin-related)	
β catenin	
Connexin43	
Fibronectin receptor β subunit	
Cadherins	
Others	
Stat 91 (transcription factor)	SH2, SH3
Stat 113 (transcription factor)	SH2, SH3
Stat3 (transcription factor)	SH2, SH3
p62 (RNA-binding protein)	
p190 (RhoGAP)	
GRB7 (RasGAP-related)	SH2, PHD
GRB10	SH2, PHD
Eps8	SH3
Eps15 (Ca^{2+} binding?)	
Tap17 (not phosphorylated)	

[a] SH2 = enzyme activity increases upon binding target P,Tyr sequence; PHD = PH domain. PTKs in the Fps, Abl, Csk, BTK, and ZAP70 subfamilies contain SH2 domains, and may be phosphorylated by other PTKs. The focal adhesion kinase, FAK, may also be a substrate for other PTKs. 3BP2, α2 chimerin and Shb are other SH2 proteins that may be substrates. A number in parentheses indicates the number of domains per molecule.

activity, and phosphorylation by RPTKs in vitro is greatly decreased when the PLCγ SH2 domains are inactivated. Second, binding of the SH2 domain to its target P.Tyr residue may result in allosteric activation of the protein that binds. Allosteric activation has been reported in the case of PI3K and SH-PTP2, where binding to an appropriate synthetic P.Tyr-containing peptide has been shown to increase enzymatic activity three- to tenfold. Third, the binding to phosphorylated RPTKs may be a mechanism for localization of SH2-containing proteins and proteins bound to SH2 domain proteins in close proximity to their substrates on the inner face of the plasma membrane. This appears to be the case for GRB2 which, through binding to an activated RPTK, brings its constitutively associated partner SOS, a Ras guanine nucleotide release factor (GNRF), to the membrane, thus allowing it to act on Ras.

Another protein-protein interaction domain found in many signaling proteins, including several SH2-containing substrates for RPTKs, is the Src-homology 3 (SH3) domain. The SH3 domain, which is about 60 residues long, was discovered as a region of homology between v-Crk, PLCγ, and Src (Mayer et al 1988; Pawson 1988; Stahl et al 1988). Three-dimensional structures have been solved for several SH3 domains (Booker et al 1993; Kohda et al 1993; Koyama et al 1993; Musacchio et al 1992; Yu et al 1992). SH3 domains also have a globular structure, and like the SH2 domain, their N- and C-terminal ends of the polypeptide come together so the folded domain can protrude and function independently. SH3 domains are also involved in protein-protein interactions, and they bind short, ten amino acid Pro-rich motifs in a sequence-specific fashion (Bar Sagi et al 1993; Cicchetti et al 1992; Liu et al 1993c; Ren et al 1993, 1994). The solution structure of a SH3 domain bound to a preferred Pro-rich peptide shows that the prolines form an extended helix that allows interactions of interspersed Arg and Leu with residues in the binding groove of the SH3 domain, thus providing ligand specificity (Yu et al 1994).

A third type of domain found in many signaling proteins that is probably involved in protein-protein interaction is the pleckstrin-homology (PH) domain, which is about 100 residues long (Musacchio et al 1993) (Table 3). The best evidence that the PH domain is involved in protein-protein interaction is that the G protein βγ subunit binding site on the βARK protein kinase is within the C-terminal PH domain (Koch et al 1993). To date, however, the role of PH domains in signaling by RPTK substrates has not been established.

RPTK Substrates and Targets

Signaling by an activated RPTK is mediated by the phosphorylation or binding of cytoplasmic proteins, which in turn propagate the signal in various ways (for recent reviews see Carpenter 1992; Glenney 1992). There are three main classes of RPTK targets (Table 3). First, there are enzymes whose activity can be altered directly by phosphorylation or who can gain access to their substrates by translocation to the plasma membrane. Second, there are proteins that lack an obvious catalytic domain. These proteins commonly contain SH2 and SH3 domains and have become known as adaptors because they are believed to serve as intermediates between RPTKs and downstream signaling molecules. This role has been convincingly established for GRB2, and the binding of GRB2 to a phosphorylated RPTK leads to activation of the Ras signaling pathway (see below). In this category there are also docking proteins such as IRS-1 that can bind multiple SH2 proteins following phosphorylation and act as an RPTK surrogate to trigger signaling via SH2 protein-mediated pathways. Third, there are structural proteins whose phosphorylation could be responsible for the rapid membrane and cytoskeletal rearrangements that commonly occur following activation of a RPTK. Because the binding of SH2 proteins to activated RPTKs can result in activation of an associated activity, not all the proteins that associate functionally with RPTKs need be substrates. A case in point is GRB2, which is not phosphorylated on Tyr in stimulated cells, yet clearly binds functionally to activated RPTKs. Such proteins can be called targets rather than substrates. Finally, it should be stressed that although many RPTK substrates contain SH2 domains, which facilitate their phosphorylation by RPTKs, not all substrates do. Moreover, SH2-mediated binding is not essential for phosphorylation of all SH2 proteins in vivo. Consequently, one should realize that like other protein kinases, activated RPTKs can phosphorylate physiological substrates without stable interaction.

RPTK substrates have been identified in a number of ways. Early studies focused on identifying proteins whose P.Tyr content increased following RPTK activation. This approach led to the identification of several substrates in the third category, such as ezrin and annexin I. A more recent variation on this approach has been to isolate P.Tyr-containing proteins after RPTK activation and raise antibodies against them, which are then used to identify the corresponding substrates by affinity purification or expression library screening. Another fruitful approach has been to test whether newly characterized cytoplasmic signaling proteins are substrates for RPTKs, and this led to the identification of PLCγ1 as a substrate. The fact that substrates often bind stably to activated RPTKs via their SH2 domains has been used to identify several substrates in the enzyme and adaptor categories including PI3K. Novel SH2 domain proteins are continuing to be identified in a variety of ways. For instance, screening of expression libraries with an autophosphorylated EGF receptor fragment resulted in the identification of a series of proteins, known as growth factor receptor-bound proteins (GRB), which include known substrates as well as potentially novel targets such as GRB2. New SH2 domain proteins, including SHC and SH-PTP2, have also been found by PCR-based and low stringency screening strategies. Despite the large number of substrates

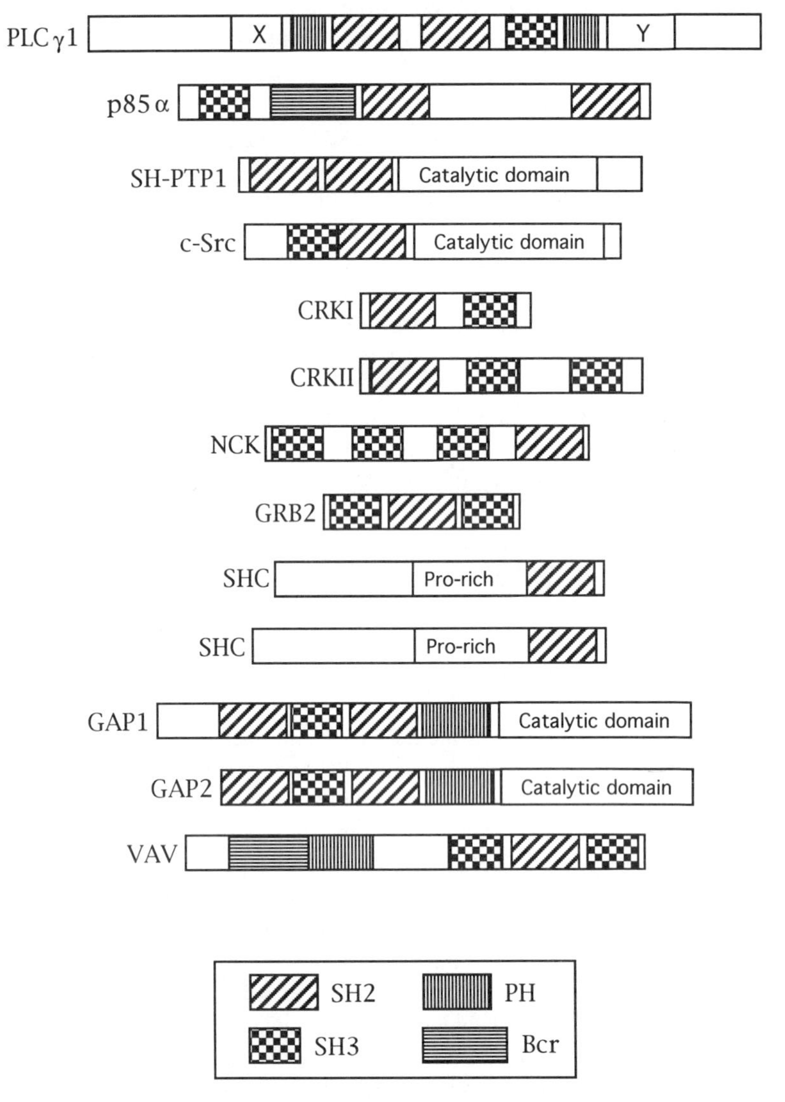

Figure 4 Structure of RPTK targets. The domain structures of several signaling molecules are shown schematically.

already identified, it seems likely that many more substrates remain to be discovered for RPTKs and nonreceptor PTKs activated by receptors lymphokine.

Genetic analysis has elucidated which substrates and targets are important for signaling by a particular RPTK. In vertebrates, this has involved the mutation of specific RPTK autophosphorylation sites and the expression of the resultant mutant RPTKs in cells lacking endogenous receptors of this type. Examples of this analysis are included as individual substrates and are discussed below. Genetic approaches are also providing fundamental insights into the signal pathways that are regulated by RPTKs. Cellular programs for differentiation that involve the induction of perhaps hundreds of genes are controlled by these receptors. The signaling pathways that traverse the cytoplasm and regulate transcription in the nucleus are still elusive, but the realization that RPTK signaling pathways are highly conserved has led to rapid progress in the elucidation of a transcytoplasmic signaling pathway that involves GRB2/Drk/Sem-5, SOS (a Ras guanine nucleotide release factor), Ras, Raf-1 (a protein-serine kinase), MAP kinase kinase (MEK) (a dual specificity kinase), and MAP kinase (a protein-serine kinase) (Marshall 1994). Activated MAP kinase can translocate into the nucleus, where it is able to phosphorylate and activate transcription factors such as Elk1 and c-Myc. An RPTK-activated pathway of this sort has been found in *C. elegans, Drosophila,* and vertebrates (see below) (for review, see Dickson & Hafen 1994). For instance, in *C. elegans* genetic analysis has linked the Let-23 RPTK, which is related to the EGF receptor, to a SH2-containing protein, Sem-5, which is the homologue of GRB2, to a Ras homologue called Let-60, and a Raf-1 homologue called Lin-45.

It is beyond the scope of this chapter to cover in detail all known RPTK substrates or the signaling pathways that are activated through these substrates (for other recent reviews of RPTK signaling and the role of SH2 and SH3 proteins see Fantl et al 1993; Kazlauskas 1994; Mayer & Baltimore 1993; Pawson & Gish 1992; Schlessinger 1994). Therefore, we review in depth only the best-understood substrates in the enzyme and adaptor categories. The domain topology of these substrates is shown in Figure 4, and some of the signaling pathways activated by these substrates and targets are shown in Figure 5. It should be borne in mind that many of the substrates for activated RPTKs are also substrates for cytoplasmic nonreceptor PTKs that are directly activated through association with lymphokine receptors lacking a PTK catalytic domain or indirectly by RPTKs.

Targets with Enzymatic Activity

PHOSPHOLIPASES Phospholipid turnover plays an important role in signal transduction. At least three classes of phospholipases can be regulated by RPTKs: phospholipase C (PLC), phospholipase A_2 (PLA_2), and phospholipase D (PLD).

Phosphatidylinositol (PI)-specific PLC preferentially hydrolyzes phosphatidylinositol 4,5-diphosphate (PIP_2) to diacylglycerol (DAG) and inositol

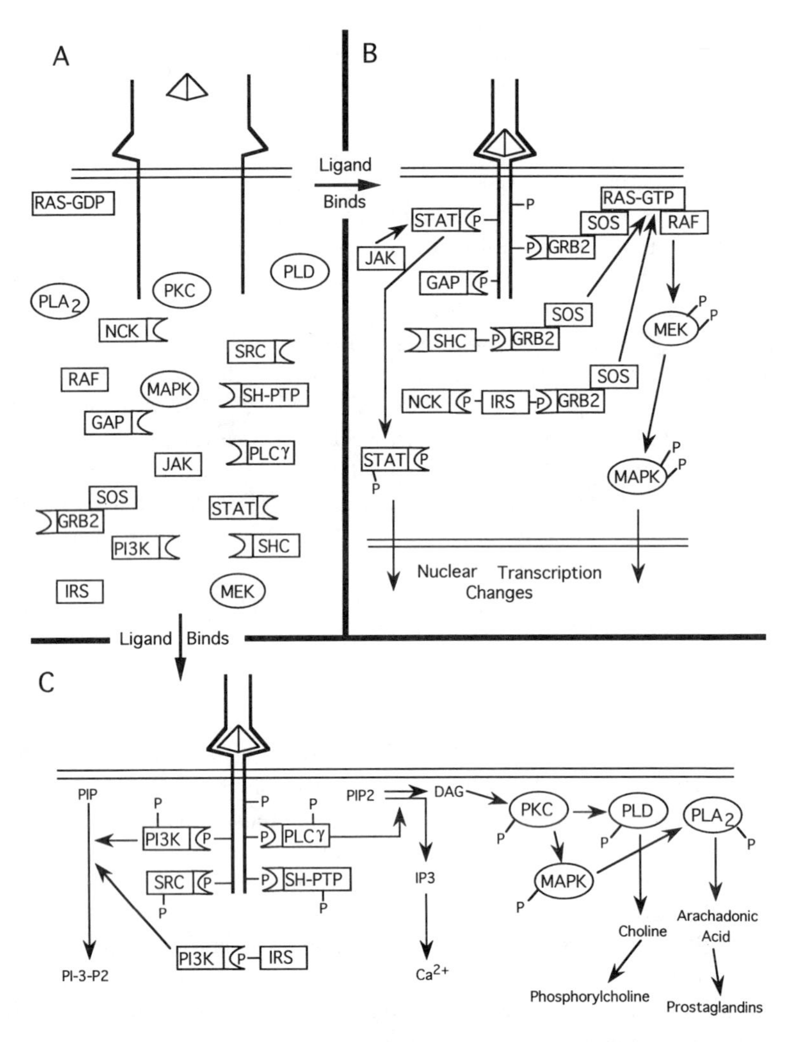

Figure 5 RPTK signaling pathways. Several pathways that become stimulated upon RPTK activation are shown. (*A*) In an unstimulated cell, the RPTK exists as a kinase-inactive monomer, is not phosphorylated on tyrosine, and has few associated proteins. (*B*) Ligand binding results in receptor dimerization and autophosphorylation, which leads to target association and/or phosphorylation. The RAS and STAT pathways lead to transcriptional changes. (*C*) RPTK activation stimulates many additional pathways that involve phospholipid messengers.

Table 4 Interaction of PLCγ with RPTKs

Receptors	Association in vivo	Association in vitro	RPTK binding site	Activation in vivo	PLCγ Tyr phosphorylation
EGF receptor	yes	yes	Y992, Y1068	yes	yes
PDGF β receptor	yes	yes	Y1021	yes	yes
FGF receptor	yes	?	Y766	yes	yes
SCF receptor	yes	?	?	yes	yes
CSF-1 receptor	no	?	—	no	no
NGF receptor	yes	?	Y785	yes	yes
TrkB	yes	?	Y785?	yes	yes
HGF receptor	?	yes	Y1349, Y1356	yes	yes
Flk2/Flt3	yes	?	?	?	yes
ERBB3	yes	?	?	?	weak
RET	?	?	?	?	yes

1,4,5 trisphosphate (IP_3). DAG is the physiological activator of protein kinase C (PKC), which has been implicated in growth control and tumorigenesis because it is a receptor for tumor-promoting phorbol esters such as TPA. IP_3 mobilizes Ca^{2+} from intracellular stores, thereby affecting Ca^{2+}-regulated processes in the cell. Three families of PI-PLC (β, γ, δ) are known (Cockcroft & Thomas 1992; Rhee & Choi 1992). The PLCγ subfamily is regulated by RPTKs, whereas the PLCβ subfamily is regulated by G protein-coupled receptors. There are two known isoforms of PLCγ, PLCγ1 and PLCγ2. PLCγ1 is widely expressed, whereas PLCγ2 is restricted to thymus, spleen, and lung tissue. Both forms have a similar structure. The only homology with other PLC isozymes is contained in the X and Y boxes, which are thought to be essential for enzymatic activity. Two SH2 domains and one SH3 domain lie between the X and Y boxes (Emori et al 1989; Stahl et al 1988; Suh et al 1988) (Figure 4). The SH2 domains direct interaction with activated RPTKs. The SH3 domain localizes PLCγ to the cytoskeleton (Bar Sagi et al 1993). Many, but not all, activated RPTKs interact with PLCγ (Table 4).

EGF stimulates PI turnover in some cells, and the activated EGF receptor associates with and phosphorylates PLCγ1 in vitro and in vivo on the same Tyr residues (Margolis et al 1989; Meisenhelder et al 1989). Tyr 771, Tyr 783, and Tyr 1254 are the major sites of Tyr phosphorylation in response to EGF in vivo (Wahl et al 1990). Phosphorylation of Tyr 783 is essential for PLCγ1 activation, and phosphorylation of Tyr 1254 may contribute to activation (Kim et al 1991). Activated ERBB2 also associates with and phosphorylates PLCγ1 on Tyr (Fazioli et al 1991; Peles et al 1991; Segatto et al 1992). Tyrosine phosphorylation of PLCγ1 stimulates phospholipase activity in vitro only when tested in the presence of profilin or under special detergent conditions (Goldschmidt-Clermont et al 1991; Nishibe et al 1990). Profilin binds to PIP_2 thereby blocking enzyme-substrate interactions. Tyrosine-phosphorylated PLCγ may compete better with profilin for PIP_2 binding than the unphosphorylated enzyme. EGF stimulation results in translocation of a major fraction of PLCγ1 to the plasma membrane (Kim et al 1990), yet binding to autophosphorylated RPTKs accounts for translocation of only a small fraction of PLCγ, meaning that other PLCγ-docking proteins must exist in the plasma membrane that remain to be identified. The PLCγ1 SH2 domains bind the EGF receptor in vitro, and binding requires Tyr phosphorylation of the receptor (Anderson et al 1990; Margolis et al 1990). Tyr 992 and Tyr 1068 in the EGF receptor C-terminus have been identified as binding sites for PLCγ1 (Margolis et al 1990; Rotin et al 1992; Vega et al 1992). Binding to the EGF receptor or ERBB2 is not necessary for PLCγ1 Tyr phosphorylation in vivo, but is essential for PLCγ1 activation (Li et al 1991; Segatto et al 1992; Vega et al 1992).

PDGF receptor activation also results in PI turnover and association of PLCγ1 with the receptor (Hasegawa 1985; Morrison et al 1990). PLCγ1 is a substrate for the PDGF receptor in vivo and in vitro (Kim et al 1991; Meisenhelder et al 1989). Tyr 1021 near the C-terminus of the PDGF β receptor has been identified as the binding site for PLCγ1 (Kashishian & Cooper 1993; Larose et al 1993; Rönnstrand et al 1992; Valius et al 1993). Binding to the activated PDGF receptor appears to be needed for efficient PLCγ1 Tyr phosphorylation (Kashishian & Cooper 1993; Rönnstrand et al 1992; Valius et al 1993). Mutation of Tyr 1021 results in a modest effect on PDGF-dependent DNA synthesis (Rönnstrand et al 1992; Valius et al 1993). Restoration of the PLCγ1 binding site in a mutant PDGF receptor that lacks five autophosphorylation sites (Tyr 740, 751, 771, 1009, and 1021) partially restores Ras activation and DNA synthesis in response to PDGF (Valius & Kazlauskas 1993).

Activation of the SCF receptor cytoplasmic domain also results in PLCγ1 Tyr phosphorylation in vivo (Hallek et al 1992; Lev et al 1991; Rottapel et al 1991). Association with the activated SCF receptor was detected in normal mast cells but not in NIH3T3 cells (Lev et al 1991; Rottapel et al 1991). However, the CSF-1 receptor, which is in the same RPTK subfamily, does not bind, phosphorylate, or activate PLCγ1 (Downing et al 1989).

Activation of the FGF receptor results in its association with PLCγ1 and Tyr phosphorylation of PLCγ1 (Burgess et al 1990; Mohammadi et al 1991). Tyr 766 is the binding site for PLCγ1 (Mohammadi et al 1991). Mutation of this site reduces PLCγ1 binding to the receptor and reduces PLCγ1 Tyr phosphorylation and activation in vivo (Mohammadi et al 1992; Peters et al 1992). Phe 766 mutant FGF receptors are unable to stimulate PI turnover or increase the intracellular Ca^{2+} concentration in response to FGF, but mutation of Tyr 766 has no effect on the induction of DNA synthesis by the FGF receptor in L6 myoblasts (Mohammadi et al 1992; Peters et al 1992).

Activation of TrkA (the NGF receptor) leads to association with and phosphorylation of PLCγ1 (Vetter et al 1991). Tyr 785 in TrkA has been identified as the binding site for PLCγ1 (Obermeier et al 1993a). TrkB has also been found to phosphorylate and associate with PLCγ1 upon activation (Middlemas et al 1994). Activation of TrkC K1, but not of TrkC, K2, or K3, leads to Tyr phosphorylation of PLCγ1 (Lamballe et al 1993). Stimulation of primary embryonic rat brain cultures with NT3 or BDNF also results in phosphorylation and activation of PLCγ1 (Widmer et al 1993). Baculovirus-expressed HGF receptors bind and phosphorylate PLCγ1 in vitro. Binding depends on autophosphorylation of the HGF receptor (Bardelli et al 1992). Experiments with chimeric receptors indicate that Flt3 and RET, but not ERBB3, utilize PLCγ as a substrate (Dosil et al 1993; Fedi et al 1994; Santoro et al 1994).

Table 4 summarizes PLCγ1 interaction with RPTKs after autophosphorylation in response to ligand binding. In general, binding to the activated receptor and Tyr phosphorylation of PLCγ1 are essential for activation. Binding of PLCγ1 to activated receptors appears to facilitate PLCγ1 Tyr-phosphorylation in response to ligand stimulation, but may not be absolutely essential. The need for association may be different for individual receptors. The possibility exists that binding to activated receptors per se could contribute to the increase in PLCγ1 activity in the cell either by bringing the enzyme into close proximity with its substrates in the plasma membrane or by allosteric activation. In many cases, activation of PLCγ1 does not appear to be necessary for a RPTK to induce mitogenesis, but activation of PLCγ1 can elicit a mitogenic response by itself (Valius & Kazlauskas 1993).

In addition to PLC, two other classes of phospholipases can be activated in response to RPTK activation. Treatment of cells with EGF, PDGF, CSF-1, or TPA stimulates PLA_2 activity (Bonventre et al 1990; Goldberg et al 1990; Lin et al 1992; Nakamura et al 1992). PLA_2 releases fatty acids, preferentially arachidonic acid, from the 2 position of phospholipids and uses phosphatidylcholine, phosphatidylethanolamine, and phosphatidylinositol as substrates (Axelrod et al 1988). Arachidonic acid is a precursor for the synthesis of prostaglandins and leukotrienes, which are important intracellular and extracellular signaling molecules (Axelrod et al 1988). Activation of the 85-kd cytoplasmic PLA_2 by RPTK or PKC pathways appears to involve MAP kinase, which phosphorylates PLA_2 at a single Ser and results in its activation (Lin et al 1993).

PDGF, EGF, lysophosphatidic acid (LPA), and TPA are all able to stimulate PLD (Ben-Av & Liscovitch 1989; Bierman et al 1990; Cook & Wakelam 1992; Fisher et al 1991; Kiss 1992; Plevin et al 1991; van der Bend et al 1992). PLD hydrolyzes phosphatidylcholine (PtdChl), which produces phosphatidic acid and choline (Billah et al 1991). Phosphatidic acid can be converted into DAG by phosphatidic acid hydrolase, and DAG activates PKC. Choline can be phosphorylated to phosphorylcholine, and there is some evidence that phosphorylcholine can function as a second messenger (Cuadrado et al 1993). The mechanism by which RPTKs activate PLD is not known, but PLD activation is dependent on PKC in some but not all systems (Cook & Wakelam 1992; Kiss 1992; Plevin et al 1991).

PHOSPHATIDYLINOSITOL 3-KINASE Interest in phosphatidylinositol (PI) kinase was first sparked by the observation that PI-kinase activity associates with middle T antigen, the main transforming protein of polyomavirus. The association with PI-kinase activity is essential, but not sufficient, for transformation by polyoma middle T. Fibroblasts express two PI kinases that can be distinguished on the basis of their sensitivity to inhibitors and that can be separated by anion-exchange chromatography (Carpenter & Cantley 1990; Whitman et al 1987). Type I PI-kinase phosphorylates PI on the D3 position; it is this enzyme that is associated with polyoma middle T (Whitman et al 1987). Type I PI kinase is now generally known as PI 3-kinase (PI3K) (Whitman et al 1988). Type II or PI 4-kinase phosphorylates PI on the D4 position, and this is the first step in the generation of PIP_2 from PI.

PI3K associates not only with polyoma middle T, but also with a number of retroviral oncogene products. Mutational analysis of these oncogenes shows a strong correlation between transformation and the ability of the oncoprotein to bind PI3K and increase the level of 3′ phosphoinositides in the cell. It was quickly established that activation of certain RPTKs resulted in stimulation of PI3K activity, association of PI3K with the activated receptors, and the generation of 3′ phosphoinositides, with $PI3,4P_2$, and $PI3,4,5P_2$ (PIP_3) showing the greatest increases (Auger et al 1989; Kaplan et al 1987; Varticovski et al 1989). The presence of PI3K activity in immunoprecipitates of oncogenic PTKs and RPTKs was found to correlate with the presence of an 81–85-kd polypeptide (Courtneidge & Heber 1987; Kaplan et al 1987).

PI3K is a heterodimer containing an 85-kd regulatory subunit and a 110-kd catalytic subunit (Carpenter et al 1990). The 85-kd subunit proves to be a family of proteins including p85α and p85β (Escobedo et al 1991b; Otsu et al 1991; Skolnik et al 1991). Both isoforms are 724 amino acids long and contain an N-terminal SH3 domain, a central (N-SH2) domain, and a C-terminal SH2 (C-SH2) domain (Figure 4). The two SH2 domains in each isoform are only 37% identical. The region between the SH3 domains and the central SH2 domain shows some homology with the C-terminal region of Bcr. This region of Bcr has GTPase-activating activity for the Rac small G protein (Diekmann et al 1991). The region between the two SH2 domains contains the binding site for the p110 subunit (Klippel et al 1993). Two SH3 domain-binding sites have been identified in the p85 subunit (Kapeller et al 1994; Liu et al 1993c). The binding of the SH3 domain of the Src family PTKs Lyn or

Fyn to purified PI3K results in activation (Pleiman et al 1994). This suggests the possibility that PI3K may interact with the SH3 domains of other signaling molecules. p110 contains the catalytic domain (Hiles et al 1992), and recent evidence suggests that there is a family of p110-related proteins. The 1068 residue p110 catalytic subunit is related to the *S. cerevisiae* PI 3-kinase Vps34p, which is involved in vesicular protein sorting (Hiles et al 1992; Schu et al 1993). However, the yeast enzyme produces only PI 3-monophosphate, whereas the mammalian enzyme generates higher order 3′ phosphoinositides. In addition, Vps34p has no p85 subunit, but instead is activated as a result of phosphorylation by the Vps15p protein kinase.

p85 mediates binding of the holoenzyme to Tyr-phosphorylated RPTKs through its SH2 domains (Escobedo et al 1991b; Otsu et al 1991). The p85 subunit binds to the sequence P.Tyr-X-X-Met (Songyang et al 1993). The two SH2 domains have slightly different binding specificities. In vitro binding studies indicate that high affinity binding in solution depends entirely on the p85 C-terminal SH2 domain (Klippel et al 1992), but in vivo both SH2 domains may be required for stable binding (Cooper & Kashishian 1993). Many RPTKs have been shown to interact with PI3K (Table 5). RPTKs have been tested for their ability to associate with the active enzyme in vivo or in vitro. In vivo association is detected by assaying RPTK immunoprecipitates from control or ligand-treated cells for associated PI3K activity. In vitro association is detected by incubating autophosphorylated RPTKs with cell lysates containing PI3K, followed by detection of associated PI3K activity. PI3K activity has commonly been detected in anti-P.Tyr immunoprecipitates from ligand-stimulated cells (Table 5). However, it is important to realize that the presence of PI3K in anti-P.Tyr immunoprecipitates can be the result of Tyr phosphorylation of any

Table 5 Interaction of PI 3-kinase with RPTKs

Receptors	Association in vivo	Association in vitro	RPTK binding site	Activity in anti-P,Tyr IPs	p85 Tyr phosphorylation
EGF receptor	no	yes	?	yes	no
PDGF β receptor	yes	yes	Y740, Y751	yes	yes
FGF receptor	?	?	?	yes	?
SCF receptor	yes	yes	Y721	yes	no
CSF-1 receptor	yes	yes	Y721	yes	no
NGF receptor	yes	?	Y751	yes	yes
HGF receptor	yes	yes	Y1349, Y1356	yes	no
Insulin receptor	yes	?	?	yes	yes
Flk2/Flt3	yes	?	?	?	yes
ERBB3	yes	?	?	yes	yes
RET	?	?	?	?	no

protein directly or indirectly associated with PI3K. Recombinant p85 or fusion proteins containing individual p85 SH2 domains have also been tested for binding to wild-type and mutant RPTKs.

Stimulation of PDGF α and β receptors leads to association of the receptor with PI3K, variable Tyr phosphorylation of p85, and an increase in higher order 3′ phosphoinositides in the cell (Escobedo et al 1991a; Heidaran et al 1991; Kaplan et al 1987; Kavanaugh et al 1992; Kazlauskas & Cooper 1989, 1990; Yu et al 1991). Both p85 SH2 domains can bind to Tyr-phosphorylated PDGF receptors independently in vitro (Klippel et al 1992; McGlade et al 1992). The C-terminal SH2 domain may be responsible for the high affinity interaction in solution in vitro (Cooper & Kashishian 1993; Klippel et al 1992), and both SH2 domains are required for a stable interaction in vivo (Cooper & Kashishian 1993). Tyr 740 and Tyr 751 in the PDGF β receptor kinase insert have been identified as the binding sites for p85 (Fantl et al 1992; Heidaran et al 1991; Kashishian et al 1992; Kazlauskas & Cooper 1989, 1990; Kazlauskas et al 1992; Severinsson et al 1990; Yu et al 1991). Analyses of the functional consequences of mutating PI3K binding sites in the PDGF α and β receptors, and of the receptors ability to initiate DNA synthesis and mitogenesis, have yielded conflicting results (Fantl et al 1992; Kazlauskas et al 1992; Yu et al 1991). In some, but not all, cells the inability to associate with PI3K decreases the mitogenic response. In HepG2 cells, a PDGF β receptor containing only the Tyr 740/751 PI3K-binding autophosphorylation sites can elicit a partial mitogenic response (Valius & Kazlauskas 1993).

CSF-1 stimulation elevates 3′ phosphoinositides (Varticovski et al 1989), and PI3K associates with the CSF-1 receptor upon ligand binding (Choudhury et al 1991; Reedijk et al 1990; Shurtleff et al 1990; Varticovski et al 1989), although no Tyr phosphorylation of p85 has been reported (Reedijk et al 1990; Shurtleff et al 1990; Varticovski et al 1989). The binding site for PI3K is Tyr 721 in the kinase insert domain of the CSF-1 receptor (Choudhury et al 1991; Reedijk et al 1990, 1992; Shurtleff et al 1990). Mutation of Tyr 721 to Phe or deletion of the kinase insert domain reduces the ability of the CSF-1 receptor to transduce signals in Rat-2 fibroblasts (Reedijk et al 1990; van der Geer & Hunter 1993). The SCF receptor (c-Kit) binds PI3K upon stimulation (Lev et al 1991; Rottapel et al 1991). In vivo and in vitro binding studies indicate that PI3K binds to the SCF receptor kinase insert region (Lev et al 1992; Rottapel et al 1991). Based on sequence comparison, Tyr 721 is most likely the binding site for PI3K in the SCF receptor.

Following stimulation of PC12 cells or NIH3T3 cells expressing TrkA with NGF, PI3K can be detected in anti-P.Tyr immunoprecipitates (Ohmichi et al 1992; Raffioni & Bradshaw 1992; Soltoff et al 1992). p85 is phosphorylated on Tyr in PC12 cells in response to NGF (Soltoff et al 1992). Conflicting reports on the association of PI3K activity with TrkA have been published

(Ohmichi et al 1992; Soltoff et al 1992). Tyr 751 has been identified as the PI3K-binding site in TrkA (Obermeier et al 1993b). TrkC K1, but not TrkC K2 or K3, associates with PI3K activity upon activation (Lamballe et al 1993). Stimulation with insulin or IGF-1 activates PI3K, and PI3K activity present in anti-P.Tyr immunoprecipitates is increased (Endemann et al 1990; Giorgetti et al 1993; Ruderman et al 1990). PI3K has been shown to associate with both the insulin receptor and IRS-1 (Backer et al 1992; Giorgetti et al 1993; Myers et al 1992; Van Horn et al 1994). There is evidence for phosphorylation of p85 by the insulin receptor in vivo and in vitro (Hayashi et al 1991, 1992, 1993).

PI3K is present in anti-P.Tyr immunoprecipitates following activation of the EGF receptor, but Tyr phosphorylation of p85 could not be detected (Hu et al 1992; Raffioni & Bradshaw 1992). Association of PI3K with the EGF receptor was described in one (Bjorge et al 1990) study, but not in a second (Cochet et al 1991). PI3K activity can be coprecipitated with activated ERBB2 (Peles et al 1992). PI3K and p85 have been found to associate in vivo and in vitro with Tyr-phosphorylated HGF receptors (Bardelli et al 1992; Graziani et al 1991; Ponzetto et al 1993). Tyr 1349 and Tyr 1356 have been identified as the binding sites for PI3K, but these sites do not conform to the usual p85 SH2 binding consensus and, instead, these two sites have P.Tyr-Val-X-Val as a consensus, thus defining a new specificity for p85 SH2 binding (Ponzetto et al 1993). Activation of Flk2/Flt3 results in association with p85. Activated Flk2/Flt3 phosphorylates p85 on Tyr in hematopoietic BaF3 cells, but not in NIH3T3 cells (Dosil et al 1993). Experiments with chimeric receptors indicate that ERBB3, but not RET, phosphorylates p85 (Fedi et al 1994; Santoro et al 1994).

PI3K is activated in response to many cytokines and associates with activated RPTKs. Moreover, PI3K activity can be detected in anti-P.Tyr immunoprecipitates, which implies that some component of the active PI3K enzyme complex is phosphorylated on Tyr. For this reason, it was anticipated that PI3K activity would be regulated directly by Tyr phosphorylation of p85 and/or p110. However, Tyr phosphorylation of p110 is never detected and, thus far, only stimulation with PDGF, NGF, and insulin has been shown to induce Tyr phosphorylation of p85 (Hayashi et al 1992; Kavanaugh et al 1992; Ohmichi et al 1992). Many RPTK ligands cause an increase in intracellular 3′ phosphoinositide levels without direct phosphorylation of p85 or p110. Since the most of these receptors associate with PI3K upon activation (Table 5), an alternative is that the binding of PI3K results in activation through a combination of translocation and allosteric activation.

PI3K binding to Tyr-phosphorylated RPTKs may bring the enzyme in close proximity to its substrates, which are located in the plasma membrane. This view is supported by experiments with v-Src and v-Abl, where mutants that can no longer associate with the plasma membrane bind PI3K, but do not raise the levels of 3′ phosphoinositides in the cell and are thus nontransforming. (Fukui & Hanafusa 1989; Varticovski et al 1991). In addition, binding of PI3K to Tyr-phosphorylated proteins stimulates PI3K catalytic activity. IRS-1, the major insulin and IGF-1 receptor substrate in the cell, contains nine putative PI3K-binding sites (Sun et al 1991; White et al 1985). IRS-1 can bind and activate PI3K in vitro (Backer et al 1992; Myers et al 1992). Moreover, synthetic P.Tyr-containing peptides derived from IRS-1, the insulin receptor, the PDGF β receptor, and polyoma middle T cause a two to threefold increase in PI3K activity (Backer et al 1992; Carpenter et al 1993a; Myers et al 1992; Van Horn et al 1994). Peptides containing two P.Tyr residues are more effective than singly phosphorylated peptides, consistent with the notion that occupancy of both p85 SH2 domains is important. Detailed studies on the interaction of p85 with P.Tyr-containing peptides indicate that binding of activating peptides results in a characteristic change in the structure of p85 (Shoelson et al 1993).

What then is the role of Tyr phosphorylation of p85? In vitro studies indicate that Tyr phosphorylation of p85 inhibits the binding of the PDGF β receptor (Kavanaugh et al 1992). Thus Tyr phosphorylation of p85 may be important for negative regulation and could also be involved in releasing PI3K from the receptor in the vicinity of the membrane. Three phosphorylation sites have been mapped in p85, but their function has not been tested (Hayashi et al 1993). PI3K is also negatively regulated by phosphorylation of p85 at Ser 608 as a result of the intrinsic protein-serine kinase activity of p110 (Carpenter et al 1993b; Dhand et al 1994). Phosphorylation of Ser 608 occurs efficiently in vitro in the presence of Mn^{2+}, which results in decreased PI3K activity, but the physiological significance of this phosphorylation, which can be detected in vivo, is currently unclear.

Despite the excellent correlation between the activation of PI3K—the generation of higher order 3′ phosphoinositides and mitogenesis—the role of PI3K in signal transduction is poorly understood. Data obtained with PDGF and CSF-1 receptor mutants that lack PI3K-binding sites suggest that in certain cell types PI3K binding to the receptor, and the consequent increase in 3′ phosphoinositides, may be important for the induction of cell division and changes in cell morphology in response to ligand binding. Vps34p, a yeast homologue of PI3K, is involved in the regulation of vesicle transport during protein sorting (Herman et al 1992; Schu et al 1993). Although Vps34p only generates PI3P, it is possible that PI3K regulates events that take place at the plasma membrane in response to RPTK activation, like membrane ruffling or receptor down-regulation. Indeed, activation of PI3K is apparently required for PDGF-induced membrane ruffling and chemotaxis by the PDGF β receptor (Wennstrom et al 1994). This would fit with the observations that p85 has

GAP activity toward Rac, which is involved in membrane ruffling (Ridley et al 1992), and that activation of PI3K is required for PDGF β receptor internalization (Joly et al 1994). There is also evidence that PI3K may lie on the pathway for mitogen activation of the 70-kd S6 kinase (Chung et al 1994). It is possible that the PDGF receptor has to be internalized in order to gain access to important substrates, and a defect in receptor internalization in cells expressing mutant PDGF receptors unable to activate PI3K could explain the pleiotropic effects that appear to depend on activation of PI3K. The fact that phosphoinositides produced by PI3K are poor substrates for all known PLCs (Serunian et al 1989) suggests that 3′ phosphoinositides themselves have a second messenger function. In this connection, it has recently been shown that PIP_3 can activate PKCζ in vitro (Nakanishi et al 1993). Nevertheless, it is clear that the roles of 3′ phosphoinositides in signal transduction remain to be established.

PROTEIN-TYROSINE PHOSPHATASES Research on signal transduction by RPTKs has focused mainly on the protein kinases and the phosphorylation of their substrates, even though the down-regulation of signals transduced by RPTKs requires dephosphorylation of the activated RPTKs themselves and their substrates. Protein-tyrosine phosphatases (PTPs) that catalyze the dephosphorylation of RPTKs and their substrates were, until recently, poorly characterized. When the first PTP was purified to homogeneity from human placenta (Tonks et al 1988), it was found to be related to the cytoplasmic domain of the leukocyte common antigen CD45 (Charbonneau et al 1988). CD45 was subsequently shown to have phosphatase activity and to be essential for T cell activation by the T cell receptor (Koretzky et al 1990, 1991; Pingel & Thomas 1989). This indicated that PTPs can contribute positively to signal transduction. Over the last several years a large number of PTPs have been molecularly cloned and characterized. They all contain a related catalytic domain with a number of absolutely conserved residues, including a cysteine that is essential for catalytic activity. They fall into two broad groups: receptor-like PTPs that contain a transmembrane domain and an extracellular putative ligand binding domain, and cytoplasmic PTPs (Mourey & Dixon 1994).

Recently two SH2 domain-containing PTPs, SH-PTP1 and SH-PTP2 (SH2-containing PTPs 1 and 2), have been cloned (Figure 4). SH-PTP1 has been identified by several groups and is also known as PTP1C, SHP, and HCP (Matthews et al 1992; Plutzky et al 1992; Shen et al 1991; Yi et al 1992). SH-PTP1 contains 595 amino acids, and has two SH2 domains present in the N-terminal half and a catalytic domain in the C-terminal half of the protein. SH-PTP1 is exclusively expressed in hematopoietic cells and some epithelial cell lines. Bacterially expressed SH-PTP1 has PTP activity (Matthews et al 1992; Plutzky et al 1992; Shen et al 1991; Yi et al 1992), and dephosphorylates the CSF-1 and SCF receptors in vitro (Pei et al 1993; Zhao et al 1993). When overexpressed in 293 cells, SH-PTP1 dephosphorylates P.Tyr residues in the cytoplasmic domains of the PDGF α, PDGF β, SCF, insulin, IGF-1, and EGF receptors, and ERBB2 (Vogel et al 1993; Yi & Ihle 1993). A fusion protein containing both SH-PTP1 SH2 domains binds in vitro to Tyr phosphorylated EGF and SCF receptors, but not to the CSF-1 receptor (Shen et al 1991; Yi & Ihle 1993). SH-PTP1 associates with activated ERBB2 cytoplasmic domain and with the SCF receptor, but not with the CSF-1 receptor in vivo (Vogel et al 1993; Yi & Ihle 1993). SH-PTP1 SH2 domain binding sites have not been identified. SH-PTP1 is a major substrate for the CSF-1 receptor in immortalized BAC1.2F5 macrophages (Yeung et al 1992). In addition, CSF-1-independent proliferation is observed when macrophages are isolated from mice homozygous for mutations at the *motheaten* locus, which encodes SH-PTP1 (Shultz et al 1993). The consequence of this mutational inactivation of SH-PTP1 suggests that it is involved in attenuation of mitogenic responses in macrophages.

SH-PTP2 has been identified by several groups and is also known as PTP1D PTP2C, SYP, and SH-PTP3 (Adachi et al 1992; Ahmad et al 1993; Bastien et al 1993; Feng et al 1993; Freeman et al 1992; Vogel et al 1993). The structure of SH-PTP2 is similar to that of SH-PTP1. In contrast to SH-PTP1, SH-PTP2 appears to be widely expressed and is present during all stages of development (Adachi et al 1992; Ahmad et al 1993; Bastien et al 1993; Feng et al 1993; Freeman et al 1992). Both immunoprecipitated and recombinant SH-PTP2 have PTP activity (Adachi et al 1992; Ahmad et al 1993; Bastien et al 1993; Feng et al 1993; Freeman et al 1992; Vogel et al 1993).

In contrast to what has been found with SH-PTP1, SH-PTP2 does not dephosphorylate the cytoplasmic domains of the PDGF α, PDGF β, EGF, insulin, and IGF-1 receptors or ERBB2 in vivo, although it shows some activity towards the SCF receptor cytoplasmic domain (Vogel et al 1993). Fusion proteins that contain both SH-PTP2 SH2 domains associate in vitro with the EGF and PDGF receptors and with IRS-1 (Feng et al 1993; Kuhne et al 1993; Lechleider et al 1993a). This association is most likely mediated by the N-terminal SH2 domain (Lechleider et al 1993a). In vivo SH-PTP2 binds to the PDGF receptor and to ERBB2, and with lower affinity binds to the EGF and SCF receptors (Feng et al 1993; Lechleider et al 1993a). SH-PTP2 is phosphorylated in vivo by the activated EGF and PDGF receptors and in v-Src transformed cells (Feng et al 1993; Lechleider et al 1993a). In the C-terminus of the PDGF receptor Tyr 1009 has been identified as a binding site for SH-PTP2 (Kazlauskas et al 1993; Lechleider et al 1993b). This would fit with P.Tyr-containing peptide-binding studies using the SH-PTP2 N-terminal SH2 in which P.Tyr-Ile-X-Val was identified as a binding site (Songyang et al 1993). Recently it was shown that SH-PTP2 mediates binding of GRB2 to the

PDGF receptor (Li et al 1994). This suggests that SH-PTP2 may be upstream of Ras and Raf-1. This conclusion is supported by mutagenesis studies in the PDGF receptor and genetic studies in *Drosophila* (Perkins et al 1992; Valius & Kazlauskas 1993). SH-PTP2 is activated by a P.Tyr-containing peptide based on the sequence around Tyr 1009 in the PDGF receptor (Lechleider et al 1993b). One study concludes that SH-PTP2 is activated following phosphorylation by the PDGF receptor (Vogel et al 1993), but this activation could instead be due to the fact that SH-PTP2 is stimulated by being bound to the phosphorylated PDGF receptor. There is evidence that SH-PTP2 is required for efficient transformation by ERBB2.

SH-PTP2 may be the mammalian homologue of the *Drosophila* PTP Corkscrew (Csw) because its amino acid sequence is more similar to Csw than SH-PTP1 (Feng et al 1993; Freeman et al 1992). Csw, however, contains an insert in the phosphatase domain that is missing in SH-PTP1 and SH-PTP2. *Csw* was identified as a gene involved in induction of terminal differentiation during *Drosophila* embryonal development (St. Johnston & Nüsslein-Volhard 1992). Activation of the Torso RPTK by the localized Torso-like ligand initiates a signal transduction cascade that results in the localized expression of the transcription factors Tailless and Huckebein. Genetic experiments indicate that Csw functions downstream of Torso and upstream of Tailless and Huckebein (Perkins et al 1992). Csw functions in concert with Polehole, a *Drosophila* homologue of the Raf-1 protein kinase (Ambrosio et al 1989; Perkins et al 1992). Csw mutant embryos have reduced levels of the transcription factors Tailless and Huckebein and are missing terminal structures (Perkins et al 1992).

In summary, both SH-PTP1 and SH-PTP2 appear to play roles in signal transduction. They may be activated through binding to Tyr-phosphorylated proteins and perhaps as a result of Tyr phosphorylation. However, they have opposite roles in RPTK signaling. SH-PTP1 dephosphorylates activated RPTKs and may be involved in suppression of the mitogenic response. In contrast, SH-PTP2 associates with activated receptors and may be involved in the activation of signal transduction cascades downstream of growth factor RPTKs. Physiological substrates remain to be identified, but PTKs in the Src family are potential substrates for SH-PTP2 and Csw since their activity is negatively regulated by C-terminal Tyr phosphorylation. They also associate with activated RPTKs.

C-SRC PTK FAMILY MEMBERS Src is the prototype of a family of cytoplasmic PTKs. At least nine family members have been identified, all of which have very similar structures. They contain a catalytic domain in the C-terminal half of the protein, followed by a short tail, which contains a regulatory Tyr phosphorylation site. A SH3 domain followed by a SH2 domain is located

N-terminal to the kinase domain (Figure 4). All Src family members contain a Gly residue at position 2 that is required for myristylation and membrane attachment. The 70–80 amino acids between the N-terminus and the SH3 domain are unique in the different Src family members. Src family members may play a role in signal transduction by cell surface receptors and, in several cases, the interaction with cell surface receptors is directed, at least in part, by the unique N-terminus (Bolen et al 1992; Bolen & Veillette 1989). There is good evidence that a number of Src family members are involved in signal transduction following stimulation of cell surface immunoglobulins, the IgE receptor, the T cell receptor complex, and the IL-2 and the IL-7 receptors (Bolen et al 1992; Bolen & Veillette 1989; Eisenman & Bolen 1990).

c-Src is regulated by a number of different protein kinases. The c-Src C-terminus contains Tyr 527, the major site of Tyr phosphorylation in vivo. Dephosphorylation of Tyr 527 increases c-Src PTK activity, and mutation of Tyr 527 increases both kinase activity and transforming potential. This indicates that c-Src kinase activity is negatively regulated by phosphorylation at Tyr 527. This regulation involves binding of the SH2 domains to the phosphorylated Tyr residue in their C-terminus (Gervais et al 1993; Liu et al 1993b). As a result of this intramolecular interaction, the catalytic domain may be sterically occluded. This fits with the observation that suppression of Src activity by the C-terminal kinase CSK in *S. cerevisiae* and *S. pombe* is dependent on the presence of the Src SH2 domain (Murphy et al 1993; Okada et al 1993; Superti-Furga et al 1993). The SH3 domain also appears to be necessary for binding of the Src SH2 domain to P.Tyr 527, possibly by modulating the affinity of the SH2 domain for its binding site (Murphy et al 1993; Okada et al 1993; Superti-Furga et al 1993). Like v-Src, c-Src can also be phosphorylated on Tyr 416, and this may be important for catalytic domain activation.

PDGF stimulation of fibroblasts leads to activation of c-Src, and additional N-terminal Ser and Tyr phosphorylation of c-Src is observed (Gould & Hunter 1988; Kypta et al 1990; Ralston & Bishop 1985). c-Yes and Fyn are also activated by PDGF treatment (Kypta et al 1990), and Src family kinases are activated by the CSF-1 receptor as well (Courtneidge et al 1993). c-Src, Fyn, and c-Yes associate with the PDGF receptor upon ligand binding (Kypta et al 1990). A fusion protein containing the Fyn SH2 domain associates with the PDGF receptor in vivo in response to PDGF (Twamley et al 1992). The association is dependent on the presence of Tyr 857 in the PDGF receptor kinase domain (Courtneidge et al 1991). However, it is unlikely that the Src SH2 domain binds directly to this Tyr, since it is not present within a consensus-binding site for the Src SH2 domain. Indeed, recent evidence indicates that c-Src binds to Tyr 579 and Tyr 581 in the juxtamembrane region of the PDGF receptor (Mori et al 1993), which are in a sequence related to the Src SH2 domain binding consensus P.Tyr-Glu-Glu-Ile (Songyang et al 1993). Thus Tyr

857 may be involved in activation of the PDGF receptor catalytic domain and required for phosphorylation of Tyr 579 and Tyr 581. The Src family kinases c-Src, Fyn, and c-Yes also associate with the activated CSF-1 receptor when it is expressed in NIH3T3 fibroblasts (Courtneidge et al 1993). Tyr 809, the autophosphorylation site in the human CSF-1 receptor that is homologous to Tyr 857 in the PDGF β receptor, is essential for Src family kinase binding and association (Courtneidge et al 1993).

How are Src family PTKs activated by PDGF and CSF-1? Tyrosine phosphorylation of c-Src at the N-terminus, which occurs following its binding to the activated PDGF and CSF-1 receptors and is probably RPTK mediated, correlates with an increase in PTK activity (Gould & Hunter 1988; Kypta et al 1990; Ralston & Bishop 1985), but there is no evidence that this phosphorylation is required for c-Src activation. Since the Src SH2 domain binds to the Tyr 579/581 site on the activated PDGF receptor with higher affinity than to P.Tyr 527, c-Src activation may simply be a result of the dissociation of the SH2 domain from P.Tyr 527, which allows substrates access to the catalytic domain. This is consistent with the fact that the activated c-Src molecules are bound to the RPTK. There is no evidence that dephosphorylation of P.Tyr 527 is involved in c-Src activation. However, the cytoplasmic PTPs that associate with activated RPTKs could in principle dephosphorylate P.Tyr 527. The Csk PTK specifically phosphorylates c-Src at Tyr 527 (Okada & Nakagawa 1988, 1989), and the phenotype of Csk knock out mice indicates that Csk is an important negative regulator of Src family PTK activity (Imamoto & Soriano 1993; Nada et al 1993). However, there is no evidence that CSK activity changes in response to activation of RPTKs.

The activation of Src family members appears to be essential for PDGF-induced mitogenesis since expression of a dominant-negative, kinase-inactive form of c-Src or Fyn inhibits PDGF-stimulated DNA synthesis in NIH3T3 cells (Twamley-Stein et al 1993). Src also appears to be essential for signaling by the NGF and FGF receptors in PC12 cells (Kremer et al 1991). However, the nature of the targets for the activated Src family PTKs and the role of the N-terminal Tyr phosphorylation remain unclear. The N-terminal Tyr phosphorylation could regulate the interaction of Src with its substrates through its SH2 and SH3 domains. There is evidence that v-Src signals through Ras both for transformation and for induction of PC12 cell differentiation (Kremer et al 1991), which could occur through v-Src-mediated Tyr phosphorylation of SHC, and consequent activation of GRB2/SOS. However, whether c-Src normally signals through Ras is unclear. Paxillin, FAK, p110, and p130 all interact physically with Src through its SH2 or SH3 domain (Cobb et al 1994; Kanner et al 1991; Weng et al 1993). PI3K activity can be detected in anti-Lyn and anti-Fyn immunoprecipitates (Prasad et al 1993a; Yamanashi et al 1992). PI3K and p85 bind to the SH3 domains of v-Src, Lyn, Fyn, and Lck in vitro (Kapeller et al 1994; Liu et al 1993c; Pleiman et al 1993; Prasad et al 1993a,b; Vogel et al 1993; Yamanashi et al 1992). Two SH3-binding sites have been identified between the SH3 and the first SH2 domain in p85 (Kapeller et al 1994; Liu et al 1993c), and SH3 binding activates PI3K (Pleiman et al 1994). Although the physiological function of these interactions is not clear, it is possible that Src family members are involved in the activation of PI3K in response to a large variety of stimuli.

SH2 Domain-Containing Adaptor Proteins

The adaptor proteins are a novel type of signaling proteins that lack a catalytic domain. Most of the adaptor proteins contain a SH2 domain and one or more additional protein interaction domains, most commonly a SH3 domain. Adaptor proteins bind to activated RPTKs via their SH2 domains, and this causes an effector protein bound to the second interaction domain to be cotranslocated. The cotranslocated protein may have a catalytic activity, as is the case for SOS, which binds to the SH3 domains in GRB2. Translocation of SOS to the membrane may in itself be sufficient to activate its target (Ras). In principle, however, proteins cotranslocated with adaptors could be substrates for RPTK phosphorylation. A second type of adaptor-docking protein is exemplified by IRS-1, which is one of the primary substrates for the insulin receptor. IRS-1 lacks a SH2 domain, but is multiply Tyr phosphorylated and, as a result, is able to bind several SH2 proteins. Some of the substrates in the enzyme category may also have an adaptor function. For instance, SH-PTP2 not only has PTP activity, but when Tyr phosphorylated, also binds GRB2, which thereby allows indirect association of GRB2 with the PDGF β receptor (Li et al 1994). Some of the best-characterized adaptor proteins are discussed below.

CRK CRK was first isolated as the viral transforming gene *v-Crk* that is present in the CT10 and ASV-1 avian sarcoma viruses (Mayer et al 1988; Tsuchie et al 1989). *v-Crk* contains viral Gag sequences fused to cellular CRK sequences that encode a SH2 domain followed by a SH3 domain (Mayer et al 1988; Tsuchie et al 1989). CT10-infected cells show increased levels of P.Tyr in several proteins (Mayer et al 1988). The ability of *v-Crk* mutants to transform cells correlates with their ability to increase cellular P.Tyr levels (Mayer & Hanafusa 1990b) and depends on the integrity of the SH2 and SH3 domains (Mayer & Hanafusa 1990b). *v-Crk* associates with a number of proteins in CT10-transformed cells, some of which appear to have protein kinase activity (Matsuda et al 1990, 1991, 1993; Mayer & Hanafusa 1990a). *v-Crk* expressed as a fusion protein in *E. coli* binds to the EGF and PDGF receptors in vitro in a manner dependent on Tyr phosphorylation (Anderson et al 1990; Birge et al 1992; Moran et al 1990), and to p62, p70, p130, and the GAP-associated p190 (Koch et al 1992; Moran et al 1990). p70 has now been identified as paxillin

(Birge et al 1993). Recently, the c-Abl protein has been found to associate with the v-Crk SH3 domain through a proline-rich sequence downstream of the catalytic domain (Ren et al 1994; H Hanafusa, personal communication), and this may account for the PTK activity associated with v-Crk. CRK is phosphorylated on Tyr in CRK-transformed cells.

There are two forms of CRK, CRKI (204 amino acids) and CRKII (304 amino acids) (Matsuda et al 1992; Reichman et al 1992) (Figure 4). CRKII contains one SH2 domain that is followed by two SH3 domains and extends about 100 amino acids beyond the end of the v-Crk protein, which lacks the second SH3 domain (Reichman et al 1992). CRKII is a phosphoprotein that is ubiquitously expressed (Reichman et al 1992). CRKI, which is not as widely expressed, lacks the second SH3 domain and is encoded by an alternatively spliced mRNA (Matsuda et al 1992). When overexpressed, CRKI has biological properties similar to v-Crk (Matsuda et al 1992). Microinjection of either CRKI or CRKII induces neuronal differentiation in PC12 cells in a Ras-dependent fashion (Tanaka et al 1993). Both the SH2 and SH3 domains are essential for this activity. CRKL, second CRK-related protein that has a similar structure to CRKII, has been identified recently (ten Hoeve et al 1993).

By analogy with GRB2, when CRK binds to activated RPTKs or other P.Tyr-containing proteins through its SH2 domain, any effector protein bound to one or both of its SH3 domains will be relocalized. The affinity of the CRK SH2 domain for its binding sites appears to be regulated by an N-terminal extension that results from alternative initiation at an upstream CUG codon (Fajardo et al 1993), which is also present in *v-Crk*. The CRK SH2 domain binds P.Tyr residues followed by a Pro at +2 (Birge et al 1993; Songyang et al 1993). CRK SH3 domains have been found to bind specifically to 136-, 125-, and 118-kd proteins (Tanaka et al 1993), whose identities are presently unknown. Another ~ 150-kd protein that associates with CRK SH3 domains has recently been shown to be related to SOS, a Ras GNRF (H Hanafusa, personal communication), which suggests that CRK signaling may be coupled to the Ras pathway.

NCK This protein of 377 amino acids contains three N-terminal SH3 domains followed by a SH2 domain (Figure 4). NCK was first molecularly cloned by fortuitous cross-reactivity with a monoclonal antibody that recognizes the melanoma-associated antigen MUC18 (Lehmann et al 1990). NCK was also cloned by a method that uses the autophosphorylated EGF receptor C-terminus as a probe to screen an expression library (Margolis et al 1992). Interest in NCK was generated when it was found that a 47-kd protein, which is immunoprecipitated by certain anti-PLCγ1 antibodies, was NCK (Meisenhelder & Hunter 1992; Meisenhelder et al 1989; Park & Rhee 1992). NCK is widely expressed in all tissues and during all stages of development.

NCK is phosphorylated on Ser in quiescent cells (Li et al 1992; Meisenhelder & Hunter 1992). Stimulation with EGF, PDGF, and orthovanadate, and activation of the low affinity IgG receptor, cell surface IgM and the T cell receptor complex, result in a rapid increase in phosphorylation on Ser and Tyr (Li et al 1992; Meisenhelder & Hunter 1992; Park & Rhee 1992). Stimulation with TPA and forskolin results in the increase of Ser phosphorylation on a number of different sites (Li et al 1992; Meisenhelder & Hunter 1992; Park & Rhee 1992). The role of these phosphorylations in NCK function has not been addressed. NCK is also phosphorylated on Tyr in v-Src-transformed cells (Meisenhelder & Hunter 1992), consistent with the observed phosphorylation upon stimulation of the low affinity IgG receptor, cell surface IgM, and the T cell receptor, which are known to activate members of the Src PTK family (Bolen et al 1992; Eiseman & Bolen 1990).

In response to receptor activation, NCK binds to the EGF and PDGF receptors via its SH2 domain (Li et al 1992). The NCK SH2 domain binds to P.Tyr in the sequence P.Tyr-Asp-Glu-X (Songyang et al 1993). Tyr 751 in the PDGF β receptor has been identified as the binding site for NCK, and it has been proposed that residues on the N-terminal side of the P.Tyr play a role in recognition of P.Tyr 751 by the NCK SH2 domain (Nishimura et al 1993). Since Tyr 751 is also the binding site for the p85 subunit of PI3K, this suggests that multiple signaling proteins can compete for binding to the same site on an activated RPTK. By using the Phe 751 mutant PDGF β receptor, receptor association was shown to be essential for the increase in both Tyr and Ser phosphorylation of NCK (Nishimura et al 1993). NCK is not phosphorylated on Tyr, nor does it associate with the receptor following activation of the CSF-1 receptor or Flk2/Flt3 (Dosil et al 1993; van der Geer & Hunter 1993). NCK also binds to the insulin receptor substrate IRS-1 in insulin-treated cells (Lee et al 1993).

The best evidence that NCK plays a role in mitogenic signal transduction is that NCK overexpression results in morphologically transformed rodent fibroblasts, which are tumorigenic (Chou et al 1992; Li et al 1992). However, unlike v-Crk, overexpression of NCK does not result in the increase in Tyr phosphorylation of cellular proteins (Chou et al 1992). Since NCK binds to activated EGF and PDGF receptors, it is likely that NCK functions as an adaptor molecule bringing effectors that bind to its SH3 domains to the activated RPTK. However, the nature of the proteins that bind to the NCK SH3 domains is unknown.

GRB2 This 217 amino acid, 24-kd protein contains nothing more than a SH2 domain located between two SH3 domains (Figure 4). GRB2 was cloned by using the CORT method and by screening a cDNA library with degenerate oligonucleotides that were designed to recognize SH2 domains (Lowenstein et al 1992; Matuoka et al 1992). GRB2 mRNA is detected in all tissues (Lowenstein et al 1992; Matuoka et al 1992; Suen et al 1993). *Drosophila* (Drk, for downstream of receptor kinases) (Olivier et al 1993; Simon et al 1993) and *C. elegans* (Sem-5) (Sternberg & Horvitz 1991) homologues of GRB2 have been identified.

GRB2 associates with Tyr phosphorylated EGF, PDGF, and CSF-1 receptors and with activated Flk2/Flt3 in vivo and in vitro (Lowenstein et al 1992; Matuoka et al 1992; Suen et al 1993; van der Geer & Hunter 1993). Tyr 1068 and Tyr 1086 have been identified as GRB2-binding sites in the EGF receptor (Batzer et al 1994). No direct GRB2-binding site has been identified in the PDGF receptor, and instead GRB2 binding may be mediated by SH-PTP2, which binds to Tyr 1009 (Li et al 1994). GRB2 also binds to the insulin receptor substrate IRS-1, to SHC, and to the receptor-like PTP, RPTPα (den Hertog et al 1994; Rozakis-Adcock et al 1992; Skolnik et al 1993b; Tobe et al 1993). A GST-Drk fusion protein binds to the mammalian EGF receptor and to the autophosphorylated Sevenless RPTK cytoplasmic domain (Olivier et al 1993; Simon et al 1993). The GRB2 SH2 domain is sufficient for binding to Tyr-phosphorylated RPTKs (Lowenstein et al 1992; Matuoka et al 1992; Suen et al 1993). The preferred binding site for the GRB2 SH2 domain is P.Tyr-Val/Leu-Asn-X (Songyang et al 1993), which is consistent with the sequences of the known binding sites in the EGF and CSF-1 receptors and in SHC. GRB2 is neither a phosphoprotein nor a substrate for activated RPTKs (Lowenstein et al 1992; Rozakis-Adcock et al 1992; Suen et al 1993). The GRB2 SH3 domain determines GRB2 localization to membrane ruffles (Bar Sagi et al 1993). In resting Rat-2 fibroblasts, GRB2 binds to a number of P.Tyr-containing proteins. None of these proteins has been identified (Matuoka et al 1993; van der Geer & Hunter 1993).

Sem-5, the *C. elegans* functional homologue of GRB2, is required for signal transduction by the Let-23 RPTK (Sternberg & Horvitz 1991). Activation of Let-23 initiates a signal transduction cascade that results in vulval development and sexual differentiation. Genetic experiments indicate that Sem-5 is downstream of Let-23 and upstream of the Ras protein, Let-60 (Sternberg & Horvitz 1991). A number of Sem-5 mutant alleles that are deficient in vulval development have been isolated; some of them have point mutations in highly conserved residues within either the SH2 or SH3 domain (Clark et al 1992). This indicates that the SH2 and SH3 domains are both essential for GRB2/Sem-5 function upstream of Ras.

Further evidence for a role of GRB2 upstream of Ras comes from work done in *Drosophila,* where Drk was originally identified as an enhancer of a weak *sevenless* allele (Simon et al 1993). Genetic evidence suggests that Drk plays a role in signal transduction downstream of Sevenless and DER, the *Drosophila* EGF receptor. The *Drosophila SOS* gene product, a GNRF for Ras

(Bonfini et al 1992), contains a number of Pro-rich sequences in its C-terminal domain, and Drk can bind directly to the SOS C-terminus in vitro (Olivier et al 1993; Simon et al 1993). GRB2 binds to mammalian SOS, and a GRB2/SOS complex exists constitutively (Buday & Downward 1993a; Chardin et al 1993; Egan et al 1993; Li et al 1993; Rozakis-Adcock et al 1993). Based on these findings, it has been suggested that when GRB2 binds to an activated RPTK, it brings the associated SOS to the cytoplasmic face of the plasma membrane, where SOS can act on Ras to catalyze GDP-GTP exchange, thus activating Ras. However, it has not been established that translocation of SOS is sufficient to activate Ras. SOS is extensively phosphorylated on Ser in response to RPTK activation (Li et al 1993; Rozakis-Adcock et al 1993), but the consequences of phosphorylation are unknown.

Consistent with a role for GRB2 in regulating Ras, overexpression of GRB2 enhances the EGF-induced activation of Ras (Gale et al 1993). Coinjection of GRB2 and c-H-Ras protein into quiescent fibroblasts leads to DNA synthesis, while microinjection of GRB2 or c-H-Ras alone does not (Lowenstein et al 1992). Overexpression of GRB2 increases the MAP kinase activation in response to insulin (Skolnik et al 1993a). Microinjection of anti-GRB2 antibodies inhibits entry into S-phase and cytoskeletal rearrangement in response to EGF or PDGF (Matuoka et al 1993). Mutation of the GRB2-binding site in the CSF-1 receptor, Tyr 697, decreases growth stimulation in response to CSF-1 (van der Geer & Hunter 1993). In summary, GRB2 plays a role as an adaptor in Ras activation linking SOS to activated RPTKs, IRS-1, or SHC. Ras activation will in turn lead to activation of the MAP kinase pathway and nuclear signaling (see below).

SHC Pelicci et al (1992) recently identified this SH2-containing protein. The SHC cDNA contains two in-frame translational start sites and encodes two overlapping proteins of 473 and 428 amino acids. These correspond to the 52- and 46-kd SHC proteins detected in cells. An additional 66-kd protein cross-reacts with anti-SHC antibodies, and this may be an alternatively spliced product of the SHC gene (Pelicci et al 1992). SHC contains a C-terminal SH2 domain and a central glycine/proline-rich sequence (Figure 4). Overexpression of SHC induces NIH3T3 fibroblast transformation (Pelicci et al 1992). SHC is Tyr-phosphorylated in resting cells, and its phosphorylation on Tyr increases upon stimulation with EGF, PDGF, CSF-1 SCF, insulin, NGF, erythropoietin, IL2, or IL3 (Burns et al 1993; Cutler et al 1993; Damen et al 1993; Pelicci et al 1992; Pronk et al 1993; Ruff-Jamison et al 1993b; Suen et al 1993; van der Geer & Hunter 1993). SHC phosphorylation also increases upon activation of ERBB2, Flk2/Flt3, and the T cell receptor, and in v-Src- and v-Fps-transformed cells (Dosil et al 1993; McGlade et al 1992; Ravichandran et al 1993; Segatto et al 1993). SHC associates with the Tyr-phosphorylated EGF receptor,

ERBB2, erythropoietin receptor, and T cell receptor ζ chain via its SH2 domain (Damen et al 1993; Pelicci et al 1992; Ravichandran et al 1993; Ruff-Jamison et al 1993b). Tyr 490 has been identified as the binding site for SHC in Trk (Obermeier et al 1993b), Tyr 1173 as the SHC-binding site in the EGF receptor (Batzer et al 1994), and Tyr 250 as the binding site in polyoma middle T (Dilworth et al 1994). The similarity between these sites is an Asn at position-2, and it is possible that SHC SH2 domain recognition requires sequences on the N-terminal side of the P.Tyr. Tyrosine-phosphorylated SHC associates with GRB2 (Cutler et al 1993; Damen et al 1993; Pronk et al 1993; Ravichandran et al 1993; Suen et al 1993; van der Geer & Hunter 1993). Another putative SH2-containing adaptor protein, SHB, has recently been reported. SHB has a topology similar to that of SHC and can also bind to the autophosphorylated PDGF β receptor (Welsh et al 1994).

In summary, SHC is phosphorylated on Tyr in response to numerous cytokines and associates with many autophosphorylated RPTKs. In addition, Tyr-phosphorylated SHC also associates with GRB2 through the GRB2 SH2 domain, and this may be another mechanism for binding GRB2 to RPTKs that lack direct binding sites. Even though SHC is a soluble protein, its phosphorylation and association with GRB2 may provide an alternative mechanism for activating the Ras pathway. This could explain why the EGF receptor mutant lacking all the autophosphorylation sites, but still capable of inducing SHC Tyr phosphorylation (Gotoh et al 1994), retains mitogenic activity (Vega et al 1992). The fact that overexpression of SHC results in transformation of NIH3T3 fibroblasts is consistent with a role for SHC in signal transduction, but its exact function remains unclear.

Docking Proteins

The insulin receptor substrate IRS-1 is the first example of what will probably prove to be a class of RPTK substrates that can be termed docking proteins. Docking proteins are substrates that are specifically phosphorylated at multiple sites by an activated RPTK and then bind a selection of SH2 proteins, which in turn propagate the signal. In principle, each docking protein will be a specific substrate for a single RPTK or a subfamily of RPTKs. IRS-1 is the only example so far, but it seems likely that there will be others.

IRS-1 The most prominent insulin RPTK substrate is IRS-1. This 185-kd protein is multiply phosphorylated on Tyr, and its sequence predicts a large number of potential SH2 protein-binding sites (White 1994). Several SH2 proteins, including PI3K, SH-PTP2, NCK, and GRB2 (Backer et al 1992; Baltensperger et al 1993; Lechleider et al 1993a; Lee et al 1993; Myers et al 1992; Skolnik et al 1993b), have been shown to bind to IRS-1 following insulin stimulation. Although there are reports of SH2-containing proteins binding to the autophosphorylated insulin receptor, it appears as though the insulin receptor uses an alternate strategy to recruit SH2 proteins, which involves phosphorylation of a large adaptor (docking) protein at multiple sites thus generating a signaling complex. Some fraction of IRS-1 is localized at the membrane and may associate with the insulin receptor. This strategy may allow insulin to elicit its characteristic metabolic response, rather than a typical RPTK growth response, despite the fact that many of the same proteins are activated. 4PS, an IRS-1-related protein, is Tyr phosphorylated in IL-4-treated hematopoietic cells and may serve a similar function in IL-4 signaling (White 1994). Mutation of Tyr 960 in the insulin receptor decreases IRS-1 phosphorylation, and a short sequence in the IL-4 receptor related to that around Tyr 960 is required for IL-4-induced phosphorylation of IRS-1 in nonhematopoietic cells (Keegan et al 1994), which implies that this may be a region involved in recognition of IRS-1 and 4PS that is required for their phosphorylation.

Ras Pathway

The small GTP-binding protein Ras, originally identified as a transforming gene present in several oncogenic retroviruses, is activated in a variety of human tumors. Like other G proteins, Ras cycles between an active GTP-bound state and an inactive GDP-bound state, thereby functioning as a molecular switch or timer. Like all small G proteins, Ras has low intrinsic GTPase activity and because GDP is bound with high affinity, Ras tends to accumulate in the GDP-bound inactive state (Bourne et al 1991). Several independent approaches, including microinjection of neutralizing antibodies, Ras overexpression, and genetic analyses in *Drosophila* and *C. elegans,* have implicated Ras in signal transduction downstream of PTKs (Burgering et al 1989, 1991; Mulcahy et al 1985; Rubin 1991; Smith et al 1986; Sternberg & Horvitz 1991). Activation of Ras can be measured directly as an increase in the GTP/GDP ratio of guanine nucleotides bound to Ras. Dividing cells have increased levels of Ras•GTP compared to quiescent cells (Satoh et al 1990a). There is a two to threefold increase in the level of Ras•GTP in response to serum, PDGF, EGF, CSF-1, and insulin (Burgering et al 1989, 1991; Gibbs et al 1990; Heidaran et al 1992; Satoh et al 1990a,b); these levels stay high for at least 30 min. Stimulation with a wide variety of lymphokines also induces an increase in Ras•GTP (Duronio et al 1992).

The levels of Ras•GTP are regulated through GNRFs and GTPase activators (GAPs) (for review, see Boguski & McCormick 1993). Ras GTPase activity is stimulated by the Ras GTPase-activating protein (RasGAP) (Gibbs et al 1988; Trahey & McCormick 1987). Two alternatively spliced forms of RasGAP exist (Trahey et al 1988; Vogel et al 1988). The longer form of RasGAP is 1047 amino acids, and contains at its C-terminus a 400 amino acid region of homology with mammalian neurofibromin (NF-1), *Drosophila* Gap-1, and the *S. cerevisiae* proteins IRA1 and IRA_2, which is sufficient for catalytic activity. Downstream of an N-terminal hydrophobic domain of ~ 180 residues, which is lacking in the short form, are two SH2 domains separated by a SH3 domain (Figure 4). NF-1 contains a domain that is related to RasGAP, and this domain is sufficient for RasGAP activity towards Ras in vitro (Martin et al 1990; Xu et al 1990). Studies of fibroblasts that lack the NF-1 protein suggest that NF-1 is the major RasGAP in most cells.

RasGAP interacts with a number of RPTKs via its SH2 domains. In response to PDGF treatment, RasGAP is Tyr phosphorylated and approximately 10% of the cellular RasGAP associates with the PDGF receptor (Kaplan et al 1990; Kazlauskas et al 1990; Molloy et al 1989). Using bacterially expressed RasGAP SH2 domains, it was found that the N-terminal SH2 domain binds with higher affinity to the PDGF receptor. There is evidence that both SH2 domains are needed for binding of RasGAP to the activated PDGF receptor in intact cells (Cooper & Kashishian 1993). Tyr 771 is the binding site for RasGAP in the PDGF receptor (Fantl et al 1992; Kazlauskas et al 1992). Mutation of this site to Phe has no effect on the induction of DNA synthesis by PDGF, but prevents RasGAP association with the activated receptor and Tyr phosphorylation in response to PDGF (Fantl et al 1992; Kashishian et al 1992; Kazlauskas et al 1992). Ras•GTP levels are regulated normally by PDGF in cells expressing mutant Phe 771 PDGF receptors.

RasGAP and the RasGAP-associated proteins, p62 and p190, are rapidly phosphorylated on Tyr in response to EGF (Ellis et al 1990; Liu & Pawson 1991). Phosphorylation of RasGAP and p62 is maximal 3–5 min after addition of EGF, whereas phosphorylation of p190 is maximal 1–2 hr after stimulation. RasGAP SH2 domains bind to the EGF receptor upon autophosphorylation in vitro. The N-terminal SH2 domain binds better than the C-terminal SH2 domain. A fusion protein containing both SH2 domains binds even better (Anderson et al 1990; Moran et al 1990). RasGAP SH2 domains bind poorly to EGF receptors that are missing the C-terminal region, which suggests that the EGF receptor C-terminus contains the RasGAP binding site(s).

RasGAP also associates with the insulin receptor and becomes Tyr phosphorylated in response to insulin, but only in NIH3T3 cells that are over-expressing the insulin receptor and that are treated with phenylarsine oxide, an inhibitor of PTPs (Pronk et al 1992). Autophosphorylated HGF receptors also associate with RasGAP in vitro (Bardelli et al 1992). RasGAP does not associate with the CSF-1 and SCF receptors upon ligand binding, nor does it become Tyr phosphorylated (Heidaran et al 1992; Reedijk et al 1990; Rottapel et al 1991). The RasGAP-associated proteins p62 and p190, however, are Tyr phosphorylated in response to CSF-1 and in v-Fms-transformed fibroblasts (Heidaran et al 1992; Reedijk et al 1990).

Tyr 460 in RasGAP has been identified as the major in vivo and in vitro site of phosphorylation by the EGF receptor, v-Src and Lck (Amrein et al 1992; Ellis et al 1990; Liu & Pawson 1991; Park et al 1992). Tyrosine phosphorylation has no effect on RasGAP activity in vitro, but may affect RasGAP's ability to associate with p62 (Park & Jove 1993). p62 is an RNA-binding protein, which is Tyr phosphorylated in response to activation of many PTKs. Indeed, its phosphorylation is often one of the most sensitive indicators of PTK activation. Moreover, P.Tyr-containing p62 binds to several SH2 proteins, which implies that it may play a role in signaling, but its true function is unknown.

Speculation about an effector role for RasGAP in Ras signaling was initially fueled by the observation that RasGAP action on Ras requires the region of Ras that interacts with its effector (Adari et al 1988; Calés et al 1988). A number of studies, however, suggest that RasGAP is a negative regulator of signal transduction by Ras. Overexpression of full length RasGAP, or the C-terminal region containing the catalytic domain, suppresses focus formation and morphological transformation of fibroblasts by Ras, or oncogenically activated Src (DeClue et al 1991; Huang et al 1993; Nori et al 1991; Zhang et al 1990). The RasGAP C-terminus also inhibits transcription from the Ras-responsive NVL3 LTR, which is induced by v-Fms. Overexpression of the RasGAP C-terminus suppresses morphological transformation by v-Fms and expression of Ras-responsive genes in response to CSF-1. It also reverses CSF-1-dependent anchorage-independent growth (Bortner et al 1991).

Genetic evidence in *Drosophila* and *S. cerevisiae* supports the conclusion that RasGAP is a negative regulator of Ras. The RAS1 and RAS2 gene products regulate cAMP levels in *S. cerevisiae* by activating adenylyl cyclase. *IRA1* and *IRA2,* two GAP-related genes, have been identified in *S. cerevisiae.* Individual mutations result in an increase in RAS-GTP levels. Deletion of both the *IRA1* and *IRA2* genes in *S. cerevisiae* has the same phenotype as that of expression of activated RAS (Tanaka et al 1990). Expression of the catalytic domain of bovine RasGAP can suppress this phenotype. The Ras GNRF, encoded by the *CDC25* gene, activates RAS by increasing the exchange of GDP for GTP. *CDC25* mutants have a severe cell cycle defect and arrest in G1. This can be rescued by activated RAS (Broek et al 1987).

In *Drosophila,* Ras acts downstream of the DER and the Sevenless RPTKs. Sevenless is required for the induction of the R7 photoreceptor cell differentiation in ommatidia, the units that together form the insect's compound eye (Rubin 1991). Several components of the Sevenless signal transduction cascade have been identified, one of which is the *Drosophila* Ras1 (Simon et al 1991). Ras1 is regulated by Son of Sevenless (Sos), a GNRF that is related to the *S. cerevisiae* CDC25 protein, and by *Drosophila* Gap1 (Bonfini et al 1992; Gaul et al 1992; Simon et al 1991). Homozygous mutations in the *GAP1* gene result in additional R7 cells, which indicates that Gap1 is not only dispensable for Ras signaling, but also acts as a negative regulator of Ras signaling (Gaul et al 1992). Gap1-negative flies are otherwise normal, which indicates that there must be other *GAP* genes in *Drosophila.* In summary, there is no evidence for an effector function of GAP downstream of Ras from genetic studies in *S. cerevisiae* or *Drosophila.* It is important to note, however, that RasGAP is only homologous to *S. cerevisiae* and *Drosophila* GAP in its catalytic domain, and that the *S. cerevisiae* and *Drosophila* GAPs lack SH2 and SH3 domains. This suggests that mammalian RasGAP could perform functions that are not performed by *S. cerevisiae* and *Drosophila* GAPs.

A number of studies support the hypothesis that RasGAP contributes positively to signal transduction. RasGAP and Ras inhibit G protein-mediated ligand-dependent K^+ channel opening in atrial cell membranes. This effect is dependent on the presence of endogenous Ras. The Ras effect is itself dependent on RasGAP present in these membranes (Yatani et al 1990). The RasGAP N-terminal region, containing two SH2 domains and a SH3 domain, also exerts this effect, but this is independent of endogenous Ras (Martin et al 1992). Transient expression of the RasGAP N-terminal domain induces the expression of a c-*fos* promoter-driven luciferase gene (Medema et al 1992). Overexpression of the RasGAP N-terminus also results in a change in cell morphology and reduced adhesion to fibronectin. This effect may be mediated by p190, a GTPase activator of Rho that associates constitutively with the RasGAP N-terminus (McGlade et al 1993). The RasGAP N-terminus also appears to cooperate in cell transformation with v-Src mutants that have lost the ability to associate with the plasma membrane (DeClue et al 1993).

Although RasGAP was purified as a negative regulator of Ras, and the RasGAP C-terminal catalytic domain may normally negatively regulate Ras, the N-terminus may have a positive signaling function. The SH2 domains in the N-terminus interact with activated RPTKs and perhaps other proteins that are Tyr phosphorylated in response to ligand treatment. The RasGAP N-terminal domain binds constitutively to p190, perhaps through its SH3 domain, and possibly to other targets that remain to be identified. The signaling activity of the N-terminus in the full length RasGAP protein is dependent on active Ras, but independent of Ras when expressed separately from the C-terminus. Based on these properties, one model for RasGAP function is that Ras•GTP binding to the C-terminus frees the N-terminal domain and allows it to participate in a signal transduction pathway at the same time that the catalytic C-terminus is inactivating Ras. Alternatively, the binding of RasGAP SH2 domains to Tyr-phosphorylated RPTKs may allow the catalytic domain to interact with Ras and the SH3 domain to signal.

Since the Ras•GDP complex is stable, a GNRF is required to generate the active Ras•GTP complex. GNRFs were first identified genetically in *S. cerevisiae* and *Drosophila.* A number of Ras GNRFs have been purified recently based on their ability to increase the nucleotide exchange rate of recombinant Ras (Downward et al 1990b; Huang et al 1990; West et al 1990; Wolfman & Macara 1990). Several distinct Ras GNRFs have been cloned. These GNRFs share a region of ~ 310 amino acids with homology to CDC25 and SOS. This region most likely represents the catalytic domain (Bowtell et al 1992; Cen et al 1992; Martegani et al 1992; Shou et al 1992; Wei et al 1992). Mammalian SOS associates constitutively with GRB2 (Buday & Downward 1993a; Chardin et al 1993; Egan et al 1993; Li et al 1993; Rozakis-Adcock et al 1993). Little is known about the regulation of these GNRFs, but they are likely to be targets for regulation by surface receptors, as is the case for SOS.

The protooncoprotein VAV has recently been shown to be a GNRF for Ras. VAV is a 95-kd cytoplasmic protein that contains an N-terminal region with weak homology to helix-loop-helix and leucine zipper domains, a domain that is homologous to the GNRF domains of Bcr, Dbl, and the yeast CDC24, a Cys-rich domain with homology to PKC (Figure 4), DAG kinase and Raf-1, which may be involved in lipid binding, and at its C-terminus a SH2 domain localized between two SH3 domains (Adams et al 1992; Katzav et al 1989, 1991). Oncogenic activation of VAV involves overexpression and deletion of the first ~ 60 amino acid residues, which suggests that the N-terminus negatively regulates VAV (Coppola et al 1991; Katzav et al 1991). The SH2 domain is essential for transforming activity of VAV (Katzav 1993). VAV is exclusively expressed in hematopoietic cells during all stages of development and in all cell lineages (Adams et al 1992; Katzav et al 1989). When VAV is ectopically expressed in fibroblasts, it is phosphorylated on Ser and Tyr in quiescent cells (Bustelo et al 1992; Margolis et al 1992). Tyrosine phosphorylation increases dramatically in response to EGF and PDGF (Bustelo et al 1992; Margolis et al 1992). VAV also associates with the EGF and PDGF receptors upon activation, and the VAV SH2 domain binds to these RPTKs in vitro (Bustelo et al 1992; Margolis et al 1992). No increase in the Tyr phosphorylation of VAV was seen after stimulation of BAC1.2F5 macrophages with CSF-1 (Margolis et al 1992). Although the association of VAV with and phosphorylation of VAV by the EGF and PDGF receptors is clearly non-physiological since these receptors are not expressed in hematopoietic cells, the activation of B and T lymphocytes and mast cells, which naturally express VAV, results in VAV Tyr phosphorylation (Bustelo & Barbacid 1992; Bustelo et al 1992; Margolis et al 1992). In B cells, VAV associates transiently with a 70-kd protein upon activation (Bustelo & Barbacid 1992). Targeted disruption of the *vav* gene in the mouse shows that VAV is essential for early embryogenesis. However, apart from its homology with Bcr and Dbl, there were no clues to the function of VAV. Recently, however, it was shown that VAV immunoprecipitates contained GNRF activity that is specific for Ras

(Gulbins et al 1993, 1994). The GNRF activity was higher in lysates from stimulated T cells than in those from unstimulated T cells, and activity correlated with Tyr phosphorylation. VAV GNRF activity appears to be stimulated directly by Tyr phosphorylation and by binding to diacylglycerols (Gulbins et al 1993, 1994).

Ras•GTP levels can in principle be regulated through changes in GNRF or GAP activity. The question is what happens in the cell? In fibroblasts, serum or RPTK ligands activate Ras through an increase in nucleotide exchange, whereas GAP activity appears to be regulated by cell density in fibroblasts (Buday & Downward 1993b; Medema et al 1993; Zhang et al 1992a). In contrast, activation of T- and B-lymphocytes is associated with decreased GAP activity. In T-lymphocytes, regulation of GAP depends on PKC (Downward et al 1990a; Lazarus et al 1993). Stimulation of HEL cells with erythropoietin results in reduced GAP activity, whereas stimulation of PC12 cells with NGF results in activation of both guanine nucleotide exchange and GAP activity (Li et al 1992; Torti et al 1992).

Ras activation appears to play a central role in signal transduction by a large number of RPTKs. What are the targets for activated Ras? Recent work indicates that one of the targets that lies directly downstream of Ras is the MAP kinase pathway (Figure 5). Expression of a dominant-negative Ras mutant (Asn 17) inhibits activation of Raf-1 or MAP kinase in response to EGF, PDGF, insulin, NGF, FGF, and TPA (de Vries-Smits et al 1992; Thomas et al 1992; Wood et al 1992). Expression of activated Ras results in the activation of both the Raf-1 protein-serine kinase and MAP kinase (Leevers & Marshall 1992; Wood et al 1992). Recently, Ras•GTP was shown to physically interact with the N-terminal regulatory region of Raf-1 (Moodie et al 1993; Van Aelst et al 1993; Vojtek et al 1993; Warne et al 1993; Zhang et al 1993), which is necessary, but probably not sufficient, to activate Raf-1. Most likely a second event, such as phosphorylation, is required for Raf-1 activation, and this probably occurs once Raf-1 is translocated to the membrane as a result of its binding to Ras•GTP. Raf-1 can phosphorylate and activate MAP kinase kinase (also known as MEK), a dual specificity protein kinase, which in turn phosphorylates and activates MAP kinase at its Thr and Tyr activating sites (Dent et al 1992; Howe et al 1992; Kyriakis et al 1992). MAP kinase kinase can also be activated by Mos or by MEK kinase (Lange-Carter et al 1993; Posada et al 1993). MAP kinase is a family of enzymes, and at least one form can translocate into the nucleus, where it can phosphorylate and activate transcription factors such as Elk-1 (Treisman 1994). Other proteins that interact with Ras•GTP have been identified by the yeast two-hybrid screen, but it remains to be established that they are true downstream targets for Ras.

RPTK-Activated Nuclear Signaling Pathways

Thus far we have focused on the primary targets for RPTK action. However, the wide variety of cellular responses triggered by RPTK activation have to be explained as a consequence of the actions of these primary targets. It is beyond the scope of this review to discuss the signaling pathways activated by these proteins, but transcytoplasmic nuclear signaling leading to induction of gene expression obviously plays a crucial role in the mitogenic responses. We have touched on the ability of RPTKs to stimulate the MAP kinase pathway through Ras, which is known to activate specific transcription factors. RPTKs probably activate multiple MAP kinase pathways (Figure 5). For instance, the Jun N-terminal kinase, JNK1, which activates the Jun transcription factor, is a novel MAP kinase that is also activated by Ras (Dérijard et al 1994).

A second RPTK-activated transcytoplasmic nuclear signaling pathway is beginning to emerge. This pathway involves members of the Jak nonreceptor PTK family and the Stat (signal transducer and activator of transcription) transcription factor family. This pathway was first uncovered as an interferon (IFN)-mediated signal pathway in which interferon α binding to its receptor activates components of the transcriptional activator ISGF3 (interferon stimulated gene factor)(Schindler et al 1992a), which binds to a specific DNA sequence and mediates IFNα-stimulated gene expression (Levy et al 1988). ISGF3 is composed of a regulatory subunit ISGF3α and a DNA binding subunit ISGFγ (Kessler et al 1990; Levy et al 1989). ISGF3α is composed of three proteins Stat 84, Stat 91, and Stat 113 (Kessler et al 1990). ISGFγ is composed of a single 48-kd protein that directly binds the ISRE (IFN-stimulated response element). The individual components of ISGFα are present in the cytoplasm, and upon IFNα binding to its receptor they are activated by Tyr phosphorylation and then translocate to the nucleus, where they form an active complex with ISGFγ (Kessler et al 1990; Levy et al 1989). The affinity of ISGFγ for the ISRE increases 25-fold upon association with activated ISGFα (Kessler et al 1990). Stat 91 contains three heptad leucine repeats in the N-terminal region and a SH3 domain followed by a SH2 domain in the C-terminus (Fu 1992; Schindler et al 1992a). Stat 84 is derived from the same gene as Stat 91 by alternative splicing (Schindler et al 1992a) and lacks the 38 C-terminal residues, but retains the SH3 and SH2 domains. Stat 113 shows homology with Stat 84/91 (Fu et al 1992). Stat 91 is also present in GAF (IFNγ-activated factor), which binds to GAS (IFNγ activated site), a response element that is different from the ISRE (Lew et al 1991; Shuai et al 1992). Like ISGF3α, GAF is activated by Tyr phosphorylation following binding to a PTK (Fu 1992; Schindler et al 1992b; Shuai et al 1992). Activation and translocation to the nucleus of GAF/Stat 91 depend on Tyr phosphorylation of a single Tyr, Tyr 701 (Shuai et al

1993a), which facilitates dimerization through mutual interactions between P.Tyr701 on one p91 molecule and the SH2 domain of the other p91 molecule (Shuai et al 1994). In the case of IFNα and IFNγ, members of the Jak/Tyk family of PTKs are required for activation of the relevant Stat proteins, although it is not clear whether these PTKs are directly responsible for their phosphorylation (Müller et al 1993; Watling et al 1993).

PDGF-induced activation of the c-*fos* transcription is in part mediated by SIF (sis-inducible factor) binding to the SIE (sis-inducible element) in the c-*fos* promoter (Hayes et al 1987; Wagner et al 1990). EGF also induces a SIE-binding activity that resembles GAF. Following stimulation with EGF, several SIE-binding factors can be detected, some of which contain Stat 91 (Fu & Zhang 1993; Ruff-Jamison et al 1993a; Sadowski et al 1993). Stat 91 can be activated in vitro in crude system by activated EGF receptors and ATP (Sadowski & Gilman 1993). Activation is inhibited by anti-P.Tyr antibodies or SH2-containing fusion proteins. The Stat 91 SH2 domain binds to the activated EGF receptor (Fu & Zhang 1993). In response to EGF, Stat 91 becomes Tyr phosphorylated in the cytoplasm after which it appears in the nucleus (Fu & Zhang 1993; Ruff-Jamison et al 1993a; Sadowski et al 1993). EGF also stimulates Tyr phosphorylation and nuclear translocation of a novel 92-kd Stat protein, Stat3 (Ruff-Jamison et al 1993a; Sadowski et al 1993; Zhong et al 1994). Evidence is mounting that PDGF and CSF-1 can also induce activation of Stat 91, whereas FGF does not (Silvennoinen et al 1993).

It is not clear which PTKs are responsible for the Tyr phosphorylation of Stat 91 and related transcription factors after RPTK activation. In the case of EGF, the receptor itself may be the PTK involved (Fu & Zhang 1993; Sadowski & Gilman 1993), since Stat 91 has been shown to bind to the activated EGF receptor via its SH2 domain (Fu & Zhang 1993). However, EGF also activates Jak1 (Shuai et al 1993b), and it is possible that the critical Stat Tyr phosphorylation is carried out by an activated Jak PTK. A growing family of Stat proteins can form specific heterodimers with individual DNA-binding specificities. The combinations of Stat proteins activated by each RPTK will allow induction of a specific subset of GAS-regulated genes. One important feature of this nuclear signaling pathway is that it is Ras independent (Silvennoinen et al 1993).

In all likelihood PTKs, in addition to the Jak and Src family kinases, are activated downstream of the RPTKs. In B and T cells, the antigen receptors, which lack intrinsic PTK activity, activate nonreceptor PTKs in the Src family, and these in turn activate other PTKs such as Syk and ZAP70, a pair of related PTKs that each have two N-terminal SH2 domains, and BTK and ITK, another pair of related PTKs with N-terminal SH3 and SH2 domains (Cambier & Jensen 1994). It is worth noting that Tec, which is related to BTK/ITK, appears to be activated by c-Kit (J Ihle, personal communication).

Cooperation of RPTK Signaling Pathways

Signaling by RPTKs can be considered the sum of signaling pathways activated by the SH2 proteins that bind to the array of autophosphorylated sites on the RPTK itself, to phosphorylation sites on RPTK-bound adaptor proteins, and on docking proteins that are specifically phosphorylated by the RPTK. These signals are integrated to provide a coordinated nuclear response. Signal diversity can be increased by heterodimerization between different RPTK subfamily members. In addition to the specific nature of the SH2 protein-mediated signals activated by an individual RPTK, it is becoming clear that the strength and duration of the signal can alter the outcome of RPTK activation. For instance, in PC12 cells if the Ras/MAP kinase pathway is activated transiently, cell growth is the outcome, whereas if there is persistent activation, cell differentiation results (Traverse et al 1992). Signal strength in turn is regulated not only by receptor number, but also by the efficiency of the feedback pathways that normally downregulate RPTK signaling.

In principle the importance of individual RPTK-activated signaling pathways in a particular cellular response can be determined by testing whether cells expressing mutant RPTKs that lack individual autophosphorylation sites or combinations of phosphorylation sites are defective in specific responses. The PDGF, EGF, FGF, and CSF-1 RPTKs have been the most intensively studied using this approach. One difficulty is to identify a truly null test cell for the expression of the wild-type and mutant RPTKs. In some instances, use has been made of chimeric RPTK molecules in which the extracellular domain is replaced by the ligand-binding domain of an RPTK that is not expressed in a particular cell, but in most cases the RPTK has been expressed in a cell type that normally does not express that RPTK. Clearly this compromises interpretation of the results, although, since most of the major transcytoplasmic signaling pathways seem to be universally expressed, this has not proved an insurmountable difficulty. However, the fact that exogenously expressed RPTKs are usually present at significantly higher levels than endogenous receptors does pose a problem. Because of these caveats, the results obtained with ectopic RPTK expression have to be considered suspect until they can be confirmed by making mutations in situ in the relevant RPTK gene in a cell where it is normally expressed, a goal that can now be achieved through homologous recombination. The problem of using ectopic expression is underscored by the fact that the phenotypic consequences of mutating specific phosphorylation sites can differ depending on the cell type in which the RPTK is expressed. Thus the mutation of the PI3K binding sites in the PDGF β receptor decreases the mitogenic response in some cells but not in other cells. In the case of the CSF-1 receptor, mutation of the PI3K-binding site decreases mitogenic signaling in fibroblasts, but apparently not in myeloid cells, where instead it inhibits CSF-1-induced differentiation.

What general conclusions can be drawn about the requirement for individual signaling pathways in cellular responses from analysis of RPTK phosphorylation site mutants? For the PDGF β receptor, mutation of the PI3K-binding sites diminishes mitogenic signaling in at least in some cell types (Fantl et al 1992; Kazlauskas et al 1992; Yu et al 1991), decreases its ability to initiate a chemotactic response (Wennstrom et al 1994), and may reduce activation of the 70 K S6 kinase (Chung et al 1994). Mutation of the GAP-binding site in the PDGF β receptor, however, does not affect the mitogenic response (Fantl et al 1992). Likewise, mutation of the PLCγ-binding site in the PDGF β receptor has little or no effect on mitogenic signaling depending on the cell type in which the mutant receptor is expressed (Kashishian & Cooper 1993; Rönnstrand et al 1992; Valius et al 1993). However, a PDGF β receptor in which all the known sites except the PLCγ-binding site have been mutated retains some mitogenic activity (Valius & Kazlauskas 1993). Mutation of the SH-PTP2-binding sites in the PDGF β receptor also has little effect on mitogenic signaling. Mutation of the c-Src binding sites is complicated by the fact that mutation of both the required Tyr residues inactivates the receptor, but mutation of Tyr 759, which reduces c-Src binding by > 90%, does not diminish the mitogenic response (Mori et al 1993). In the case of the FGF receptor, mutation of the PLCγ binding site, which is the only reported substrate binding site, has no effect on mitogenic signaling, and it has been speculated that this is required for some other FGF receptor-mediated response (Mohammadi et al 1992; Peters et al 1992). For the CSF-1 receptor, mutation of either the PI3K-binding site or the GRB2-binding site diminishes CSF-1 stimulated growth of fibroblasts (van der Geer & Hunter 1993). Mutation of Tyr 706 in the kinase insert decreases CSF-1 induction of c-*fos* (van der Geer & Hunter 1991), and mutation of Tyr 809 in the catalytic domain diminishes c-*myc* induction (Roussel et al 1991). However, in the latter case Tyr 809 is equivalent to Tyr 416 in Src, and rather than its phosphorylation being directly required for activating a c-*myc* induction pathway, it seems likely that phosphorylation of Tyr 809 enhances overall catalytic activity and thereby increases phosphorylation at other autophosphorylation sites. Finally, for the EGF receptor, although the binding of specific SH2 proteins to individual receptor phosphorylation sites has been reported, it has been harder to establish critical roles for any of the individual EGF receptor autophosphorylation sites. Indeed, an EGF receptor lacking all known autophosphorylation sites can still deliver a mitogenic signal (Vega et al 1992).

The general conclusion is that no one signaling pathway is essential for RPTK-mediated mitogenic responses. However, the failure to see effects upon elimination of individual signaling pathways may be because in this type of analysis the mutant RPTK is generally overexpressed and ligand is used at saturating levels. Under such conditions, the RPTK may be able to achieve a mitogenic signaling threshold by activating only one or two of the pathways that it is capable of activating. In contrast, under normal circumstances in vivo, where both receptor numbers are lower and ligand is limiting, it seems likely that the activation of multiple pathways will be required for a complete cellular response. Indeed, in some cases there is evidence that cooperation between receptor-activated pathways is important in achieving maximal cellular responses in vitro. For instance, a CSF-1 receptor mutant lacking both the PI3K- and the GRB2-binding sites has no growth-promoting activity in rat fibroblasts, whereas elimination of either site alone only reduces the growth response (van der Geer & Hunter 1993). The fact that most RPTKs have the capacity to trigger multiple signaling pathways through the binding of SH2-containing proteins to the array of autophosphorylation sites also suggests that multiple RPTK-activated signal pathways are needed to achieve optimal responses.

If, as seems likely, cooperation between pathways is important for RPTK signaling, an important question is whether such cooperation occurs at the level of single RPTK dimers through the association of multiple signaling proteins. Although a single RPTK molecule probably can be phosphorylated at multiple sites, the evidence that this actually occurs is not compelling. Often the overall stoichiometry of RPTK phosphorylation is not that high, which suggests that on average each activated RPTK molecule will only be autophosphorylated at one or two sites (one exception is a RPTK like the insulin receptor, where there is processive phosphorylation of a set of three Tyr within the catalytic domain that are required for maximal PTK activity, but these residues do not appear to be binding sites for SH2 proteins). Within an activated RPTK dimer, the individual molecules can be phosphorylated at different sites, and this would allow two different SH2 proteins to bind to the same dimer. Indeed, the existence of complexes containing more than one SH2 protein is demonstrated by the fact that one can detect a second SH2 protein in immunoprecipitates of a different SH2-containing protein in PDGF-treated cells (Kaplan et al 1990). In addition, it is possible to detect both PI3K p85 and PLCγ simultaneously associated with TrkA in cells overexpressing all three proteins (Obermeier et al 1993b). However, evidence for the original concept of a signal transfer particle with multiple SH2-containing proteins bound to a single RPTK dimer has not been forthcoming (Ullrich & Schlessinger 1990). Moreover, many RPTK targets contain two SH2 domains, and stable binding to activated RPTK dimers requires binding of both SH2 domains in *cis* or in *trans* (Cooper & Kashishian 1993), which reduces the number of SH2 proteins able to bind simultaneously. It is also possible that there is competition for the binding of SH2 proteins to closely spaced phosphorylation sites. In summary, there is no compelling evidence for interactions between separate SH2 proteins bound to an activated RPTK dimer.

In this regard we need to know whether SH2 proteins signal only when bound to the activated RPTK, or whether they can dissociate in an activated form. The former must be true when target activation occurs via an allosteric mechanism, but when activation is mediated by target phosphorylation, then the phosphorylated, activated protein could dissociate from the RPTK. Only if the activated target dissociates can there be amplification of the signal pathway at the level of the primary target protein. Although phosphorylated SH2-containing substrates do not readily dissociate from RPTKs in vitro, there is some evidence that this occurs in vivo. For instance, not all the Tyr-phosphorylated PLCγ in EGF-treated A431 cells is associated with the EGF receptor. Moreover, there is rapid translocation of a significant fraction of the PLCγ population to the membrane, which may occur in part through the association of PLCγ via its SH2 domain with other membrane proteins that are phosphorylated by the EGF receptor. This suggests that despite the high binding affinity of the SH2 domain, there may be a mechanism for dissociating proteins from activated RPTKs following phosphorylation. Chaperonin-mediated unfolding might be one way to achieve this end.

Another question is whether SH2 proteins remain associated with RPTK dimers as they are internalized following ligand binding, and whether they are degraded with the receptors as they enter the lysosomal pathway. The fact that there is no evidence for ligand-activated substrate degradation implies that SH2 proteins do dissociate prior to the RPTK entering multivesicular bodies. However, annexin I and SHC (Wada et al 1992) do remain associated with endosomal vesicles containing activated RPTKs. The possibility that activation of PI3K is involved in relocalizing the activated RPTK, giving it access to different substrates, may be important in this regard.

Regardless of whether signaling pathway cooperation occurs at the level of SH2 proteins bound to the activated RPTK dimer, the SH2 recruitment mechanism provides a means of ensuring that the necessary signal pathways are activated in a microenvironment near an activated RPTK molecule in the plasma membrane. Under such conditions, even if activated SH2-containing proteins are released, they can stay close to the receptor and may facilitate pathway crosstalk that is required for specific responses. In vivo, where only a few hundred RPTK molecules per cell are sufficient to provoke a mitogenic response, it could be important to have localized activation of signal pathways in order to achieve a threshold level of signal.

RPTK Signal Down-Regulation

While it is beyond the scope of this article to thoroughly review how RPTK-activated signals are down-regulated, a number of mechanisms are known that act at the level of the RPTK itself. Ligand-induced RPTK internalization leads to degradation of the RPTK by the lysosomal pathway, thus destroying activated RPTKs and decreasing the level of available surface receptors. Based on mutant analysis, down-regulation is generally found to require RPTK activity, but this may be due to phosphorylation of a substrate such as PI3K, rather than RPTK autophosphorylation. Dephosphorylation of the activated RPTK (and its substrates) by PTPs must play an important role, and nonreceptor PTPs like SH-PTP1 have been implicated in this process. Whether receptor-like PTPs (RPTPs) are also involved is unclear, but the fact that some RPTPs are capable of homotypic interaction via their extracellular domains suggests that they might be activated when cells come into contact, thus down-regulating signals generated by RPTKs when cells are crowded under conditions that are known to inhibit growth. Finally, there is considerable evidence that RPTK activity is subject to negative feedback through phosphorylation by protein-serine/threonine kinases. For instance, PKC phosphorylates the EGF receptor at Thr 654 and decreases its activity. Several protein-serine/threonine kinases that are activated by RPTK signaling, such as PKC, MAP kinase, and Ca^{2+}/calmodulin-dependent protein kinase, have been shown to phosphorylate and regulate RPTK activity. This last mechanism also affords the possibility of heterologous desensitization of unrelated receptors. PDGF-mediated down-regulation of the EGF receptor is an example of this sort of transmodulation and occurs via activation of PKC by the PDGF receptor leading to phosphorylation at Thr 654 of the EGF receptor.

SUMMARY AND PERSPECTIVES

More than 50 RPTK genes that exist in mammals are organized into 14 distinct RPTK subfamilies. These RPTKs have roles in cell growth, cell survival, cell metabolism, and differentiation. Cognate ligand binding to RPTKs causes these receptors to dimerize and autophosphorylate at specific tyrosine residues. These phosphorylated tyrosines can then serve as docking sites for SH2 domain-containing signaling or adaptor proteins. Signaling molecules may be activated allosterically upon binding, as a result of tyrosine phosphorylation, as a result of relocalization in close proximity to their substrates at the plasma membrane, or by a combination of the above.

A large number of SH2-containing signaling and adaptor proteins are present in the cytoplasm, ready to interact with activated growth factor RPTKs. The activation of these proteins can, in turn, initiate signaling pathways leading to the nucleus or other cellular targets. Ras appears to be central in executing the response to RPTK activation, and at least four SH2-containing proteins, GRB2, SHC, RasGAP, and VAV, are either directly or indirectly involved in regulation of Ras activity. Ras activation results in stimulation of the MAP kinase pathway which, through the phosphorylation of transcription factors, can lead to gene induction. Other adaptor proteins such as CRK and NCK activate mitogenic pathways of an unknown nature. It is likely that the study of such proteins will identify novel signaling pathways. The recent discovery of a Ras-independent nuclear signaling pathway, which utilizes Jak family PTKs and which can be activated by RPTKs, emphasizes the possibility that there are novel signaling pathways waiting to be identified. RPTK activation of phospholipase C and phospholipase A_2 results in production of many second messengers, which can also activate signaling pathways. Other RPTK targets such as Src family members and the SH2-containing PTPs may initiate their own specific signaling pathways or modulate Ras activation or signal transduction downstream of Ras. The role of PI3K in RPTK signaling remains an enigma. It could activate a unique signaling pathway, or otherwise be involved in membrane-associated events such as RPTK translocation that are required for RPTK signaling. In cells that overexpress RPTKs, the activation of a single signaling pathway may be sufficient to drive cells into mitosis. However, in normal cells, which express fewer receptors, the coordinated activation of several pathways that result from the interaction of different signaling proteins with activated RPTKs may be required to elicit a complete response to ligand stimulation.

Can we use the recent explosion of information about RPTK-mediated signaling mechanisms to predict what response activation of a RPTK will induce? The sequence specificity of SH2 domain binding encourages the hope that one could predict what signaling pathways will be activated by a specific RPTK based on the sequences around its autophosphorylation sites. This, coupled with the knowledge of how the different RPTK-signaling pathways interact, might ultimately allow one to predict the nature of the cellular responses to the binding of RPTK ligand. At present, however, the fact that some SH2 domains have overlapping binding specificities, that a single SH2 domain can bind more than one consensus sequence, and that some RPTK autophosphorylation sites bind more than one SH2 protein complicates predictions of which signaling pathways can be activated by an RPTK. There is also increasing evidence that RPTKs in the same subfamily can heterodimerize in the presence of ligand. This adds further complexity, since a heterodimer, by activating different combinations of cooperating signal pathways, may generate a response distinct from that of either homodimer alone. Nevertheless, the continued study of RPTK-signaling mechanisms at the biochemical level and through genetic analysis will undoubtedly lead to a clearer picture of how an integrated cellular response to ligand binding is achieved.

ACKNOWLEDGMENTS

We would like to thank Anne Marie Quinn for help with Figures 2 and 3, and Mike Fox, Cary Lai, Joe Ruiz, Mark Henkemeyer, Saburo Hanafusa, Jim Ihle, and Kari Alitalo for sharing unpublished data.

Literature Cited

Adachi M, Sekiya M, Miyachi T, Matsuno K, Hinoda Y, et al. 1992. Molecular cloning of a novel protein-tyrosine phosphatase SH-PTP3 with sequence similarity to the src-homology region 2. *FEBS Lett.* 314:335–39

Adams JM, Houston H, Allen J, Lints T, Harvey R. 1992. The hematopoietically expressed vav proto-oncogene shares homology with the dbl GDP-GTP exchange factor, the *bcr* gene and a yeast gene (*CDC24*) involved in cytoskeletal organization. *Oncogene* 7:611–18

Adari H, Lowy DR, Willumsen BM, Der CJ, McCormick F. 1988. Guanosine triphosphatase activating protein (GAP) interacts with the p21 ras effector binding domain. *Science* 240:518–21

Ahmad S, Banville D, Zhao Z, Fischer EH, Shen SH. 1993. A widely expressed human protein-tyrosine phosphatase containing src homology 2 domains. *Proc. Natl. Acad. Sci. USA* 90:2197–201

Amaya E, Musci TJ, Kirschner MW. 1991. Expression of a dominant negative mutant of the FGF receptor disrupts mesoderm formation in Xenopus embryos. *Cell* 66:257–70

Ambrosio L, Mahowald AP, Perrimon N. 1989. Requirement of the *Drosophila raf* homologue for *torso* function. *Nature* 342:288–91

Amrein KE, Flint N, Panholzer B, Burn P. 1992. Ras GTPase-activating protein: a substrate and a potential binding protein of the protein-tyrosine kinase $p56^{lck}$ *Proc. Natl. Acad. Sci. USA* 89:3343–46

Anderson D, Koch CA, Grey L, Ellis C, Moran MF, Pawson T. 1990. Binding of SH2 domains of phospholipase C gamma 1, GAP, and Src to activated growth factor receptors. *Science* 250:979–82

Aprelikova O, Pajusola K, Partanen J, Armstrong E, Alitalo R, et al. 1992. *FLT4*, a novel class III receptor tyrosine kinase in chromosome 5q33-qter. *Cancer Res.* 52:746–48

Aroian RV, Koga M, Mendel JE, Ohshima Y, Sternberg PW. 1990. The *let*-23 gene necessary for *Caenorhabditis elegans* vulval induction encodes a tyrosine kinase of the EGF receptor subfamily. *Nature* 348:693–99

Aroian RV, Lesa GM, Sternberg PW. 1994. Mutations in the *Caenorhabditis elegans let-23* EGFR-like gene define elements important for cell-type specificity and function. *EMBO J.* 13:360–66

Auger KR, Serunian LA, Soltoff SP, Libby P, Cantley LC. 1989. PDGF-dependent tyrosine phosphorylation stimulates production of novel polyphosphoinositides in intact cells. *Cell* 57:167–75

Avivi A, Zimmer Y, Yayon A, Yarden Y, Givol D. 1991. Flg-2, a new member of the family of fibroblast growth factor receptors. *Oncogene* 6:1089–92

Axelrod J, Burch RM, Jelsema CL. 1988. Receptor mediated activation of phospholipase A2 via GTP-binding proteins: arachidonic acid and its metabolites as second messengers. *Trends Neurosci.* 11:117–23

Backer JM, Myers MG Jr, Shoelson SE, Chin DJ, Sun XJ, et al. 1992. Phosphatidylinositol 3′-kinase is activated by association with IRS-1 during insulin stimulation. *EMBO J.* 11:3469–79

Baker NE, Rubin GM. 1989. Effect on eye development of dominant mutations in *Drosophila* homologue of the EGF receptor. *Nature* 340:150–53

Baltensperger K, Kozma LM, Cherniack AD, Klarlund JK, Chawla A, et al. 1993. Binding of the Ras activator son of sevenless to insulin receptor substrate-1 signaling complexes. *Science* 260:1950–52

Barbacid M. 1993. Nerve growth factor: a tale of two receptors. *Oncogene* 8:2033–44

Bardelli A, Maina F, Gout I, Fry MJ, Waterfield MD, et al. 1992. Autophosphorylation promotes complex formation of recombinant hepatocyte growth factor receptor with cytoplasmic effectors containing SH2 domains. *Oncogene* 7:1973–78

Bar Sagi D, Rotin D, Batzer A, Mandiyan V, Schlessinger J. 1993. SH3 domains direct cellular localization of signaling molecules. *Cell* 74:83–91

Bartley TD, Hunt R, Welcher AA, Boyle W, Parker V, et al. 1994. B61 is a ligand for the ECK receptor protein-tyrosine kinase. *Nature* 368:558–60

Bastien L, Ramachandran C, Liu S, Adam M. 1993. Cloning, expression and mutational analysis of SH-PTP2, human protein-tyrosine phosphatase. *Biochem. Biophys. Res. Commun.* 196:124–33

Batzer AG, Rotin D, Ureña JM, Skolnik EY, Schlessinger J. 1994. Hierarchy of binding sites for Grb2 and Shc on the epidermal growth factor receptor. *Mol. Cell. Biol.* In press

Bauskin AR, Alkalay I, Ben-Neriah Y. 1991.

Redox regulation of a protein tyrosine kinase in the endoplasmic reticulum. *Cell* 66:685–96

Bazan JF. 1991. Genetic and structural homology of stem cell factor and macrophage colony-stimulating factors. *Cell* 65:9–10

Ben-Av P, Liscovitch M. 1989. Phospholipase D activation by the mitogens platelet-derived growth factor and 12-*O*-tetradecanoylphorbol 13-acetate in NIH-3T3 cells. *FEBS. Lett.* 259:64–66

Bierman AJ, Koenderman L, Tool AJ, de Laat SW. 1990. Epidermal growth factor and bombesin differ strikingly in the induction of early responses in Swiss 3T3 cells. *J. Cell. Physiol.* 142:441–48

Billah MM, Anthes JC, Mullmann TJ. 1991. Receptor-coupled phospholipase D: regulation and functional significance. *Biochem. Soc. Trans.* 19:324–29

Birge RB, Fajardo JE, Mayer BJ, Hanafusa H. 1992. Tyrosine-phosphorylated epidermal growth factor receptor and cellular p130 provide high affinity binding substrates to analyze Crk-phosphotyrosine-dependent interactions in vitro. *J. Biol. Chem.* 267:10588–95

Birge RB, Fajardo JE, Reichman C, Shoelson SE, Songyang Z, et al. 1993. Identification and characterization of a high-affinity interaction between v-Crk and tyrosine-phosphorylated paxillin in CT10-transformed fibroblasts. *Mol. Cell. Biol.* 13:4648–56

Bjorge JD, Chan T-O, Antczak M, Kung H-J, Fujita DJ. 1990. Activated type I phosphatidylinositol kinase is associated with the epidermal growth factor (EGF) receptor following EGF stimulation. *Proc. Natl. Acad. Sci. USA* 87:3816–20

Boguski MS, McCormick F. 1993. Proteins regulating Ras and its relatives. *Nature* 366: 643–54

Bohme B, Holtrich U, Wolf G, Luzius H, Grzeschik KKS, Rubsamen-Waigmann H. 1993. PCR mediated detection of a new human receptor-tyrosine-kinase, HEK 2. *Oncogene* 8:2857–62

Bolen JB, Rowley RB, Spana C, Tsygankov A. 1992. The Src family of tyrosine protein kinases in hemopoietic signal transduction. *FASEB J.* 6:3403–9

Bolen JB, Veillette A. 1989. A function for the *lck* proto-oncogene. *Trends Biol. Sci.* 14: 404–7

Bonfini L, Karlovich CA, Dasgupta C, Banerjee U. 1992. The Son of sevenless gene product: a putative activator of Ras. *Science* 255:603–6

Bonventre JV, Gronich JH, Nemenoff RA. 1990. Epidermal growth factor enhances glomerular mesangial cell soluble phospholipase A_2 activity. *J. Biol. Chem.* 265: 4934–38

Booker GW, Breeze AL, Downing AK, Panayotou G, Gout I, et al. 1992. Structure of an SH2 domain of the p85 alpha subunit of phosphatidylinositol-3-OH kinase. *Nature* 358:684–87

Booker GW, Gout I, Downing AK, Driscoll PC, Boyd J, et al. 1993. Solution structure and ligand-binding site of the SH3 domain of the p85α subunit of phosphatidylinositol 3-kinase. *Cell* 73:813–22

Bortner DM, Ulivi M, Roussel MF, Ostrowski MC. 1991. The carboxy-terminal catalytic domain of the GTPase-activating protein inhibits nuclear signal transduction and morphological transformation mediated by the CSF-1 receptor. *Genes Dev.* 5:1777–85

Bossemeyer D, Engh RA, Kinzel V, Ponstingl H, Huber R. 1993. Phosphotransferase and substrate binding mechanism of the cAMP-dependent protein kinase catalytic subunit from porcine heart as deduced from the 2.0 Å structure of the complex with Mn^{2+} adenyl imidodiphosphate and inhibitors peptide PKI(5–24). *EMBO J.* 12:849–59

Bottaro DP, Rubin JS, Faletto DL, Chen AML, Kmiecick TE, et al. 1991. Identification of the hepatocyte growth factor receptor as the c-*met* proto-oncogene. *Science* 251:802–4

Bourne HR, Sanders DA, McCormick F. 1991. The GTPase superfamily: conserved structure and molecular mechanism. *Nature* 349: 117–27

Bowtell D, Fu P, Simon M, Senior P. 1992. Identification of murine homologues of the *Drosophila son of sevenless* gene: potential activators of ras. *Proc. Natl. Acad. Sci. USA* 89:6511–15

Brachmann R, Lindquist PB, Nagashima M, Kohr W, Lipari T, et al. 1989. Transmembrane TGF-alpha precursors activate EGF/TGF-alpha receptors. *Cell* 56:691–700

Brannan CI, Lyman SD, Williams DE, Eisenman J, Anderson DM, et al. 1991. Steel-Dickie mutation encodes a c-kit ligand lacking transmembrane and cytoplasmic domains. *Proc. Natl. Acad. Sci. USA* 88:4671–74

Broek D, Toda T, Michaeli T, Levin L, Birchmeier C, et al. 1987. The S. cerevisiae *CDC25* gene product regulates the *RAS*/adenylate cyclase pathway. *Cell* 48:789–99

Buday L, Downward J. 1993a. Epidermal growth factor regulates p21ras through the formation of a complex of receptor, Grb2 adapter protein, and Sos nucleotide exchange factor. *Cell* 73:611–20

Buday L, Downward J. 1993b. Epidermal growth factor regulates the exchange rate of guanine nucleotides on $p21^{ras}$ in fibroblasts. *Mol. Cell. Biol.* 13:1903–10

Burgering BMT, Medema RH, Maassen JA, van de Wetering ML, van der Eb AJ, et al. 1991. Insulin stimulation of gene expression mediated by p21ras activation. *EMBO J.* 10: 1103–9

Burgering BMT, Snijders AJ, Maassen JA, van der Eb AJ, Bos JL. 1989. Possible involvement of normal p21 H-*ras* in the insulin/insulin-like growth factor 1 signal transduction pathway. *Mol. Cell. Biol.* 9:4312–22

Burgess WH, Dionne CA, Kaplow J, Mudd R, Friesel R, et al. 1990. Characterization and cDNA cloning of phospholipase C-gamma, a major substrate for heparin-binding growth factor 1 (acidic fibroblast growth factor)-activated tyrosine kinase. *Mol. Cell. Biol.* 10: 4770–77

Burns LA, Karnitz LM, Sutor SL, Abraham RT. 1993. Interleukin-2-induced tyrosine phosphorylation of p52shc in T lymphocytes. *J. Biol. Chem.* 268:17659–61

Bustelo XR, Barbacid M. 1992. Tyrosine phosphorylation of the vav proto-oncogene product in activated B cells. *Science* 256: 1196–99

Bustelo XR, Ledbetter JA, Barbacid M. 1992. Product of vav proto-oncogene defines a new class of tyrosine protein kinase substrates. *Nature* 356:68–71

Calesac C, Hancock JF, Marshall CJ, Hall A. 1988. The cytoplasmic protein GAP is implicated as the target for regulation by the *ras* gene product. *Nature* 332:548–50

Cambier JC, Jensen WA. 1994. The hetero-oligomeric antigen receptor complex and its coupling to cytoplasmic effectors. *Curr. Opin. Genet. Dev.* 4:55–63

Campos-Gonzalez R, Glenney JR. 1992. Tyrosine phosphorylation of mitogen-activated protein kinase in cells with tyrosine kinase-negative epidermal growth factor receptors. *J. Biol. Chem.* 267:14535–38

Cantley LC, Auger KR, Carpenter C, Duckworth B, Graziani A, et al. 1991. Oncogenes and signal transduction. *Cell* 64: 281–302

Carpenter CL, Auger KR, Chanudhuri M, Yoakim M, Schaffhausen B, et al. 1993a. Phosphoinositide 3-kinase is activated by phosphopeptides that bind to the SH2 domains of the 85-kDa subunit. *J. Biol. Chem.* 268:9478–83

Carpenter CL, Auger KR, Duckworth BC, Huo W-M, Schaffhausen B, Cantley LC. 1993b. A tightly associated serine/threonine protein kinase regulates phosphoinositide 3-kinase activity. *Mol. Cell. Biol.* 13:1657–65

Carpenter CL, Cantley LC. 1990. Phosphoinositide kinases. *Biochemistry* 29: 11147–56

Carpenter CL, Duckworth BC, Auger KR, Cohen B, Schaffhausen BS, Cantley LC. 1990. Purification and characterization of phosphoinositide 3-kinase from rat liver. *J. Biol. Chem.* 265:19704–11

Carpenter G. 1992. Receptor tyrosine kinase substrates: *src* homology domains and signal transduction. *FASEB J.* 6:3283–89

Cen H, Papageorge AG, Zippel R, Lowy DR, Zhang K. 1992. Isolation of multiple mouse cDNAs with coding homology to *Saccharomyces cerevisiae CDC25:* identification of a region related to Bcr, Vav, Dbl and CDC24. *EMBO J.* 11:4007–15

Chan J, Watt V. 1991. *eek* and *erk*, new members of the eph subclass of receptor protein-tyrosine kinases. *Oncogene* 6:1057–61

Chao MV. 1992. Neurotrophin receptors: a window into neuronal differentiation. *Neuron* 9:583–93

Charbonneau H, Tonks NK, Walsh KA, Fischer EH. 1988. The leukocyte common antigen (CD45): A putative receptor-linked protein tyrosine phosphatase. *Proc. Natl. Acad. Sci. USA* 85:7182–86

Chardin P, Camonis JH, Gale NW, van Aelst L, Schlessinger J, et al. 1993. Human Sos1: a guanine nucleotide exchange factor for Ras that binds to GRB2. *Science* 260:1338–43

Cheon H-G, LaRochelle WJ, Bottaro DP, Burgess WH, Aaronson SA. 1994. High-affinity binding sites for related fibroblast growth factor ligands reside within different receptor immunoglobulin-like domains. *Proc. Natl. Acad. Sci. USA* 91:989–93

Chou MM, Fajardo JE, Hanafusa H. 1992. The SH2- and SH3-containing Nck protein transforms mammalian fibroblasts in the absence of elevated phosphotyrosine levels. *Mol. Cell. Biol.* 12:5834–42

Chou Y-H, Hayman MJ. 1991. Characterization of a member of the immunoglobulin gene superfamily that possibly represents an additional class of growth factor receptor. *Proc. Natl. Acad. Sci. USA* 88:4897–901

Choudhury GG, Wang LM, Pierce J, Harvey SA, Sakaguchi AY. 1991. A mutational analysis of phosphatidylinositol-3-kinase activation by human colony-stimulating factor-1 receptor. *J. Biol. Chem.* 266:8068–72

Chung J, Grammer TC, Lemon KP, Kazlauskas A, Blenis J. 1994. PDGF-dependent regulation of pp70-S6 kinase is coupled to receptor-dependent binding and activation of phosphatidylinositol 3-kinase. *Nature.* In press

Cicchetti P, Mayer BJ, Thiel G, Baltimore D. 1992. Identification of a protein that binds to the SH3 region of Abl and is similar to Bcr and GAP-rho. *Science* 257:803–6

Clark SG, Stern MJ, Horvitz HR. 1992. *C. elegans* cell-signalling gene sem-5 encodes a protein with SH2 and SH3 domains. *Nature* 356:340–44

Cobb BS, Schaller MD, Leu T-H, Parsons JT. 1994. Stable association of $pp60^{src}$ and $pp59^{fyn}$ with the focal adhesion-associated protein tyrosine kinase, $pp125^{FAK}$. *Mol. Cell. Biol.* 14:147–55

Cochet C, Filhol O, Payrastre B, Hunter T, Gill GN. 1991. Interaction between the epidermal growth factor receptor and phosphoinositide kinases. *J. Biol. Chem.* 266:637–44

Cockcroft S, Thomas GMH. 1992. Inositol-lipid-specific phospholipase C isoenzymes and their differential regulation by receptors. *Biochem. J.* 288:1–14

Cook SJ, Wakelam MJ. 1992. Epidermal growth factor increases sn-1,2-diacylglycerol levels and activates phospholipase D-catalysed phosphatidylcholine breakdown in Swiss 3T3 cells in the absence of inositol-lipid hydrolysis. *Biochem. J.* 285:247–53

Cooper JA, Kashishian A. 1993. In vivo binding properties of SH2 domains from GTPase-activating proteins and phosphatidylinositol 3-kinase. *Mol. Cell. Biol.* 13:1737–45

Coppola J, Bryant S, Koda T, Conway D, Barbacid M. 1991. Mechanism of activation of the vav protooncogene. *Cell Growth Differ.* 2:95–105

Courtneidge S, Dhand R, Pilat D, Twamley GM, Waterfield MD, Roussel MF. 1993. Activation of Src family kinases by colony simulating factor-1, and their association with its receptor. *EMBO J.* 12:943–50

Courtneidge SA, Heber A. 1987. An 81 kd protein complexed with middle T antigen and pp60c-src: a possible phosphatidylinositol kinase. *Cell* 50:1031–37

Courtneidge SA, Kypta RM, Cooper JA, Kazlauskas A. 1991. Platelet-derived growth factor receptor sequences important for binding of src family tyrosine kinases. *Cell Growth Differ.* 2:483–86

Crowley C, Spencer SD, Nishimura MC, Chen KS, Pitts-Meek S, et al. 1994. Mice lacking nerve growth factor display perinatal loss of sensory and sympathetic neurons yet develop basal forebrain cholinergic neurons. *Cell* 76:1001–11

Crumley G, Bellot F, Kaplow JM, Schlessinger J, Jaye M, Dionne CA. 1991. High-affinity binding and activation of a truncated FGF receptor by both aFGF and bFGF. *Oncogene* 6:2255–62

Cuadrado A, Carnero A, Dolfi F, Jiménez B, Lacal JC. 1993. Phosphorylcholine: a novel second messenger essential for the mitogenic activity of growth factors and oncogenes. *Oncogene* 8:2959–68

Culouscou JM, Plowman GD, Carlton GW, Green JM, Shoyab M. 1993. Characterization of a breast cancer cell differentiation factor that specifically activates the HER4/p180erbB4 receptor. *J. Biol. Chem.* 268: 18407–10

Cunningham BC, Ultsch M, De Vos AM, Mulkerrin MG, Clauser KR, Wells JA. 1991. Dimerization of the extracellular domain of the human growth hormone receptor by a single hormone molecule. *Science* 254:821–25

Cutler RL, Liu L, Damen JE, Krystal G. 1993. Multiple cytokines induce the tyrosine phosphorylation of Shc and its association with Grb2 in hemopoietic cells. *J. Biol. Chem.* 268:21463–65

Dai W, Pan H, Hassanain H, Gupta S, Murphy MJ. 1994. Molecular cloning of a novel receptor tyrosine kinase, tif, highly expressed in human ovary and testis. *Oncogene* 9:975–79

Damen JE, Liu L, Cutler RL, Krystal G. 1993. Erythropoietin stimulates the tyrosine phosphorylation of Shc and its association with Grb2 and a 145-Kd tyrosine phosphorylated protein. *Blood* 82:2296–303

De Bondt HL, Rosenblatt J, Jancarik J, Jones HD, Morgan DO, Kim S. 1993. Crystal structure of cyclin-dependent kinase 2. *Nature* 363:595–602

DeClue JE, Vass WC, Johnson MR, Stacey DW, Lowy DR. 1993. Functional role of GTPase-activating protein in cell transformation by pp60v-src. *Mol. Cell. Biol.* 13:6799–809

DeClue JE, Zhang K, Redford P, Vass WC, Lowy DR. 1991. Suppression of src transformation by overexpression of full-length GTPase-activating protein (GAP) or of the GAP C terminus. *Mol. Cell. Biol.* 11:2819–25

den Hertog J, Tracy S, Hunter T. 1994. Phosphorylation of receptor protein-tyrosine phosphatase a on tyrosine 789, a binding site for the SH3-SH2-SH3 adaptor protein GRB-2 in vivo. *EMBO J.* In press

Dent P, Haser W, Haystead TA, Vincent LA, Roberts TM, Sturgill TW. 1992. Activation of mitogen-activated protein kinase kinase by v-Raf in NIH 3T3 cells and in vitro. *Science* 257:1404–7

Dérijard B, Hibi M, Wu I-H, Barrett T, Su B, et al. 1994. JNK1: A protein kinase stimulated by UV light and Ha-Ras that binds and phosphorylates the c-Jun activation domain. *Cell* 76:1025–37

de Vries C, Escobedo JA, Ueno H, Houck K, Ferrara N, Williams LT. 1992. The fms-like tyrosine kinase, a receptor for vascular endothelial growth factor. *Science* 255:989–91

de Vries-Smits AMM, Burgering BM, Leevers SJ, Marshall CJ, Bos JL. 1992. Involvement of p21ras in activation of extracellular signal-regulated kinase 2. *Nature* 357:602–4

Dhand R, Hiles I, Panayotou G, Roche S, Fry MJ, et al. 1994. PI 3-kinase is a dual specificity enzyme: autoregulation by an intrinsic protein-serine kinase activity. *EMBO J.* 13: 522–33

Dickson B, Hafen E. 1994. Genetics of signal transduction in invertebrates. *Curr. Opin. Genet. Dev.* 4:64–70

Diekmann D, Brill S, Garret MD, Totty N, Hsuan J, et al. 1991. Bcr encodes a GTPase-activating protein for p21^{rac}. *Nature* 351: 400–2

Dilworth SM, Brewster CEP, Jones MD, Lanfrancone L, Pelicci G, Pelicci PG. 1994. Transformation by polyoma virus middle T-antigen involves the binding and tyrosine phosphorylation of Shc. *Nature* 367:87–90

Di Marco E, Cutuli N, Guerra L, Cancedda R, De Luca M. 1993. Molecular cloning of trkE, a novel trk-related putative tyrosine kinase receptor isolated from normal human keratinocytes and widely expressed by normal human tissues. *J. Biol. Chem.* 268: 24290–95

Dosil M, Wang S, Lemischka IR. 1993. Mitogenic signalling and substrate specificity of the Flk2/Flt3 receptor tyrosine kinase in fibroblasts and interleukin 3-dependent hematopoietic cells. *Mol. Cell. Biol.* 13:6572–85

Downing JR, Margolis BL, Zilberstein A, Ashmun RA, Ullrich A, et al. 1989. Phospholipase C-γ, a substrate for PDGF receptor kinase, is not phosphorylated on tyrosine during the mitogenic response to CSF-1. *EMBO J.* 8:3345–50

Downward J, Graves JD, Warne PH, Rayter S, Cantrell DA. 1990a. Stimulation of p21ras upon T-cell activation. *Nature* 346:719–23

Downward J, Riehl R, Wu L, Weinberg RA. 1990b. Identification of a nucleotide exchange-promoting activity for p21ras. *Proc. Natl. Acad. Sci. USA* 87:5998–6002

Dumont DJ, Yamaguchi TP, Conlon RA, Rossant J, Breitman ML. 1992. *tek,* a novel tyrosine kinase gene located on mouse chromosome 4, is expressed in endothelial cells and their presumptive precursors. *Oncogene* 7:1471–80

Duronio V, Welham MJ, Abraham S, Dryden P, Schrader JW. 1992. p21ras activation via hemopoietin receptors and c-kit requires tyrosine kinase activity but not tyrosine phosphorylation of p21ras GTPase-activating protein. *Proc. Natl. Acad. Sci. USA* 89:1587–91

Egan SE, Giddings BW, Brooks MW, Buday L, Sizeland AM, Weinberg RA. 1993. Association of Sos Ras exchange protein with Grb2 is implicated in tyrosine kinase signal transduction and transformation. *Nature* 363:45–51

Eichmann A, Marcelle C, Breant C, Le Douarin NM. 1993. Two molecules related to the VEGF receptor are expressed in early endothelial cells during avian embryonic development. *Mech. Dev.* 42:33–48

Eiseman E, Bolen JB. 1990. Src-related tyrosine protein kinases as signalling components in hematopoietic cells. *Cancer Cells* 2:303–10

Ellis C, Moran M, McCormick F, Pawson T. 1990. Phosphorylation of GAP and GAP-associated proteins by transforming and mitogenic tyrosine kinases. *Nature* 343:377–81

Ellis L, Clauser E, Morgan DO, Edery M, Roth RA, Rutter WJ. 1986. Replacement of insulin receptor tyrosine residues 1162 and 1163 compromises insulin-stimulated kinase activity and uptake of 2-deoxyglucose. *Cell* 45: 721–32

Emori Y, Homma Y, Sorimachi H, Kawasaki H, Nakanishi O, et al. 1989. A second type of rat phosphoinositide-specific phospholipase C containing a *src*-related sequence not essential for phosphoinositide-hydrolysing activity. *J. Biol. Chem.* 264:21885–90

Endemann G, Yonezawa K, Roth RA. 1990. Phosphatidylinositol kinase or an associated protein is a substrate for the insulin receptor tyrosine kinase. *J. Biol. Chem.* 265:396–400

Ernfors P, Lee K, Jaenisch R. 1994. Mice lacking brain-derived neurotrophic factor develop with sensory deficits. *Nature* 368: 147–50

Escobedo JA, Kaplan DR, Kavanaugh WM, Turck CW, Williams LT. 1991a. A phosphatidylinositol-3 kinase binds to platelet-derived growth factor receptors through a specific receptor sequence containing phosphotyrosine. *Mol. Cell. Biol.* 11:1125–32

Escobedo JA, Navankasattusas S, Kavanaugh WM, Milfay D, Fried VA, Williams LT. 1991b. cDNA cloning of a novel 85 kd protein that has SH2 domains and regulates binding of PI3-kinase to the PDGF beta-receptor. *Cell* 65:75–82

Fajardo JE, Birge RB, Hanafusa H. 1993. A 31-amino-acid N-terminal extension regulates c-Crk binding to tyrosine-phosphorylated proteins. *Mol. Cell. Biol.* 13:7295–302

Falls DL, Rosen KM, Corfas G, Lane WS, Fischbach GD. 1993. ARIA, a protein that stimulates acetylcholine receptor synthesis, is a member of the Neu ligand family. *Cell* 72:801–15

Fantl WJ, Escobedo JA, Martin GA, Turck CW, del Rosario M, et al. 1992. Distinct phosphotyrosines on a growth factor receptor bind to specific molecules that mediate different signaling pathways. *Cell* 69:413–23

Fantl WJ, Escobedo JA, Williams LT. 1989. Mutations of the platelet-derived growth factor receptor that cause a loss of ligand-induced conformational change, subtle changes in kinase activity, and impaired ability to stimulate DNA synthesis. *Mol. Cell. Biol.* 9:4473–78

Fantl WJ, Johnson DE, Williams LT. 1993. Signalling by receptor tyrosine kinases. *Annu. Rev. Biochem.* 62:453–81

Fazioli F, Kim UH, Rhee SG, Molloy CJ, Segatto O, Di Fiore PP. 1991. The erbB-2 mitogenic signaling pathway: tyrosine phosphorylation of phospholipase C-gamma and

GTPase-activating protein does not correlate with erbB-2 mitogenic potency. *Mol. Cell. Biol.* 11:2040–48

Fedi P, Pierce JH, Di Fiore PP, Kraus MH. 1994. Efficient coupling with phosphatidylinositol 3-kinase, but not phospholipase Cγ or GTPase-activating protein, distinguishes ErbB3 signaling from that of other ErbB/EGFR family members. *Mol. Cell. Biol.* 14:492–500

Felder S, Zhou M, Hu P, Urena J, Ullrich A, et al. 1993. SH2 domains exhibit high-affinity binding to tyrosine phosphorylated peptides yet also exhibit rapid dissociation and exchange. *Mol. Cell. Biol.* 13:1449- 55

Feng GS, Hui CC, Pawson T. 1993. SH2-containing phosphotyrosine phosphatase as a target of protein-tyrosine kinases. *Science* 259: 1607–11

Fisher GJ, Henderson PA, Voorhees JJ, Baldassare JJ. 1991. Epidermal growth factor-induced hydrolysis of phosphatidylcholine by phospholipase D and phospholipase C in human dermal fibroblasts. *J. Cell. Physiol.* 146:309–17

Freeman RJ, Plutzky J, Neel BG. 1992. Identification of a human src homology 2-containing protein-tyrosine-phosphatase: a putative homolog of *Drosophila* corkscrew. *Proc. Natl. Acad. Sci. USA* 89:11239–43

Friesel R, Dawid IB. 1991. cDNA cloning and developmental expression of fibroblast growth factor receptors from *Xenopus laevis*. *Mol. Cell. Biol.* 11:2481–88

Fu XY. 1992. A transcription factor with SH2 and SH3 domains is directly activated by an interferon alpha-induced cytoplasmic protein tyrosine kinase(s). *Cell* 70:323–35

Fu XY, Schindler C, Improta T, Aebersold R, Darnell JE. 1992. The proteins of ISGF-3, the interferon alpha-induced transcriptional activator, define a gene family involved in signal transduction. *Proc. Natl. Acad. Sci. USA* 89:7840–43

Fu XY, Zhang JJ. 1993. Transcription factor p91 interacts with the epidermal growth factor receptor and mediates activation of the c-fos gene promoter. *Cell* 74:1135–45

Fujimoto J, Yamamoto T. 1994. *brt*, a mouse gene encoding a novel receptor-type protein-tyrosine kinase, is preferentially expressed in the brain. *Oncogene* 9:693–98

Fukui Y, Hanafusa H. 1989. Phosphatidylinositol kinase activity associates with viral $p60^{src}$ protein. *Mol. Cell. Biol.* 9: 1651–58

Gale NW, Kaplan S, Lowenstein EJ, Schlessinger J, Bar Sagi D. 1993. Grb2 mediates the EGF-dependent activation of guanine nucleotide exchange on Ras. *Nature* 363:88–92

Gaul U, Mardon G, Rubin GM. 1992. A putative Ras GTPase activating protein acts as a negative regulator of signaling by the Sevenless receptor tyrosine kinase. *Cell* 68:1007–19

Geissler EN, Ryan EN, Housman DE. 1988. The dominant-white spotting (W) locus of the mouse encodes the c-*kit* proto-oncogene. *Cell* 55:185–92

Gervais FG, Chow LM, Lee JM, Branton PE, Veillette A. 1993. The SH2 domain is required for stable phosphorylation of $p56^{lck}$ at tyrosine 505, the negative regulatory site. *Mol. Cell. Biol.* 13:7112–21

Gherardi E, Stoker M. 1991. Hepatocyte growth factor-scatter factor: mitogen, motogen, and Met. *Cancer Cells* 3:227–32

Gibbs JB, Marshall MS, Scolnick EM, Dixon RAF, Vogel US. 1990. Modulation of guanine nucleotides bound to Ras in NIH3T3 cells by oncogenes, growth factors, and the GTPase activating protein (GAP). *J. Biol. Chem.* 265:20437–42

Gibbs JB, Schaber MD, Allard WJ, Sigal IS, Scolnick EM. 1988. Purification of ras GTPase activating protein from bovine brain. *Proc. Natl. Acad. Sci. USA* 85:5026–30

Gilardi-Hebenstreit P, Nieto MA, Frain M, Mattei M-G, Chestier A, et al. 1992. An Eph-related receptor protein tyrosine kinase gene segmentally expressed in the developing mouse hindbrain. *Oncogene* 7:2499–506

Giordano S, Di Renzo MF, Ferracini R, Chiado Piat L, Comiglio PM. 1988. p145, a protein with associated tyrosine activity in a human gastric carcinoma cell line. *Mol. Cell. Biol.* 8:3510–17

Giorgetti S, Ballotti R, Kowalski-Chauvel A, Tartare S, Van Obberghen E. 1993. The insulin and insulin-like growth factor-1 receptor substrate IRS-1 associates with and activates phosphatidylinositol 3-kinase in vitro. *J. Biol. Chem.* 268:7358–64

Glass DJ, Yancopoulos GD. 1993. The neurotropins and their receptors. *Trends Cell Biol.* 3:263–68

Glazer L, Shilo BZ. 1991. The *Drosophila* FGF-R homolog is expressed in the embryonic tracheal system and appears to be required for directed tracheal cell extension. *Genes Dev.* 5:697–705

Glenney JRJ. 1992. Tyrosine-phosphorylated proteins: mediators of signal transduction from the tyrosine kinases. *Biochim. Biophys. Acta* 1134:113–27

Goldberg HJ, Viegas MM, Margolis BL, Schlessinger J, Skorecki KL. 1990. The tyrosine kinase activity of the epidermal-growth-factor receptor is necessary for phospholipase A2 activation. *Biochem. J.* 267:461–65

Goldschmidt-Clermont PJ, Kim JW, Machesky LM, Rhee SG, Pollard TD. 1991. Regulation of phospholipase C-γ1 by profilin and tyrosine phosphorylation. *Science* 251:1231–33

Gotoh N, Tojo A, Muroya K, Hashimoto Y,

Hattori S, et al. 1994. Epidermal growth factor-receptor mutant lacking the autophosphorylation sites induces phosphorylation of the Shc protein and Shc-Grb2/ASH association and retains mitogenic activity. *Proc. Natl. Acad. Sci. USA* 91:167–71

Gould KL, Hunter T. 1988. Platelet-derived growth factor induces multisite phosphorylation of pp60c-src and increases its protein-tyrosine kinase activity. *Mol. Cell. Biol.* 8: 3345–56

Graziani A, Gramaglia D, Cantley LC, Comoglio PM. 1991. The tyrosine-phosphorylated hepatocyte growth factor/scatter factor receptor associates with phosphatidylinositol 3-kinase. *J. Biol. Chem.* 266:22087–90

Grieco M, Santoro M, Beringieri MT, Melillo RM, Donghi R, et al. 1990. PTC is a novel rearranged form of the ret proto-oncogene and is frequently detected in vivo in human thyroid papillary carcinomas. *Cell* 60:557–63

Gulbins E, Coggeshall KM, Baier G, Katzav S, Burn P, Altman A. 1993. Tyrosine kinase-stimulated guanine nucleotide exchange activity of Vav in T cell activation. *Science* 260:822–25

Gulbins E, Coggeshall KM, Langlet C, Baier G, Bonnefoy-Berard N, et al. 1994. Activation of Ras in vitro and in intact fibroblasts by the Vav guanine nucleotide exchange protein. *Mol. Cell. Biol.* 14:906–13

Hack N, Sue-A-Quan A, Mills GB, Skorecki KL. 1993. Expression of human tyrosine kinase-negative epidermal growth factor receptor amplifies signaling through endogenous murine epidermal growth factor receptor. *J. Biol. Chem.* 268:26441–46

Hafen E, Basler K, Edstroem J-E, Rubin GM. 1987. *Sevenless,* a cell-specific homeotic gene of *Drosophila,* encodes a putative transmembrane receptor with a tyrosine kinase domain. *Science* 236:55–63

Hallek M, Druker B, Lepisto EM, Wood KW, Ernst TJ, Griffin JD. 1992. Granulocyte-macrophage colony-stimulating factor and steel factor induce phosphorylation of both unique and overlapping signal transduction intermediates in a human factor-dependent hematopoietic cell line. *J. Cell. Physiol.* 153: 176–86

Hammacher A, Mellstrom K, Heldin C, Westermark B. 1989. Isoform-specific induction of actin reorganization by platelet-derived growth factor suggests that the functionally active receptor is a dimer. *EMBO J.* 8:2489–95

Han S, Stuart LA, Degen SJF. 1991. Characterization of the DNF15S2 locus on human chromosome 3: identification of a gene coding for four kringle domains with homology to hepatocyte growth factor. *Biochemistry* 30:9768–80

Hanks SK. 1991. Eukaryotic protein kinases. *Curr. Opin. Struct. Biol.* 1:369–83

Hanks SK, Quinn AM. 1991. Protein kinase catalytic domain sequence database: identification of conserved features of primary structure and classification of family members. *Methods Enzymol.* 200:38–62

Hanks SK, Quinn AM, Hunter T. 1988. The protein kinase family: conserved features and deduced phylogeny of the catalytic domains. *Science* 241:42–52

Hart AC, Kramer H, Zipersky SL. 1993. Extracellular domain of the boss transmembrane ligand acts as an antagonist of the sev receptor. *Nature* 361:732–36

Hasegawa H. 1985. Early changes in inositol lipids and their metabolites induced by platelet-derived growth factor in quiescent Swiss mouse 3T3 cells. *Biochem. J.* 232:99–109

Hayashi H, Kamohara S, Nishioka Y, Kanai F, Miyake N, et al. 1992. Insulin treatment stimulates the tyrosine phosphorylation of the alpha-type 85-kDa subunit of phosphatidylinositol 3-kinase in vivo. *J. Biol. Chem.* 267:22575–80

Hayashi H, Miyake N, Kanai F, Shibasaki F, Takenawa T, Ebina Y. 1991. Phosphorylation in vitro of the 85 kDa subunit of phosphatidylinositol 3-kinase and its possible activation by insulin receptor tyrosine kinase. *Biochem. J.* 280:769–75

Hayashi H, Nishioka Y, Kamohara S, Kanai F, Ishii K, et al. 1993. The α-type 85 kda subunit of phosphatidylinositol 3-kinase is phosphorylated at tyrosines 368, 580, and 607 by the insulin receptor. *J. Biol. Chem.* 268: 7107–17

Hayes TE, Kitchen AM, Cochran BH. 1987. Inducible binding of a factor to the c-*fos* regulatory region. *Proc. Natl. Acad. Sci. USA* 84:1272–76

Heidaran MA, Molloy CJ, Pangelinan M, Choudhury GG, Wang LM, et al. 1992. Activation of the colony-stimulating factor 1 receptor leads to the rapid tyrosine phosphorylation of GTPase-activating protein and activation of cellular $p21^{ras}$. *Oncogene* 7: 147–52

Heidaran MA, Pierce JH, Lombardi D, Ruggiero M, Gutkind JS, et al. 1991. Deletion or substitution within the alpha platelet-derived growth factor receptor kinase insert domain: effects on functional coupling with intracellular signaling pathways. *Mol. Cell. Biol.* 11:134–42

Henkemeyer M, Marengere LEM, McGlade J, Olivier JP, Conlon RA, et al. 1994. Immunolocalization of the Nuk receptor tyrosine kinase suggests roles in segmental patterning of the brain and axonogenesis. *Oncogene* 9:1001–14

Herman PK, Stack JH, Emr SD. 1992. An essential role for a protein and lipid kinase

complex in secretory protein sorting. *Trends Cell Biol.* 2:363–68
Hiles ID, Otsu M, Volinia S, Fry MJ, Gout I, et al. 1992. Phosphatidylinositol 3-kinase: structure and expression of the 110 kd catalytic subunit. *Cell* 70:419–29
Hirai H, Maru Y, Hagiwara K, Nishida J, Takaku F. 1987. A novel putative tyrosine kinase receptor encoded by the *eph* gene. *Science* 238:1717–20
Hofstra RMW, Landsvater RM, Ceccherini I, Stulp RP, Stelwagen T, et al. 1994. A mutation in the RET proto-oncogene associated with multiple endocrine neoplasia type 2B and sporadic medullary thyroid carcinoma. *Nature* 367:375–76
Holmes WE, Sliwkowski MX, Akita RW, Henzel WJ, Lee J, et al. 1992. Identification of heregulin, a specific activator of $p185^{erbB2}$. *Science* 256:1205–10
Holzman LB, Marks RM, Dixit VM. 1990. A novel immediate-early response gene of endothelium is induced by cytokines and encodes a secreted protein. *Mol. Cell. Biol.* 10: 5830–38
Houssaint E, Blanquet PR, Champion-Arnaud P, Gesnel MC, Torriglia A, Courtois Y. 1990. Related fibroblast growth factor receptor genes exist in the human genome. *Proc. Natl. Acad. Sci. USA* 87:8180–84
Hovens CM, Stacker SA, Andres A-C, Harpur AG, Ziemiecki A, Wilks AF. 1992. RYK, a receptor tyrosine kinase-related molecule with unusual kinase domain motifs. *Proc. Natl. Acad. Sci. USA* 89:11818–22
Howe LR, Leevers SJ, Gomez N, Nakielny S, Cohen P, Marshall CJ. 1992. Activation of the MAP kinase pathway by the protein kinase raf. *Cell* 71:335–42
Hu P, Margolis B, Skolnik EY, Lammers R, Ullrich A, Schlessinger J. 1992. Interaction of phosphatidylinositol 3-kinase-associated p85 with epidermal growth factor and platelet-derived growth factor receptors. *Mol. Cell. Biol.* 12:981–90
Huang DC, Marshall CJ, Hancock JF. 1993. Plasma membrane-targeted ras GTPase-activating protein is a potent suppressor of p21ras function. *Mol. Cell. Biol.* 13:2420–31
Huang YK, Kung HF, Kamata T. 1990. Purification of a factor capable of stimulating the guanine nucleotide exchange reaction of ras proteins and its effect on ras-related small molecular mass G proteins. *Proc. Natl. Acad. Sci. USA* 87:8008–12
Huff JL, Jelinek MA, Borgman CA, Lansing TJ, Parsons JT. 1993. The protooncogene c-sea encodes a transmembrane protein-tyrosine kinase related to the Met/hepatocyte growth factor/scatter factor receptor. *Proc. Natl. Acad. Sci. USA* 90:6140–44
Hunter T. 1991. Protein kinase classification. *Methods Enzymol.* 200:3–37
Imamoto A, Soriano P. 1993. Disruption of the *csk* gene, encoding a negative regulator of Src family tyrosine kinases, leads to neural tube defects and embryonic lethality in mice. *Cell* 73:1117–24
Imamoto A, Soriano P, Stein PL. 1994. Genetics of signal transduction: tales from the mouse. *Curr. Opin. Genet. Dev.* 4:40–46
Janssen JWG, Schulz AS, Steenvoorden ACM, Schmidberger M, Strehl S, et al. 1991. A novel putative tyrosine kinase receptor with oncogenic potential. *Oncogene* 6:2113–20
Jennings CGB, Dyer SM, Burden SJ. 1993. Muscle specific trk-related receptor with a kringle domain defines a distinct class of receptor tyrosine kinases. *Proc. Natl. Acad. Sci. USA* 90:2895–99
Jia R, Hanafusa H. 1994. The proto-oncogene of v-*eyk* (v-*ryk*) is a novel receptor-type protein tyrosine kinase with extracellular Ig/FN-III domains. *J. Biol. Chem.* 269:1839–44
Johnson JD, Edman JC, Rutter WJ. 1993. A receptor tyrosine kinase found in breast carcinoma cells has an extracellular discoidin I-like domain. *Proc. Natl. Acad. Sci. USA* 90:10891–95
Joly M, Kazlauskas A, Fay FS, Corvera S. 1994. Disruption of PDGF receptor trafficking by mutation of its PI-3 kinase binding sites. *Science* 263:684–87
Jones KR, Fariñas I, Backus C, Reichardt LF. 1994. Targeted disruption of the *BDNF* gene perturbs brain and sensory neuron development but not motor neuron development. *Cell* 76:989–99
Kan M, Wang F, Xu J, Crabb JW, Hou J, McKeehan WL. 1993. An essential heparin-binding domain in the fibroblast growth factor receptor kinase. *Science* 259:1918
Kanner SB, Reynolds AB, Wang HC, Vines RR, Parsons JT. 1991. The SH2 and SH3 domains of pp60src direct stable association with tyrosine phosphorylated proteins p130 and p110. *EMBO J.* 10:1689–98
Kapeller R, Prasad KVS, Janssen O, Hou W, Schaffhausen BS, et al. 1994. Identification of two SH3-binding motifs in the regulatory subunit of phosphatidylinositol 3-kinase. *J. Biol. Chem.* 269:1927–33
Kaplan DR, Morrison DK, Wong G, McCormick F, Williams LT. 1990. PDGF beta-receptor stimulates tyrosine phosphorylation of GAP and association of GAP with a signaling complex. *Cell* 61:125–33
Kaplan DR, Whitman M, Schaffhausen B, Pallas DC, White M, et al. 1987. Common elements in growth factor stimulation and oncogenic transformation: 85 kd phosphoprotein and phosphatidylinositol kinase activity. *Cell* 50:1021–29
Karn T, Holtrich U, Brauninger A, Bohme B, Wolf G, Rubsamen-Waigmann H, Strebhardt K. 1993. Structure, expression and chromo-

somal mapping of TKT from man and mouse: a new subclass of receptor tyrosine kinases with a factor VIII-like domain. *Oncogene* 8:3433–40

Kashishian A, Cooper JA. 1993. Phosphorylation sites at the C-terminus of the platelet-derived growth factor receptor bind phospholipase Cγl. *Mol. Biol. Cell* 4:49–57

Kashishian A, Kazlauskas A, Cooper JA. 1992. Phosphorylation sites in the PDGF receptor with different specificities for binding GAP and PI3 kinase in vivo. *EMBO J.* 11:1373–82

Kashles O, Yarden Y, Fischer R, Ullrich A, Schlessinger J. 1991. A dominant negative mutation suppresses the function of normal epidermal growth factor receptors by heterodimerization. *Mol. Cell. Biol.* 11: 1454–63

Katzav S. 1993. Single point mutations in the SH2 domain impair the transforming potential of vav and fail to activate proto-vav. *Oncogene* 8:1757–63

Katzav S, Cleveland JL, Heslop HE, Pulido D. 1991. Loss of the amino-terminal helix-loop-helix domain of the vav proto-oncogene activates its transforming potential. *Mol. Cell. Biol.* 11:1912–20

Katzav S, Martin ZD, Barbacid M. 1989. vav, a novel human oncogene derived from a locus ubiquitously expressed in hematopoietic cells. *EMBO J.* 8:2283–90

Kavanaugh WM, Klippel A, Escobedo JA, Williams LT. 1992. Modification of the 85-kilodalton subunit of phosphatidylinositol-3 kinase in platelet-derived growth factor-stimulated cells. *Mol. Cell. Biol.* 12: 3415–24

Kazlauskas A. 1994. Receptor tyrosine kinases and their targets. *Curr. Opin. Genet. Dev.* 4:5–14

Kazlauskas A, Cooper JA. 1989. Autophosphorylation of the PDGF receptor in the kinase insert region regulates interactions with cell proteins. *Cell* 58:1121–33

Kazlauskas A, Cooper JA. 1990. Phosphorylation of the PDGF receptor beta subunit creates a tight binding site for phosphatidylinositol 3 kinase. *EMBO J.* 9: 3279–86

Kazlauskas A, Durden DL, Cooper JA. 1991. Functions of the major tyrosine phosphorylation site of the PDGF receptor beta subunit. *Cell Regul.* 2:413–25

Kazlauskas A, Ellis C, Pawson T, Cooper JA. 1990. Binding of GAP to activated PDGF receptors. *Science* 247:1578–81

Kazlauskas A, Feng GS, Pawson T, Valius M. 1993. The 64-kDa protein that associates with the platelet-derived growth factor receptor beta subunit via Tyr-1009 is the SH2-containing phosphotyrosine phosphatase Syp. *Proc. Natl. Acad. Sci. USA* 90:6939–43

Kazlauskas A, Kashishian A, Cooper JA, Valius M. 1992. GTPase-activating protein and phosphatidylinositol 3-kinase bind to distinct regions of the platelet-derived growth factor receptor beta-subunit. *Mol. Cell. Biol.* 12:2534–44

Keck PJ, Hauser SD, Krivi G, Sanzo K, Warren T, Feder J, Connolly DT. 1989. Vascular permeability factor, an endothelial cell mitogen related to PDGF. *Science* 246:1309–12

Keegan AD, Nelms K, White M, Wang L-M, Pierce JH, Paul WE. 1994. An IL-4 receptor region containing an insulin receptor motif is important for IL-4-mediated IRS-1 phosphorylation and cell growth. *Cell* 76:811–20

Keegan K, Johnson DE, Williams LT, Hayman MJ. 1991. Isolation of an additional member of the fibroblast growth factor receptor family, FGFR-3. *Proc. Natl. Acad. Sci. USA* 88: 1095–99

Kelman Z, Simon-Chazottes D, Guenet J-L., Yarden Y. 1993. The murine *vik* gene (chromosome 9) encodes a putative receptor with unique protein kinase motifs. *Oncogene* 8: 37–44

Kessler DS, Veals SA, Fu XY, Levy DE. 1990. Interferon-alpha regulates nuclear translocation and DNA-binding affinity of ISGF3, a multimeric transcriptional activator. *Genes Dev.* 4:1753–65

Kim HK, Kim JW, Zilberstein A, Margolis B, Kim JG, et al. 1991. PDGF stimulation of inositol phospholipid hydrolysis requires PLC-gamma 1 phosphorylation on tyrosine residues 783 and 1254. *Cell* 65:435–41

Kim UH, Kim HS, Rhee SG. 1990. Epidermal growth factor and platelet-derived growth factor promote translocation of phospholipase C-gamma from cytosol to membrane. *FEBS Lett.* 270:33–36

Kimura H. 1993. Schwannoma-derived growth factor must be transported into the nucleus to exert mitogenic activity. *Proc. Natl. Acad. Sci. USA* 90:2165–69

Kiss Z. 1992. Differential effects of platelet-derived growth factor, serum and bombesin on phospholipase D-mediated hydrolysis of phosphatidylethanolamine in NIH 3T3 fibroblasts. *Biochem. J.* 285:229–33

Klein R, Silos-Santiago I, Smeyne RJ, Lira SA, Brambilla R, et al. 1994. Disruption of the neurotrophin-3 receptor gene *trk*C eliminates Ia muscle afferents and results in abnormal movements. *Nature* 368:249–51

Klein R, Smeyne RJ, Wurst W, Long LK, Auerbach BA, et al. 1993. Targeted disruption of the trkB neurotropin receptor gene results in nervous system lesions and neonatal death. *Cell* 75:113–22

Klippel A, Escobedo JA, Fantl WJ, Williams LT. 1992. The C-terminal SH2 domain of p85 accounts for the high affinity and specificity of the binding of phosphatidylinositol 3-kinase to phosphorylated platelet-derived growth factor beta receptor. *Mol. Cell. Biol.* 12:1451–59

Klippel A, Escobedo JA, Hu Q, Williams LT. 1993. A region of the 85-kilodalton (kDa) subunit of phosphatidylinositol 3-kinase binds the 110-kDa catalytic subunit in vivo. *Mol. Cell. Biol.* 13:5560–66

Kmiecik TE, Shalloway D. 1987. Activation and suppression of pp60c-^{src} transforming ability by mutation of its primary sites of tyrosine phosphorylation. *Cell* 49:65–73

Knighton DR, Cadena DL, Zheng J, Ten Eyck LF, Taylor SS, et al. 1993. Structural features that specify tyrosine kinase activity deduced from homology modeling of the epidermal growth factor receptor. *Proc. Natl. Acad. Sci. USA* 90:5001–5

Knighton DR, Zheng J, Ten Eyck LF, Ashford VA, Xuong N-H, et al. 1991a. Crystal structure of the catalytic subunit of cyclic adenosine monophosphate-dependent protein kinase. *Science* 253:407–20

Knighton DR, Zheng J, Ten Eyck LF, Xuong N-H, Taylor SS, Sowadski JM. 1991b. Structure of a peptide inhibitor bound to the catalytic subunit of cyclic adenosine monophosphate-dependant protein kinase. *Science* 253:414–20

Koch CA, Anderson D, Moran MF, Ellis C, Pawson T. 1991. SH2 and SH3 domains: elements that control interactions of cytoplasmic signaling proteins. *Science* 252:668–74

Koch CA, Moran MF, Anderson D, Liu XQ, Mbamalu G, Pawson T. 1992. Multiple SH2-mediated interactions in v-src-transformed cells. *Mol. Cell. Biol.* 12:1366–74

Koch WJ, Inglese J, Stone WC, Lefkowitz RJ. 1993. The binding site for the beta gamma subunits of heterotrimeric G proteins on the beta-adrenergic receptor kinase. *J. Biol. Chem.* 268:8256–60

Kohda D, Hatanaka H, Odaka M, Maddiyan V, Ullrich A, et al. 1993. Solution structure of the SH3 domain of phospholipase Cγ. *Cell* 72:953–60

Koretzky GA, Picus J, Schultz T, Weiss A. 1991. Tyrosine phosphatase CD45 is required for T-cell antigen receptor and CD2-mediated activation of a protein tyrosine kinase and interleukin 2 production. *Proc. Natl. Acad. Sci. USA* 88:2037–41

Koretzky GA, Picus J, Thomas ML, Weiss A. 1990. Tyrosine phosphatase CD45 is essential for coupling T-cell antigen receptor to the phosphatidylinositol pathway. *Nature* 346:66–69

Koyama S, Yu H, Dalgarno DC, Shin TB, Zydowsky LD, Schreiber SL. 1993. Structure of the PI3K SH3 domain and analysis of the SH3 family. *Cell* 72:945–52

Kraus MH, Fedi P, Starks V, Muraro R, Aaronson SA. 1993. Demonstration of ligand-dependent signaling by the *erb*B-3 tyrosine kinase and its constitutive activation in human breast tumor cells. *Proc. Natl. Acad. Sci. USA* 90:2900–4

Kremer NE, D'Arcangelo G, Thomas SM, DeMarco M, Brugge JS, Halegoua S. 1991. Signal transduction by nerve growth factor and fibroblast growth factor in PC12 cells requires a sequence of Src and Ras actions. *J. Cell. Biol.* 115:809–19

Kuhne MR, Pawson T, Lienhard GE, Feng GS. 1993. The insulin receptor substrate 1 associates with the SH2-containing phosphotyrosine phosphatase Syp. *J. Biol. Chem.* 268:11479–81

Kypta RM, Goldberg Y, Ulug ET, Courtneidge SA. 1990. Association between the PDGF receptor and members of the *src* family of tyrosine kinases. *Cell* 62:481–92

Kyriakis JM, App H, Zhang XF, Banerjee P, Brautigan DL, et al. 1992. Raf-1 activates MAP kinase-kinase. *Nature* 358:417–21

Lai C, Lemke G. 1991. An extended family of protein-tyrosine kinase genes differentially expressed in the vertebrate nervous system. *Neuron* 6:691–704

Lai C, Lemke G. 1994. Structure and expression of the Tyro 10 receptor tyrosine kinase. *Oncogene* 9:877–83

Lamballe F, Klein R, Barbacid M. 1991. trkC, a new member of the trk family of tyrosine protein kinases, is a receptor for neurotrophin-3. *Cell* 66:967–79

Lamballe F, Tapley P, Barbacid M. 1993. trkC encodes multiple neurotrophin-3 receptors with distinct biological properties and substrate specificities. *EMBO J.* 12:3083–94

Lange-Carter C, Pleiman CM, Gardner AM, Blumer KJ, Johnson GL. 1993. A divergence in the MAP kinase regulatory network defined by MEK kinase and Raf. *Science* 260: 315–19

Larose L, Gish G, Shoelson S, Pawson T. 1993. Identification of residues in the beta platelet-derived growth factor receptor that confer specificity for binding to phospholipase C-gamma 1. *Oncogene* 8:2493–99

Lax I, Bellot F, Howk R, Ullrich A, Givol D, Schlessinger J. 1989. Functional analysis of the ligand binding site of EGF-receptor utilizing chimeric chicken/human receptor molecules. *EMBO J.* 8:421–27

Lax I, Burgess WH, Bellot F, Ullrich A, Schlessinger J, Givol D. 1988. Localization of a major receptor-binding domain for epidermal growth factor by affinity labeling. *Mol. Cell. Biol.* 8:1831–34

Lazarus AH, Kawauchi K, Rapoport MJ, Delovitch TL. 1993. Antigen-induced B lymphocyte activation involves the p21ras and ras•GAP signaling pathway. *J. Exp. Med.* 178:1765–69

Lechleider RJ, Freeman RMJ, Neel BG. 1993a.

Tyrosyl phosphorylation and growth factor receptor association of the human *corkscrew* homologue, SH-PTP2. *J. Biol. Chem.* 268: 13434–38

Lechleider RJ, Sugimoto S, Bennett AM, Kashishian AS, Cooper JA, et al. 1993b. Activation of the SH2-containing phosphotyrosine phosphatase SH-PTP2 by its binding site, phosphotyrosine 1009, on the human platelet-derived growth factor receptor. *J. Biol. Chem.* 268:21478–81

Lee C-H, Li W, Nishimura R, Zhou M, Batzer A, et al. 1993. Nck associates with the SH2 domain-docking protein IRS-1 in insulin-stimulated cells. *Proc. Natl. Acad. Sci. USA* 90:11713–17

Leevers SJ, Marshall CJ. 1992. Activation of extracellular signal-regulated kinase, ERK2, by p21ras oncoprotein. *EMBO J.* 11:569–74

Lehmann JM, Riethmuller G, Johnson JP. 1990. Nck, a melanoma cDNA encoding a cytoplasmic protein consisting of the src homology units SH2 and SH3. *Nucleic Acids Res.* 18:1048

Letwin K, Yee SP, Pawson T. 988. Novel protein-tyrosine kinase cDNAs related to *fps-fes* and *eph* cloned using anti-phosphotyrosine antibodies. *Oncogene* 3:621–27

Leung DW, Cachianes G, Kuang W, Goeddel DV, Ferrara N. 1989. Vascular endothelial growth factor is a secreted angiogenic mitogen. *Science* 246:1306–9

Lev S, Blechman J, Nishikawa S-I, Givol D, Yarden Y. 1993. Interspecies molecular chimeras of Kit help define the binding site of the stem cell factor. *Mol. Cell. Biol.* 13: 2224–34

Lev S, Givol D, Yarden Y. 1991. A specific combination of substrates is involved in signal transduction by the kit-encoded receptor. *EMBO J.* 10:647–54

Lev S, Givol D, Yarden Y. 1992. Interkinase domain of kit contains the binding site for phosphatidylinositol 3′ kinase. *Proc. Natl. Acad. Sci. USA* 89:678–82

Levy DE, Kessler DS, Pine R, Darnell JE. 1989. Cytoplasmic activation of ISGF3, the positive regulator of interferon-alpha-stimulated transcription, reconstituted in vitro. *Genes Dev.* 3:1362–71

Levy DE, Kessler DS, Pine R, Reich N, Darnell JE. 1988. Interferon-induced nuclear factors that bind a shared promotor element correlate with positive and negative transcriptional control. *Genes Dev.* 2:383–93

Lew DJ, Decker T, Strehlow I, Darnell JE. 1991. Overlapping elements in the guanylate-binding protein gene promotor mediate transcriptional induction by alpha and gamma interferons. *Mol. Cell. Biol.* 11:182–91

Li BQ, Kaplan D, Kung HF, Kamata T. 1992. Nerve growth factor stimulation of the Ras-guanine nucleotide exchange factor and GAP activities. *Science* 256:1456–59

Li N, Batzer A, Daly R, Yajnik V, Skolnik E, et al. 1993. Guanine-nucleotide-releasing factor hSos1 binds to Grb2 and links receptor tyrosine kinases to Ras signalling. *Nature* 363:85–88

Li W, Hack N, Margolis B, Ullrich A, Skorecki K, Schlessinger J. 1991. Carboxy-terminal truncations of epidermal growth factor (EGF) receptor affect diverse EGF-induced cellular responses. *Cell Regul.* 2:641–49

Li W, Hu P, Skolnik EY, Ullrich A, Schlessinger J. 1992. The SH2 and SH3 domain-containing Nck protein is oncogenic and a common target for phosphorylation by different surface receptors. *Mol. Cell. Biol.* 12: 5824–33

Li W, Nishimura R, Kashishian A, Batzer AG, Kim WJH, et al. 1994. A new function for a phosphotyrosine phosphatase: linking GRB2-Sos to a receptor tyrosine kinase. *Mol. Cell. Biol.* 14:509–17

Lin L-L, Wartmann M, Lin AY, Knopf JL, Seth A, Davis RJ. 1993. cPLA$_2$ is phosphorylated and activated by MAP kinase. *Cell* 72:269–78

Lin L-L, Lin AY, Knopf JL. 1992. Cytosolic phospholipase A2 is coupled to hormonally regulated release of arachidonic acid. *Proc. Natl. Acad. Sci. USA* 89:6147–51

Lindberg R, Hunter T. 1990. cDNA cloning and characterization of *eck*, an epithelial cell receptor protein-tyrosine kinase in the *eph/elk* family of protein kinases. *Mol. Cell. Biol.* 10:6316–24

Liu J-P, Baker J, Perkins AS, Robertson EJ, Efstratiadis A. 1993a. Mice carrying null mutations of the genes encoding insulin-like growth factor I (*Igf-1*) and Type 1 IGF receptor (*Igf1r*). *Cell* 75:59–72

Liu X, Brodeur SR, Gish G, Zhou S, Cantley LC, et al. 1993b. Regulation of c-Src tyrosine kinase activity by the Src SH2 domain. *Oncogene* 8:1119–26

Liu X, Marengere LE, Koch CA, Pawson T. 1993c. The v-Src SH3 domain binds phosphatidylinositol 3′-kinase. *Mol. Cell. Biol.* 13:5225–32

Liu XQ, Pawson T. 1991. The epidermal growth factor receptor phosphorylates GTPase-activating protein (GAP) at Tyr-460, adjacent to the GAP SH2 domains. *Mol. Cell. Biol.* 11:2511–16

Livneh E, Glazer L, Segal D, Schlessinger J, Shilo B-Z. 1985. The Drosophila EGF receptor gene homolog: conservation of both hormone binding and kinase domains. *Cell* 40: 599–607

Longati P, Bardelli A, Ponzetto C, Naldini L, Comoglio PM. 1994. Tyrosines$^{1234-1235}$ are critical for activation of the tyrosine kinase encoded by the MET proto-oncogene (HGF receptor). *Oncogene* 9:49–57

Lowenstein EJ, Daly RJ, Batzer AG, Li W, Margolis B, et al. 1992. The SH2 and SH3 domain-containing protein GRB2 links receptor tyrosine kinases to ras signaling. *Cell* 70:431–42

Luetteke NC, Phillips HK, Qiu TH, Copeland NG, Earp HS, et al. 1994. The mouse *waved-2* phenotype results from a point mutation in the EGF receptor tyrosine kinase. *Genes Dev.* 8:399–413

Lyman SD, James L, Bos TV, de Vries P, Brasel K, et al. 1993. Molecular cloning of a ligand for the flt3/flk-2 tyrosine kinase receptor: a proliferative factor for primitive hematopoietic cells. *Cell* 75:1157–67

Maglione D, Guerriero V, Viglietto G, Delli-Bovi P, Persico MG. 1991. Isolation of a human placenta cDNA coding for a protein related to the vascular permeability factor. *Proc. Natl. Acad. Sci. USA* 88:9267–71

Maisonpierre PC, Barrezueta NX, Yancopoulos GD. 1993. Ehk-1 and Ehk-2: two novel members of the Eph receptor-like tyrosine kinase family with distinctive structures and neuronal expression. *Oncogene* 8:3277–88

Maminta MLD, Williams KL, Nakagawara A, Enger KT, Guo C, et al. 1992. Identification of a novel tyrosine kinase receptor-like molecule in neuroblastomas. *Biochem. Biophys. Res. Commun.* 189:1077–83

Marchionni MA, Goodearl ADJ, Chen MS, Bermingham-McDonogh O, Kirk C, et al. 1993. Glial growth factors are alternatively spliced erbB2 ligands expressed in the nervous system. *Nature* 362:312–18

Margolis B, Bellot F, Honegger AM, Ullrich A, Schlessinger J, Zilberstein A. 1990. Tyrosine kinase activity is essential for the association of phospholipase C-gamma with the epidermal growth factor receptor. *Mol. Cell. Biol.* 10:435–41

Margolis B, Hu P, Katzav S, Li W, Oliver JM, et al. 1992. Tyrosine phosphorylation of vav proto-oncogene product containing SH2 domain and transcription factor motifs. *Nature* 356:71–74

Margolis B, Rhee SG, Felder S, Mervic M, Lyall R, et al. 1989. EGF induces tyrosine phosphorylation of phospholipase C-II: a potential mechanism for EGF receptor signaling. *Cell* 57:1101–7

Margolis B, Silvennoinen O, Comoglio F, Roonprapunt C, Skolnik E, et al. 1992. High-efficiency expression/cloning of epidermal growth factor-receptor-binding proteins with Src homology 2 domains. *Proc. Natl. Acad. Sci. USA* 89:8894–98

Marshall CJ. 1994. MAP kinase kinase kinase, MAP kinase kinase and MAP kinase. *Curr. Opin. Genet. Dev.* 4:82–89

Martegani E, Vanoni M, Zippel R, Coccetti P, Brambilla R, et al. 1992. Cloning by functional complementation of a mouse cDNA encoding a homologue of CDC25, a *Saccharomyces cerevisiae* RAS activator. *EMBO J.* 11:2151–57

Martin GA, Viskochil D, Bollag G, McCabe PC, Crosier WJ, et al. 1990. The GAP-related domain of the neurofibromatosis type 1 gene product interacts with ras p21. *Cell* 63:843–49

Martin GA, Yatani A, Clark R, Conroy L, Polakis P, et al. 1992. GAP domains responsible for ras p21-dependent inhibition of muscarinic atrial K^+ channel currents. *Science* 255:192–94

Maru Y, Hirai H, Takaku F. 1990. Overexpression confers an oncogenic potential upon the *eph* gene. *Oncogene* 5:199–204

Masiakowski P, Carroll RD. 1992. A novel family of cell surface receptors with tyrosine kinase-like domain. *J. Biol. Chem.* 267: 26181–90

Matsuda M, Mayer BJ, Fukui Y, Hanafusa H. 1990. Binding of transforming protein, P47gag-crk, to a broad range of phosphotyrosine-containing proteins. *Science* 248:1537–39

Matsuda M, Mayer BJ, Hanafusa H. 1991. Identification of domains of the v-crk oncogene product sufficient for association with phosphotyrosine-containing proteins. *Mol. Cell. Biol.* 11:1607–13

Matsuda M, Nagata S, Tanaka S, Nagashima K, Kurata T. 1993. Structural requirement of CRK SH2 region for binding to phosphotyrosine-containing proteins. Evidence from reactivity to monoclonal antibodies. *J. Biol. Chem.* 268:4441–46

Matsuda M, Tanaka S, Nagata S, Kojima A, Kurata T, Shibuya M. 1992. Two species of human CRK cDNA encode proteins with distinct biological activities. *Mol. Cell. Biol.* 12: 3482–89

Matsuda S, Kadowaki Y, Ichino M, Akiyama T, Toyoshima K, Yamamoto T. 1993. 17β-Estradiol mimics ligand activity of the c-*er*B2 protooncogene produce. *Proc. Natl. Acad. Sci. USA* 90:10803–7

Matthews RJ, Bowne DB, Flores E, Thomas ML. 1992. Characterization of hematopoietic intracellular protein tyrosine phosphatases: description of a phosphatase containing an SH2 domain and another enriched in proline-, glutamic acid-, serine-, and threonine-rich sequences. *Mol. Cell. Biol.* 12:2396–405

Matthews W, Jordan CT, Gavin M, Jenkins NA, Copeland NG, Lemischka IR. 1991a. A receptor tyrosine kinase cDNA isolated from a population of enriched primitive hematopoietic cells and exhibiting close genetic linkage to c-kit. *Proc. Natl. Acad. Sci. USA* 88:9026–30

Matthews W, Jordan CT, Wiegand GW, Pardoll

D, Lemischka IR. 1991b. A receptor tyrosine kinase specific to hematopoietic stem and progenitor cell-enriched populations. *Cell* 65:1143–52

Matuoka K, Shibasaki F, Shibata M, Takenawa T. 1993. Ash/Grb-2, a SH2/SH3-containing protein, couples to signaling for mitosis and cytoskeletal reorganisation by EGF and PDGF. *EMBO J.* 12:3467–73

Matuoka K, Shibata M, Yamakawa A, Takenawa T. 1992. Cloning of ASH, a ubiquitous protein composed of one Src homology region (SH) 2 and two SH3 domains, from human and rat cDNA libraries. *Proc. Natl. Acad. Sci. USA* 89:9015–19

Mayer BJ, Baltimore D. 1993. Signalling through SH2 and SH3 domains. *Trends Cell Biol.* 3:8–13

Mayer BJ, Hamaguchi M, Hanafusa H. 1988. A novel oncogene with a structural similarity to phospholipase C. *Nature* 332:272–75

Mayer BJ, Hanafusa H. 1990a. Association of the v-crk oncogene product with phosphotyrosine-containing proteins and protein kinase activity. *Proc. Natl. Acad. Sci. USA* 87:2638–42

Mayer BJ, Hanafusa H. 1990b. Mutagenic analysis of the v-crk oncogene: requirement for SH2 and SH3 domains and correlation between increased cellular phosphotyrosine and transformation. *J. Virol.* 64:3581–89

Mayer BJ, Jackson PK, Baltimore D. 1991. The non-catalytic src homology region 2 segment of abl tyrosine kinase binds to tyrosine-phosphorylated cellular proteins with high affinity. *Proc. Natl. Acad. Sci. USA* 88:627–31

McGlade CJ, Ellis C, Reedijk M, Anderson D, Mbamalu G, et al. 1992. SH2 domains of the p85 alpha subunit of phosphatidylinositol 3-kinase regulate binding to growth factor receptors. *Mol. Cell. Biol.* 12:991–97

McGlade J, Brunkhorst B, Anderson D, Mbamalu G, Settleman J, et al. 1993. The N-terminal region of GAP regulates cytoskeletal structure and cell adhesion. *EMBO J.* 12:3073–81

McGlade J, Cheng A, Pelicci G, Pelicci PG, Pawson T. 1992. Shc proteins are phosphorylated and regulated by the v-Src and v-Fps protein-tyrosine kinases. *Proc. Natl. Acad. Sci. USA* 89:8869–73

Medema RH, de Laat WL, Martin GA, McCormick F, Bos JL. 1992. GTPase-activating protein SH2-SH3 domains induce gene expression in a Ras-dependent fashion. *Mol. Cell. Biol.* 12:3425–30

Medema RH, de Vries-Smits AMM, van der Zon GCM, Maassen JA, Bos JL. 1993. Ras activation by insulin and epidermal growth factor through enhanced exchange of guanine nucleotides on p21ras. *Mol. Cell. Biol.* 13: 155–62

Meisenhelder J, Hunter T. 1992. The SH2/SH3 domain-containing protein Nck is recognized by certain anti-phospholipase C-gamma 1 monoclonal antibodies, and its phosphorylation on tyrosine is stimulated by platelet-derived growth factor and epidermal growth factor treatment. *Mol. Cell. Biol.* 12:5843–56

Meisenhelder J, Suh P-G, Rhee SG, Hunter T. 1989. Phospholipase C-γ is a substrate for the PDGF and EGF receptor protein-tyrosine kinases in vivo and in vitro. *Cell* 57:1109–22

Middlemas DS, Meisenhelder J, Hunter T. 1994. Identification of TrkB autophosphorylation sites and evidence that phospholipase C-γl is a substrate of the Trk B receptor. *J. Biol. Chem.* 269:5458–66

Millauer B, Shawver L, Plate KH, Risau W, Ullrich A. 1994. Glioblastoma growth inhibited in vivo by a dominant-negative Flk-1 mutant. *Nature* 367:576–79

Millauer G, Wizigmann-Voos S, Schnürch H, Martinez R, Møller NPH, et al. 1993. High affinity VEGF binding and developmental expression suggest Flk-1 as a major regulator of vasculogenesis and angiogenesis. *Cell* 72:835–46

Mohammadi M, Dionne CA, Li W, Li N, Spivak T, et al. 1992. Point mutation in FGF receptor eliminates phosphatidylinositol hydrolysis without affecting mitogenesis. *Nature* 358:681–84

Mohammadi M, Honegger AM, Rotin D, Fischer R, Bellot F, et al. 1991. A tyrosine-phosphorylated carboxy-terminal peptide of the fibroblast growth factor receptor (Flg) is a binding site for the SH2 domain of phospholipase C-gamma 1. *Mol. Cell. Biol.* 11:5068–78

Molloy CJ, Bottaro DP, Fleming TP, Marshall MS, Gibbs JB, Aaronson SA. 1989. PDGF induction of tyrosine phosphorylation of GTPase activating protein. *Nature* 342:711–14

Moodie SA, Willumsen BM, Weber MJ, Wolfman A. 1993. Complexes of Ras•GTP with Raf-1 and mitogen-activated protein kinase kinase. *Science* 260:1658–61

Moran MF, Koch CA, Anderson D, Ellis C, England L, et al. 1990. Src homology region 2 domains direct protein-protein interactions in signal transduction. *Proc. Natl. Acad. Sci. USA* 87:8622–26

Morgan WR, Greenwald I. 1993. Two novel transmembrane protein tyrosine kinases expressed during *Caenorabditis elegans* hypodermal development. *Mol. Cell Biol.* 13: 7133–43

Mori S, Rönnstrand L, Yokote Y, Engström Å, Courtneidge SA, et al. 1993. Identification of two juxtamembrane autophosphorylation sites in the PDGF β-receptor; involvement in the interaction with Src tyrosine kinases. *EMBO J.* 12:2257–64

Morrison DK, Kaplan DR, Rhee SG, Williams

LT. 1990. Platelet-derived growth factor (PDGF)-dependent association of phospholipase C-gamma with the PDGF receptor signaling complex. *Mol. Cell. Biol.* 10:2359–66

Mourey RJ, Dixon JE. 1994. Protein tyrosine phosphatases: characterization of extracellular and intracellular domains. *Curr. Opin. Genet. Dev.* 4:31–39

Mulcahy LS, Smith MR, Stacey DW. 1985. Requirement for *ras* proto-oncogene function during serum-stimulated growth of NIH3T3 cells. *Nature* 313:241–43

Müller M, Laxton C, Briscoe J, Schindler C, Improta T, et al. 1993. Complementation of a mutant cell line: central role of the 91 kDa polypeptide of ISGF3 in the interferon-α and -γ signal transduction pathway. *EMBO J.* 12: 4221–28

Mulligan LM, Kwok JB, Healy CS, Elsdon MJ, Eng C, et al. 1993. Germline mutations of the RET proto-oncogene in multiple endocrine neoplasia type 2A. *Nature* 363:458–60

Murphy SM, Bergman M, Morgan DO. 1993. Suppression of c-Src activity by C-terminal Src kinase involves the c-Src SH2 and SH3 domains: analysis with *Saccharomyces cerevisiae*. *Mol. Cell. Biol.* 13:5290–300

Musacchio A, Gibson T, Rice P, Thompson J, Saraste M. 1993. The PH domain: a common piece in the structural pathwork of signalling proteins. *Trends Biochem. Sci.* 18:343–48

Musacchio A, Noble M, Pauptit R, Wierenga R, Saraste M. 1992. Crystal structure of a Src-homology 3 (SH3) domain. *Nature* 359:851–55

Myers MG, Backer JM, Sun XJ, Shoelson S, Hu P, et al. 1992. IRS-1 activates phosphatidylinositol 3′-kinase by associating with *src* homology 2 domain of p85. *Proc. Natl. Acad. Sci. USA* 89:10350–54

Nada S, Yagi T, Takeda H, Tokunaga T, Nakagawa H, et al. 1993. Constitutive activation of Src family kinases in mouse embryos that lack Csk. *Cell* 73:1125–35

Nakamura T, Lin L-L, Kharbanda S, Knopf J, Kufe D. 1992. Macrophage colony stimulating factor activates phosphatidylcholine hydrolysis by cytoplasmic phospholipase A_2. *EMBO J.* 11:4917–22

Nakamura T, Nishizawa T, Hagiya M, Seki T, Shimonishi M, et al. 1989. Molecular cloning and expression of human hepatocyte growth factor. *Nature* 342:440–43

Nakanishi H, Brewer KA, Exton JH. 1993. Activation of the ζ isozyme of protein kinase C by phosphatidylinositol 3,4,5-trisphosphate. *J. Biol. Chem.* 268:13–16

Naldini L, Vigna E, Narshiman RP, Gaudino G, Zarnigar R, et al. 1991. Hepatocyte growth factor (HGF) stimulates the tyrosine kinase activity of the receptor encoded by the proto-oncogene c-MET. *Oncogene* 6:501–4

Nishibe S, Wahl MI, Hernandez-Sotomayor SMT, Tonks NK, Rhee SG, Carpenter G. 1990. Increase of the catalytic activity of phospholipase C-γ1 by tyrosine phosphorylation. *Science* 250:1253–56

Nishida Y, Hata M, Nishizuka Y, Rutter WJ, Ebina Y. 1986. Cloning of a *Drosophila* cDNA encoding a polypeptide similar to the human insulin receptor precursor. *Biochem. Biophys. Res. Commun.* 141:474–81

Nishimura R, Li W, Kashishian A, Mondino A, Zhou M, et al. 1993. Two signaling molecules share a phosphotyrosine-containing binding site in the platelet-derived growth factor receptor. *Mol. Cell. Biol.* 13:6889–96

Nocka K, Tan JC, Chiu E, Chu TY, Ray P, et al. 1990. Molecular bases of dominant negative and loss of function mutations at the murine c-kit/white spotting locus: W37, Wv, W41 and W. *EMBO J.* 9:1805–13

Nori M, Vogel US, Gibbs JB, Weber MJ. 1991. Inhibition of v-src-induced transformation by a GTPase-activating protein. *Mol. Cell. Biol.* 11:2812–18

O'Bryan JP, Frye RA, Cogswell PC, Neubauer A, Kitch B, et al. 1991. *Axl*, a transforming gene isolated from primary human myeloid leukemia cells, encodes a novel receptor tyrosine kinase. *Mol. Cell. Biol.* 11:5016–31

Obermeier A, Halfter H, Wiesmüller K-H, Jung G, Schlessinger J, Ullrich A. 1993a. Tyrosine 785 is a major determinant of Trk-substrate interaction. *EMBO J.* 12:933–41

Obermeier A, Lammers R, Wiesmüller K-H, Jung G, Schlessinger J, Ullrich A. 1993b. Identification of Trk binding sites for SHC and phosphatidylinositol 3′-kinase and formation of a multimeric signaling complex. *J. Biol. Chem.* 268:22963–66

Ohashi K, Mizuno K, Kuma K, Miyata T, Nakamura T. 1994. Cloning of the cDNA for a novel receptor tyrosine kinase, Sky, predominantly expressed in brain. *Oncogene* 9: 699–705

Ohmichi M, Decker SJ, Saltiel AR. 1992. Activation of phosphatidylinositol-3 kinase by nerve growth factor involves indirect coupling of the trk proto-oncogene with src homology 2 domains. *Neuron* 9:769–77

Okada M, Howell BW, Broome MA, Cooper JA. 1993. Deletion of the SH3 domain of Src interferes with regulation by the phosphorylated carboxyl-terminal tyrosine. *J. Biol. Chem.* 268:18070–75

Okada M, Nakagawa H. 1988. Identification of a novel protein tyrosine kinase that phosphorylates $pp60^{c\text{-}src}$ and regulates its activity in neonatal rat brain. *Biochem. Biophys. Res. Commun.* 154:796–802

Okada M, Nakagawa H. 1989. A protein tyrosine kinase involved in regulation of $pp60^{c\text{-}src}$ function. *J. Biol. Chem.* 264:20886–93

Olivier JP, Raabe T, Henkemeyer M, Dickson

B, Mbamalu G, et al. 1993. A Drosophila SH2-SH3 adaptor protein implicated in coupling the sevenless tyrosine kinase to an activator of Ras guanine nucleotide exchange, Sos. *Cell* 73:179–91

Otsu M, Hiles I, Gout I, Fry MJ, Ruiz LF, et al. 1991. Characterization of two 85 kd proteins that associate with receptor tyrosine kinases, middle-T/pp60c-src complexes, and PI3-kinase. *Cell* 65:91–104

Overduin M, Rios CB, Mayer BJ, Baltimore D, Cowburn D. 1992. Three-dimensional solution structure of the src homology 2 domain of c-abl. *Cell* 70:697–704

Panayotou G, Gish G, End P, Truong O, Gout I, et al. 1993. Interactions between SH2 domains and tyrosine phosphorylated platelet-derived growth factor βreceptor sequences: analysis of kinetic parameters by a novel biosensor-based approach. *Mol. Cell. Biol.* 13: 3567–76

Park D, Rhee SG. 1992. Phosphorylation of Nck in response to a variety of receptors, phorbol myristate acetate, and cyclic AMP. *Mol. Cell. Biol.* 12:5816–23

Park S, Jove R. 1993. Tyrosine phosphorylation of Ras GTPase-activating protein stabilizes its association with p62 at membranes of v-Src transformed cells. *J. Biol. Chem.* 268: 25728–34

Park S, Liu X, Pawson T, Jove R. 1992. Activated Src tyrosine kinase phosphorylates Tyr-457 of bovine GTPase-activating protein (GAP) in vitro and the corresponding residue of rat GAP in vivo. *J. Biol. Chem.* 267: 17194–200

Partanen J, Armstrong E, Makela TP, Korhonen J, Sandberg M, et al. 1992. A novel endothelial cell surface receptor tyrosine kinase with extracellular epidermal growth factor homology domains. *Mol. Cell. Biol.* 12:1698–707

Partanen J, Makela TP, Eerola E, Korhonen J, Hirvonen H, et al. 1991. FGFR-4, a novel acidic fibroblast growth factor receptor with a distinct expression pattern. *EMBO J.* 10: 1347–54

Pasquale EB. 1991. Identification of chicken embryo kinase 5, a developmentally regulated receptor-type tyrosine kinase of the Eph family. *Cell Regul.* 2:523–34

Paul SR, Merberg D, Finnerty H, Morris GE, Morris JC, et al. 1992. Molecular cloning of the cDNA encoding a receptor tyrosine kinase-related molecule with a catalytic region homologous to c-*met*. *Int. J. Cell Cloning* 10: 309–14

Pawson T. 1988. Non-catalytic domains of cytoplasmic protein-tyrosine kinases: regulatory domains in signal transduction. *Oncogene* 3:491–95

Pawson T, Bernstein A. 1990. Receptor tyrosine kinases: genetic evidence for their role in *Drosophila* and mouse development. *Trends Genet.* 6:350–56

Pawson T, Gish GD. 1992. SH2 and SH3 domains: from structure to function. *Cell* 71: 359–62

Pei D, Neel BG, Walsh CT. 1993. Overexpression, purification, and characterization of SHPTP1, a Src homology 2-containing protein-tyrosine-phosphatase. *Proc. Natl. Acad. Sci. USA* 90:1092–96

Peles E, Lamprecht R, Ben LR, Tzahar E, Yarden Y. 1992. Regulated coupling of the Neu receptor to phosphatidylinositol 3′-kinase and its release by oncogenic activation. *J. Biol. Chem.* 267:12266–74

Peles E, Levy RB, Or E, Ullrich A, Yarden Y. 1991. Oncogenic forms of Neu/HER2 tyrosine kinase are permanently coupled to PLCγ. *EMBO J.* 10:2077–86

Pelicci G, Lanfrancone L, Grignani F, McGlade J, Cavallo F, et al. 1992. A novel transforming protein (SHC) with an SH2 domain is implicated in mitogenic signal transduction. *Cell* 70:93–104

Perez JL, Shen X, Finkernagel S, Sciorra L, Jenkins NA, et al. 1994. Identification and chromosomal mapping of a receptor tyrosine kinase with a putative phospholipid binding sequence in its ectodomain. *Oncogene* 9: 211–19

Perkins LA, Larsen I, Perrimon N. 1992. Corkscrew encodes a putative protein tyrosine phosphatase that functions to transduce the terminal signal from the receptor tyrosine kinase torso. *Cell* 70:225–36

Peters KG, Marie J, Wilson E, Ives HE, Escobedo J, et al. 1992. Point mutation of an FGF receptor abolishes phosphatidylinositol turnover and Ca^{2+} flux but not mitogenesis. *Nature* 358:678–81

Pingel JT, Thomas ML. 1989. Evidence that the leukocyte-common antigen is required for antigen-induced T lymphocyte proliferation. *Cell* 58:1055–65

Piwnica-Worms H, Saunders KB, Roberts TM, Smith AE, Cheng SH. 1987. Tyrosine phosphorylation regulates the biochemical and biological properties of pp60c-src. *Cell* 49: 75–82

Pleiman CM, Clark MR, Gauen LK, Winitz S, Coggeshall KM, et al. 1993. Mapping of sites on the Src family protein tyrosine kinases p55blk, p59fyn, and p56lyn which interact with the effector molecules phospholipase C-gamma 2, microtubule-associated protein kinase, GTPase-activating protein, and phosphatidylinositol 3-kinase. *Mol. Cell. Biol.* 13: 5877–87

Pleiman CM, Hertz WM, Cambier JC. 1994. Activation of phosphatidylinositol-3′ kinase by Src-family kinase SH3 binding to the p85 subunit. *Science* 263:1609–12

Plevin R, Cook SJ, Palmer S, Wakelam MJ.

1991. Multiple sources of sn-1,2-diacylglycerol in platelet-derived-growth-factor-stimulated Swiss 3T3 fibroblasts. Evidence for activation of phosphoinositidase C and phosphatidylcholine-specific phospholipase D. *Biochem. J.* 279: 559–65

Plowman GD, Culouscou J-M, Whitney GS, Green JM, Carlton GW, et al. 1993a. Ligand-specific activation of HER4/p180^{erbB4}, a fourth member of the epidermal growth factor receptor family. *Proc. Natl. Acad. Sci. USA* 90:1746–50

Plowman GD, Green JM, Culouscou J, Carlton GW, Rothwell VM, Buckley S. 1993b. Heregulin induces tyrosine phosphorylation of HER4/p180^{erbB4}. *Nature* 366:473–75

Plutzky J, Neel BG, Rosenberg RD. 1992. Isolation of a src homology 2-containing tyrosine phosphatase. *Proc. Natl. Acad. Sci. USA* 89:1123–27

Ponzetto C, Bardelli A, Maina F, Longati P, Panayotou G, et al. 1993. A novel recognition motif for phosphatidylinositol 3-kinase binding mediates its association with the hepatocyte growth factor/scatter factor receptor. *Mol. Cell. Biol.* 13:4600–8

Posada J, Yew N, Ahn NG, Vande WG, Cooper JA. 1993. Mos stimulates MAP kinase in *Xenopus* oocytes and activates a MAP kinase kinase in vitro. *Mol. Cell. Biol.* 13:2546–53

Prasad KVS, Janssen O, Kapeller R, Raab M, Cantley LC, Rudd CE. 1993a. Src-homology 3 domain of protein kinase p59fyn mediates binding to phosphatidylinositol 3-kinase in T cells. *Proc. Natl. Acad. Sci. USA* 90:7366–70

Prasad KVS, Kapeller R, Janssen O, Repke H, Duke-Cohan JS, et al. 1993b. Phosphatidylinositol (PI) 3-kinase and PI 4-kinase binding to the CD4-p56^{lck} complex: the p56^{lck} SH3 domain binds to PI 3-kinase but not PI 4-kinase. *Mol. Cell. Biol.* 13:7708–17

Price JV, Clifford RJ, Schüpbach T. 1989. The maternal ventralizing locus *torpedo* is allelic to *faint little ball*, an embryonic lethal, and encodes the Drosophila EGF receptor homolog. *Cell* 56:1085–92

Pronk GJ, McGlade J, Pelicci G, Pawson T, Bos JL. 1993. Insulin-induced phosphorylation of the 46- and 52-kDa Shc proteins. *J. Biol. Chem.* 268:5748–53

Pronk GJ, Medema RH, Burgering BM, Clark R, McCormick F, Bos JL. 1992. Interaction between the p21ras GTPase activating protein and the insulin receptor. *J. Biol. Chem.* 267:24058–63

Pulido D, Campuzano S, Koda T, Modolell J, Barbacid M. 1992. *Dtrk*, a *Drosophila* gene related to the trk family of neurotrophin receptors, encodes a novel class of neural cell adhesion molecule. *EMBO J.* 11:391–404

Raffioni S, Bradshaw RA. 1992. Activation of phosphatidylinositol 3-kinase by epidermal growth factor, basic fibroblast growth factor, and nerve growth factor in PC12 pheochromocytoma cells. *Proc. Natl. Acad. Sci. USA* 89:9121–25

Ralston R, Bishop JM. 1985. The product of the protooncogene c-*src* is modified during the response to platelet-derived growth factor. *Proc. Natl. Acad. Sci. USA* 82:7845–49

Ravichandran KS, Lee KK, Songyang Z, Cantley LC, Burn P, Burakoff SJ. 1993. Interaction of SHC with the ζ chain of the T cell receptor upon T cell activation. *Science* 262: 902–5

Reedijk M, Liu XQ, van der Geer P, Letwin K, Waterfield MD, et al. 1992. Tyr721 regulates specific binding of the CSF-1 receptor kinase insert to PI 3′-kinase SH2 domains: a model for SH2-mediated receptor-target interactions. *EMBO J.* 11:1365–72

Reedijk M, Liu XQ, Pawson T. 1990. Interactions of phosphatidylinositol kinase, GTPase-activating protein (GAP), and GAP-associated proteins with the colony-stimulating factor 1 receptor. *Mol. Cell. Biol.* 10: 5601–8

Reichman CT, Mayer BJ, Keshav S, Hanafusa H. 1992. The product of the cellular *crk* gene consists primarily of SH2 and SH3 regions. *Cell Growth Differ.* 3:451–60

Reith AD, Ellis C, Maroc N, Pawson T, Bernstein A, Dubreuil P. 1993. 'W' mutant forms of the Fms receptor tyrosine kinase act in a dominant manner to suppress CSF-1 dependent cellular transformation. *Oncogene* 8: 45–53

Ren R, Mayer BJ, Cicchetti P, Baltimore D. 1993. Identification of a ten-amino acid proline-rich SH3 binding site. *Science* 259: 1157–61

Ren R, Ye Z-S, Baltimore D. 1994. Abl protein-tyrosine kinase selects the Crk adapter as a substrate using SH3 binding sites. *Genes Dev.* 8:783–95

Rescigno J, Mansukhani A, Basilico C. 1991. A putative receptor tyrosine kinase with unique structural topology. *Oncogene* 6: 1909–13

Rhee SG, Choi KD. 1992. Regulation of inositol phospholipid-specific phospholipase C isozymes. *J. Biol. Chem.* 267:12393–96

Ridley AJ, Paterson HF, Johnston CL, Diekmann D, Hall A. 1992. The small GTP-binding protein rac regulates growth factor-induced membrane ruffling. *Cell* 70:401–10

Rodrigues GA, Park M. 1993. Dimerization mediated through a leucine zipper activates the oncogenic potential of the *met* receptor tyrosine kinase. *Mol. Cell. Biol.* 13:6711–22

Rodrigues GA, Park M. 1994. Oncogenic activation of tyrosine kinases. *Curr. Opin. Genet. Dev.* 4:15–24

Rönnstrand L, Mori S, Arridson A-K, Eriksson A, Wernstedt C, et al. 1992. Identification of two C-terminal autophosphorylation sites in the PDGF βreceptor: involvement in the interaction with phospholipase C-γ. *EMBO J.* 11:3911–19

Ronsin C, Muscatelli F, Mattei MG, Breathnach R. 1993. A novel putative receptor protein tyrosine kinase of the met family. *Oncogene* 8:1195–202

Rosnet O, Marchetto S, deLapeyriere O, Birnbaum D. 1991. Murine *Flt3,* a gene encoding a novel tyrosine kinase receptor of the PDGFR/CSF1R family. *Oncogene* 6:1641–50

Rotin D, Margolis B, Mohammadi M, Daly RJ, Daum G, et al. 1992. SH2 domains prevent tyrosine dephosphorylation of the EGF receptor: identification of Tyr992 as the high-affinity binding site for SH2 domains of phospholipase C gamma. *EMBO J.* 11:559–67

Rottapel R, Reedijk M, Williams DE, Lyman SD, Anderson DM, et al. 1991. The Steel/W transduction pathway: kit autophosphorylation and its association with a unique subset of cytoplasmic signaling proteins is induced by the Steel factor. *Mol. Cell. Biol.* 11:3043–51

Roussel MF, Cleveland JL, Shurtleff SA, Sherr CJ. 1991. *Myc* rescue of a mutant CSF-1 receptor impaired in mitogenic signalling. *Nature* 363:361–63

Rozakis-Adcock M, Fernley R, Wade J, Pawson T, Bowtell D. 1993. The SH2 and SH3 domains of mammalian Grb2 couple the EGF receptor to the Ras activator mSos1. *Nature* 363:83–85

Rozakis-Adcock M, McGlade J, Mbamalu G, Pelicci G, Daly R, et al. 1992. Association of the Shc and Grb2/Sem5 SH2-containing proteins is implicated in activation of the Ras pathway by tyrosine kinases. *Nature* 360: 689–92

Rubin GM. 1991. Signal transduction and the fate of the R7 photoreceptor in *Drosophila. Trends Genet.* 7:372–77

Ruderman NB, Kapeller R, White MF, Cantley LC. 1990. Activation of phosphatidylinositol 3-kinase by insulin. *Proc. Natl. Acad. Sci. USA* 87:1411–15

Ruff-Jamison S, Chen K, Cohen S. 1993a. Induction by EGF and interferon-γ of tyrosine phosphorylated DNA binding proteins in mouse liver nuclei. *Science* 261:1733–36

Ruff-Jamison S, McGlade J, Pawson T, Chen K, Cohen S. 1993b. Epidermal growth factor stimulates the tyrosine phosphorylation of SHC in the mouse. *J. Biol. Chem.* 268:7610–12

Ruiz JC, Robertson EJ. 1994. The expression of the receptor-protein tyrosine kinase gene, *eck* is highly restricted during early mouse development. *Mech. Dev.* 46:87–100

Sadowski HB, Gilman MZ. 1993. Cell free activation of a DNA-binding protein by epidermal growth factor. *Nature* 362:79–83

Sadowski HB, Shuai K, Darnell JE, Gilman MZ. 1993. A common nuclear signal transduction pathway activated by growth factor and cytokine receptors. *Science* 261:1739–44

Sadowski I, Stone JC, Pawson T. 1986. A noncatalytic domain conserved among cytoplasmic tyrosine kinases modifies the kinase function and transforming activity of the Fujinami sarcoma virus p130$^{gag\text{-}fps}$. *Mol. Cell. Biol.* 6:4396–408

Sajjadi FG, Pasquale EB. 1993. Five novel avian Eph-related tyrosine kinases are differentially expressed. *Oncogene* 8:1807–13

Sajjadi FG, Pasquale EB, Subramani S. 1991. Identification of a new *eph*-related receptor tyrosine kinase gene from mouse and chicken that is developmentally regulated and encodes at least two forms of the receptor. *New Biol.* 3:769–78

Santoro M, Wong WT, Aroca P, Santos E, Matoskova B, et al. 1994. An epidermal growth factor receptor/*ret* chimera generates mitogenic and transforming signals: evidence for a *ret*-specific signaling pathway. *Mol. Cell. Biol.* 14:663–75

Satoh T, Endo M, Nakafuku M, Akiyama T, Yamamoto T, Kaziro Y. 1990a. Accumulation of p21ras•GTP in response to stimulation with epidermal growth factor and oncogene products with tyrosine kinase activity. *Proc. Natl. Acad. Sci. USA* 87:7926–29

Satoh T, Endo M, Nakafuku M, Nakamura S, Kaziro Y. 1990b. Platelet-derived growth factor stimulates formation of active p21ras•GTP complex in Swiss mouse 3T3 cells. *Proc. Natl. Acad. Sci. USA* 87:5993–97

Schejter ED, Shilo B-Z. 1989. The Drosophilia EGF receptor homolog (*DER*) gene is allelic to *faint little ball,* a locus essential for embryonic development. *Cell* 56:1093–104

Schindler C, Fu XY, Improta T, Aebersold R, Darnell JE. 1992a. Proteins of transcription factor ISGF-3: one gene encodes the 91-and 84-kDa ISGF-3 proteins that are activated by interferon alpha. *Proc. Natl. Acad. Sci. USA* 89:7836–39

Schindler C, Shuai K, Prezioso VR, Darnell JE. 1992b. Interferon-dependent tyrosine phosphorylation of a latent cytoplasmic transcription factor. *Science* 257:809–13

Schlessinger J. 1994. SH2/SH3 signaling proteins. *Curr. Opin. Genet. Dev.* 4:25–30

Schlessinger J, Ullrich A. 1992. Growth factor signaling by receptor tyrosine kinases. *Neuron* 9:383–91

Schneider R. 1992. The human protooncogene *ret*: a communicative cadherin? *Trends Biochem. Sci.* 17:468–69

Schneider R, Schweiger M. 1991. A novel modular mosaic of cell adhesion motifs in the extracellular domains of the neurogenic trk and trkB tyrosine kinase receptors. *Oncogene* 6:1807–11

Schu PV, Takegawa K, Fry MJ, Stack JH, Waterfield MD, Emr SD. 1993. Phosphatidylinositol 3-kinase encoded by yeast VPS34 gene is essential for protein sorting. *Science* 260:88–91

Schuchardt A, D'Agati V, Larsson-Blomberg L, Costantini F, Pachis V. 1994. Defects in the kidney and enteric nervous system of mice lacking the tyrosine receptor Ret. *Nature* 367:380–83

Segatto O, Lonardo F, Helin K, Wexler D, Fazioli F, et al. 1992. erbB-2 autophosphorylation is required for mitogenic action and high-affinity substrate coupling. *Oncogene* 7:1339–46

Segatto O, Pelicci G, Giuli S, Digiesi G, Di FP, et al. 1993. Shc products are substrates of erbB-2 kinase. *Oncogene* 8:2105–12

Selva E, Raden DL, Davis RJ. 1993. Mitogen-activated protein kinase stimulation by a tyrosine kinase-negative epidermal growth factor receptor. *J. Biol. Chem.* 268:2250–54

Serunian LA, Haber MT, Fukui T, Kim JW, Rhee SG, et al. 1989. Polyphosphoinositides produced by phosphatidylinositol 3-kinase are poor substrates for phospholipase C from rat liver and bovine brain. *J. Biol. Chem.* 264:17809–15

Severinsson L, Ek B, Mellstrom K, Claesson WL, Heldin C-H. 1990. Deletion of the kinase insert sequence of the platelet-derived growth factor β receptor affects receptor kinase activity and signal transduction. *Mol. Cell. Biol.* 10:801–9

Shen SH, Bastien L, Posner BI, Chretien P. 1991. A protein-tyrosine phosphatase with sequence similarity to the SH2 domain of the protein-tyrosine kinases. *Nature* 352: 736–39

Shishido E, Higashijima S, Emori Y, Saigo K. 1993. Two FGF-receptor homologues of *Drosophilia*: one is expressed in mesodermal primordium in early embryos. *Development* 117:751–61

Shoelson SE, Sivaraja M, Williams KP, Hu P, Schlessinger J, Weiss MA. 1993. Specific phosphopeptide binding regulates a conformational change in the PI 3-kinase SH2 domain associated with enzyme activation. *EMBO J.* 12:795–802

Shou C, Farnsworth CL, Neel BG, Feig LA. 1992. Molecular cloning of cDNAs encoding a guanine-nucleotide-releasing factor for Ras p21. *Nature* 358:351–54

Shuai K, Horvath CM, Tsai Huang LH, Qureshi SA, Cowburn D, Darnell JE. 1994. Interferon activation of the transcription factor Stat91 involves dimerization through SH2-phosphotyrosyl peptide interactions. *Cell* 76:821–28

Shuai K, Schindler C, Prezioso VR, Darnell JE. 1992. Activation of transcription by IFN-gamma: tyrosine phosphorylation of a 91-kD DNA binding protein. *Science* 258:1808–12

Shuai K, Stark GR, Kerr IM, Darnell JE. 1993a. A single phosphotyrosine residue of Stat91 required for gene activation by interferon-gamma. *Science* 261:1744–46

Shuai K, Ziemiecki A, Wilks AF, Harpur AG, Sadowski HB, et al. 1993b. Polypeptide signalling through tyrosine phosphorylation of Jak and Stat proteins. *Nature* 366:580–83

Shultz LD, Schweitzer PA, Rajan TV, Yi T, Ihle JN, et al. 1993. Mutations at the murine *motheaten* locus are within the hematopoietic cell protein-tyrosine phosphatase (*Hcph*) gene. *Cell* 73:1445–54

Shurtleff SA, Downing JR, Rock CO, Hawkins SA, Roussel MF, Sherr CJ. 1990. Structural features of the colony-stimulating factor 1 receptor that affect its association with phosphatidylinositol 3-kinase. *EMBO J.* 9:2415–21

Silvennoinen O, Schindler C, Schlessinger J, Levy DE. 1993. Ras-independent growth factor signaling by transcription factor tyrosine phosphorylation. *Science* 261:1736–39

Simon MA, Bowtell DD, Dodson GS, Laverty TR, Rubin GM. 1991. Ras1 and a putative guanine nucleotide exchange factor perform crucial steps in signaling by the sevenless protein tyrosine kinase. *Cell* 67:701–16

Simon MA, Dodson GS, Rubin GM. 1993. An SH3-SH2-SH3 protein is required for p21Ras1 activation and binds to sevenless and Sos proteins in vitro. *Cell* 73:169–77

Skolnik EY, Batzer A, Li N, Lee CH, Lowenstein E, et al. 1993a. The function of GRB2 in linking the insulin receptor to Ras signaling pathways. *Science* 260:1953–55

Skolnik EY, Lee CH, Batzer A, Vicentini LM, Zhou M, et al. 1993b. The SH2/SH3 domain-containing protein GRB2 interacts with tyrosine-phosphorylated IRS1 and Shc: implications for insulin control of ras signalling. *EMBO J.* 12:1929–36

Skolnik EY, Margolis B, Mohammadi M, Lowenstein E, Fischer R, et al. 1991. Cloning of PI3 kinase-associated p85 utilizing a novel method for expression/cloning of target proteins for receptor tyrosine kinases. *Cell* 65: 83–90

Smeyne RJ, Klein R, Schnapp A, Long LK, Bryant S, et al. 1994. Severe sensory and sympathetic neuropathies in mice carrying a disrupted Trk/NGF receptor gene. *Nature* 368:246–49

Smith DR, Vogt PK, Hayman MJ. 1989. The v-sea oncogene of the avian retrovirus S13: another member of the protein-tyrosine kinase family. *Proc. Natl. Acad. Sci. USA* 86: 5291–96

Smith MR, DeGudicibus SJ, Stacey DW. 1986.

Requirement for c-ras proteins during viral oncogene transformation. *Nature* 320:540–43

Snijders AJ, Haase VH, Bernards A. 1993. Four tissue-specific mouse *ltk* mRNAs predict tyrosine kinases that differ upstream of their transmembrane segment. *Oncogene* 8:27–35

Soltoff SP, Rabin SL, Cantley L, Kaplan DR. 1992. Nerve growth factor promotes the activation of phosphatidylinositol 3-kinase and its association with the *trk* tyrosine kinase. *J. Biol. Chem.* 267:17472–77

Songyang Z, Shoelson SE, Chaudhuri M, Gish G, Pawson T, et al. 1993. SH2 domains recognize specific phosphopeptide sequences. *Cell* 72:767–78

Songyang Z, Shoelson SE, McGlade J, Olivier P, Pawson T, et al. 1994. Specific motifs recognized by the SH2 domains of Csk, 3BP2, fps/fes, GRB-2, HCP, SHC, Syk, and Vav. *Mol. Cell. Biol.* 14:2777–85

Sprenger F, Stevens LM, Nüsslein-Volhard C. 1989. The *Drosophila* gene *torso* encodes a putative receptor tyrosine kinase. *Nature* 338:478–83

Stahl ML, Ferenz CR, Kelleher KL, Kriz RW, Knopf JL. 1988. Sequence similarity of phospholipase C with the non-catalytic region of src. *Nature* 332:269–72

Stephenson DA, Mercola M, Anderson E, Wang C, Stiles CD, et al. 1991. Platelet-derived growth factor receptor α-subunit gene (*Pdgfa*) is deleted in the mouse patch (*Ph*) mutation. *Proc. Natl. Acad. Sci. USA* 88:6–10

Sternberg PW, Horvitz HR. 1991. Signal transduction during *C. elegans* vulval induction. *Trends Genet.* 7:366–71

St. Johnston D, Nüsslein-Volhard C. 1992. The origin of pattern and polarity in the Drosophila embryo. *Cell* 68:201–19

Suen KL, Bustelo XR, Pawson T, Barbacid M. 1993. Molecular cloning of the mouse *grb2* gene: differential interaction of the Grb2 adaptor protein with epidermal growth factor and nerve growth factor receptors. *Mol. Cell. Biol.* 13:5500–12

Suh P-G, Ryu SH, Moon KH, Suh HW, Rhee SG. 1988. Inositol phospholipid-specific phospholipase C: complete cDNA and protein sequence homology to tyrosine kinase-related oncogene products. *Proc. Natl. Acad. Sci. USA* 85:5419–23

Sun XJ, Rothenberg P, Kahn CR, Backer JM, Araki E, et al. 1991. Structure of the insulin receptor substrate IRS-1 defines a unique signal transduction protein. *Nature* 352:73–77

Superti-Furga G, Fumagalli S, Koegl M, Courtneidge SA, Draetta G. 1993. Csk inhibition of c-Src activity requires both the SH2 and SH3 domains of Src. *EMBO J.* 12:2625–34

Takahashi M, Cooper GM. 1987. *ret* transforming gene encodes a fusion protein homologous to tyrosine kinases. *Mol. Cell. Biol.* 7: 1378–85

Tanaka K, Nakafuku M, Satoh T, Marshall MS, Gibbs JB, et al. 1990. S. cerevisiae genes *IRA1* and *IRA2* encode proteins that may be functionally equivalent to mammalian ras GTPase activating protein. *Cell* 60:803–7

Tanaka S, Hattori S, Kurata T, Nagashima K, Fukui Y, et al. 1993. Both the SH2 and SH3 domains of human CRK protein are required for neuronal differentiation of PC12 cells. *Mol. Cell. Biol.* 13:4409–15

ten Hoeve J, Morris C, Heisterkamp N, Groffen J. 1993. Isolation and chromosomal localization of *CRKL*, a human *crk*-like gene. *Oncogene* 8:2469–74

Terman BI, Carrion ME, Kovacs E, Rasmussen BA, Eddy RL, Shows TB. 1991. Identification of a new endothelial cell growth factor receptor tyrosine kinase. *Oncogene* 6:1677–83

Terman BI, Dougher-Vermazen M, Carrion ME, Dimitrov D, Armellino DC, et al. 1992. Identification of the KDR tyrosine kinase as a receptor for vascular endothelial cell growth factor. *Biochem. Biophys. Res. Commun.* 187:1579–86

Thomas SM, De Marco M, D'Arcangelo G, Halegoua S, Brugge JS. 1992. Ras is essential for nerve growth factor- and phorbol ester-induced tyrosine phosphorylation of MAP kinases. *Cell* 68:1031–40

Tobe K, Matuoka K, Tamemoto H, Ueki K, Kaburagi Y, et al. 1993. Insulin stimulates association of insulin receptor substrate-1 with the protein abundant Src homology/growth factor receptor-bound protein 2. *J. Biol. Chem.* 268:11167–71

Tonks NK, Diltz CD, Fischer EH. 1988. Purification of the major protein-tyrosine-phosphatases of human placenta. *J. Biol. Chem.* 263:6722–30

Torti M, Marti KB, Altschuler D, Yamamoto K, Lapetina EG. 1992. Erythropoietin induces p21ras activation and p120GAP tyrosine phosphorylation in human erythroleukemia cells. *J. Biol. Chem.* 267:8293–98

Toyoshima H, Kozutsumi H, Maru Y, Hagiwara K, Furuya A, et al. 1993. Differently spliced cDNAs of human leukocyte tyrosine kinase receptor tyrosine kinase predict receptor proteins with and without a tyrosine kinase domain and a soluble receptor protein. *Proc. Natl. Acad. Sci. USA* 90:5404–8

Trahey M, McCormick F. 1987. A cytoplasmic protein stimulates normal N-ras p21 GTPase, but does not affect oncogenic mutants. *Science* 238:542–45

Trahey M, Wong G, Halenbeck R, Rubinfeld B, Martin GA, et al. 1988. Molecular cloning of two types of GAP complementary DNA from human placenta. *Science* 242:1697–700

Treisman R. 1994. Ternary complex factors: growth factor regulated transcriptional activators. *Curr. Opin. Genet. Dev.* 4:96–101

Tsoulfas P, Soppet D, Escandon E, Tessarollo L, Mendoza-Ramirez JL, et al. 1993. The rat trkC locus encodes multiple neurogenic receptors that exhibit differential response to neurotrophin-3 in PC12 cells. *Neuron* 10: 975–90

Tsuchie H, Chang CH, Yoshida M, Vogt PK. 1989. A newly isolated avian sarcoma virus, ASV-1, carries the crk oncogene. *Oncogene* 4:1281–84

Twamley GM, Kypta RM, Hall B, Courtneidge SA. 1992. Association of Fyn with the activated platelet-derived growth factor receptor: requirements for binding and phosphorylation. *Oncogene* 7:1893–901

Twamley-Stein GM, Pepperkok R, Ansorge W, Courtneidge SA. 1993. The Src family tyrosine kinases are required for platelet-derived growth factor-mediated signal transduction in NIH 3T3 cells. *Proc. Natl. Acad. Sci. USA* 90:7696–700

Ueno H, Colbert H, Escobedo JA, Williams LT. 1991. Inhibition of PDGF beta receptor signal transduction by coexpression of a truncated receptor. *Science* 252:844–48

Ullrich A, Schlessinger J. 1990. Signal transduction by receptors with tyrosine kinase activity. *Cell* 61:203–12

Valenzuela DM, Maisonpierre PC, Glass DJ, Rojas E, Nunez L, et al. 1993. Alternative forms of rat TrkC with different functional capabilities. *Neuron* 10:963–74

Valius M, Bazenet C, Kazlauskas A. 1993. Tyrosines 1021 and 1009 are phosphorylation sites in the carboxy terminus of the platelet-derived growth factor receptor β subunit and are required for binding of phospholipase Cγ and a 64-kilodalton protein, respectively. *Mol. Cell. Biol.* 13:133–43

Valius M, Kazlauskas A. 1993. Phospholipase C-gamma 1 and phosphatidylinositol 3 kinase are the downstream mediators of the PDGF receptor's mitogenic signal. *Cell* 73: 321–34

Van Aelst L, Barr M, Marcus S, Polverino A, Wigler M. 1993. Complex formation between RAS and RAF and other protein kinases. *Proc. Natl. Acad. Sci. USA* 90:6213–7

van der Bend RL, de Widt J, van Corven EJ, Moolenaar WH, van Blitterswijk WJ. 1992. The biologically active phospholipid, lysophosphatidic acid, induces phosphatidylcholine breakdown in fibroblasts via activation of phospholipase D. Comparison with the response to endothelin. *Biochem. J.* 285:235–40

van der Geer P, Hunter T. 1991. Tyrosine 706 and 807 phosphorylation site mutants in the murine colony-stimulating factor-1 receptor are unaffected in their ability to bind or phosphorylate phosphatidylinositol-3 kinase but show differential defects in their ability to induce early response gene transcription. *Mol. Cell. Biol.* 11:4698–709

van der Geer P, Hunter T. 1993. Mutation of Tyr697, a GRB2-binding site, and Tyr 721, a PI 3-kinase binding site, abrogates signal transduction by the murine CSF-1 receptor expressed in Rat-2 fibroblasts. *EMBO J.* 12:5161–72

Van Horn DJ, Myers MG, Backer JM. 1994. Direct activation of the phosphatidylinositol 3′-kinase by the insulin receptor. *J. Biol. Chem.* 269:29–32

Varticovski L, Daley GQ, Jackson P, Baltimore D, Cantley LC. 1991. Activation of phosphatidylinositol 3-kinase in cells expressing abl oncogene variants. *Mol. Cell. Biol.* 11: 1107–13

Varticovski L, Druker B, Morrison D, Cantley L, Roberts T. 1989. The colony stimulating factor-1 receptor associates with and activates phosphatidylinositol-3 kinase. *Nature* 342:699–702

Vega QC, Cochet C, Filhol O, Chang C-P, Rhee SG, Gill GN. 1992. A site of tyrosine phophorylation in the C-terminus of the epidermal growth-factor receptor is required to activate phospholipase C. *Mol. Cell. Biol.* 12: 128–35

Vetter ML, Martin ZD, Parada LF, Bishop JM, Kaplan DR. 1991. Nerve growth factor rapidly stimulates tyrosine phosphorylation of phospholipase C-gamma 1 by a kinase activity associated with the product of the trk protooncogene. *Proc. Natl. Acad. Sci. USA* 88:5650–54

Vogel US, Dixon RAF, Schaber MD, Diehl RE, Marshall MS, et al. 1988. Cloning of bovine GAP and its interaction with oncogenic *ras* p21. *Nature* 335:90–93

Vogel W, Lammers R, Huang J, Ullrich A. 1993. Activation of a phosphotyrosine phosphatase by tyrosine phosphorylation. *Science* 259:1611–14

Vojtek AB, Hollenberg SM, Cooper JA. 1993. Mammalian Ras interacts directly with the serine/threonine kinase Raf. *Cell* 74:205–14

Wada I, Lai WH, Posner BI, Bergeron JJ. 1992. Association of the tyrosine phosphorylated epidermal growth factor receptor with a 55-kD tyrosine phosphorylated protein at the cell surface and in endosomes. *J. Cell Biol.* 116:321–30

Wagner BJ, Hayes TE, Hoban CJ, Cochran BH. 1990. The SIF binding element confers *sis*/PDGF inducibility onto the c-*fos* promoter. *EMBO J.* 9:4477–84

Wahl M, Nishibe S, Kim JW, Kim H, Rhee SG, Carpenter G. 1990. Identification of two epidermal growth factor-sensitive tyrosine phosphorylation sites of phospholipase C-γ in intact HSC-1 cells. *J. Biol. Chem.* 265: 3944–48

Waksman G, Kominos D, Robertson SC, Pant N, Baltimore D, et al. 1992. Crystal structure of the phosphotyrosine recognition domain SH2 of v-*src* complexed with tyrosine-phosphorylated peptides. *Nature* 358:646–53

Wang Z, Myles GM, Brandt CS, Lioubin MN, Rohrschneider L. 1993. Identification of the ligand-binding regions in the macrophage colony-stimulating factor receptor extracellular domain. *Mol. Cell. Biol.* 13:5348–59

Warne PH, Viciana PR, Downward J. 1993. Direct interaction of Ras and the amino-terminal region of Raf-1 in vitro. *Nature* 364: 352–55

Watling D, Guschin D, Müller M, Silvennoinen O, Witthuhn BA, et al. 1993. Complementation of a mutant cell line defective in the interferon-γ signal transduction pathway by the protein tyrosine kinase JAK2. *Nature* 366:166–70

Wei W, Mosteller RD, Sanyal P, Gonzales E, McKinney D, et al. 1992. Identification of a mammalian gene structurally and functionally related to the *CDC25* gene of *Saccharomyces cerevisiae*. *Proc. Natl. Acad. Sci. USA* 89:7100–4

Weiner DB, Liu J, Cohen JA, Williams WV, Greene MI. 1989. A point mutation in the *neu* oncogene mimics ligand induction of receptor aggregation. *Nature* 339:230–31

Welsh M, Mares J, Karlsson T, Lavergne C, Breant B, Claesson-Welsh L. 1994. Shb is a ubiquitously expressed Src homology 2 protein. *Oncogene* 9:19–27

Wen D, Peles E, Cupples R, Suggs SV, Bacus SS, et al. 1992. Neu differentiation factor: a transmembrane glycomembrane containing an EGF domain and an immunoglobulin homology unit. *Cell* 69:559–72

Weng Z, Taylor JA, Turner CE, Brugge JS, Seidel-Dugan C. 1993. Detection of Src homology 3-binding proteins, including paxillin, in normal and v-Src-transformed Balb/c 3T3 cells. *J. Biol. Chem.* 268:14956–63

Wennstrom S, Siegbahn A, Yokote K, Avridsson A-K, Mori S, et al. 1994. Membrane ruffling and chemotaxis transduced by the PDGF-β receptor require the binding site for phosphatidylinositol 3′ kinase. *Oncogene* 9:651–60

Werner S, Duan DR, De Vries C, Peters KG, Johnson DE, Williams LT. 1992. Differential splicing in the extracellular region of fibroblast growth factor receptor 1 generates receptor variants with different ligand-binding specificities. *Mol. Cell. Biol.* 12:82–88

Werner S, Weinberg W, Liao X, Peters KG, Blessing M, et al. 1993. Targeted expression of a dominant-negative FGF receptor mutant in the epidermis of transgenic mice reveals a role of FGF in keratinocyte organizaton and differentiation. *EMBO J.* 12:2635–43

West M, Kung HF, Kamata T. 1990. A novel membrane factor stimulates guanine nucleotide exchange reaction of ras proteins. *FEBS Lett.* 259:245–48

White MF. 1994. The IRS-1 signaling system. *Curr. Opin. Genet. Dev.* 4:47–54

White MF, Maron R, Kahn CR. 1985. Insulin rapidly stimulates tyrosine phosphorylation of a M_r-185,000 protein in intact cells. *Nature* 318:183–86

Whitman M, Downes PC, Keeler M, Keller T, Cantley L. 1988. Type I phosphatidylinositol kinase makes a novel inositol phospholipid, phosphatidylinositol-3-phosphate. *Nature* 332: 644–46

Whitman M, Kaplan D, Roberts T, Cantley L. 1987. Evidence for two distinct phosphatidylinositol kinases in fibroblasts. *Biochem. J.* 247:165–74

Wicks IP, Wilkinson D, Salvaris E, Boyd AW. 1992. Molecular cloning of *HEK*, the gene encoding a receptor tyrosine kinase expressed by human lymphoid tumor cell lines. *Proc. Natl. Acad. Sci. USA* 89:1611–15

Widmer HR, Kaplan DR, Rabin SJ, Beck KD, Hefti F, Knusel B. 1993. Rapid phosphorylation of phospholipase C gamma 1 by brain-derived neurotrophic factor and neurotrophin-3 in cultures of embryonic rat cortical neurons. *J. Neurochem.* 60:2111–23

Wiedlocha A, Falnes PO, Madshus IH, Sandvig K, Olsnes S. 1994. Dual mode of signal transduction by externally added acidic fibroblast growth factor. *Cell* 76:1039–51

Wiktor-Jedrezejczak W, Bartocci A, Ferrante AW, Ahmed-Ansari A, Sell AW, et al. 1990. Total absence of colony-stimulating factor 1 in the macrophage-deficient osteopetrotic (op/op) mouse. *Proc. Natl. Acad. Sci. USA* 87:4828–32

Williams LT. 1989. Signal transduction by the platelet-derived growth factor receptor. *Science* 243:1564–70

Wilson C, Goberdhan DCI, Steller H. 1993. *Dror*, a potential neurotrophic receptor gene, encodes a *Drosophila* homolog of the vertebrate Ror family of Trk-related receptor tyrosine kinases. *Proc. Natl. Acad. Sci. USA* 90:7109–13

Wittbrodt J, Adam D, Malitschek B, Mäueler W, Raulf F, et al. 1989. Novel putative receptor tyrosine kinase encoded by the melanoma-inducing *Tu* locus in *Xiphophorus*. *Nature* 341:415–21

Witte ON. 1990. Steel locus defines new multipotent growth factor. *Cell* 63:5–6

Wolfman A, Macara IG. 1990. A cytosolic protein catalyzes the release of GDP from p21ras. *Science* 248:67–69

Woltjer RL, Lukas TJ, Staros JV. 1992. Direct identification of residues of the epidermal growth factor receptor in close proximity to the amino terminus of bound epidermal growth factor. *Proc. Natl. Acad. Sci. USA* 89:7801–5

Wong ST, Winchell LF, McCune BK, Earp HS, Teixido JJM, et al. 1989. The TGF-alpha precursor expressed on the cell surface binds to the EGF receptor on adjacent cells, leading to signal transduction. *Cell* 56:495–506

Wood KW, Sarnecki C, Roberts TM, Blenis J. 1992. Ras mediates nerve growth factor receptor modulation of three signal-transducing protein kinases: MAP kinase, Raf-1, and RSK. *Cell* 68:1041–50

Xu GF, O'Connell P, Viskochil D, Cawthon R, Robertson M, et al. 1990. The neurofibromatosis type 1 gene encodes a protein related to GAP. *Cell* 62:599–608

Xu Q, Holder N, Patient R, Wilson SW. 1994. Spatially regulated expression of three receptor tyrosine kinase genes during gastrulation in the zebrafish. *Development.* 120:287–99

Yamanashi Y, Fukui Y, Wongsasant B, Kinoshita Y, Ichimori Y, et al. 1992. Activation of Src-like protein-tyrosine kinase Lyn and its association with phosphatidylinositol 3-kinase upon B-cell antigen receptor-mediated signaling. *Proc. Natl. Acad. Sci. USA* 89:1118–22

Yarden Y, Ullrich A. 1988. Growth factor receptor tyrosine kinases. *Annu. Rev. Biochem.* 57:443–78

Yatani A, Okabe K, Polakis P, Halenbeck R, McCormick F, Brown AM. 1990. *ras* p21 and GAP inhibit coupling of muscarinic receptors to atrial K^+ channels. *Cell* 61:769–76

Yee K, Bishop TR, Zon LI. 1993. Isolation of a novel receptor tyrosine kinase cDNA expressed by developing erythroid progenitors. *Blood* 82:1335–43

Yeung Y-G, Berg KL, Pixley FJ, Angeletti RH, Stanley ER. 1992. Protein tyrosine phosphatase-1C is rapidly phosphorylated on tyrosine in macrophages in response to colony stimulating factor-1. *J. Biol. Chem.* 267:23447–50

Yi T, Ihle JN. 1993. Association of hematopoietic cell phosphatase with c-Kit after stimulation with c-Kit ligand. *Mol. Cell. Biol.* 13: 3350–58

Yi TL, Cleveland JL, Ihle JN. 1992. Protein tyrosine phosphatase containing SH2 domains: characterization, preferential expression in hematopoietic cells, and localization to human chromosome 12p12-p13. *Mol. Cell. Biol.* 12:836–46

Yoshida H, Hayashi S, Kunisada T, Ogawa M, Nishikawa S, et al. 1990. The murine mutation osteopetrosis is in the coding region of the macrophage colony stimulating factor gene. *Nature* 345:442–44

Yu H, Chen JK, Feng S, Dalgarno DC, Brauer AW, Schreiber SL. 1994. Structural basis for the binding of proline-rich peptides to SH3 domains. *Cell* 76:933–45

Yu H, Rosen MK, Shin TB, Seidel-Dugan C, Brugge JS, Schreiber SL. 1992. Solution structure of the SH3 domain of Src and identification of its ligand binding site. *Science* 258:1665–68

Yu JC, Heidaran MA, Pierce JH, Gutkind JS, Lombardi D, et al. 1991. Tyrosine mutations within the alpha platelet-derived growth factor receptor kinase insert domain abrogate receptor-associated phosphatidylinositol-3 kinase activity without affecting mitogenic or chemotactic signal transduction. *Mol. Cell. Biol.* 11:3780–85

Zerlin M, Julius MA, Goldfarb M. 1993. NEP: a novel receptor-like tyrosine kinase expressed in proliferating neuroepithelia. *Oncogene* 8:2731–39

Zhang F, Strand A, Robbins D, Cobb MH, Goldsmith EJ. 1994. Atomic structure of the MAP kinase ERK2 at 2.3 Å resolution. *Nature* 367:704–11

Zhang K, DeClue JE, Vass WC, Papageorge AG, McCormick F, Lowy DR. 1990. Suppression of c-ras transformation by GTPase-activating protein. *Nature* 346:754–56

Zhang K, Papageorge AG, Lowy DR. 1992a. Mechanistic aspects of signaling through Ras in NIH 3T3 cells. *Science* 257:671–74

Zhang XF, Settleman J, Kyriakis JM, Takeuchi-Suzuki E, Elledge SJ, et al. 1993. Normal and oncogenic p21ras proteins bind to the amino-terminal regulatory domain of c-Raf-1. *Nature* 364:308–13

Zhang XK, Egan JO, Huang DP, Sun ZL, Chien VKY, Chiu JF. 1992b. Hepatitis B virus DNA integration and expression of an ERB B-like gene in human hepatocellular carcinoma. *Biochem. Biophys. Res. Commun.* 188:344–51

Zhao Z, Bouchard P, Diltz CD, Shen SH, Fischer EH. 1993. Purification and characterization of a protein tyrosine phosphatase containing SH2 domains. *J. Biol. Chem.* 268:2816–20

Zhong Z, Wen Z, Darnell JE. 1994. Stat3: A STAT family member activated by tyrosine phosphorylation in response to epidermal growth factor and interleukin 6. *Science* 264: 95–98

Annu. Rev. Cell Biol. 1994. 10:339–72

DYNEINS: Molecular Structure and Cellular Function

E. L. F. Holzbaur
University of Pennsylvania, Philadelphia, Pennsylvania 19104–6046

R. B. Vallee
Worcester Foundation for Experimental Biology, Shrewsbury, Massachusetts 01545

KEY WORDS: dynein, microtubules, flagella, axonal transport, organelle transport

---

CONTENTS

0743–4634/94/1115–0339$05.00

INTRODUCTION

Dynein was first identified in eukaryotic axonemes as the ATPase required for flagellar and ciliary beating (Gibbons & Rowe 1965). By analogy to the acto-myosin cross-bridge cycle, dynein was proposed to couple ATP hydrolysis to sliding between the adjacent outer doublet microtubules of the axoneme. Over two decades later, a cytoplasmic form of the enzyme was isolated from brain tissue as the motor molecule responsible for transport along microtubules toward their minus ends (Paschal & Vallee 1987).

While the role of axonemal dynein in mediating flagellar and ciliary beating seems reasonably clear, the complete range of cytoplasmic dynein functions remains an issue of considerable interest. The enzyme is thought to be responsible for the retrograde or minus-end directed transport of organelles within the cell, including retrograde axonal transport, as well as the centripetal transport of endosomes, lysosomes, and the elements of the Golgi apparatus in cells in general. However, there is evidence that also points to a role for cytoplasmic dynein in mitosis, although the precise stage of mitosis during which the enzyme may act remains uncertain.

Axonemal and cytoplasmic dyneins appear to be similar to kinesin and myosin in the general mechanism for coupling the energy of ATP hydrolysis to the production of mechanical force. However, the dyneins are structurally distinct from kinesin and myosin, as well as from the newly discovered superfamilies of kinesin- and myosin-related proteins (see Goldstein 1991; Bloom 1992; Endow & Titus 1992; Hammer 1991; Cheney et al 1993; Pollard et al 1991; Goodson & Spudich 1993). Some distinctive aspects of the mechanism of force production by axonemal dyneins have already been described (Porter & Johnson 1983; Johnson 1985; Holzbaur & Johnson 1989a,b), and other differences will surely be discovered as analysis of cytoplasmic dynein mechanochemistry proceeds.

In this chapter, we review recent developments in our understanding of the molecular structure of axonemal and cytoplasmic dyneins, and the insights this information has provided into the cellular function and mechanochemistry of these motor proteins. This chapter extends our own recent reviews of dynein molecular biology (Holzbaur et al 1994; Vallee 1993) and updates earlier reviews of dynein structure and function (Gibbons 1988; Porter & Johnson 1989; Vallee & Shpetner 1990).

Diversity of Axonemal and Cytoplasmic Dyneins

The dyneins represent a structurally similar but functionally diverse family of proteins. Three general classes of dyneins can be distinguished.

Axonemal dyneins may be classified as either outer arm or inner arm dyneins, based on their distribution within the axoneme. Outer arm dyneins were the first to be isolated and characterized because of their relative ease of purification from species such as sea urchin, *Tetrahymena*, and *Chlamydomonas*. They remain the best characterized of the axonemal dyneins. In *Chlamydomonas*, the outer row of dynein arms is composed of a single three-headed complex that is uniformly distributed at 24-nm intervals over the length of the outer doublet microtubules.

Inner arm dyneins are more diverse in both composition and distribution. Biochemical isolation and fractionation of inner arm dyneins from an outer arm-deficient *Chlamydomonas* mutant strain led to the identification of seven distinct inner arm subtypes (Kagami & Kamiya 1992; Piperno et al 1990). The inner arms are thought to be arranged in two rows in a complex pattern of alternating isoforms (Mastronarde et al 1992). The composition of the inner arm dyneins has been found to differ along the length of the axoneme (Piperno & Ramanis 1991).

Cytoplasmic dyneins represent a third structural class. In contrast to the axonemal dyneins, the cytoplasmic enzyme has been implicated in a relatively wide range of intracellular functions, but may be simpler and more highly conserved in structure than the axonemal forms.

Dynein Structure

Axonemal and cytoplasmic dyneins are morphologically similar as analyzed by a variety of electron microscopic techniques. Perhaps the oddest feature of the dyneins is the difference in the number of force-producing "heads." Of several axonemal dyneins that have been investigated, *Chlamydomonas* flagellar outer arm dynein, *Tetrahymena* ciliary outer arm dynein, and *Paramecium* ciliary outer arm dynein have three heads. In contrast, sea urchin flagellar outer arm dynein, trout and bull sperm outer arm dynein, and pig tracheal ciliary dynein have two heads. Biochemical and structural analysis of flagellar inner arms dyneins from *Chlamydomonas* revealed a two-headed structure (Piperno et al 1990). All true cytoplasmic dyneins that have been examined (as distinct from soluble ciliary precursor forms) have two heads as well. The existing data suggest that the third head is unique to protist axonemal dyneins.

The two or three large globular dynein heads, 10–14 nm in diameter, are connected by slender stalks to a common base (Johnson & Wall 1983; Goodenough & Heuser 1985; Vallee et al 1988). Cytoplasmic dyneins have a mass of 1.2×10^6 Daltons (Vallee et al 1988), and axonemal dyneins range in mass from 1.2 to 1.9×10^6 Daltons, depending on the number of globular heads in the complex (reviewed in Johnson et al 1986). The base of the molecule appears to be about as massive as each of the heads (Johnson & Wall 1983; Vallee et al 1988). Some images suggest that the base is composed of several small globular domains (Sale et al 1985) that could represent individual dynein accessory subunits (see below). Some differences in the overall shape of individual heads have been detected (for example, pear-shaped vs round; Sale et al 1985). In addition, a variety of protrusions, such as filamentous stalks, appear to be associated with the heads of both axonemal and cytoplasmic dyneins. These protrusions may play role in microtubule binding (Goodenough & Heuser 1982, 1984; Amos 1989).

In the axoneme, the globular dynein heads interact in a nucleotide-dependent manner with the B-tubule of the outer microtubule doublets (Johnson & Wall 1983). The base of the dynein complex binds to the A-tubule in an ATP-insensitive manner. By analogy, the base of cytoplasmic dynein is thought to anchor the molecule to organelles and, possibly, to kinetochores, while the globular heads interact in an ATP-dependent manner with the microtubules of the cytoskeleton (Vallee et al 1988).

Subunit Composition of Axonemal and Cytoplasmic Dyneins

Dyneins contain from one to three heavy chain polypeptides, which range in size from 471 to 540 kDa, and varying numbers of intermediate chains (57–140 kDa) and light chains (6–22 kDa). The well-characterized outer arm axonemal dynein from *Chlamydomonas* is composed of three heavy chains (α, β, and γ) of ~ 500 kDa, two intermediate chains of 70 and 78 kDa, and at least eight light chains (Table 1; Pfister et al 1982; Piperno & Luck 1979). The inner row of dynein arms in the axoneme is more heterogeneous. Biochemical and morphological analysis of inner arm dyneins from mutants lacking either outer or inner dynein arms indicate three distinct species separable by sucrose density gradient centrifugation (I1, I2, and I3; Piperno et al 1990). However, these inner arm dyneins may be further fractionated by FPLC to reveal seven distinct biochemical species that vary in their polypeptide composition, as well as their ATPase activities and rates of microtubule translocation (Kagami & Kamiya 1992). In all, eight distinct inner arm dynein heavy chains have been identified by SDS-gel electrophoresis and vanadate/UV-mediated photocleavage (Kagami & Kamiya 1992). The intermediate and light chain content of each of the inner arm dyneins differ markedly (Kagami & Kamiya 1992). It has also been suggested that the subunit composition of inner arm dynein I2 varies with its position along the length of the axoneme (Piperno et al 1990; Kagami & Kamiya 1992).

The subunit composition of cytoplasmic dynein has been defined by sucrose density gradient centrifugation, chromatography, and immunoprecipitation. The bovine brain enzyme is composed of two high molecular weight polypeptides of 532 K, three 74 K intermediate chains, and a quartet of eleven light intermediate chains of 59, 57, 55, and 53 K (Paschal et al 1987b, 1993; Vallee et al 1988; Hughes et al 1993). Polypeptides corresponding in size to the intermediate chains have also been seen to co-purify with the heavy chain in

Table 1 Subunit composition of cytoplasmic and axonemal dyneins

	Cytoplasmic dynein	Axonemal dyneins	
	Bovine brain[a] (kDa)	*Chlamydomonas* outer arm dynein[b]	*Chlamydomonas* inner arm dynein, I1[c] (kDa)
Heavy chains	530	480 (α) 440 (β) 415 (γ)	433 418
Intermediate chains	74	77 (IC80) 64 (IC70)	140 97
Light intermediate chains	57/59 53/55		
Light chains	—	22, 20, 19 18, 16, 14, 14, 11, 8, 8	—

[a] Paschal et al 1987; [b] Pfister et al 1982; Witman et al 1994; [c] Piperno et al, 1990; Kagami & Kamiya, 1992.

cytoplasmic dyneins isolated from other tissues (Collins & Vallee 1989; Neely et al 1990) and species (Steuer et al 1990; Pfarr et al 1990).

Force Generation in Vitro

Axonemal and cytoplasmic dyneins are capable of driving microtubule gliding in vitro (Paschal et al 1987a,b; Lye et al 1987; Vale & Toyoshima 1988), and both the cytoplasmic (Paschal & Vallee 1987) and axonemal enzymes (Vale & Toyoshima 1988) generate force directed toward the minus end of the microtubule. However, the average rate of microtubule gliding powered by cytoplasmic dynein is slower (1.25–2.0 μm/sec; Paschal et al 1987b; Lye et al 1987) than that for axonemal outer arm dyneins (4–10 μm/sec: Paschal et al 1987a; Vale & Toyoshima 1988). Purified inner arm dyneins also support rapid microtubule gliding rates (2–12 μm/sec; Kagami & Kamiya 1992), all minus end-directed.

Some forms of axonemal dynein (Vale & Toyoshima 1988; Kagami & Kamiya 1992) have caused microtubules to rotate as they glide, which was interpreted to indicate an off-axial component to force production. Recently, kinesin has also been shown to produce microtubule rotation that correlates strongly with the pitch of the microtubule lattice supertwist (Sanghamitra et al 1993). A physiological requirement for off-axial force production is unclear, and it remains to be determined whether this behavior contributes significantly to intracellular motility.

In order to address the question of whether a single dynein head is capable of producing sustained force, or whether the combined activity of multiple heads is required, Sale & Fox (1988) and Vale et al (1989) measured the motility characteristics of sea urchin outer arm dynein that had been separated into single-headed particles by low ionic strength dialysis. Under these conditions, a discrete peak consisting of the beta heavy chain and the intermediate chains can be isolated, while the alpha heavy chain cannot be isolated as a discrete biochemical species. The beta/IC particle supported microtubule gliding at rapid rates (4–11 μm/sec), which suggests that a single head is sufficient for motility in this assay.

Rates of gliding of microtubules on dynein-coated coverslips and of movement of dynein-coated beads along microtubules in general appear to be considerably greater than those observed for kinesin or its related proteins. The patterns of kinesin-coated and cytoplasmic dynein-coated bead movement also differ. Kinesin-coated latex beads move in a linear pattern, apparently along a single microtubule protofilament, while dynein wanders across the tubulin lattice (Wang & Sheetz 1993). Whether this reflects a difference in the mechanism by which the two classes of protein interact with microtubules, or is a consequence of their different size and flexibility, remains to be resolved.

SEQUENCE AND STRUCTURE OF THE DYNEIN HEAVY CHAINS

Conservation of Multiple Consensus ATP-Binding Motifs within Both Axonemal and Cytoplasmic Dynein Sequences

The full-length sequences of three axonemal and six cytoplasmic dynein heavy chains have been obtained from analysis of cDNA clones: the sea urchin outer arm β-heavy chain (Gibbons et al 1991; Ogawa 1991); and the *Chlamydomonas* outer arm β- and γ-heavy chains (Mitchell & Brown 1994; Wilkerson et al 1994). Six cytoplasmic dynein heavy chains have been sequenced from *Dictyostelium* (Koonce et al 1992); rat (Mikami et al 1993; Zhang et al 1993); *Saccharomyces cerevisiae* (Li et al 1993; Eschel et al 1993); *Aspergillus nidulans* (Xiang et al 1994); *Neurospora crassa* (Plamann et al 1994); *Drosophila melanogaster* (Y Li et al, unpublished results); and *Caenorhabditis elegans* (R Lye et al, unpublished results). In addition, many partial sequences have been obtained. Comparisons of these sequences reveal a highly conserved polypeptide structure with some unexpected features.

Overall, the central third of the dynein heavy chain is the most highly conserved portion, followed by the C-terminal third. The N-terminal third of the sequence is the most highly divergent and shows almost no relatedness in comparisons of axonemal and cytoplasmic dynein sequences (Figure 1).

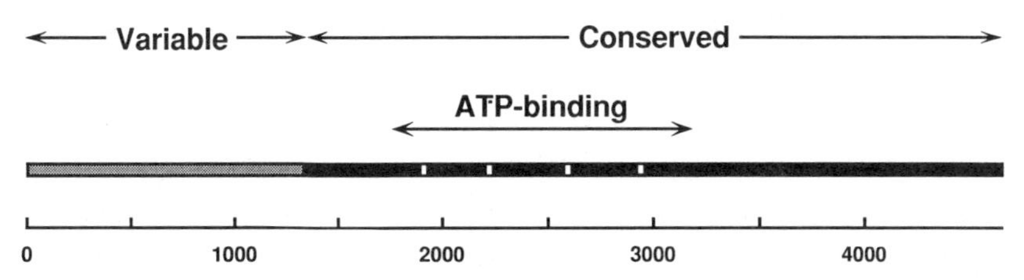

Figure 1 Schematic diagram of the dynein heavy chain, which indicates the location of the four conserved P-loop motifs (*open boxes*), as well as the domains that are most highly conserved and most divergent (variable) among the axonemal and cytoplasmic dyneins. Amino acid numbers are shown at the bottom (from Vallee 1993).

An intriguing feature of the heavy chain is the presence of multiple P-loop elements involved in γ-phosphate binding and hydrolysis in ATPases and GTPases (Walker et al 1982). In the sea urchin axonemal dynein β-heavy chain, a total of five P-loop elements were observed (Gibbons et al 1991; Ogawa 1991). One P-loop element, found near the N-terminus, has not been seen in other forms of dynein heavy chain. The other four elements are evenly spaced at ~ 35 to 40 kDa intervals within the central region of the polypeptide (Figure 1). The primary sequence and spacing between the four P-loop elements have been highly conserved among all forms of dynein examined to date.

These observations suggest that dynein may have as many as four 35–40 kDa ATPase or ATP-binding domains. However, from previous work there is little reason to believe that more than one domain is functional. A major site of ATP hydrolysis had been previously mapped by vanadate-mediated photocleavage of the heavy chain of two forms of axonemal dynein. In the presence of ATP and vanadate, UV-irradiation of dyneins leads to specific cleavage at a single site (Lee-Eiford et al 1986; Paschal et al 1987b; Lye et al 1987; Euteneuer et al 1988). Comparison of the vanadate cleavage site with the primary sequence information indicates that the principal site for ATP hydrolysis is nearest to the first P-loop in the β dynein heavy chain (Gibbons et al 1991). The identification of this P-loop motif as the site of vanadate-mediated photocleavage within the dynein heavy chain has been further substantiated by epitope mapping analysis using monoclonal antibodies to different domains within the heavy chain (Wilkerson et al 1994). Further support of a functional role for the first P-loop is its absolute conservation within all known dynein sequences (Figure 2).

While P-loops 2, 3, and 4 may be clearly identified in all full-length dynein sequences available, there is more variation between these sequences (Figure 2). However, P-loop 4 is as well conserved as P-loop 1 among the axonemal dyneins characterized to date (Figure 2). P-loop 3 is almost as highly conserved among the cytoplasmic dyneins—the only difference within this element be-

	P-Loop 1	P-Loop 2	P-Loop 3	P-Loop4
Rat	GPAGTGKT	GPSGSGKS	GPPGSGKT	GVSGAGKT
Drosophila	GPAGTGKT	GPSGSGKS	GPPGGGKT	GVSGAGKT
Dictyostelium	GPAGTGKT	GPSGGGKT	GPPGSGKT	GVSGGGKS
Aspergillus	GPAGTGKT	GKSGSGKS	GPPGSGKT	GVSGSGKT
Neurospora	GPAGTGKT	GNSGSGKS	GPPGSGKT	GVSGSGKT
Yeast	GPAGTGKT	GKAGCGKT	GPPGSGKT	GASRTGKT
CONSENSUS, CD	GPAGTGKT	G G GK	GPPG GKT	G S GKT
Sea Urchin, β	GPAGTGKT	GNADTGKS	GNAGLGKS	GVGGSGKQ
Chlamy, β	GPAGTGKT	GAAGCGKT	GNTGTGKS	GVGGSGKQ
Chlamy, α	GPAGTGKT	GPTGTGRT	GGAGVGKT	GVGGSGKQ
Chlamy, γ	GPAGTGKT	GPSGSGKS	GGPGTAKT	GVGGSGKQ
CONSENSUS, AX	GPAGTGKT	G GK	G G KT	GVGGSGKQ

Figure 2 P-Loop motif sequences from cytoplasmic and axonemal dynein heavy chains. Cytoplasmic and axonemal consensus sequences indicate which amino acid residues are absolutely conserved among the known full-length dynein heavy chain sequences.

tween the diverse species examined is a substitution of Gly for Ser within the *Drosophila* sequence. P-loop 2 is the most variable of these elements. In the *Chlamydomonas* flagellar outer arm dynein α-heavy chain, P-loop 2 may be argued to have lost any residual ability to bind or hydrolyze ATP because of the substitution of an Arg for Lys (Wilkerson et al 1994), which has been implicated in a direct interaction with the phosphates of the bound nucleotide in other ATP-binding proteins (Saraste et al 1990). Similarly it has been argued that P-loop 3 of the *Chlamydomonas* γ-heavy chain may also have lost its function because of the substitution of an Ala for the strongly conserved Gly of the P-loop motif (Wilkerson et al 1994).

A similar pattern of sequence conservation is also evident in the amino acid sequences that flank the P-loop motifs. These amino acids presumably are involved in the folding of the nucleotide-binding site. Again, the sequences that flank P-loop 1 are the most highly conserved among the characterized dynein heavy chains. In contrast, the sequences flanking P-loop 2 are the least well conserved.

The ability of P-loops 2, 3, and 4 to bind nucleotide in either cytoplasmic or axonemal dynein remains to be tested. Biochemical studies on the three-headed *Tetrahymena* ciliary outer arm dynein are consistent with a single ATP hydrolytic site for each globular dynein head and, therefore, each dynein heavy chain (Johnson 1983). Vanadate-mediated photocleavage also occurs at a single major site per dynein heavy chain (Gibbons et al 1987). Therefore, the meaning of the strong conservation of multiple potential nucleotide-binding sites within the dynein heavy chain sequence is unclear. Conceivably, the other P-loops may have some nucleotide-binding function, for example, in the hydrolysis of related nucleotides. While axonemal dyneins are highly specific for ATP (Gibbons 1966), cytoplasmic dynein was found to hydrolyze CTP, TTP, UTP, GTP, and ITP at rates greater than that for ATP (Shpetner et al 1988; Collins & Vallee 1989). However, only ATP was capable of supporting microtubule gliding in vitro (Paschal & Vallee 1987). Thus the role of the general nucleotidase activity in cytoplasmic dynein function remains unclear, but could be associated with P-loops 2–4.

It is noteworthy that other kinetic parameters of axonemal and cytoplasmic dynein differ significantly and may be associated with differences in the primary structure of the heavy chains. While the observed extents of microtubule activation of the basal rates of ATP hydrolysis for axonemal and cytoplasmic dyneins are similar (five to sixfold stimulation), the basal rate of ATP hydrolysis for the cytoplasmic enzyme (0.09 μmol/mg-min: Paschal et al 1987b; Shpetner et al 1988), is much lower than that for axonemal dynein, measured at 0.4 μmol/mg-min for the *Tetrahymena* ciliary outer arm dynein (Johnson 1983). Cytoplasmic dynein was also found to be much more sensitive to activation by microtubules than was the axonemal enzyme. Microtubules stimulated the ATPase activity of bovine brain cytoplasmic dynein with a K_m of 0.16 mg/ml tubulin, with saturation observed at ~ 1 mg/ml (10 μM) tubulin (Shpetner et al 1988). For the ciliary enzyme, saturation was not observed even at tubulin concentrations up to 500 μM. This may reflect differences in functional requirements for the two forms of dynein. The axonemal enzyme is structurally constrained within the axoneme where the effective local concentration of tubulin is high (Shimizu et al 1989). This contrasts with cytoplasmic dynein-driven movements along the less well-ordered cytoplasmic micro- tubules, where significant activation of the enzyme at lower effective tubulin concentrations may be required.

Other Features of the Dynein Heavy Chain Sequence

As noted above, the amino terminal third of the dynein heavy chain sequence appears to be almost completely divergent between cytoplasmic and axonemal dyneins (Mikami et al 1993). Clear sequence conservation within this region is observed among cytoplasmic dyneins (e.g. Mikami et al 1993) and among the available axonemal heavy chains (Mitchell & Brown 1994; Wilkerson et al 1994), although the β-heavy chain N-terminal sequences are more closely related to each other than to the one known γ-heavy chain sequence. These observations suggest that the amino acid sequence of the first third of the dynein heavy chain specifies an isoform-specific function, such as mediating the association of the heavy chain with the other subunits of the dynein complex.

The overall folding of the dynein heavy chain to form the characteristic globular head and stalk of the molecule is not well understood. The central domain in which the four P-loops are located is flanked by regions predicted to form short α-helical coiled-coils in both axonemal and cytoplasmic dynein sequences (Mikami et al 1993; Mitchell & Brown 1994). Whether these regions actually take on a coiled-coil configuration in the folded heavy chain remains to be determined. However, it is conceivable that such a structure would comprise a neck region separating the globular head domain from the N- and C-terminal segments of the heavy chain.

Some existing data suggest an involvement of the amino terminal domain in the binding interaction of the heavy chain with the other subunits of the dynein complex. The *Chlamydomonas* mutant *oda*4-s7 produces a truncated β dynein heavy chain encompassing only the N-terminal 160 kDa. However, the axoneme exhibits a fully assembled two-headed outer arm that includes all three heavy chains and a full complement of accessory subunits. It is not yet clear which of these polypeptides bind directly to the truncated β-heavy chain (Sakakibara et al 1993).

The amino terminal region has also been implicated in anchoring dynein within the axoneme. Mocz & Gibbons (1993) investigated the microtubule-binding characteristics of axonemal dynein fragments generated by limited proteolysis in situ. Peptides derived from the carboxyl terminal region of the sea urchin β chain were released by ATP, but peptides generated from the amino-terminal region could only be released by elevated ionic strength. These data suggest that the amino terminal region of the heavy chain may also participate in linking the base of the dynein molecule to the A-microtubule of the axonemal outer doublet.

These limited data support a model in which the amino terminal domain of the heavy chain mediates interactions with at least one other subunit and, perhaps, with the axonemal A-microtubule as well. The polypeptide backbone may then extend to form the stalk observed by electron microscopy. The central P-loop-containing domain would be expected to fold to form the globular head. Whether the C-terminal domain folds back to participate in the formation of the stalk, as suggested by earlier models based on proteolytic degradation and epitope mapping, or instead comprises some other structural feature of the molecule, remains to be determined.

Extent of the Dynein Heavy Chain Gene Family

Current data support a single cytoplasmic dynein heavy chain gene in rat, mouse, *Dictyostelium, Aspergillus,* and yeast (Koonce et al 1992; Mikami et al 1993; Xiang et al 1994; Li et al 1993; Eschel et al 1993). In contrast, each organism may have multiple axonemal dynein heavy chain genes.

A number of laboratories have devised PCR screens for expressed dynein mRNAs, which use primers that hybridize to the most highly conserved sequences of the heavy chains. The resulting PCR fragments have been cloned and sequenced, yielding a large data set of partial dynein sequences spanning the first, and most highly conserved, P-loop motif. Gibbons et al (1994) identified 14 distinct dynein cDNA sequences in sea urchin, one of which corresponded to the previously characterized flagellar β heavy chain, and another of which is thought to be a cytoplasmic dynein sequence based on its homology to the *Dictyostelium* and rat sequences. Most of the remaining partial sequences are believed to encode distinct isoforms of axonemal dynein heavy chains based on their up-regulation following deciliation of the embryo. One of these sequences could conceivably encode a cytoplasmic dynein based on its similarity to other cytoplasmic sequences, but it is unclear why its expression should be affected by deciliation. A similar experiment in *Paramecium tetraurelia* led to the characterization of eight partial dynein cDNAs, one of which was proposed to be the cytoplasmic form (Asai et al 1994). Wilkerson et al (1994) identified seven PCR products from *Chlamydomonas*, three of which correspond to the α, β, and γ sequences, and four of which are apparently novel. A similar PCR screen of *Drosophila* yielded seven partial sequences, one of which encoded the cytoplasmic dynein heavy chain (Rasmusson et al 1994).

These partial sequences have been used to estimate the probable extent of the dynein heavy chain gene family. In sea urchin, Southern blotting analysis has suggested at least 15 distinct genes (Gibbons et al 1994), and in *Paramecium,* evidence was found for the expression of at least 12 dynein heavy chain genes (Asai et al 1994). These results are consistent with the biochemical analysis of the diversity of axonemal dynein isoforms, many of which are likely to correspond to the less well-characterized inner arm dyneins (Kagami & Kamiya 1992).

These sequences have also been used to identify conserved motifs outside the P-loops, diagnostic for either axonemal or cytoplasmic dynein isoforms (Asai et al 1994). Two highly conserved elements, CFDEFNR and FITMNP, were found in the vicinity of P-loop 1 (Gibbons et al 1994; Wilkerson et al 1994). The potential contribution of these motifs to either dynein structure or function is not clear, although the second motif is thought to be involved in microtubule-binding based on its similarity to a short sequence in kinesin (Witman 1992; Wilkerson et al 1994).

The multiple dynein heavy chain genes are to date best characterized as a family, rather than a superfamily. Kinesin-related proteins form a superfamily because although they each encode a conserved domain homologous to the

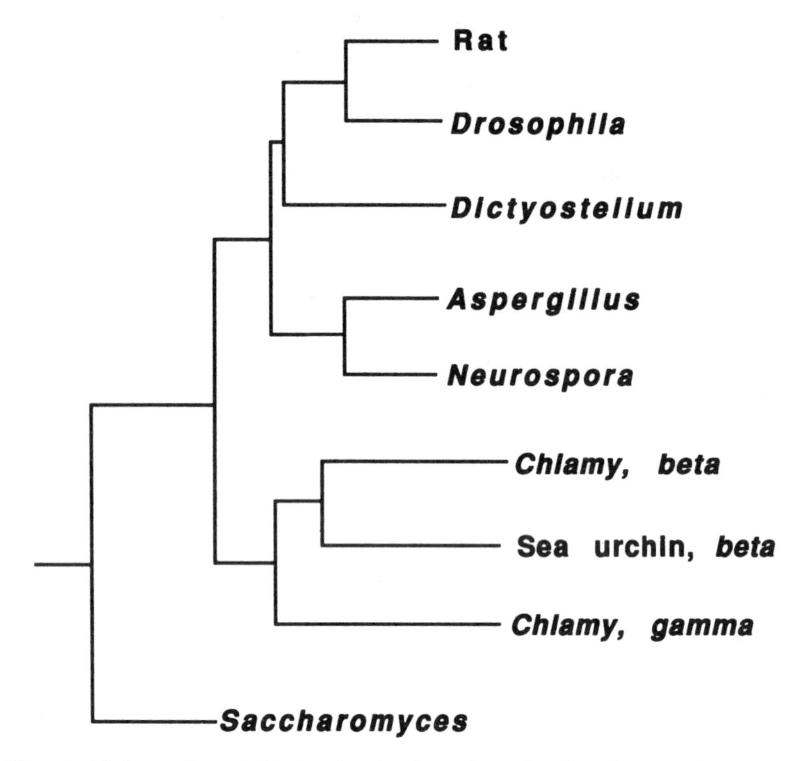

Figure 3 Phylogenetic tree indicating the relatedness of cytoplasmic and axonemal dynein heavy chain sequences. This cladogram was generated with the Clustal V algorithm of Higgins & Sharp (1989), using the MegAlign program in the LASERGENE analysis software package from DNAStar on a Macintosh computer.

motor domain of kinesin, the position of this domain within the overall primary sequence, as well as the sequences outside this motor domain, differ among the superfamily members (Goldstein 1991). In contrast, for those dynein genes for which full-length sequences are available, the domain organization is conserved.

The availability of several full and partial dynein sequences provides some insight into the evolution of the dynein gene family. The available full-length dynein heavy chain sequences were used to construct a phylogenetic tree using the Clustal method (Higgins & Sharp 1989), as shown in Figure 3. As expected, the cytoplasmic and axonemal dyneins form separate branches of this tree, with divergence between the different forms of dynein apparently a relatively ancient evolutionary event. Unexpectedly, the cytoplasmic dynein heavy chain from yeast appears to be most divergent from other known dynein sequences.

It will be of interest to determine whether this observation reflects the evolutionary distance between yeast and the other organisms or, rather, a functional difference between yeast cytoplasmic dynein and other forms of dynein.

Analysis of multiple partial dynein sequences obtained from a single organism also indicates that the duplication and divergence of the dynein genes were relatively ancient events. Using PCR, Gibbons et al (1994) identified at least six distinct classes of dynein genes in sea urchin. Comparisons of partial sequences from *Drosophila* dynein genes indicated that each of the seven *Drosophila* dyneins so far identified is apparently more closely related to a homologous sea urchin sequence than to the other *Drosophila* genes (Rasmusson et al 1994; Gibbons et al 1994). Taken together the existing data suggest that the divergence of an axonemal dynein progenitor was an ancient event and that subsequently there has been strong evolutionary pressure to maintain the specialized sequences required for this complex type of cellular motility.

INTERMEDIATE CHAINS

While all dyneins characterized so far have heavy chains of comparable size, they have been found to contain a somewhat baffling array of accessory proteins. *Chlamydomonas* flagellar outer arm dynein has as many as ten accessory polypeptides: two intermediate chains of ~ 78 and 70 kDa (IC78 and IC70), and eight light chains of M_r 19.6–7.8 K (Pfister et al 1982). The equivalent form of axonemal dynein from sea urchin contains polypeptides of 122, 90, 76 kDa and a series of light chains of 14–24 kDa (Bell et al 1979). Trout sperm outer arm dynein contains polypeptides of 85, 73, 65, 63, 57, 22 19, 11.5, 9, 7.5, and 6 kDa (Gatti et al 1989). Pig tracheal ciliary outer arm dynein also contained polypeptides in the intermediate chain range; the light chain content was not assessed (Hastie et al 1986). Each of three inner arm dyneins from *Chlamydomonas* was reported to have a distinctive complement of accessory polypeptides (Goodenough & Heuser 1985; Piperno et al 1990; Kagami & Kamiya 1992). Cytoplasmic dyneins are again different in composition with a group of polypeptides of 74 kDa and four electrophoretic species in the 53–57 kDa range (Paschal et al 1987b; Collins & Vallee 1989). Light chains associated with cytoplasmic dynein have not been detected.

The compositional variation among axonemal outer arm dyneins from different species suggests that some of the electrophoretic complexity among the dyneins may be more apparent than real. Perhaps the dynein accessory subunits have been functionally conserved during evolution, but have diverged in length. An indication that this hypothesis may be valid has comc from the use of monoclonal antibodies, which have revealed cross-reactivity between the *Chlamydomonas* outer arm dynein intermediate chains and polypeptides of somewhat different size in sea urchin and trout-sperm outer arm dyneins (King et al 1985, 1990). Similarly, a monoclonal antibody raised against a cytoplasmic dynein intermediate chain was found to react with a somewhat smaller polypeptide in cilia and flagella, possibly representing a mammalian axonemal dynein intermediate chain (Paschal et al 1992).

Molecular Characterization of the Dynein Intermediate Chains

Molecular characterization of the dynein accessory polypeptides has begun to clarify the relationship between accessory subunits found in different forms of dynein. Genomic clones encoding the *Chlamydomonas* IC70 were obtained by screening an expression library using a monoclonal antibody (Williams et al 1986). Subsequent analysis mapped IC70 to the *ODA*6 gene, known to be involved in flagellar motility, and provided full-length deduced amino acid sequence (Mitchell & Kang 1991).

cDNAs encoding the 74 kDa polypeptide of rat cytoplasmic dynein were selected using oligonucleotide probes designed on the basis of extensive amino acid sequence (Paschal et al 1992). Weak but clearly significant homology with *Chlamydomonas* flagellar IC70 was detected (23.9% identity and 47.5% similarity). Relatedness was restricted to the C-terminal half of the two sequences, while the N-terminal regions were completely divergent (Figure 4).

The primary structures of additional intermediate chains, which include IC78 from the *Chlamydomonas* flagellar outer arm (King et al 1992), a cytoplasmic dynein intermediate chain from *Dictyostelium discoideum* (Trivi-

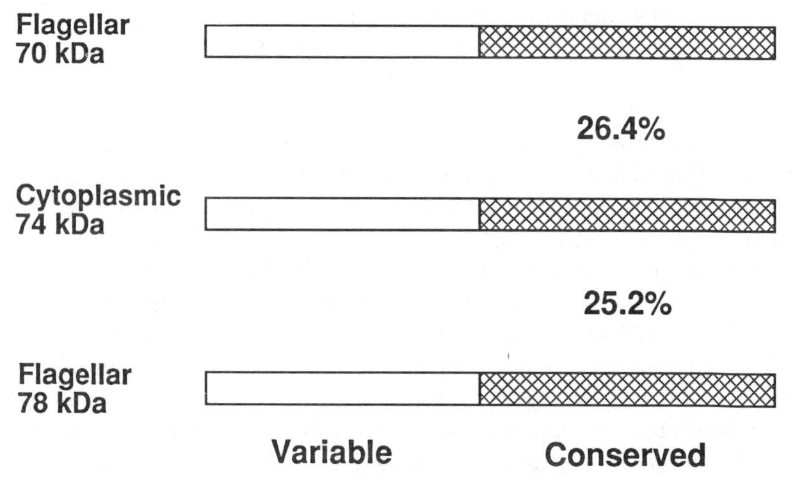

Figure 4 Schematic diagram of the relationship among the dynein intermediate chains from *Chlamydomonas* (70 and 78 kDa) and the 74 kDa intermediate chain from rat cytoplasmic dynein (this figure was contributed by B Paschal, Scripps Research Institute, La Jolla, CA; see Paschal et al 1992). The numbers refer to % sequence identity with rat cytoplasmic sequences over the C-terminal region. N-terminal values were comparable, but significance scores were much lower.

nos-Lagos et al 1993), and what appears to be a second intermediate chain gene from rat (Vaughan & Vallee 1993), have since been determined. All the new sequences show the same general pattern observed from the comparison of the IC70 and the rat cytoplasmic dynein intermediate chain sequence, with the greatest extent of homology being within the C-terminal domains. Overall sequence identity between rat and *Dictyostelium* intermediate chains was 39%, considerably higher than between any of the cytoplasmic and axonemal sequences. The two rat sequences were even more closely related, showing 76% identity overall (Vaughn et al 1993). Again, homology was highest within the C-terminal domain (92 vs 68% within the N-terminal portions). Because the rat intermediate chain sequences differ throughout their lengths, they are thought to derive from different genes, but this contention has to be proven. Multiple products of each of the rat genes are also suggested by comparison of peptide sequences, the sequences of PCR products spanning portions of the intermediate chains, and the sequences of numerous partial cDNA clones (Paschal et al 1992; Vaughan & Vallee 1993). Alternative sequences occur at least one site within the initially described cytoplasmic dynein intermediate chain (Paschal et al 1992), and at two sites within the newly identified intermediate chain (Vaughan & Vallee 1993). In all cases, the alternative sequences are located within the variable N-terminal portion of the polypeptide.

Biochemical evidence suggests an important role for the intermediate chains in dynein function. IC78 was found to lie at the base of the flagellar dynein molecule by immunoelectron microscopy of the purified protein using a monoclonal anti-IC78 antibody (King & Witman 1990). Cross-linking studies showed a direct interaction between IC70 and tubulin within the axoneme (King et al 1991), and microtubule binding by in vitro translated IC70 has also been observed (King et al 1992). IC78 has not itself been shown to be at the base of the dynein molecule. However, it appears to be directly associated with IC70 in experiments involving the immunoprecipitation of dynein subunits under conditions that partially dissociate the complex (Mitchell & Rosenbaum 1986). In the presence of either 0.5% Nonidet P40 or low levels of sodium dodecyl sulfate, immunoprecipitation using anti-IC70 monoclonal antibody resulted in the selective co-precipitation of IC78. Together these data suggest that the intermediate chains lie at the base of the dynein molecule in contact with each other and with tubulin. Presumably, they are at least in part responsible for anchoring outer arms to the A-microtubules of the axoneme. The need for two different intermediate chains to accomplish this task is unknown.

By analogy with the *Chlamydomonas* flagellar outer arm dynein, the cytoplasmic dynein intermediate chains are proposed to play a role in targeting the enzyme complex within the cell (Paschal et al 1992). In view of the potential roles of cytoplasmic dynein, the intermediate chains possibly link the enzyme to the surface of membranous organelles and/or to kinetochores. No indication of membrane-spanning or lipid-binding sites was seen in the primary sequence. Thus it seems likely that the intermediate chains bind to other organelle surface proteins that have not been identified.

In view of the pattern of sequence conservation between the intermediate chains, it seems reasonable to expect that the C-terminal region is responsible for a common intermediate chain function, such as heavy chain binding (Paschal et al 1992). The N-terminal region, by contrast, would confer functional specificity. Thus this region might be involved in segregating inner and outer arm dyneins to their respective binding sites on the A-microtubule of the axonemal outer doublet, and to directing cytoplasmic dynein to membrane surfaces or to the kinetochore.

In vitro translation of full-length and truncated forms of IC78 has been performed to define the tubulin-binding domain, using purified brain tubulin or axonemes from mutant strains of *Chlamydomonas* lacking outer dynein arms (King et al 1992). Intact IC78 failed to bind to microtubules. However, using the truncated constructs, two tubulin-binding regions were identified, one near the center of the IC78 polypeptide within the variable region, and the other within the more highly conserved C-terminal region. Use of the zero-length cross-linker EDC supported a functional role for the variable domain site (King et al 1991, 1992).

CHARACTERIZATION OF LOWER MOLECULAR WEIGHT ACCESSORY POLYPEPTIDES

Relatively little is known about the structure and function of the myriad lower molecular weight dynein polypeptides. Clues to the possible function of some of the axonemal dynein light chains have come from biochemical analysis of flagellar outer arm dynein. Immunoprecipitation of IC70 from partially dissociated *Chlamydomonas* flagellar outer arm dynein preparations (see above) revealed co-precipitation of not only IC78, but of light chains of 21 and 12.5 kDa as well (Mitchell & Rosenbaum 1986). In addition, cross-linking of dynein molecules produced intermediate chain-containing products of a size consistent with an intermediate chain/light chain complex (King et al 1991). Together these studies suggest that at least some of the light chains are directly associated with the intermediate chains and, presumably, lie at the base of the dynein molecule. Whether this model holds for all of the light chains is not known, nor is it clear whether the light chains represent a family of related polypeptides. However, the existing data raise the possibility that some of these polypeptides might modulate the interaction of axonemal dynein with the A-microtubule.

Thus far corresponding low molecular weight polypeptides have not been reported in cytoplasmic dynein preparations. Instead, a series of four electrophoretic species of 59, 57, 55, and 53 kDa were found to co-purify with cytoplasmic dynein to reproducible and near integral stoichiometry under a variety of purification conditions (Paschal et al 1987b). This behavior suggests that the 53-59 kDa species are true subunits of the dynein molecule, but they are apparently distinct from the axonemal dynein light chains.

Two-dimensional gel electrophoresis revealed the 53–59 kDa region of cytoplasmic dynein to be complex, exhibiting numerous isoelectric as well as size variants (Gill et al 1992; Hughes et al 1993). Alkaline phosphatase treatment reduced the four one-dimensional electrophoretic species to two of 53 and 57 kDa, and substantially simplified the two-dimensional electrophoretic pattern (Hughes et al 1993). Proteolytic digestion of the 53/55 and 57/59 kDa electrophoretic species yielded distinct peptide patterns. However, homology between the sequences of peptides derived from p53/55 and p57/59 was detected (Hughes et al 1993).

cDNA clones corresponding to members of this group of polypeptides have been selected from chicken (Gill et al 1992) and rat libraries (Hughes et al 1993). RNA blot analysis revealed two transcripts of 4.4 and 2.5 kb in rat. The larger transcript may encode p53, while the smaller appears to encode p57 (Hughes et al 1993). The deduced amino acid sequence revealed no apparent homology with the intermediate chains or other proteins in the sequence databases. For this reason, and to distinguish these polypeptides from the light chains, we have proposed the term light intermediate chains (LICs).

Together the existing data are consistent with separate but related genes encoding p53 and p57, the products of which are extensively modified by phosphorylation. Further work will clearly be needed to learn more about the function of this polypeptide group. However, in view of their apparent extent of phosphorylation, they are excellent candidates for a role in regulation of cytoplasmic dynein activity in the cell.

FUNCTIONAL ANALYSIS OF DYNEINS

Genetic Analysis of Axonemal Dyneins

Classical genetic analysis in *Chlamydomonas* has identified a number of loci involved in flagellar function. Mutations at several of these loci affect flagellar motility. Nonmotile mutants have been identified that either lack outer arms, such as pf13 and pf22, or have defective inner arms, such as pf23 (Huang et al 1979). Dynein mutants with decreased or aberrant motility include the *oda* mutants 1–10 (named *oda* for outer dynein assembly) and the inner arm dynein mutants *ida*1–3, *idb*1, and *pf*30 (reviewed by Kamiya 1991). In general, motile mutants with defects in outer arm dyneins exhibit decreased flagellar beat frequency, while defects in inner arm dyneins may alter the characteristics of the flagellar waveform (Brokaw & Kamiya 1987). The dynein polypeptide cDNA sequences from *Chlamydomonas* described above have now been mapped to genetic loci corresponding to some of these previously characterized mutations. The α, β, and γ heavy chains of *Chlamydomonas* dyneins map to the *oda*11, *oda*4, and *oda*2 loci, respectively (Sakakibara et al 1991; Kamiya 1988; Wilkerson et al 1994). The 70 and 78 kDa *Chlamydomonas* intermediate chains have been mapped to the *oda*6 and *oda*9 loci, respectively (Mitchell & Kang 1991; Witman et al 1994).

Some insight has been provided by these mutations into the assembly of the dynein complex within the axoneme. The absence of outer dynein arms in *oda*6 strains supports a role for IC70 in binding to the axonemal microtubules. An noted earlier, analysis of the *oda*4-s7 allele of the *oda*4 locus has revealed that a truncated form of the β-heavy chain, corresponding to the amino terminal 160 kDa of this polypeptide, is sufficient for the assembly of an α-(truncated β)-γ dynein complex (Sakakibara et al 1993), which suggests that the determinants for assembly and association of the heavy chains within the dynein complex are encoded in the amino-terminal sequences of the heavy chain polypeptide. In contrast, however, loss of the α-heavy chain from the axoneme, as seen in the *oda*11 mutant, does not prevent the assembly of an apparently functional, albeit less active, outer arm dynein.

The recent characterization of two distinct alleles of *sup-pf*-1, identified as extragenic suppressors of a radial spoke mutation in *Chlamydomonas*, and now known to map to the *oda*4 locus encoding the β-heavy chain, indicates that two small, in-frame deletions in the highly conserved small α-helical coiled-coil domain of the heavy chain resulted in a relatively significant effect on flagellar motility (Porter et al 1993). Further analyses of this type to functionally map the domains of the heavy chain of axonemal dynein may lead to the identification of regions likely to be involved in protein-protein associations required for cellular motility.

In *Drosophila,* one of the partial sequences for an axonemal dynein heavy chain identified by PCR has been mapped to the *kl*-5 locus within the fertility region of the Y chromosome (Gepner & Hays 1993). Mutations at this locus result in immotile sperm with a loss of dynein outer arms.

Genetic Analysis of Cytoplasmic Dyneins

To examine the cellular role of cytoplasmic dynein in the yeast *Saccharomyces cerevisiae,* Li et al (1993) and Eschel et al (1993) constructed deletion disruption mutants in the heavy chain gene. Cytoplasmic dynein was found not to be essential for vegetative growth, meiosis, or karyogamy, although slowed growth was observed for some mutant constructs (Li et al 1993), but not for others (Eschel et al 1993). What was clearly defective in the mutants was the process of nuclear segregation. In synchronized cells, 38% of those cells that completed nuclear division showed improper segregation of the nuclei between the mother and the bud (Eschel et al 1993). Many cells exhibited abnormal spindle orientation, which was apparently responsible for the lack of complete fidelity in chromosome segregation. These results suggest a role for cytoplasmic dynein not in chromosome-to-pole movement or spindle elongation, but in proper positioning of the nucleus. This process presumably involves force production in association with the extranuclear cytoplasmic microtubules rather than in association with the intranuclear spindle microtubules.

Nonetheless, analysis of a strain made null for the kinesin-related protein Kip1p and cytoplasmic dynein and with a temperature-sensitive mutation in the kinesin-related protein Cin8p suggests an additional role for dynein in mitosis (Saunders et al 1993). A shift to elevated temperature during anaphase in double mutants for Kip1p and Cin8p had no effect on continued spindle elongation. However, in the triple mutant, further spindle elongation was blocked. This behavior may reflect the role of dynein in pulling along the extranuclear cytoplasmic microtubules. However, it could signal a novel role for the enzyme in producing force within the spindle, which is degenerate with other microtubule-based motors.

Cytoplasmic dynein has been implicated in a potentially related process in the filamentous fungi *Aspergillus nidulans* and *Neurospora crassa.* In *Aspergillus* the *nud*A gene has recently been shown to encode the cytoplasmic dynein heavy chain (Xiang 1994). *nud*A is one of a series of mutations in *Aspergillus* that is defective in nuclear migration (*nud* = nuclear distribution mutants; Morris 1976). This process is known to involve translocation of the nuclei along cytoplasmic microtubules (Morris 1976). While the orientation of microtubules within the hypha has not been determined, the identification of *nud*A as a cytoplasmic dynein polypeptide implies that movement of the nucleus along the microtubules is minus-end directed. Nuclear division is unaffected in *nud*A.

In *Neurospora,* the cytoplasmic dynein heavy chain has been found to be encoded by the *Ro*1 gene (*Ro* = *ropy*; Plamann et al 1994). A series of at least eight *ropy* genes were identified in a search for suppressors of mutations in the *cot*-1 gene, which encodes a protein kinase proposed to be involved in the transport of precursors to the hyphal tip (Steele & Trinci 1977; Yarden et al 1992). Mutations in the *ropy*A genes result in a defect in nuclear migration similar to that seen in the *Aspergillus nud* mutations.

Thus mutations in the cytoplasmic dynein heavy chain produce a similar phenotype in two different filamentous fungi. It seems likely that this phenotype will prove to be related to that observed in *S. cerevisiae.* In both yeast and the filamentous fungi, cytoplasmic dynein may be involved in producing force along microtubules emanating from the cytoplasmic face of the pole body. A simple model would have the motor protein associated with the cortex of the cell, pulling on the microtubule plus ends. It will, therefore, be of interest to determine the distribution of cytoplasmic dynein in these organisms.

In higher eukaryotes, dynein has been observed by immunofluorescence microscopy to localize within the spindle and, more specifically, to the kinetochores of prometaphase chromosomes (Pfarr et al 1990; Steuer et al 1990; Wordeman et al 1991), although this distribution has not been seen in all studies (Lin & Collins 1992). An association of dynein with kinetochores is supported by observations of prometaphase mitotic movements in newt lung cells (Rieder et al 1990). Kinetochores were observed to attach tangentially to astral microtubules and migrate poleward (toward the minus microtubule end) at rates corresponding to those observed in in vitro assays for cytoplasmic dynein force production. Hyman & Mitchison (1991) also observed microtubule gliding across kinetochores of isolated prometaphase chromosomes. Together these results point strongly to a role for cytoplasmic dynein during prometaphase. Whether cytoplasmic dynein might also participate in chromosome-to-pole movement during anaphase A is unclear.

To address these issues more directly, Vaisberg et al (1993) examined the effects of microinjecting antibodies directed against the ATPase domain of cytoplasmic dynein into PtK1 cells. The antibodies blocked mitosis in prophase. However, rather than interfering with poleward chromosome movement, they prevented the formation of the spindle or, when injected during mitosis, caused the spindle to collapse. These results appear to indicate a role for cytoplasmic dynein in spindle pole separation.

It is difficult to envision how the protein might accomplish this task if it were located within the spindle. However, dynein could serve to separate spindle poles if it were located toward the outer region of the spindle, perhaps associated with the astral microtubules. This model necessitates that cytoplasmic dynein be anchored to some cellular structure outside the spindle. The protein would then be in position to generate outward-directed force on the centrosomes, which would lead to the separation of the spindle poles and the development of the bipolar spindle. This model suggests that force-generation along the astral microtubules is key to the separation of the poles, rather than force-production along the inter-digitating spindle microtubules. Such a model is consistent with the observations of Waters et al (1993), who noted that the force production mechanism responsible for centrosome separation during vertebrate spindle formation is intrinsic to each aster, in that each centrosome apparently moves autonomously. Furthermore, it suggests some relationship between the role for dynein in fungi and vertebrates in that in both cases dynein could be operating on the spindle from the outside.

Contradicting this model are the results of immunolocalization studies that have so far shown cytoplasmic dynein to be distributed within the spindle. It is also worth noting that the effects of the microinjected anti-dynein antibodies on mitosis are inconsistent with a role for the enzyme in prometaphase, as suggested by these and other earlier studies (see above), and conceivably in anaphase. It is possible that any effect of the antibodies on these other processes was simply obscured by the collapse of the spindle. Alternatively, cytoplasmic dynein may have a degenerate role during prometaphase and anaphase that is compensated for by other motor proteins.

Cytoplasmic Dynein in Intracellular Organelle Transport

Cytoplasmic dynein was initially proposed to be responsible for retrograde organelle transport (Paschal & Vallee 1987; Vallee et al 1989). Such a role has been supported by pharmacological analysis of organelle movements in vitro (Schnapp & Reese 1989; Schroer et al 1989). Immunocytochemistry in cultured cells with antibodies to cytoplasmic dynein has also clearly shown co-localization with lysosomes and endosomes (Lin & Collins 1992). In vitro reconstitution assays suggest that cytoplasmic dynein is required for the perinuclear localization of the Golgi apparatus as well (Corthesy-Theulaz et al 1992) and for vesicular transport from early to late endosomes (Aniento et al 1993). The extent of cellular functions involving cytoplasmic dynein has yet to be fully defined, but this effort should be greatly facilitated by continued genetic and molecular genetic analysis in a variety of organisms.

CELLULAR REGULATORY MECHANISMS

Regulation of Axonemal Dyneins

Coordination of the activity of the diverse forms of dynein within the axoneme to produce a coherent waveform, and the overall regulation of flagellar beating, are complex problems that will probably require a combination of genetic and other approaches for solution. Existing evidence points to a regulatory role for radial spokes and to a recently described "dynein regulatory complex" in the coordination of bend propagation.

Mutations resulting in defects in radial spoke assembly have a paralyzed flagellar phenotype. In order to test the hypothesis that paralysis results from the disruption of a direct interaction between the radial spokes and dynein, Smith & Sale (1992) investigated the motor properties of inner arm dyneins isolated from a mutant strain lacking radial spokes. It was determined that the velocity of microtubule translocation catalyzed by dynein isolated from the spoke-less mutant was significantly slower than that of dynein isolated from the wild-type. Reconstitution experiments suggested that radial spokes activate dynein by a mechanism involving a relatively stable modification of the dynein, rather than by a continued direct interaction between the radial spokes and the dynein. The nature of dynein modification has yet to be identified.

Analysis of extragenic suppressors of radial spoke mutations has also led to insight into the regulation of dynein within the axoneme. Demonstration that the primary defect in one of these suppressor mutations, *sup-pf-*1, is in the β-heavy chain suggests a direct interaction between the radial spokes and outer arm dyneins. Other radial spoke suppressor mutants have been used to define the dynein regulatory complex as a set of six polypeptides of 108, 83, 65, 60, 40, and 29 kDa (Piperno et al 1992; Huang et al 1982). These polypeptides may form a complex that mediates regulatory signals between the radial spokes and inner dynein arms and is localized in close proximity to these structures in the axoneme (O'Toole et al 1993). Further studies will be required to define the complex biochemically and to examine its possible regulatory role.

Regulation in the axoneme also occurs at the level of control of waveform and beat frequency. Numerous studies have shown that calcium may be involved in the regulation of beat parameters. For example, the Ca^{2+} concentration appears to mediate the avoidance response in *Paramecium* (Eckert et al 1976) and the changes in flagellar waveform during the phototactic response of *Chlamydomonas* (Schmidt & Eckert 1976; Bessen et al 1980). cAMP also has a demonstrated effect on flagellar motility, but this effect may be species specific—in some systems cAMP may activate and in others inhibit flagellar beating. The polypeptides involved in the Ca^{2+}-mediated and cAMP-dependent signal transduction pathways have not been clearly identified, although there are some candidates. Calmodulin has been observed to co-purify with some preparations of axonemal dynein (Jamieson et al 1979), and a direct effect of calcium and calmodulin on activating the ATPase activity of axonemal dynein was demonstrated by Blum et al (1980). Piperno et al (1992) demonstrated an association of the Ca^{2+}-binding protein caltractin/centrin with inner arm dyneins from *Chlamydomonas*, and Chilcote & Johnson (1990) identified a co-purifying axonemal cAMP-dependent protein kinase from *Tetrahymena* that can phosphorylate outer arm dynein in vitro.

Regulation of Axonemal Dynein by Phosphorylation

Several studies have investigated the phosphorylation of dynein within the axoneme. Chilcote & Johnson (1990) determined that isolated *Tetrahymena* 22S outer arm dynein typically contains six phosphates, localized to the 78, 76, 47, and 23 kDa intermediate and light chains. Removal of two of the phosphates from the purified dynein resulted in an ~ 30% decrease in the ATPase and microtubule-activated ATPase activities of the preparation. In contrast, metabolic labeling studies indicate that in *Chlamydomonas* only the α dynein heavy chain is phosphorylated in vivo (Piperno & Luck 1981). Phosphorylation sites have been mapped to at least six distinct sites along the α-heavy chain (King & Witman; 1994); however, the effect of phosphorylation on activity was not examined. The role of dynein phosphorylation in the regulation of dynein function within the axoneme is not clear. This mechanism is likely to be too slow to be involved in the coordination of dynein activation during flagellar beating, but may have a role in a more global regulation of dynein function.

It remains to be demonstrated what effect the direct phosphorylation of axonemal dyneins may have on the force generation properties of the enzymes. In *Paramecium,* thiophosphorylation of a 29 kDa polypeptide, which co-purifies with outer arm dynein, results in an increased rate of dynein-mediated microtubule gliding in vitro (Hamasaki et al 1991). The phosphorylation of this polypeptide, proposed to be a dynein regulatory light chain, is both cAMP and Ca^{2+} dependent (Barkalow et al 1993). This result suggests that axonemal motility may be regulated by the specific phosphorylation of subunits of the dynein complex. Further analysis of this mechanism is clearly required, but may be hindered by the species-specific differences that have been observed to date.

Regulation of Cytoplasmic Dynein

For cytoplasmic dynein, regulation is likely to be critical at two levels. Force production by cytoplasmic dynein was found to be exclusively minus end directed in vitro (Paschal & Vallee 1987). There is little reason to believe that the directionality of force production by dynein is regulated, although a dynein-like ATPase purified from the giant amoeba *Reticulomyxa* was reported to be responsible for bidirectional motility (Euteneuer et al 1989; Schliwa et al 1991).

There is good reason to expect that dynein motor activity may be turned on and off in vivo. It is difficult to understand how the protein reaches the axon terminus to serve in retrograde transport unless it is delivered there in an inactive state. It is also uncertain whether organelles thought to be minus-end directed, such as late endosomes, lysosomes, and the Golgi apparatus, remain under the continued control of cytoplasmic dynein once they have reached their destination within the cell. Regulation of the association of cytoplasmic dynein with organelles may also be significant in the normal physiology of the cell and could be involved in specifying the distinctive distributions of membranous organelles observed among different cell types.

Phosphorylation of Cytoplasmic Dynein

Regulation of cytoplasmic dynein motor activity is of particular interest in the neuron because the extended length of the axon may require directed transport of inactive dynein to the synaptic region, where it may be specifically activated. Hirokawa et al (1990) found by using immunocytochemistry of rat peripheral neurons that cytoplasmic dynein accumulated at both the proximal and distal sides of ligations, which indicated bidirectional transport. Presumably the two dynein fractions differ in their motor activity, but it is difficult to test this possibility directly.

Dillman & Pfister (1993) compared the phosphorylation of total rat brain cytoplasmic dynein to the anterogradely transported fraction of cytoplasmic dynein in rat optic nerve. Phosphate was observed to be incorporated into the heavy chain as well the intermediate chains and lower molecular weight subunits in both fractions. However, a lower extent of heavy chain phosphorylation was observed in the anterograde pool. These data were consistent with a role for phosphorylation in converting cytoplasmic dynein to a more active form capable of driving retrograde transport.

Recent observations of Lin & Collins (1993a,b) indicate that reversible phosphorylation of cytoplasmic dynein may also be involved in the regulation of its activity in nonneuronal cells. Treatment of cultured fibroblasts with the phosphatase inhibitor okadaic acid resulted in an apparent redistribution of cytoplasmic dynein, observable as a decreased labeling of lysosomes with dynein-specific antibody. Immunoprecipitation of cytoplasmic dynein from the labeled NRK cells following treatment with okadaic acid revealed increased levels of dynein heavy chain phosphorylation in comparison to untreated control cells. Exposure to serum or okadaic acid has also been reported to increase the total number of outward and inward organelle movements observed in cultured CV-1 endothelial cells (Hamm-Alvarez et al 1992), consistent with an effect of phosphorylation to increase motor activity. Nonetheless, considerably more work will be needed to assess the relative effects of phosphorylation on ATPase activity, motor activity, and/or subcellular targeting. Also remaining to be investigated is the possible regulatory role of phosphorylation of the cytoplasmic dynein accessory subunits, which appear to be phosphorylated as isolated (see above; Gill et al 1992; Hughes et al 1993; Dillman & Pfister 1993).

Interacting Factors within the Cell: the Glued/Dynactin Complex

Cytoplasmic dyneins purified from rat liver and rat testis were found to copurify with additional polypeptides of 150 and 45 kDa (Collins & Vallee 1989). Similar polypeptides were subsequently observed to co-purify with cytoplasmic dyneins from other sources through microtubule-affinity and sucrose gradient density centrifugation (Steuer et al 1990; Holzbaur et al 1991). To investigate a potential role for these polypeptides in the regulation of cytoplasmic dynein activity in vivo, Holzbaur et al (1991) cloned cDNAs encoding the 150 kDa polypeptide. Analysis of the predicted protein sequence of this polypeptide revealed significant homology with the product of the *Drosophila* gene *Glued* (Holzbaur et al 1991). Gl^1, the first *Glued* allele to be identified

(Plough & Ives 1935), has a dominant phenotype that results in the aberrant development of the compound eye and optic lobe in the heterozygous fly. The null mutation is lethal early in development. Studies on genetic mosaics in *Drosophila* suggest that the expression of the *Glued* gene is essential in all cells (Harte & Kankel 1982).

Studies on a similar polypeptide identified in chick, and described as dynactin (Gill et al 1991), suggest that this polypeptide may be a component of a factor required for dynein-mediated vesicular transport along microtubules. Gill et al (1991) examined the ability of cytoplasmic dynein to drive vesicle motility along microtubules in an in vitro assay. It was determined that cytoplasmic dynein preparations that included the 150 kDa polypeptide could support movement of alkali-stripped vesicles along microtubules. However, further purification of the dynein by FPLC chromatography, which resulted in the loss of the 150 kDa polypeptide along with other polypeptides from the dynein preparation, also resulted in a substantial reduction in the frequency of dynein-mediated organelle movements in the in vitro motility assay. Vesicle motor activity could be partially restored by re-addition of the FPLC fraction that was enriched in the 150 kDa polypeptide.

Further studies have determined that the 150 kDa polypeptide is a component of a heteromeric microtubule-associated complex that sediments at 20S along with cytoplasmic dynein (Gill et al 1991; Paschal et al 1993). Immunoprecipitation and immunodepletion experiments indicate that the Glued or dynactin complex includes polypeptides of 150, 135, 62, 50, 45, 37, and 32 kDa. The stoichiometry of the major polypeptides of the complex was determined to be 1:2:7 for the 150, 50, and 45 kDa polypeptides, respectively (Paschal et al 1993). These three polypeptides are the best characterized to date; these studies have yielded the unexpected result that the 45 kDa polypeptide is a novel form of actin (described below).

The rat polypeptide $p150^{Glued}$ is 31% identical to the *Drosophila* gene product overall (Holzbaur et al 1991; Swaroop et al 1987). $p150^{Glued}$ is more highly conserved in higher eukaryotes. Both chick (Gill et al 1991) and human cDNAs (Tokito et al 1993) encoding the 150 kDa polypeptide have been isolated and characterized. Overall the chick sequence is 78% identical to rat (although the chick cDNA lacks the first 175 amino acids encoded by the rat and *Drosophila* cDNAs), and the human sequence is 94% identical to rat. $p150^{Glued}$ is characteristically expressed as a polypeptide doublet of 150 and 135 kDa. The cloning and characterization of multiple human cDNAs suggest that these isoforms result from alternative splicing (Tokito et al 1993). Existing evidence suggests only a single gene in rat, mouse, and *Drosophila* (Holzbaur et al 1991; K Moore & E Holzbaur, unpublished data; Swaroop et al 1987).

$p150^{Glued}$ (dynactin) is predicted to encode a highly α-helical polypeptide, including extended regions of heptad repeat sequence that may form coiled-coil motifs. Potentially, these α-helical coiled-coils may mediate protein-protein interactions within the Glued complex, including the possible dimerization of the polypeptide. Analysis of the predicted amino acid sequence of p150^{Glued} has also revealed a motif that is homologous to a microtubule-binding motif characterized in the endosome-microtubule linker protein CLIP-170 (Pierre et al 1992). This motif has also been identified by sequence homology in the yeast microtubule-associated protein BIK1 (Pierre et al 1992; Berlin et al 1990).

The ability of p150^{Glued} to bind to microtubules has been examined both in vitro and in transfected cells. Studies examining the binding of p150^{Glued} synthesized in an in vitro transcription/translation system to purified bovine brain microtubules have defined the microtubule-binding motif to amino acids 39–150. Cellular studies in which full-length and truncated constructs of p150^{Glued} were transfected into Rat2 cells have revealed that the over-expression of the polypeptide in vivo results in the decoration of the microtubule cytoskeleton (Waterman-Storer et al 1993).

The 45 kDa polypeptide of the Glued (dynactin) complex appears to be a novel form of actin. Full-length cDNA sequences encoding centractin, also known as actin-RPV, were isolated from both human and canine libraries; the predicted amino acid sequences were identical (Lees-Miller et al 1992a; Clark & Meyer 1992). Overall, the predicted sequence of centractin is ~ 50% identical to those of cytoplasmic actins. However, centractin is distinct from both previously characterized novel actins in yeast—*act*2 from *Schizosaccharomyces pombe* (Lees-Miller et al 1992b) and *Act*2 from *Saccharomyces cerevisiae* (Schwob & Martin 1992). The polypeptide is predicted to fold in a similar fashion to conventional actins, with most of the conserved residues mapping to the core of the folded polypeptide, while the divergent amino acids map to the surface (Clark & Meyer 1992; Lees-Miller et al 1992a). The amino acids comprising the ATP-binding site of actin are conserved in the centractin sequence, and in vitro-translated centractin has been shown to bind to an ATP-agarose column (Melki et al 1993).

Centractin synthesized in vitro has also been observed to co-sediment with polymerized actin filaments, but not with monomeric actin, which indicates that the two polypeptides may co-polymerize under certain conditions (Melki et al 1993). However, centractin appears to exist exclusively as a stoichiometric component of the Glued complex in vivo. Paschal et al (1993) fractionated brain cytosol by sucrose gradient centrifugation and found centractin to co-sediment in a single, symmetric peak at 20S with the 150 and 50 kDa components of the Glued complex. Waterman-Storer et al (1993) examined the binding interaction of p150^{Glued} to centractin by affinity chromatography and determined that these polypeptides are capable of a direct association. This result is of particular interest because if p150^{Glued} binds directly to microtubules, and if centractin may interact with actin filaments, this suggests a mechanism for the interaction of the microtubule and actin-based cellular cytoskeletons (Waterman-Storer et al 1993).

The predicted amino acid sequence of the 50 kDa polypeptide, the second most abundant component in the Glued complex following centractin, has been determined from both human and rat cDNA clones (C Echeverri et al, unpublished results). No significant homology to other sequences in the GenEMBL database was observed. Three regions are predicted to form α-helical coiled-coils, although total coiled-coil sequence is considerably less than that predicted for p150^{Glued}. Thus the 50 kDa polypeptide and p150^{Glued} may form homodimers within the Glued complex, but it will be interesting to test this hypothesis directly. p50 also contains a strongly predicted helix-turn-helix motif, characteristic of DNA-binding proteins. The significance of this motif in p50 is uncertain.

The Glued complex has also been reported to contain the α and β subunits of a barbed-end capping protein related to capZ (Schroer et al 1992). These polypeptides are likely to cap one end of the short actin-like filament revealed by electron microscopic analysis of the complex (Schroer et al 1992). Both the defined stoichiometry and the EM analysis indicate that the filaments are of uniform and defined size (equivalent to 7–8 centractin subunits). The mechanism for the polymerization of centractin filaments of defined length, as well as the identity of a capping protein for the other end of the filament, remain to be determined.

The nature of the interaction of the Glued complex with cytoplasmic dynein remains to be resolved. Only marginal levels of the complex were observed to co-immunoprecipitate with cytoplasmic dynein from brain cytosol, and vice versa. Furthermore, each of the complexes alone sediments at 20S, yet in preparations containing both, no obvious shift toward a higher S-value has been reported.

However, the components of the complex have been found by immunofluorescence to be associated with vesicle-like structures in cultured cells and to be concentrated at the centrosome (Gill et al 1991; Clark & Meyer 1992; Paschal et al 1993). The centrosomal association of the Glued complex has been found to be sensitive to nocodazole, which suggests that the complex may be associated with the minus ends of microtubules that converge at that site (Paschal et al 1993).

More recent genetic evidence suggests that the two proteins function in the same pathway. Among the ropy mutations identified as partial suppressors of mutations in the *cot*-1 protein kinase in *Neurospora* (see above), two are related to components of the Glued complex (Plamann et al 1994). *Ro*4 is related to centractin, and *Ro*3 is related to p150^{Glued}. Mutant alleles at either locus produce a defect in nuclear migration, as is seen for mutations in *Ro*1, which encodes the cytoplasmic dynein heavy chain in *Neurospora.*

Whether these data indicate a direct or indirect interaction between the Glued complex and cytoplasmic dynein is not known. Potentially, the complex might serve as an organelle membrane receptor for cytoplasmic dynein, although a substantial fraction of the complex is found in the cytosol. Since the complex may interact directly with microtubules via the microtubule-binding motif in $p150^{Glued}$, it could serve as a docking protein, similar to the function proposed for the endosome-microtubule linker protein CLIP-170 (Pierre et al 1992). In addition to a role as a docking protein, the complex might also be required constitutively during vesicular transport as a tether. This would prevent the diffusion of the vesicle away from the microtubule during the phase of the cytoplasmic dynein cross-bridge cycle in which the heads are detached from the microtubule. Further work will be required to refine and test these models for the role of the Glued complex, but the lethality of the null phenotype in *Drosophila* clearly indicates an essential cell function.

CONCLUSIONS

Dynein is the largest and most complex of the known motor proteins. For this reason conventional biochemistry has so far provided relatively limited insight into its structure. The dramatic increase in molecular data for the dynein subunits and for proteins functionally related to dynein seen during the past few years has provided exciting new insight into how the molecule is organized and how it may function. Analysis of the sequence information now available suggests that two paradigms exist for dyneins—the heavy chain of cytoplasmic dynein is encoded by a single gene in all organisms examined, but axonemal dynein heavy chains are encoded by a more complex gene family. Nonetheless, cytoplasmic dynein appears to be involved in diverse functions, while multiple heavy chains play coordinated but distinct roles in the axoneme. While PCR analysis has revealed an extensive family of dyneins, no strong evidence has come forth to suggest a superfamily of dyneins related primarily within their motor domains, in contrast to recent findings in the kinesin and myosin fields. Further exploration is required to define the full extent of the family.

The overall structure of the heavy chains is conserved among both axonemal and cytoplasmic dyneins, with the most striking feature being the conservation of four P-loop consensus motifs for nucleotide binding and hydrolysis. This observation contrasts with existing biochemical data that have so far indicated only a single functional ATP binding and hydrolysis site per polypeptide. Much work clearly remains to resolve this issue, to define the structural organization of the heavy chain, and to understand how it produces force.

Molecular cloning of the accessory subunits of dynein has already greatly simplified the otherwise baffling complexity observed in the composition of dyneins from different sources. The identification of an intermediate chain family encourages further extensive efforts to clone cDNAs encoding dynein subunits, with the promise of a dramatic simplification in understanding the relationship between different forms of dynein. Present information on the relationship between the intermediate chains of axonemal vs cytoplasmic forms of dynein strongly suggests a role for the cytoplasmic intermediate chain in intracellular targeting, and it will be of considerable interest to continue efforts to define the cellular polypeptides with which the intermediates chains interact.

Genetics may provide some of the most valuable information on both cytoplasmic and axonemal dynein. Studies in a number of organisms have already revealed unexpected roles for cytoplasmic dynein. Analysis of *Chlamydomonas* axonemal mutants has also begun to provide insight into the structural organization of the dynein molecule and how the activity of the axonemal forms may be regulated. Ultimately, genetic analysis is likely to be invaluable for understanding the remarkable level of dynein coordination required to produce effective ciliary and flagellar beating.

ACKNOWLEDGMENTS

The authors thank Drs. Christine Collins, Kevin Pfister, Michael Plamann, Ron Morris, George Witman, David Asai, David Mitchell, and Rex Chisholm for sharing their unpublished data; Dr. Paula Henthorn for her advice on DNA analysis software; and George Witman, Curt Wilkerson, and Kevin Vaughan for their helpful comments.

Literature Cited

Amos LA. 1989. Brain dynein crossbridges microtubules into bundles. *J. Cell Sci.* 93:19–28

Aniento F, Emans N, Griffiths G, Gruenberg J. 1993. Cytoplasmic dynein-dependent vesicular transport from early to late endosomes. *J. Cell Biol.* 123:1373–87

Asai DJ, Beckwith SM, Kandl KA, Keating HH, Tjandra H, Froney JD. 1994. The dynein genes of *Paramecium tetraurelia:* sequences adjacent to the catalytic P-loop identify cytoplasmic and axonemal heavy chain isoforms. *J. Cell Sci.* 107:839–47

Barkalow T, Hamasaki T, Nair S, Satir P. 1993. A 29 kDa cAMP-dependent phosphorylation is directly associated with and regulates axonemal 22S dynein. *Mol. Biol. Cell* 4:161 (Abstr.)

Bell CW, Fronk E, Gibbons IR. 1979. Polypeptide subunits of dynein I from sea urchin sperm flagella. *J. Supramol. Struct.* 11:311–17

Berlin V, Styles CA, Fink GR. 1990. BIK1, a protein required for microtubule function during mating and mitosis in *Saccharomyces cerevisiae,* colocalizes with tubulin. *J. Cell Biol.* 111:2573–86

Bessen M, Fay RB, Witman GB. 1980. Calcium control of waveform in isolated flagellar axonemes of *Chlamydomonas*. *J. Cell Biol.* 86: 446–55

Bloom GS. 1992. Motor proteins for cytoplasmic microtubules. *Curr. Opin. Cell Biol.* 4: 66–73

Blum JJ, Hayes A, Jamieson GA, Vanaman TC. 1980. Calmodulin confers calcium sensitivity on ciliary dynein ATPase. *J. Cell Biol.* 87:386–97

Brokaw CJ, Kamiya R. 1987. Bending patterns of *Chlamydomonas* flagella: IV. Mutants with defects in inner and outer dynein arms indicate differences in dynein arm function. *Cell Motil. Cytoskeleton* 8:68–75

Cheney RE, Riley MA, Mooseker MS. 1993. Phylogenetic analysis of the myosin superfamily. *Cell Motil. Cytoskeleton* 24:215–23

Chilcote TJ, Johnson KA. 1990. Phosphorylation of *Tetrahymena* 22S dynein. *J. Biol. Chem.* 265:17257–66

Clark SW, Meyer DI. 1992. Centracin is an actin homologue associated with the centrosome. *Nature* 359:246–50

Collins CA, Vallee RB. 1989. Preparation of microtubules from rat liver and testis: cytoplasmic dynein is a major microtubule associated protein. *Cell Motil. Cytoskeleton* 14:491–500

Corthesy-Theulaz I, Pauloin A, Pfeffer SR. 1992. Cytoplasmic dynein participates in the centrosomal localization of the Golgi complex. *J. Cell Biol.* 118:1333–45

Dillman JF, Pfister KK. 1993. Differential phosphorylation in vivo of cytoplasmic dynein in anterograde and total cell compartments. *Mol. Biol. Cell* 4:160 (Abstr.)

Eckert R, Naitoh Y, Machemer H. 1976. Calcium in the bioelectric and motor functions of *Paramecium. Symp. Soc. Exp. Biol.* 30: 233–55

Endow SA, Titus MA. 1992. Genetic approaches to molecular motors. *Annu. Rev. Cell Biol.* 8:29–66

Eshel D, Urrestarauzu LA, Vissers S, Jauniaux JC, van Vliet-Reedijk JC, et al. 1993. Cytoplasmic dynein is required for normal nuclear segregaton in yeast. *Proc. Natl. Acad. Sci. USA* 90:11172–76

Euteneuer U, Johnson KB, Schliwa M. 1989. Photolytic cleavage of cytoplasmic dynein inhibits organelle transport in *Reticulomyxa. Eur. J. Cell Biol.* 50:34–40

Euteneuer U, Koonce MP, Pfister KK, Schilwa M. 1988. An ATPase with properties expected for the organelle motor of the giant amoeba *Reticulomyxa. Nature* 332:176–78

Gatti JL, King SM, Moss AG, Witman GB. 1989. Outer arm dynein from trout spermatozoa. Purification, polypeptide composition, and enzymatic properties. *J. Biol. Chem.* 264: 11450–57

Gepner J, Hays TS. 1993. A fertility region on the Y chromosome of *Drosophila melanogaster* encodes a dynein microtubule motor. *Proc. Natl. Acad. Sci. USA* 90:11132–36

Gibbons BH, Asai DJ, Tang WJY, Hays TS, Gibbons IR. 1994. Phylogeny and expression of axonemal and cytoplasmic dynein genes in sea urchins. *Mol. Biol. Cell* 5:57–70

Gibbons IR. 1966. Studies on the adenosine triphosphatase activity of 14S and 30S dynein from cilia of *Tetrahymena. J. Biol. Chem.* 241:5590–96

Gibbons IR. 1988. Dynein ATPases as microtubule motors. *J. Biol. Chem.* 263:15837–40

Gibbons IR, Gibbons BH, Mocz G, Asai DJ. 1991. Multiple nucleotide-binding sites in the sequence of dynein β heavy chain. *Nature* 352:640–43

Gibbons IR, Lee-Eiford A, Mocz G, Phillipson CA, Tang WJY, Gibbons BH. 1987. Photosensitized cleavage of dynein heavy chains. *J. Biol. Chem.* 262:2780–86

Gibbons IR, Rowe A. 1965. Dynein: a protein with adenosine triphosphatase activity from cilia. *Science* 149:424

Gill SR, Salcedo AF, Cleveland DW, Schroer TA. 1992. Cloning of a cytoplasmic dynein 50 kD subunit. *Mol. Biol. Cell* 3:161 (Abstr.)

Gill SR, Schroer TA, Szilak I, Steuer ER, Sheetz MP, Cleveland DW. 1991. Dynactin, a conserved, ubiquitously expressed component of an activator of vesicle motility mediated by cytoplasmic dynein. *J. Cell Biol.* 115: 1639–50

Goldstein LSB. 1991. The kinesin superfamily: tails of functional redundancy. *Trends Cell Biol.* 1:93–98

Goodenough UW, Heuser JE. 1982. The substructure of the outer dynein arm. *J. Cell Biol.* 95:798–815

Goodenough UW, Heuser JE. 1984 Structural comparison of purified proteins with in situ dynein arms. *J. Mol. Biol.* 180:1083–1118

Goodenough UW, Heuser JE. 1985. Substructure of inner dynein arms, radial spokes, and the central pair/projection complex of cilia and flagella. *J. Cell Biol.* 100:2008–18

Goodson HV, Spudich JA. 1993. Molecular evolution of the myosin family: relationships derived from comparisons of amino acid sequences. *Proc. Natl. Acad. Sci. USA* 90:659–63

Hamasaki T, Barkalow K, Richmond J, Satir P. 1991. cAMP-stimulated phosphorylation of an axonemal polypeptide that copurifies with the 22S dynein arm regulates microtubule translocation velocity and swimming speed in *Paramecium. Proc. Natl. Acad. Sci. USA* 88:7918–22

Hamm-Alvarez SF, Kim PY, Sheetz MP. 1993. Regulation of vesicle transport in CV-1 cells and extracts. *J. Cell Sci.* 106:955–66

Hammer J. 1991. Novel myosin. *Trends Cell Biol.* 1:50–56

Harte PJ, Kankel DR. 1982. Genetic analysis

of mutations at the *Glued* locus and interacting loci in *Drosophila melanogaster. Genetics* 101:477–501
Hastie AT, Dicker DT, Hingley ST, Keuppers F, Higgins ML, Weinbaum G. 1986. Isolation of cilia from porcine tracheal epithelium and extraction of dynein arms. *Cell Motil. Cytoskeleton* 6:25–34
Higgins DG, Sharp PM. 1989. Fast and sensitive multiple sequence alignments on a microcomputer. *CABIOS Commun.* 5:151–53
Hirokawa N, Sato-Yoshitake R, Yoshida T, Kawashima T. 1990. Brain dynein (MAP1C) localizes on both anterogradely and retrogradely transported membranous organelles in vivo. *J. Cell Biol.* 111:1027–37
Holzbaur ELF, Hammarback JA, Paschal BM, Kravit NG, Pfister KK, Vallee RB. 1991. Homology of a 150K cytoplasmic dynein-associated polypeptide with the *Drosophila* gene *Glued. Nature* 351:579–83
Holzbaur ELF, Johnson KA. 1989a. ADP release is rate limiting in steady-state turnover by the dynein adenosinetriphosphatase. *Biochemistry* 28:5577–85
Holzbaur ELF, Johnson KA. 1989b. Microtubules accelerate ADP release by dynein. *Biochemistry* 28:7010–16
Holzbaur ELF, Mikami A, Paschal BM, Vallee RB. 1994. Molecular characterization of cytoplasmic dynein. In *Microtubules,* ed. J Hyams, C LLoyd, pp. 251–68. New York: Wiley-Liss
Huang B, Piperno G, Luck DJL. 1979. Paralyzed flagella mutants of *Chlamydomonas reinhardtii* defective for axonemal doublet microtubule arms. *J. Biol. Chem.* 254:3091–99
Huang B, Ramanis Z, Luck DJL. 1982. Suppressor mutations in Chlamydomonas reveal a regulatory mechanism for flagellar function. *Cell* 28:115–24
Hughes SM, Herskovits JS, Vaughan KT, Vallee RB. 1993. Cloning and characterization of cytoplasmic dynein 53/55 and 57/59 subunits. *Mol. Biol. Cell* 4:47 (Abstr.)
Hyman AA, Mitchison TJ. 1991. Two different microtubule-base motor activities with opposite polarities in kinetochores. *Nature* 351: 187–88
Jamieson GA, Vanaman TC, Blum JJ. 1979. Presence of calmodulin in *Tetrahymena. Proc. Natl. Acad. Sci. USA* 76:6471–75
Johnson KA. 1983. The pathway of ATP hydrolysis by dynein. *J. Biol. Chem.* 258: 13825–32
Johnson KA. 1985. Pathway of the microtubule-dynein ATPase and the structure of dynein: a comparison with actomyosin. *Annu. Rev. Biophys. Biophys. Chem.* 14: 161–88
Johnson KA, Marchese-Ragona SP, Clutter DB, Holzbaur ELF, Chilcote TJ. 1986. Dynein structure and function. *J. Cell Sci. Suppl.* 5:189–96
Johnson KA, Wall JS. 1983. Structure and molecular weight of the dynein ATPase. *J. Cell Biol.* 96:669–78
Kagami O, Kamiya R. 1992. Translocation and rotation of microtubules caused by multiple species of *Chlamydomonas* inner-arm dynein. *J. Cell Sci.* 103:653–64
Kamiya R. 1988. Mutations at twelve independent loci result in absence of outer dynein arms in *Chlamydomonas reinhardtii. J. Cell Biol.* 107:2253–58
Kamiya R. 1991. Selection of *Chlamydomonas* dynein mutants. *Methods Enzymol.* 196:348–55
King SM, Gatti JL, Moss AG, Witman GB. 1990. Outer-arm dynein from trout spermatozoa: substructural organization. *Cell Motil. Cytoskeleton* 16:266–78
King SM, Otter T, Witman GB. 1985. Characterization of monoclonal antibodies against *Chlamydomonas* flagellar dyneins. *Proc. Natl. Acad. Sci. USA* 82:4717–21
King SM, Wilkerson CG, Witman GB. 1991. The M_r 78,000 intermediate chain of *Chlamydomonas* outer arm dynein interacts with α-tubulin in situ. *J. Biol. Chem.* 266: 8401–7
King SM, Wilkerson CG, Witman GB. 1992. Molecular cloning and domain analysis of the M_r 78,000 intermediate chain of *Chlamydomonas* outer arm dynein. *Mol. Biol. Cell* 3:2 (Abstr.)
King SM, Witman GB. 1990. Localization of an intermediate chain of outer arm dynein by immunoelectron microscopy. *J. Biol. Chem.* 265:19807–11
King SM, Witman GB. 1994. Multiple sites of phosphorylation within the α heavy chain of *Chlamydomonas* outer arm dynein. *J. Biol. Chem.* 269:5452–57
Koonce MP, Grissom PM, McIntosh JR. 1992. Dynein from *Dictyostelium:* primary structure comparisons between a cytoplasmic motor enzyme and flagellar dynein. *J. Cell Biol.* 119:1597–1604
Lee-Eiford A, Ow RA, Gibbons IR. 1986. Specific cleavage of dynein heavy chains by ultraviolet irradiation in the presence of ATP and vanadate. *J. Biol. Chem.* 261: 2337–42
Lees-Miller JP, Helfman DM, Schroer TA. 1992a. A vertebrate actin-related protein is a component of a multisubunit complex involved in microtubule-based vesicle motility. *Nature* 359:244–46
Lees-Miller JP, Henry G, Helfman DM. 1992b. Identification of *act2,* an essential gene in the fission yeast *Schizosaccharomyces pombe* that encodes a protein related to actin. *Proc. Natl. Acad. Sci. USA* 89:80–83
Li YY, Yeh E, Hays T, Bloom K. 1993. Disruption of mitotic spindle orientation in a

yeast dynein mutant. *Proc. Natl. Acad. Sci. USA* 90:10096–100

Lin SXH, Collins CA. 1992. Immunolocalization of cytoplasmic dynein to lysosomes in cultured cells. *J. Cell Sci.* 101:125–37

Lin SXH, Collins CA. 1993a. Regulation of the intracellular distribution of cytoplasmic dynein by serum factor and calcium. *J. Cell Sci.* 105:579–88

Lin SXH, Collins CA. 1993b. Cytoplasmic dynein is phosphorylated and undergoes changes in intracellular distribution in response to okadaic acid. *Mol. Cell. Biol.* 4:401 (Abstr.)

Lye RJ, Porter ME, Scholey JM, McIntosh JR. 1987. Identification of a microtubule-based cytoplasmic motor in the nematode C. elegans. *Cell* 51:309–18

Mastronarde DN, O'Toole ET, McDonald KL, McIntosh JR, Porter ME. 1992. Arrangement of inner dynein arms in wild-type and mutant flagella of *Chlamydomonas. J. Cell Biol.* 118:1145–62

Melki R, Vainberg IE, Chow RL, Cowan NJ. 1993. Chaperonin-mediated folding of vertebrate actin-related protein and gamma-tubulin. *J. Cell Biol.* 122:1301–10

Mikami A, Paschal BM, Mazumdar M, Vallee RB. 1993. Molecular cloning of the retrograde transport motor cytoplasmic dynein (MAP 1C). *Neuron* 10:787–96

Mitchell DR. 1994. Cell and molecular biology of flagellar dyneins. *Int. Rev. Cytol.* In press

Mitchell DR, Brown KS. 1994. Sequence analysis of the *Chlamydomonas* alpha and beta dynein heavy chain genes. *J. Cell Sci.* 107: 635–44

Mitchell DR, Kang Y. 1991. Identification of *oda6* as a *Chlamydomonas* dynein mutant by rescue with the wild-type gene. *J. Cell Biol.* 113:835–42

Mitchell DR, Rosenbaum JL. 1986. Protein-protein interactions in the 18S ATPase of *Chlamydomonas* outer dynein arms. *Cell Motil. Cytoskeleton* 6:510–20

Mocz G, Gibbons IR. 1993. ATP-insensitive interaction of the amino-terminal region of the β heavy chain of dynein with microtubules. *Biochemistry* 32:3456–60

Morris NR. 1976. Mitotic mutants of *Aspergillus nidulans. Genet. Res.* 26:237–54

Neely MD, Erickson HP, Boekelheide K. 1990. HMW-2, the Sertoli cell cytoplasmic dynein from rat testis, is a dimer composed of nearly identical subunits. *J. Biol. Chem.* 265:8691–98

Ogawa K. 1991. Four ATP binding sites in the midregion of the β heavy chain of dynein. *Nature* 352:643–45

O'Toole ET, Gardner LC, Giddings TH, Porter ME. 1993. The "dynein regulatory complex" resides at the junction between the radial spokes and the inner dynein arms. *Mol. Biol. Cell* 4:161 (Abstr.)

Paschal BM, Holzbaur ELF, Pfister KK, Clark S, Meyer D, Vallee RB. 1993. Characterization of a 50-kDa polypeptide in cytoplasmic dynein preparations reveals a complex with p150^{Glued} and a novel actin. *J. Biol. Chem.* 268:15318–23

Paschal BM, King SM, Moss AG, Collins CA, Vallee RB, Witman GB. 1987a. Isolated flagellar outer arm dynein translocates brain microtubules in vitro. *Nature* 330:672–74

Paschal BM, Mikami A, Pfister KK, Vallee RB. 1992. Homology of the 74-kD cytoplasmic dynein subunit with a flagella dynein polypeptide suggests an intracellular targeting function. *J. Cell Biol.* 118:1133–43

Paschal BM, Shpetner HS, Vallee RB. 1987b. MAP 1C is a microtubule-activated ATPase which translocates microtubules in vitro and has dynein-like properties. *J. Cell Biol.* 105: 1273–82

Paschal BM, Vallee RB. 1987. Retrograde transport by the microtubule-associated protein MAP 1C. *Nature* 330:181–83

Pfarr CM, Coue M, Grisson PM, Hays TS, Porter ME, McIntosh JR. 1990. Cytoplasmic dynein is localized to kinetochores during mitosis. *Nature* 345:263–65

Pfister KK, Fay RB, Witman, GB. 1982. Purification and polypeptide composition of dynein ATPases from *Chlamydomonas* flagella. *Cell Motil.* 2:525–47

Pfister KK, Haley BE, Witman GB. 1984. The photoaffinity probed 8-azidoadenosine 5′-triphosphate selectively labels the heavy chain of *Chlamydomonas* 12S dynein. *J. Biol. Chem.* 259:8499–8504

Pfister KK, Witman GB. 1984. Subfractionation of *Chlamydomonas* 18S dynein into two unique subunits containing ATPase activity. *J. Biol. Chem.* 259:12072–80

Pierre P, Scheel J, Rickard JE, Kreis TE. 1992. CLIP-170 links endocytic vesicles to microtubules. *Cell* 70:887–900

Piperno G, Luck DJL. 1979. Axonemal adenosine triphosphatases from flagella of *Chlaymdomonas reinhardtii:* purification of two dyneins. *J. Biol. Chem.* 254:3084–90

Piperno G, Luck DJL. 1981. Inner arm dyneins from flagella of Chlamydomonas reinhardtii. *Cell* 27:331–40

Piperno G, Mead K, Shestak W. 1992. The inner dynein arms I2 interact with a "dynein regulatory complex" in *Chlamydomonas* flagella. *J. Cell Biol.* 118:1455–63

Piperno G, Ramanis Z. 1991. The proximal portion of *Chlamydomonas* flagella contains a distinct set of inner dynein arms. *J. Cell Biol.* 112:701–9

Piperno G, Ramanis Z, Smith EF, Sale WS. 1990. Three distinct inner dynein arms in *Chlamydomonas* flagella: molecular compo-

sition and locationin the axoneme. *J. Cell Biol.* 110:379–89
Plamann M, Minke PF, Tinsley JH, Bruno KS. 1994. Cytoplasmic dynein and centractin are required for normal nuclear distribution in filamentous fungi. *J. Cell Biol.* Submitted
Plough HH, Ives PT. 1935. Induction of mutations by high temperature in *Drosophila. Genetics* 20:42–69
Pollard TD, Doberstein SK, Zot HG. 1991. Myosin-I. *Annu. Rev. Physiol.* 53:653–81
Porter ME, Johnson KA. 1983. Transient state kinetic analysis of the ATP-induced dissociation of the dynein-microtubule complex. *J. Biol. Chem.* 258:6582–87
Porter ME, Johnson KA. 1989. Dynein structure and function. *Annu. Rev. Cell Biol.* 5: 199–51
Porter ME, Knott JA, Gardner LC, Mitchell DR, Dutcher SK. 1993. Mutations in the structural gene for the dynein beta chain reveal the location of a regulatory domain. *Mol. Cell Biol.* 4:47 (Abstr.)
Rasmusson K, Serr M, Gepner J, Gibbons I, Hays TS. 1994. A family of dynein genes in *Drosophila melanogaster. Mol. Biol. Cell* 5: 45–55
Rieder CL, Alexander SP, Rupp G. 1990. Kinetochores are transported poleward along a single astral microtubule during chromosome attachment to the spindle in newt lung cells. *J. Cell Biol.* 110:81–95
Sakakibara H, Mitchell DR, Kamiya R. 1991. A *Chlamydomonas* outer arm dynein mutant missing the α heavy chain. *J. Cell Biol.* 113:615–22
Sakakibara H, Takada S, King SSM, Witman GB, Kamiya R. 1993. A *Chlamydomonas* outer arm dynein mutant with a truncated β heavy chain. *J. Cell Biol.* 122:653–61
Sale WS, Fox LA, 1988. Isolated β-heavy chain subunit of dynein translocates microtubules in vitro. *J. Cell Biol.* 107:1793–97
Sale WS, Goodenough UW, Heuser JE. 1985. The substructure of isolated and in situ outer dynein arms of sea urchin sperm flagella. *J. Cell Biol.* 101:400–12
Sanghamitra R, Meyhöfer E, Milligan RA, Howard J. 1993. Kinesin follows the microtubule's protofilament axis. *J. Cell Biol.* 121:1083–93
Saraste M, Sibblad PR, Wittinghofer A. 1990. The P-loop—a common motif in ATP- and GTP-binding proteins. *Trends Biochem. Sci.* 15:430–34
Saunders WS, Eshel D, Gibbons IR, Totis L, Hoyt MA. 1993. Anaphase in *Saccharomyces cerevisiae* is a cooperative effort between the kinesin-related proteins Cin8P and Kip1P and dynein-related Dyn1P. *Mol. Biol. Cell* 4:118 (Abstr.)
Schliwa M, Shimizu T, Vale RD, Euteneuer U. 1991. Nucleotide specificities of anterograde and retrograde organelle transport in *Reticulomyxa* are indistinguishable. *J. Cell Biol.* 112:1199–1203
Schmidt JA, Eckert R. 1976. Calcium couples flagellar reversal to photostimulation in *Chlamydomonas reinhardtii. Nature* 262: 713–15
Schnapp BJ, Reese TS. 1989. Dynein is the motor for retrograde axonal transport of organelles. *Proc. Natl. Acad. Sci. USA* 86: 1548–52
Schroer TA, Heuser JE, Helfman DM, Lees-Miller JP. 1992. Actin-RPV: the major component of the dynactin complex, activator of cytoplasmic dynein. *Mol. Biol. Cell* 3:2 (Abstr.)
Schroer TA, Steuer ER, Sheetz MP. 1989. Cytoplasmic dynein is a minus end-directed motor for membranous organelles. *Cell* 56: 937–46
Schwob E, Martin RP. 1992. New yeast actin-like gene required late in the cell cycle. *Nature* 355:179–82
Shimizu T, Marchese-Ragon SP, Johnson KA. 1989. Activation of the dynein adenosinetriphosphatase by cross-linking to microtubules. *Biochemistry* 28:7016–21
Shpetner HS, Paschal BM, Vallee RB. 1988. Characterization of the microtubule-activated ATPase of brain cytoplasmic dynein (MAP1C). *J. Cell Biol.* 107:1001–9
Smith EF, Sale WS. 1992. Regulation of dynein-driven microtubule sliding by the radial spokes in flagella. *Science* 257:1557–59
Steele GC, Trinci APJ. 1977. Effect of temperature and temperature-shifts on growth and branching of a wild-type and a temperature-sensitive colonial mutant (*cot*-1) of *Neurospora crassa. Arch. Microbiol.* 113:43–48
Steuer E, Wordeman L, Schroer TA. Sheetz MP. 1990. Localization of cytoplasmic dynein to mitotic spindles and kinetochores. *Nature* 345:266–68
Swaroop A, Swaroop M, Garen A. 1987. Sequence analysis of the complete cDNA and encoded polypeptide for the *Glued* gene of *Drosophila melanogaster. Proc. Natl. Acad. Sci. USA* 84:6501–5
Tokito, MK, Lee VMY, Holzbaur ELF. 1993. Characterization of the human cDNA encoding p150*Glued* and its expression in the human NTera 2 cell line. *Mol. Biol. Cell* 4:162 (Abstr.)
Trivinos-Lagos L, Collins CA, Chisholm RL. 1993. Cloning of dynein intermediate chain: multiple isoforms are expressed during *Dictyostelium* development. *Mol. Biol. Cell* 4:47 (Abstr.)
Vaisberg EA, Koonce MP, McIntosh JR. 1993. Cytoplasmic dynein plays a role in mammalian mitotic spindle formation. *J. Cell Biol.* 123:849–58
Vale RD, Soll, DR, Gibbons IR. 1989. One-di-

mensional diffusion of microtubules bound to flagellar dynein. *Cell* 59:915–25

Vale RD, Toyoshima YY. 1988. Rotation and translocation of microtubules in vitro induced by dyneins from *Tetrahymena* cilia. *Cell* 52:459–69

Vallee RB. 1993. Molecular analysis of the microtubule motor dynein. *Proc. Natl. Acad. Sci. USA* 90:8769–72

Vallee RB, Shpetner HS. 1990. Motor proteins of cytoplasmic microtubules. *Annu. Rev. Biochem.* 59:909–32

Vallee RB, Wall JS, Paschal BM, Shpetner HS. 1988. Microtubule-associated protein 1C from brain is a two-headed cystolic dynein. *Nature* 332:561–63

Vaughan KT, Vallee RB. 1993 Transfection of Cos-7 cells with cytoplasmic dynein intermediate chains. *Mol. Biol. Cell* 4:162 (Abstr.)

Walker JE, Saraste M, Runswick MJ, Gay NJ. 1982. Distantly related sequences in the α- and β-subunits of ATP synthase, myosin, kinases and other ATP-requiring enzymes and a common nucleotide binding fold. *EMBO J.* 1:945–51

Wang ZH, Sheetz MP. 1993. Cytoplasmic dynein dances while kinesin walks along the microtubule surface. *Mol. Biol. Cell* 4:48 (Abstr.)

Waterman-Storer CM, Karki S, Holzbaur ELF. 1993. Analysis of the p150^{Glued}-centractin complex reveals a microtubule-binding function in vivo and in vitro. *Mol. Cell Biol.* 4:162 (Abstr.)

Waters JC, Cole RW, Rieder CL. 1993. The force-producing mechanism for centrosome separation during spindle formation in vertebrates is intrinsic to each aster. *J. Cell Biol.* 122:361–72

Wilkerson CG, King SM, Wiman GB. 1994. Molecular analysis of the γ heavy chain of *Chlamydomonas* flagellar outer-arm dynein. *J. Cell Sci.* 107:497–506

Williams BD, Mitchell DR, Rosenbaum JL. 1986. Molecular cloning and expression of flagellar radial spoke and dynein genes of *Chlamydomonas. J. Cell Biol.* 103:1–11

Witman GB. 1992. Axonemal dyneins. *Curr. Opin. Cell Biol.* 4:74–79

Witman GB, Wilkerson CG, King SG. 1994. The biochemistry, genetics, and molecular biology of flagellar dynein. See Holzbaur et al 1994, pp. 229–49

Wordeman L, Steuer ER, Sheetz, MP, Mitchison T. 1991. Chemical subdomains within the kinetochore domain of isolated CHO mitotic chromosomes. *J. Cell Biol.* 114:285–94

Xiang X, Beckwith SM, Morris NR. 1994. Cytoplasmic dynein is involved in nuclear migration in *Aspergillus nidulans. Proc. Natl. Acad. Sci. USA* 91:2100–4

Yarden O, Plamann M, Ebbole DJ, Yanofsky C. 1992. *cot-1,* a gene required for hyphal elongation in *Neurospora crassa,* encodes a protein-kinase. *EMBO J.* 11(6):2159–66

Yu H, Toyoshima I, Steuer ER, Sheetz, MP. 1992. Kinesin and cytoplasmic dynein binding to brain microsomes. *J. Biol. Chem.* 267: 20457–64

Zhang Z, Tanaka Y, Noanaka S, Aizawa H, Kawasaki H, et al. 1993. The primary structure of rat brain (cytoplasmic) dynein heavy chain, a cytoplasmic motor enzyme. *Proc. Natl. Acad. Sci. USA* 90:7928–32

Annu. Rev. Cell Biol. 1994. 10:373–403

CELL BIOLOGY OF THE AMYLOID β-PROTEIN PRECURSOR AND THE MECHANISM OF ALZHEIMER'S DISEASE

Dennis J. Selkoe
Department of Neurology, and Program in Neuroscience, Harvard Medical School and Center for Neurologic Diseases, Brigham and Women's Hospital, Boston, Massachusetts 02115

KEY WORDS: membrane proteins, endocytosis, proteolysis, polarized cells, neurodegeneration

---

CONTENTS

0743–4634/94/1115–0373$05.00

Intensive research on the molecular pathology of Alzheimer's disease (AD) led to the cloning of the amyloid β-protein precursor (βPP) and the recognition that altered βPP metabolism and cerebral accumulation of its amyloid β-protein (Aβ) fragment are early and invariant features of AD. In turn, the study of the normal structure and function of this ubiquitously expressed type I integral membrane glycoprotein has led to the cloning of other members of the βPP gene family and the identification of several unusual properties of this molecule. The study of βPP provides a compelling example of research that began with a strictly disease-oriented focus and gave rise to insights into the normal biology of a class of macromolecules.

In this article, I review the salient structural and functional properties of βPP and what is currently known about its several alternative trafficking and processing pathways. I attempt to integrate this rapidly emerging information into a hypothetical mechanism for the involvement of βPP metabolism and progressive Aβ deposition in the pathogenetic cascade of AD. As is customary for this forum, the review is selective rather than exhaustive, and it presents a critical appraisal of current progress on the biology of βPP and its role in the most common form of age-related mental failure in humans.

The study of βPP is providing novel insights into the cell biology of certain cell surface proteins that can serve both as receptors and as secretory precursors. Conversely, progress in understanding the trafficking and metabolism of such proteins in polarized and nonpolarized cells provides important lessons for students of Alzheimer's disease.

BIOCHEMICAL ANALYSES OF ALZHEIMER BRAIN LESIONS LEAD TO THE IDENTIFICATION OF Aβ AND ITS PRECURSOR

The Histopathology of Alzheimer's Disease

Virtually all of the information we consider in this chapter emanates from the interest during the early 1980's in defining the biochemical compositions of the principal neuropathological abnormalities that define AD. Whereas progressive clinical dementia consistent with Alzheimer's disease has long been known, it was not until 1906 that the Bavarian psychiatrist, Alois Alzheimer, first called attention to the combined occurrence of innumerable neurofibrillary tangles and senile (neuritic) plaques in the cerebral cortex of such patients. Neurofibrillary tangles are nonmembrane-bound bundles of abnormal proteinaceous filaments, referred to as paired helical filaments (PHF), that accumulate in the perinuclear cytoplasm of many cortical and limbic neurons. Extensive biochemical and immunochemical studies of PHF have led to the conclusion that their principal protein subunit is an altered phosphorylated form of the neuronal microtubule-associated protein, tau (Grundke-Iqbal et al 1986a: Kondo et al 1988; Kosik et al 1986; Lee et al 1991; Nukina & Ihara 1986; Wischik et al 1988; Wood et al 1986). Bundles of PHF also accumulate within some of the dystrophic neurites that occur in and around the other hallmark lesion of the disease, the senile plaque. The latter term refers to a structurally diverse array of spherical deposits of extracellular Aβ. The first type of senile plaque to be described is now generally referred to as a classical plaque and consists of a compacted extracellular deposit (core) of 6–10 nm amyloid filaments (composed of Aβ), which is intimately surrounded by morphologically abnormal dendrites and axons. Such plaques are also characterized by the presence of reactive astrocytes ringing the plaque periphery and activated microglia in and around its central amyloid deposit. Thus the classical or "mature" plaque is a complex multicellular lesion that always contains an extracellular deposit of filamentous Aβ.

The advent of sensitive immunocytochemical methods for detecting Aβ revealed that such classical plaques with their compacted amyloid cores represent only a minority of all Aβ deposits in Alzheimer brains. Much more common than these is an amorphous form of Aβ deposit now generally referred to as a diffuse or preamyloid plaque. In these plaques, silver staining and immunocytochemistry with antibodies to tau and other proteins reveal very few altered neurites and few or no reactive astrocytes or activated microglia. Moreover, electron microscopy has generally shown far fewer amyloid filaments in diffuse than neuritic plaques, although the possibility that the diffuse plaques contain Aβ which is principally in a filamentous rather than amorphous form has recently been raised (Davies & Mann 1993). It appears that Aβ plaque formation represents a continuum from extracellular deposits with minimal or no associated neuronal and glial alteration to those with profound cytopathology. The precise sequence of evolution of amyloid plaques and the time required to develop a so-called mature neuritic plaque remain elusive in the absence of a manipulable animal model in which this process can be studied dynamically.

In addition to its deposition in various types of parenchymal plaques, Aβ is also found in highly variable amounts in the walls of meningeal and parenchymal arterioles, small arteries, capillaries and, sometimes, venules. Although such microvascular β-amyloid is present to some extent in virtually all AD cases, its abundance does not correlate closely with the density of amyloid plaques in the same tissue. Besides plaque and vascular amyloid deposits and neurofibrillary tangles, AD brain is characterized by a variable, but often marked, loss of neuronal cell bodies, a widespread alteration of cortical neurites beyond those immediately surrounding Aβ plaques, and a severe reactive astrocytosis. This complex array of lesions can be seen in the hippocampus, entorhinal cortex, amygdala, cerebral cortex, and certain subcortical nuclei that project to hippocampus and neocortex (e.g. the cholinergic nucleus basalis of Meynert). While the density and detailed topography of the lesions vary considerably from case to case, widespread Aβ deposition occurs in limbic and association cortices in virtually all cases of AD.

Characterization of Aβ and βPP

A considerable portion of the protein chemical and molecular biological information about Alzheimer's disease that has accrued during the last decade has emanated from the original isolation by Glenner & Wong of Aβ from meningovascular amyloid filaments (Glenner & Wong 1984). Subsequent purification of compacted amyloid plaque cores revealed that they too were principally composed of Aβ, but in an even less soluble form than that found in the blood vessel wall (Gorevic et al 1986; Masters et al 1985; Roher et al 1986; Selkoe et al 1986). Protein sequencing studies have demonstrated that Aβ from AD brain shows N- and C-terminal heterogeneity. The principal species in vascular amyloid is generally believed to be 40 residues long (Aβ1–40), whereas that in plaques apparently includes both Aβ1–40 and Aβ1–42 and probably other minor species that extend to residue 43 and/or begin after the asp 1 that forms the N-terminus of the major Aβ species (Kang et al 1987; Miller et al 1993; Mori et al 1992; Roher et al 1993).

Previous work on numerous other amyloid deposition diseases in humans revealed that the subunit proteins of these diverse amyloid filaments usually consisted of relatively small proteins that arose as proteolytic fragments of larger, normal or mutant gene products. The β-amyloid characteristic of AD has turned out to be no exception. Cloning of a full-length cDNA (Kang et al 1987) using degenerate oligonucleotides based on the Aβ sequence demonstrated that the latter was a 40–43 residue fragment of a single membrane-spanning glycoprotein now referred to as the amyloid β-protein precursor (βAPP or βPP) (Figure 1). The precursor has a 17–residue signal peptide, a large ectodomain containing numerous interesting motifs, a single hydrophobic transmembrane-like domain, and a small cytoplasmic tail. βPP is encoded by a single gene located on human chromosome 21, which undergoes alternative splicing to produce several membrane-associated polypeptides (Kang et al 1987; Kitaguchi et al 1988; Ponte et al 1988; Tanzi et al 1988) and at least one derivative that lacks the transmembrane domain (de Sauvage & Octave 1989). The most common alternate transcripts contain a 56–amino acid insert in the middle of the ectodomain that has homology to the Kunitz family of serine protease inhibitors (KPI) and inhibits trypsin, chymotrypsin, γ subunit of NGF, and other serine proteases in vitro. The most abundantly expressed isoform in non-neural tissues is the 751–residue polypeptide containing this KPI domain. It is expressed in neurons, glia, and other brain cells as well, but

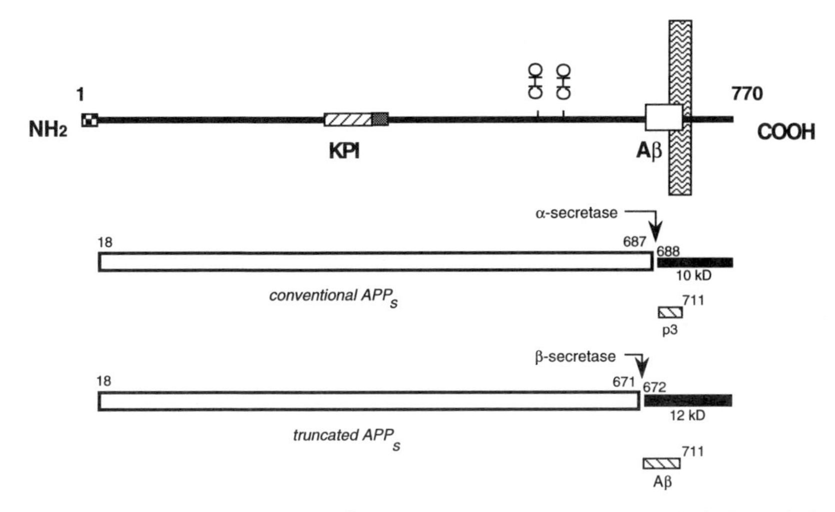

Figure 1 Schematic diagrams of the β-amyloid precursor protein and its principal metabolic derivatives. The upper diagram depicts the largest of the known βAPP alternate transcripts, comprising 770 amino acids. Regions of interest are indicated at their correct relative positions. A 17–residue signal peptide occurs at the N-terminus. Two alternatively spliced exons of 56 and 19 amino acids are inserted at residue 289; the first contains a serine protease inhibitor domain of the Kunitz type (KPI). Two sites of N-glycosylation (CHO) are found at residues 542 and 571. A single membrane-spanning domain at amino acids 700–723 is indicated by the vertical hatched bar. The amyloid β-protein (Aβ) fragment (*white box*) includes 28 residues just outside the membrane plus the first 12–15 residues of the transmembrane domain. In the middle diagram, the arrow indicates the site (after residue 687) of a constitutive proteolytic cleavage made by an unknown protease(s) designated α-secretase that enables secretion of the large, soluble ectodomain of βAPP (APP_s) into the medium and retention of the 83 residue C-terminal fragment (~ 10 kDa) in the membrane. The lower diagram depicts the alternative proteolytic cleavage after residue 671 by an unknown enzyme(s) called β-secretase that results in the secretion of a truncated APP_s molecule and the retention of a 99 residue (~ 12 kDa) C-terminal fragment. The latter may serve as an intermediate in the generation of Aβ, whereas the 10–kDa fragment is believed to give rise to the p3 peptide.

neurons (and to a lesser extent glia) also express a 695–residue form lacking the KPI domain.

In addition to heterogeneity derived from alternative splicing, βPP undergoes several posttranslational modifications, including N- and O-linked glycosylation, sulfation, phosphorylation, and a proteolytic cleavage just before the transmembrane region to release the large, soluble ectodomain as a secreted derivative (βPP_s). The principal proteolytic cleavage that generates βPP_s occurs at residues 687–688 (βPP_{770} numbering); this is equivalent to residues 16–17 of the Aβ region (Esch et al 1990) (Figure 1). As such, this scission precludes formation of intact Aβ and thus can be considered "non-amyloidogenic." The unknown enzyme(s) that affects this cleavage has been designated α-secretase. Mutagenesis experiments suggest that α-secretase is capable of cleaving many different peptide bonds and derives its specificity by cleaving at a distance (12–13 residues) N-terminal to the start of the membrane (Sisodia et al 1990). It appears that βPP needs to be membrane anchored for α-secretase cleavage to occur, which suggests that the protease itself is also membrane-associated (Sisodia 1992). Its mechanism may be similar to that of proteolytic activities known to cleave a number of other single membrane-spanning precursors just outside of the membrane to release the ectodomains (e.g. pro TGFα, colony stimulating factor-1, the TNFα receptor, etc) (Ehlers & Riordan 1991).

ALTERNATIVE PROTEOLYTIC PROCESSING OF βPP AND THE GENERATION OF Aβ

Because the intact Aβ fragment accumulates extracellularly in AD, considerable effort has been directed at defining the enzymatic processing of those precursor molecules that do not undergo cleavage by α-secretase. The existence of such alternate processing could be inferred from the fact that even cells that show robust secretion of βPP_s appear to process only a minority of their βPP molecules by this route (Weidemann et al 1989). Early evidence that some βPP molecules may be localized to lysosomes (Benowitz et al 1989) and that lysosomotropic agents alter βPP processing (Cole et al 1989) led to systematic attempts to define the role of the endosomal/lysosomal system in βPP processing. Treatment of βPP-transfected cells with the thiol protease inhibitor, leupeptin, or with ammonium chloride or chloroquine leads to accumulation of full-length βPP and various C-terminal proteolytic fragments inside the cells (Caporaso et al 1992; Golde et al 1992). In addition to the ~10–kDa C-terminal fragment (CTF) that is retained in the membrane following α-secretase-mediated release of the ectodomain, such cells contain several slightly larger CTFs whose size and immunoreactivity suggest that they contain the intact Aβ region. Similar CTFs are detectable in AD brain tissue and its microvessels (Estus et al 1992; Tamaoka et al 1992). To confirm the processing of βPP within the endosomal/lysosomal system, surface βPP molecules were labeled by biotinylation or ectodomain antibodies and allowed to undergo reinternalization in living cells (Haass et al 1992a). The tagged surface βPP was internalized; purification of late endosomes/lysosomes from such cells revealed the presence of full-length βPP, the 10–kDa CTF, and additional CTFs greater than 10 kDa that contained the intact Aβ region (Haass et al 1992a). Therefore, potentially amyloidogenic fragments of βPP can be generated in the endosomal/lysosomal system. The internalization of βPP is believed

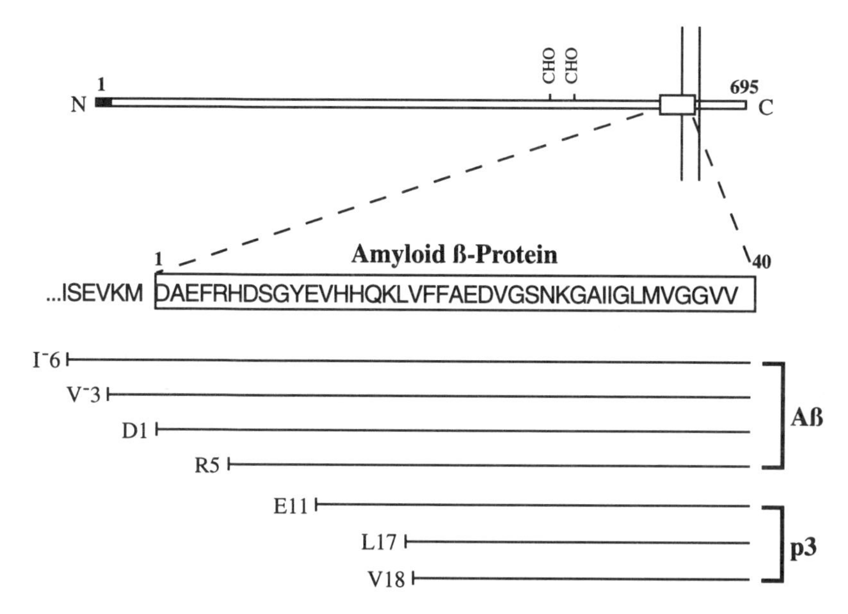

Figure 2 Results of N-terminal radiosequencing of the secreted Aβ and p3 species immunoprecipitated from the conditioned medium of human kidney 293 cells transfected with βAPP_{695}. The major Aβ peptide begins at asp_1, and the major p3 peptides begin at val_{18} or leu_{17}.

to employ the cytoplasmic asn-pro-xxx-tyr consensus sequence that mediates clathrin-coated pit binding (Chen et al 1990).

The delineation of two general pathways for βPP processing—exocytic (secretory) and endocytic (internalization and lysosomal targeting)—leaves open the question of how and where the Aβ fragment is generated. Because the Aβ peptide accumulates pathologically in AD and trisomy 21 (Down's syndrome) and because it includes a hydrophobic region that is normally anchored within the membrane, it had been widely assumed that Aβ could only be released from cells by an abnormal processing event. However, metabolic labeling of βPP-expressing primary or transfected cells, followed by immunoprecipitation of their media with sensitive Aβ antibodies, revealed the presence of the 4–kDa Aβ peptide(s) and a 3–kDa fragment thereof (designated p3) (Haass et al 1992b; Seubert et al 1992; Shoji et al 1992).

Radiosequencing demonstrated that the 4–kDa peptide begins principally at the asp_1 residue that is the major N-terminus of the Aβ purified from AD brain but shows N-terminal heterogeneity, with minor species beginning at val_{-3}, ile_{-6}, and glu_{11} (Haass et al 1992b) (Figure 2). The p3 peptide begins primarily at the α-secretase cleavage site (leu_{17}) or one residue later (val_{18}) (Figure 2).

Aβ is also present in human and rodent cerebrospinal fluid (CSF) (Seubert et al 1992; Shoji et al 1992). Combined N-terminal sequencing and mass spectrometry confirm that Aβ peptides ending at residues 40 and 42 occur in CSF (Vigo-Pelfrey et al 1993) and culture media (Dovey et al 1993). Aβ immunoreactivity has also been detected in human serum, but its molecular form has not been definitively identified (Seubert et al 1992). A wide range of βPP-expressing cells examined to date show varying levels of Aβ in conditioned media but not in lysates (Busciglio et al 1993; Haass et al 1992b; Seubert et al 1992; Shoji et al 1992) [except for one report of Aβ in lysates; (Wertkin et al 1993)]. Thus Aβ is constitutively secreted during normal cellular metabolism. Importantly, the detection of Aβ peptides in normal human CSF, which are identical to those secreted by cells in culture, indicates that studies of the processing of βPP into Aβ conducted in vitro are fully relevant to such processing in vivo.

GENETIC EVIDENCE IMPLICATES Aβ IN THE PATHOGENESIS OF FAMILIAL ALZHEIMER'S DISEASE

Molecular biological and protein chemical studies of βPP have been paralleled by substantial progress in elucidating the genetics of familial (autosomal dominant) AD. The first genetic alteration to be associated with the AD phenotype was trisomy 21. It has been known for several decades that patients with Down's syndrome invariably develop the full histological lesions of Alzheimer's disease by the sixth decade of life. Neuropathological studies of Down's subjects dying in their teens and twenties reveal the presence of some or many diffuse Aβ plaques in the absence of surrounding neuritic and glial pathology and without neurofibrillary tangles (Giaccone et al 1989; Mann & Esiri 1989; Motte & Williams 1989). This early deposition of Aβ, which appears to substantially precede any evidence of mental or behavioral deterioration in these already retarded patients, is believed to derive from the enhanced transcription of the βPP gene at a level higher than that expected from the increased gene dosage alone (Neve et al 1988). However, the precise mechanism of the premature deposition of Aβ and subsequent development of AD lesions in Down's syndrome is not understood, and the possible involvement of other genes on chromosome 21 cannot be excluded.

Direct evidence that alterations of βPP can cause AD came from the detection of several distinct missense mutations in exons 16 and 17, which encode the Aβ region (Figure 3). The first mutation to be found occurs in families with hereditary cerebral hemorrhage with amyloidosis of the Dutch type (HCHWA-D), a rare syndrome of recurrent and ultimately fatal cerebral hemorrhages caused by severe β-amyloid deposition in meningeal and cerebral microvessels (Levy et al 1990; van Broeckhoven et al 1990). The histopathol-

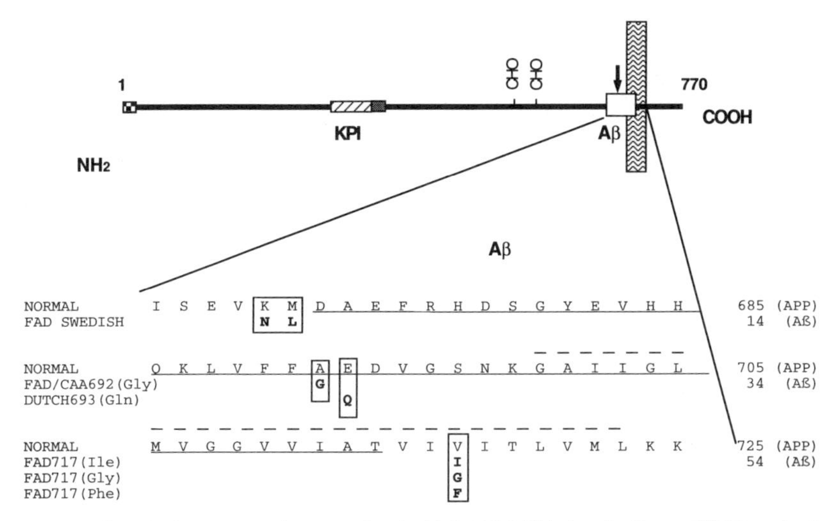

Figure 3 βAPP missense mutations associated with familial Alzheimer's disease (FAD) or closely related β-amyloidotic disorders. The sequence of βAPP containing the Aβ and transmembrane region is expanded and shown by the single-letter amino acid code. The underlined residues represent the $A\beta_{1-43}$ peptide. The broken line indicates the location of the transmembrane domain. The boxed residues depict the currently known missense mutations identified in patients with FAD or hereditary cerebral hemorrhage with amyloidosis of the Dutch-type (Dutch). FAD/CAA indicates that the family with the 692 ala → gly mutation contains individuals with the congophilic amyloid angiopathy phenotype and/or the FAD phenotype. Three-digit numbers refer to the residue number according to the βAPP_{770} isoform. Two-digit numbers refer to the residue number according to the Aβ sequence (asp_{672}=1).

ogy of HCHWA-D is closely related to that of AD, in that the former patients have large numbers of diffuse Aβ plaques in cerebral cortex in addition to the heavy vascular deposition. These plaques show little or no surrounding neuritic and glial alteration, and the patients have no progressive cortical dementia, but rather experience abrupt neurological symptoms from their cerebral hemorrhages. It is not clear why diffuse, but not neuritic, plaques accumulate in the cortices of these middle-aged patients, but it should be noted that diffuse plaques far outnumber neuritic plaques in the brains of normal elderly humans and even AD patients.

Shortly after the identification of the βPP_{693} glu → gln mutation in HCHWA-D, a val → ile mutation at βPP_{717} was found in affected members of two familial Alzheimer's disease (FAD) families (Goate et al 1991). Subsequently, AD families with substitutions of either val → gly or val → phe at this same codon were reported (Chartier-Harlin et al 1991; Murrell et al 1991) (Figure 3). A mutation at the residue immediately N-terminal to that causing HCHWA-D (βPP_{692} ala → gly) led to an interesting phenotype of progressive

AD and/or severe amyloid angiopathy causing multiple cerebral hemorrhages within the same pedigree, which supports the hypothesis that HCHWA-D and AD are pathological variants of the same disease. A double mutation (βPP_{670} lys → asn/ βPP_{671} met → leu) found in a large Swedish pedigree has aroused particular interest because it is located immediately N-terminal to the start site of Aβ (asp_{672}) (see Figure 1, 3). Expression of this mutant in transfected cells leads to a five to eightfold increase in secretion of Aβ starting at asp_1 (Cai et al 1993; Citron et al 1992). Moreover, primary cells (skin fibroblasts) cultured from members of this kindred confirm a severalfold increase in Aβ secretion in gene carriers; one of these individuals is more than a decade younger than the mean age of onset of symptoms in this family (Citron et al 1994). The finding that cells from a phenotypically unaffected peripheral tissue (skin) show elevation of Aβ production many years prior to onset of neurological symptoms strongly supports the concept that excess Aβ can arise prior to, or in the absence of, local cellular pathology rather than as a secondary effect. Mechanistically, it is assumed that the Swedish mutant βPP provides a highly favorable substrate (a leu-asp bond) for "β-secretase," the unidentified protease that generates the N-terminus of Aβ. Indeed, a ~ 12–kDa CTF of βPP that starts at or near this site is elevated in the lysates of some cells expressing the Swedish mutant βPP (Cai et al 1993).

Although such βPP missense mutations have been detected in only a handful of FAD families to date, their importance for elucidating the early pathogenesis of AD, in which severe Aβ deposition is a constant feature, is considerable. The various mutations have been strongly genetically linked to the AD phenotype in the respective families. Therefore, these βPP mutations represent the first specific molecular cause of Alzheimer's disease to be discovered.

Even prior to the detection of the βPP mutations, it was apparent that numerous families with early onset AD showed no linkage to markers on chromosome 21 (Schellenberg et al 1988; St George-Hyslop et al 1990). A sizable fraction of these early onset pedigrees have now been linked to markers on the long arm of chromosome 14 (Mullan et al 1992b; Schellenberg et al 1992; St. George-Hyslop et al 1992; Van Broeckhoven et al 1992). The identification of the responsible gene is actively being sought. These families show a severe neuropathological phenotype, including abundant Aβ plaques, and their onset of clinical disease is usually approximately a decade earlier than the families having βPP mutations (Mullan et al 1993).

The most recent chromosomal locus to be associated with AD is on the long arm of chromosome 19. Pericak-Vance and colleagues (Pericak-Vance et al 1991) had originally demonstrated an apparent linkage to chromosome 19 in families with late (> 65 years) clinical onset, but it was difficult to define this locus further because of the limitations of performing linkage analysis in very late onset diseases. Subsequently, apolipoprotein E was detected immunocytochemically in plaques and tangles in AD brain (Namba et al 1991), and synthetic Aβ immobilized on filters was found to bind ApoE from human CSF in in vitro experiments (Strittmatter et al 1993). These observations led to the discovery of allelic segregation of the *ApoE* gene, which is located on chromosome 19q, with both late onset familial and sporadic AD (Corder et al 1993; Saunders et al 1993; Strittmatter et al 1993). Numerous studies have confirmed the association of the ε4 allele of the *ApoE* gene with AD in Caucasian populations. Humans homozygous for ε4 have been reported to have as much as an eightfold elevated risk of developing AD after age 70 compared to subjects homozygous for ε3 (the much more common allele in humans) (Corder et al 1993). Apoε3/ε4 heterozygotes are found to have an intermediate elevation of risk (Corder et al 1993). It should be emphasized that ApoE4 is felt to be a risk factor for the development of AD rather than a causative trait, because individuals with one or two ε4 alleles do not necessarily develop AD and, conversely, individuals with no ε4 alleles frequently do. The biological mechanism by which the ApoE4 protein confers a greater likelihood of developing AD is under study. In this regard, it has been reported that subjects harboring one or two ε4 alleles have a significantly higher number of amyloid plaques and amyloid-laden blood vessels than those with no ε4 alleles (Rebeck et al 1993; Schmechel et al 1993). Recently, it was shown that individuals with an ApoE2 allele appear to have a reduced risk of developing AD, a finding that supports the concept that *ApoE* is the responsible gene on chromosome 19q that is implicated in the disease (Corder et al 1994).

In addition to the AD families already associated with one of the two known causative chromosomal loci (14 and 21), other FAD pedigrees show no linkage to markers on these chromosomes. Therefore, autosomal dominant AD is distinctly heterogeneous, and additional chromosomal loci will likely be found. The genetically diverse forms of the disease, however, show a highly similar neuropathological phenotype; specific histological differences have not been described other than the accentuation of Aβ deposition in patients harboring one or two ε4 alleles.

SEVERAL POTENTIAL FUNCTIONS HAVE BEEN ASCRIBED TO βPP

Experiments conducted in vitro suggest a variety of possible functional activities for the βPP holoprotein and its secreted derivatives. One putative function that became apparent as soon as alternate transcripts of βPP were cloned (Kitaguchi et al 1988; Ponte et al 1988; Tanzi et al 1988) was the role of the KPI insert in inhibiting certain serine proteases. Secreted forms of βPP_s expressing the KPI exon [referred to as protease nexin 2 (PN2)] (Oltersdorf et al 1989; van Nostrand & Cunningham 1987) appear to inhibit several different serine proteases in vitro (Oltersdorf et al 1989; Sinha et al 1990; Van Nostrand et al 1989). A previously identified serine protease inhibitor involved in coagulation, factor XIa inhibitor, appears to be identical to PN2 (Smith et al 1990), a finding that suggests one physiological role for the major secreted βPP derivative. Because PN2 is stored in the α-granules of platelets and released upon platelet activation, it has been postulated that this secreted derivative may be involved in the repair process of vascular injury and in wound healing (Smith et al 1990; Van Nostrand et al 1990).

Several laboratories have provided evidence for the function of βPP_s as an extracellular matrix molecule that can mediate the adhesion of neural or non-neural cells (Breen et al 1991; Chen & Yankner 1991; Klier et al 1990; Milward et al 1992; Schubert et al 1989). The fact that endothelial cells express βPP and polarized epithelial cells secrete βPP_s overwhelmingly from their basolateral surfaces (Haass et al 1994; see below) suggests that such cells could serve as a source of βPP_s in the basement membrane of blood vessels. In addition to a possible role in cell adhesion, βPP_s (both KPI+ and KPI- forms) appears to have a growth-promoting activity on non-neural and neural cells (Saitoh et al 1989). The region responsible for this autocrine activity has been localized to an arg-glu-arg-met-ser sequence in the middle of the ectodomain (Ninomiya et al 1993). βPP_s also provides neuroprotective properties for cultured primary neurons, in part by lowering intracellular calcium levels (Mattson et al 1993). This activity has been localized to the C-terminal portion of the ectodomain (Mattson et al 1993).

The activities just summarized relate to soluble βPP_s, the principal secreted derivative. Recent evidence indicates that the holoprotein expressed at the cell surface may confer neuronal attachment and neurite-promoting activities (Qiu et al 1994). In this work, the adhesion and neurite elongation of rat hippocampal neurons co-cultured on βPP-expressing non-neural cells could be attributed to the surface protein and not to secreted derivatives in the medium. The responsible region was localized to the C-terminal portion of the ectodomain, apparently distinct from the arg-glu-arg-met-ser motif that mediates the autocrine effects of βPP_s (Qiu et al 1994).

The realization that Aβ is also a constitutively secreted product of βPP metabolism has raised the question of whether it may have a specific function throughout life. Alternatively, it may simply represent a semi-stable proteolytic intermediate during βPP catabolism, one that could accumulate in part because of its highly hydrophobic sequence. Possible physiological activities of Aβ have been detected during studies attempting to model its neurotoxicity in culture. Soluble Aβ synthetic peptides exert some trophic activity on freshly cultured rat neurons (Whitson et al 1989; Yankner et al 1990), and neurite-promoting activity has been detected in short-term cultures (Whitson et al 1990). The effects of Aβ have also been examined by presenting it to neurons as an insoluble matrix molecule associated with conventional neuronal substrates such as laminin or fibronectin; in this paradigm, Aβ had a supportive effect on neuronal attachment and neurite outgrowth (Koo et al 1993). The neurotoxic effects of Aβ have, in general, been modeled by using relatively long exposures to high doses of synthetic peptides that have been allowed to aggregate in vitro prior to their addition to culture media (see e.g. Pike et al 1993).

CELL BIOLOGICAL STUDIES OF THE TRAFFICKING AND SECRETORY PROCESSING OF βPP

Understanding the details of the intracellular processing of βPP has become a subject of intense scrutiny. This interest arises not only from a desire to know precisely where and how the Aβ fragment is generated and released, but also from the realization that this ubiquitously expressed membrane protein has a number of intriguing and unusual characteristics. These include (*a*) βPP functions as an apparent cell-surface receptor and a secreted derivative that acts upon other cells; (*b*) proteolysis of βPP leads to regulated release of several soluble derivatives, including hydrophobic peptides that include much of the transmembrane domain; (*c*) βPP is phosphorylated solely on its ectodomain, which leads to the secretion of phosphorylated βPP_s (Hung & Selkoe 1994); (*d*) the large βPP ectodomain contains diverse structural motifs that give rise to several putative functions of the molecule; and (*e*) βPP is a member of a conserved gene family, including homologues in *Drosophila* and *C. elegans.* In view of these interesting properties, elucidating how βPP is trafficked into various subcellular compartments and then processed into several secretory fragments that may have distinct functional activities should provide answers about the biology of other membrane/secretory proteins.

After its synthesis, βPP is translocated into the ER via its signal peptide and then undergoes maturation by the addition of first N-linked and then O-linked sugars in the Golgi (Oltersdorf et al 1990; Weidemann et al 1989). Treatment of βPP-expressing cells with brefeldin A (10 μM), which leads to resorption of Golgi into ER and blockage of transport through the central vacuolar system, produces an intermediate glycosylated species that migrates on electrophoretic gels between the usual positions of the N- and O-glycosylated isoforms (Haass et al 1993). This failure of normal maturation is accompanied by a marked inhibition of production of Aβ and the smaller p3 peptide. Treatment of cells with another non-specific toxin, the monovalent ionophore monensin (10 μM), also results in an intermediate glycosylated holoprotein and a sharp reduction in formation of Aβ and p3 (Haass et al 1993). Treatment with other agents that non-specifically alter vesicular pH, e.g. ammonium chloride (5 mM) or chloroquine, likewise inhibits production of Aβ and p3. These findings suggest that production of the Aβ fragment follows the maturation of the precursor through the Golgi and requires an acidic intracellular vesicle. However, no Aβ can be detected in isolated late endosome/lysosome fractions, which suggests that it is not the lysosome itself that is the principal site of Aβ formation (Haass et al 1993). Although production of βPP_s is also inhibited by agents that block the full maturation of βPP (e.g. brefeldin A, monensin), ammonium chloride and chloroquine have only modest and variable effects on βPP_s secretion.

Pulse-chase experiments have shown that βPP molecules are rapidly turned over by secretory cleavage. The half-life of the mature O-glycosylated holoprotein, which is the substrate for α-secretase, is in the range of 30–60 min, and βPP_s is initially detected in the medium after 5–10 min in several cell lines expressing endogenous or transfected βPP (Haass et al 1993; Oltersdorf et al 1990; Weidemann et al 1989) In such experiments, Aβ and p3 appear in the medium simultaneously, but at a rate considerably slower than that of βPP_s secretion (Haass et al 1993). The relative levels of Aβ and p3 in medium remain constant during a prolonged chase, thus suggesting that there is no precursor/product relationship between these two peptides. Instead, p3 appears to be derived from the ~ 10–kDa CTF, which arises when α-secretase cleavage of βPP releases βPP_s, whereas Aβ must arise from alternative proteolytic cleavages of precursor molecules that do not undergo α-secretase processing. Prolonged pulse-chase experiments reveal that the Aβ and p3 peptides are far more stable than the βPP_s derivative, in that they have a half-life in βPP-transfected 293 cells that may exceed 12–18 h (C Haass & D Selkoe, unpublished data). In general, the levels of p3 in the medium parallel the levels of βPP_s when secretory processing is modulated by a variety of factors (e.g. activation of protein kinase C) (see below). Thus the amount of p3 in the medium can serve as an indicator of the level of α-secretory cleavage under many circumstances.

Although the enzymes that cleave βPP have not been definitively identified, the properties of the substrate that are required for these cleavages to occur have been defined in part. As mentioned previously, cleavage by α-secretase requires βPP to be membrane inserted and occurs following its full maturation, which leads to the postulate that the cleavage occurs at the cell surface by a membrane-anchored protease (Sisodia 1992). Direct evidence that cell surface βPP can serve as the immediate precursor of βPP_s has been obtained by labeling surface molecules with biotin or sulfate and then detecting the labeled βPP_s in medium (Haass et al 1992a; Sisodia 1992). Quantitation of βPP_s release from surface molecules has been undertaken in at least one cell type, namely βPP-transfected CHO cells (Koo et al 1993). When monoclonal antibodies or Fab fragments are used to label surface βPP in living cells, approximately 30–50% of the precursor is converted to βPP_s during the first 30 min following labeling; another 40–60% undergoes rapid reinternalization into the cell. Although surface βPP is a principal source of secreted βPP_s, evidence indicates that the α-secretase cleavage can also occur intracellularly (De Strooper et al 1993; Sambamurti et al 1992; C Haass & D Selkoe, unpublished data). Detailed analyses of the relative amounts of intracellular vs cell-surface processing by α-secretase in various cell types await identification of the enzyme(s) and its localization within the cell.

Information about the characteristics of the β-secretase processing event has begun to emerge. A truncated βPP_s species that ends at met_{670}, i.e. just prior to the start of Aβ, has been detected immunochemically in the medium of βPP-transfected cells (Seubert et al 1993) (Figure 1). This finding, coupled with the detection in cell lysates of an ~ 12–kDa CTF beginning at or near asp_{671} (Figure 1) under some experimental circumstances (Cai et al 1993; Golde et al 1992), suggests that alternative secretory processing of the holoprotein could serve as an initiating step in Aβ formation. The amount of the truncated βPP_s species appears to be much less than that of conventional βPP_s in the cell types examined to date. As in the case of α-secretase processing, cell-surface βPP can serve as a substrate for the β-secretase cleavage (P Seubert, personal communication). This finding suggests that surface-inserted βPP could be one source of Aβ. Given the evidence reviewed above that agents that alter intravesicular pH block Aβ formation, we have postulated that one site for Aβ generation may be within early endosomes following reinternalization of full-length βPP or an ~ 12–kDa CTF from the cell surface (Haass & Selkoe 1993). Such a mechanism has now been demonstrated directly by labeling of surface βPP in living cells with radioiodine followed by incubation at 37°C to allow internalization and processing (Koo & Squazzo 1994). Radioiodinated Aβ was released into the medium. The use of potassium depletion to inhibit internalization led to a marked decrease in the production of iodinated Aβ in pulse-chase experiments; restoration of normal potassium levels restored Aβ secretion (Koo & Squazzo 1994). If the YENPTY sequence in the cytoplasmic tail was deleted, Aβ release was markedly decreased (Koo & Squazzo 1994), consistent with a similar decrease observed previously when the entire cytoplasmic domain was deleted (Haass et al 1993). These experiments provide the first demonstration of a specific pathway for Aβ generation: internalization of surface βPP into early endosomes followed by excision and release of the Aβ peptide.

That endosomes containing βPP (and fragments thereof) can rapidly recycle to the cell surface has been demonstrated directly by labeling of the surface precursor with monoclonal antibodies at 4°C, warming briefly to initiate internalization, then acid-stripping the residual surface label and incubating at 37°C for varying times in the presence of a fluorescent secondary antibody. In this paradigm, any internalized βPP tagged with primary antibody that recycles to the cell surface will pick up secondary antibody and reinternalize, and this can then be demonstrated by fixation and fluorescent immunocytochemistry. These experiments showed that βPP is internalized and rapidly recycled to the cell surface within 5–10 min (Yamazaki et al 1993). A portion of the internalized βPP moved progressively from smaller vesicles near the cell surface to larger vesicles localized immediately perinuclear, which suggests that some internalized molecules traffic to lysosomes, as expected. Repetition of these studies with cells expressing a derivative of βPP lacking virtually the entire cytoplasmic domain showed far less βPP internalization and recycling (Yamazaki et al 1993).

With regard to the substrate requirements for β-secretase cleavage, mutagenesis around the N-terminus of Aβ suggests a high sequence specificity of the enzyme(s) (Citron et al 1993). Besides the methionine at βPP_{670} that occurs in wild-type molecules, only the leu substitution of this site, which occurs in the Swedish FAD kindred, allowed proper cleavage and Aβ generation. All other substitutions at this position examined to date led to little or no detectable Aβ. Similarly, a variety of substitutions at asp_{671} resulted in a loss of the 4–kDa Aβ peptide and variable production of an intermediate 3.5–kDa species, which begins at glu_{11} (Citron et al 1993). Mutations at βPP_{669} or $_{672}$ also interfered markedly with Aβ generation. Deletion of four to five residues between asp_{671} and the membrane did not significantly alter β-secretase cleavage, which still occurred at the met-asp bond. Therefore, in contrast to α-secretase, β-secretase does not appear to cleave at a specified distance from the membrane, at least in the cell types studied (Citron et al 1993). Importantly, cleavage by β-secretase and generation of Aβ required βPP to be membrane-inserted. When stop codons were introduced either 40 or 51 residues after the Aβ start site, only the latter truncated molecule underwent membrane insertion and could generate Aβ; the former showed no membrane insertion and no Aβ or p3 production (Citron et al 1993). These experiments suggest that β-secretase is a highly sequence-specific enzyme(s) with properties distinct from α-secretase and that it may itself be membrane-associated. The N-terminal heterogeneity of Aβ species seen in vitro (Haass et al 1992b; Haass et al 1994) and in vivo (Masters et al 1985; Miller et al 1993; Vigo-Pelfrey et al 1993) probably represents cleavages by distinct but related proteases because the mutagenesis experiments just reviewed make it unlikely that a single β-secretase enzyme could effect these varied cleavages.

The generation of the C-terminus of Aβ is an unusual event because it requires access by the protease to the hydrophobic transmembrane region. Aβ may be the first demonstrated example of a protein fragment released by such intramembranous proteolysis in normal cells. It will be interesting to determine whether peptides containing the transmembrane domains of other proteins are secreted in a similar fashion. Exactly how and when the protease(s) accesses the Aβ C-terminal region remains a mystery. What we know is that a protease(s) capable of generating this cleavage in cultured cells shows considerable nonspecificity because mutagenesis of residue 42 of the Aβ sequence to various other hydrophobic residues, deletion of residues 39–42, or insertion of four leucines between positions 39 and 40 does not substantially alter the generation of an ~ 4–kDa Aβ peptide (Selkoe et al 1993).

Considerable effort is being directed at identifying enzymes that can cleave βPP at the N- and C-termini of Aβ. A number of such studies have begun by using synthetic peptide substrates spanning these cleavage sites (see e.g. Araham et al 1992). However, the use of small synthetic peptides can lead to identification of activities that are not necessarily relevant to the cleavage of the holoprotein. The finding that truncated βPP molecules that are not membrane-anchored do not undergo β-secretase cleavage (Citron et al 1993) suggests that the correct identification of relevant proteases requires testing on full-length, membrane-associated βPP substrates. Biochemical strategies for isolating and purifying the Aβ-generating proteases must also grapple with the issue of the appropriate source of enzyme. Although postmortem human cerebral cortex would be one obvious choice, this tissue presents formidable technical obstacles. Moreover, because all βPP-expressing cells within brain (neurons, astrocytes, microglia, endothelial cells, smooth muscle cells, etc) are capable of generating Aβ, it remains unclear which cell type(s) is actually contributing to the deposition of Aβ in AD brain and thus which cell should be considered the optimal source for identification of the proteases. Once candidate proteases that cleave βPP at the correct position are identified, it still may be difficult to confirm that they are related to Aβ production in human brain because it is unclear what the topographic distribution of such an enzyme should be and what level of expression is necessary, given the apparent slow accrual of Aβ deposits over the course of decades in AD subjects. One criterion for correct identification will be the ability of the candidate enzyme to generate Aβ having the proper sequence when transfected into both neural and non-neural human cells.

βPP Trafficking in Polarized Epithelial Cells

Virtually all of the information about βPP processing described above has been generated in nonpolarized cell types. Such studies have recently been extended to analyses conducted in Madin-Darby canine kidney (MDCK) cells, which display a polarized phenotype with distinct apical and basolateral plasma membranes when grown as monolayers on polycarbonate or nitrocellulose filters (Matlin & Simons 1984; Matter et al 1992). In addition to the benefit of extensive comparative data on the polarized trafficking of other surface proteins in MDCK cells, studies of βPP processing in these epithelial cells could be relevant to the preferential deposition of the Aβ fragment in the abluminal basement membrane of endothelial cells in AD brain (Yamaguchi et al 1992).

The endogenous expression of βPP in MDCK cells is sufficiently high that initial observations about its polarized secretion could be made without the need for transfection. The conventional βPP_s generated by α-secretase cleavage was released ~ 80–90% from the basolateral surface, as were the Aβ and p3 peptides (Haass et al 1994). Stable transfection of wild-type βPP into these cells resulted in the same strong basolateral secretion of βPP_s, Aβ, and p3 as occurs for the endogenous molecule. The enhanced βPP signal produced by transfection enabled surface antibody labeling studies, which demonstrated ~ 80–90% insertion of the holoprotein on the basolateral membrane. Moreover, surface biotinylation of the transfectants revealed an overwhelming release of biotinylated βPP_s from the basolateral membrane (Haass et al 1994).

Treatment of the transfected MDCK cells with 10 mM ammonium chloride abolishes the polarized secretion of βPP_s so that it is released almost equally from apical and basolateral surfaces (Haass et al 1994). Nonetheless, surface βPP distribution still remains ~ 90% basolateral. These findings suggest that βPP_s may exist in part in intracellular acidic vesicles whose trafficking can be regulated independently of the routing of holoβPP to the cell surface. Further support for this hypothesis derives from the detection of intraluminal βPP_s in membrane vesicles purified from MDCK cells and extracted in sodium carbonate buffer (Haass et al 1994). The signals that could direct the polarized trafficking of the nonmembrane-inserted βPP_s molecule are not known. In this regard, the expression of the Swedish mutant βPP molecule in MDCK cells has led to the finding that the truncated βPP_s species (ending at leu_{670}) released abundantly from this precursor by β-secretase is principally secreted apically rather than basolaterally (S Sisodia et al, unpublished data). Similarly, the small amount of truncated βPP_s released from wild-type transfected MDCK cells is also predominantly released apically (S Sisodia et al, unpublished data). These results indicate that β-secretase-generated βPP_s traffics differently than α-secretase-generated βPP_s and raise the possibility that there is a signal in the last 16 residues of the latter derivative that helps direct it basolaterally.

Information about the signals that regulate the polarized trafficking of holoβPP has begun to emerge. A series of increasing truncations of the cytoplasmic domain, which remove the last 12, 22, or 32 residues, caused little detectable change in the polarized basolateral sorting of βPP (Haass et al 1994). However, removal of the last 42 residues, i.e. almost the entire cytoplasmic tail, led to a shift in βPP polarity: 40–50% of the molecules were now inserted on the apical membrane (Haass et al 1994). This result suggests that a signal in or near the region 729–738 of βPP_{770} participates in directing the precursor basolaterally. Indeed, mutagenesis of the tyrosine at residue 728 also altered polarization so that ~ 40% of molecules now trafficked apically. Similar results were obtained when this Y728A mutant was combined with a deletion of the last 12 residues of the cytoplasmic tail, including the NPTY motif. The fact that these particular mutants did not switch almost entirely to an apical distribution, as has been shown with several other basolateral surface receptors undergoing cytoplasmic deletion, suggests that the Y728 and surrounding residues serve as only a partial signal for basolateral sorting. Additional mutagenesis experiments will be needed to elucidate more fully the signals that mediate polarized trafficking of βPP. Importantly, in all of the MDCK studies to date, the pattern of Aβ secretion has generally followed the trafficking of the holoprotein, e.g. Aβ is ~ 40% apically released from molecules bearing the Y728A mutation (Haass et al 1994).

These various findings have led to a model for βPP trafficking in polarized epithelial cells that includes cleavage of a small population of precursor molecules by β-secretase at a relatively early point in βPP processing to generate a truncated βPP_s form that is principally secreted apically. In contrast, conventional βPP_s generated by α-secretase may arise somewhat later during βPP sorting and is directed basolaterally. The holoprotein is normally trafficked preferentially to the basolateral membrane, but alteration of signals in the N-terminal part of the cytoplasmic tail can partially redirect it to the apical surface. Aβ secretion appears to depend on the pattern of trafficking of holoβPP to the surface and thus may arise following endocytosis of the cell-surface molecule, consistent with studies in nonpolarized cells discussed above.

SECRETORY PROCESSING OF βPP IS UNDER COMPLEX REGULATION BY FIRST MESSENGERS

The secretory processing of a number of single membrane-spanning proteins including pro-TGFα (Pandiella & Massagué 1991), the TNF receptor (Lantz et al 1990), and the CSF 1 receptor (Downing et al 1989) is regulated by protein kinase C (PKC)-mediated pathways (reviewed in Ehlers & Riordan 1991). Activation of PKC by treatment of cells with phorbol esters was similarly shown to enhance α-secretase cleavage of βPP and thus to generate increased βPP_s (Buxbaum et al 1990). This increased secretory processing of βPP was postulated to result from the direct phosphorylation of serines or threonines in the cytoplasmic tail, based on in vitro phosphorylation of a synthetic peptide comprising that region (Gandy et al 1988) and on the ability of exogenous PKC to phosphorylate the βPP cytoplasmic domain in permeabilized cells (Suzuki et al 1992). Agonist stimulation of cell-surface receptors whose actions are mediated by the phospholipase C/PKC pathway, e.g. muscarinic m1 and m3 receptors, likewise leads to increased βPP_s secretion. As expected, this enhanced α-secretase activity is accompanied by a parallel decline in Aβ release (Buxbaum et al 1993; Hung et al 1993). Subsequent studies have shown that stimulation of bradykinin receptors or activation of hippocampal slices by electrical depolarization can also lead to increased βPP_s release (Nitsch et al 1993). Therefore, it is postulated that a number of first messengers that activate the phospholipase C/PKC signal transduction pathway can enhance non-amyloidogenic cleavage of βPP.

Although one postulated mechanism for these findings was a direct effect of PKC activation on βPP phosphorylation (Gandy et al 1988; Suzuki et al 1992), the basal phosphorylation state of βPP in intact cells was not known and needed to be established. Metabolic labeling of βPP-transfected 293 cells with ^{32}P orthophosphate followed by phosphoamino acid analysis demonstrated exclusive phosphorylation of serines (Hung & Selkoe 1994). There are two serines in the cytoplasmic domain, but mutagenesis of both to alanines still allowed phosphorylation of βPP. This result suggests that βPP undergoes ectodomain phosphorylation. Immunoprecipitation of $βPP_s$ from the conditioned medium of phosphate-labeled cells confirmed that βPP was phosphorylated on its ectodomain, which resulted in secretion of phosphorylated $βPP_s$ (Hung & Selkoe 1994; Knops et al 1993). The lack of phosphate incorporation into the 10-kDa CTF arising from α-secretase cleavage indicated that βPP was solely phosphorylated on ectodomain serines. Similar results were obtained in other cell types including HS683 glioma cells, CHO cells, and primary rat hippocampal neurons (Hung & Selkoe 1994).

Treatment of transfected 293 cells with phorbol ester in the presence of radiophosphate did not lead to increased phosphorylation of βPP (Hung & Selkoe 1994). Instead, the amount of phosphorylated βPP declined during a chase in PDBu, consistent with the increase in turnover of the holoprotein into $βPP_s$. The use of various deletion constructs has enabled the phospho-acceptor region to be localized to the N-terminal half of the βPP ectodomain (residues −78–305 of $βAPP_{695}$), a stretch that contains 10 serines (Hung & Selkoe 1994). The precise serine(s) involved is not defined. Phorbol ester stimulation of 293 cells transfected with a construct lacking this region still led to the expected increase in $βPP_s$ secretion, thus confirming that the PKC-mediated regulation of βPP processing did not involve direct phosphorylation of the precursor. Deletion of most of the cytoplasmic tail as well as of other portions of the ectodomain narrowed the region of βPP necessary for PKC-mediated regulation of secretion to a 64–amino acid stretch (residues 590–653 of $βPP_{695}$) surrounding the α-secretase cleavage site. On the basis of these findings, it appears that several first messenger systems acting via PKC regulate the α-secretase cleavage event indirectly, perhaps by activating α-secretase itself, or by altering the trafficking of βPP in a way that enhances its interaction with this protease (Hung & Selkoe 1994).

It is likely that the processing of βPP via amyloidogenic vs non-amyloidogenic pathways is under complex, multifactorial regulation, only one example of which is the PKC-mediated pathway. For example, increases in intracellular calcium produced by the ionophore A23187 led to increased Aβ production, which is dependent upon extracellular calcium levels (Querfurth & Selkoe

1994). The mechanism of this effect is unknown but could involve increased activity of calcium-activated protease(s). Further clarification of the factors that regulate alternative βPP proteolysis are of interest not only for understanding the physiological generation of βPP_s and Aβ but also for devising strategies to pharmacologically down-regulate Aβ production in AD.

Other topics for future study are the mechanisms and function of the selective ectodomain phosphorylation of βPP. The precursor may be a substrate for an ecto-protein kinase (A Hung & D Selkoe, unpublished data), but it could also be phosphorylated by an intraluminal kinase in an intracellular compartment. One must also determine whether both phosphorylated and non-phosphorylated species of βPP_s exist in extracellular fluid and, if so, whether these have distinct functional properties. There are a few examples of transmembrane proteins in which selective ectodomain phosphorylation modulates the ability of the protein to interact with other molecules, e.g. the CD36 surface protein in platelets, which binds to collagen in its phosphorylated form but to thrombospondin after dephosphorylation (Asch et al 1993). It will be interesting to see whether specific phosphatases regulate the functions of the ectodomain of either cell-surface βPP or secreted βPP_s in a manner similar to that described for CD36.

UNDERSTANDING THE ROLE OF βPP PROCESSING IN THE PATHOGENETIC CASCADE OF AD

Although many aspects of βPP trafficking and processing remain to be clarified, recent advances in this field allow one to postulate a general framework for the contribution of amyloidogenic βPP processing to AD (Figure 4). The identification of at least two chromosomal loci for genes that underlie autosomal dominant forms of the disease and the prospect that several additional loci will be identified indicate that AD is a syndrome that can be initiated by several distinct molecular alterations that trigger a common pathological cascade. Such a model predicts that the earliest molecular events in AD pathogenesis will differ among different genetic forms of the disease, but that these will converge to induce an alteration of the expression, intracellular processing, or degradative metabolism of βPP that ultimately results in a critical imbalance between Aβ production and clearance in cerebral tissue. This imbalance, the degree of which is likely to be regulated by numerous other gene products besides βPP, results in the gradual cerebral accumulation of Aβ peptides, a portion of which polymerizes over time into insoluble, particulate Aβ deposits (including amyloid filaments). The accrual of these deposits and their gradual association with a variety of extracellular molecules that have biological activity lead to subtle and then more profound neuronal, synaptic, and glial cytopathology that ultimately produces the symptoms of progressive dementia.

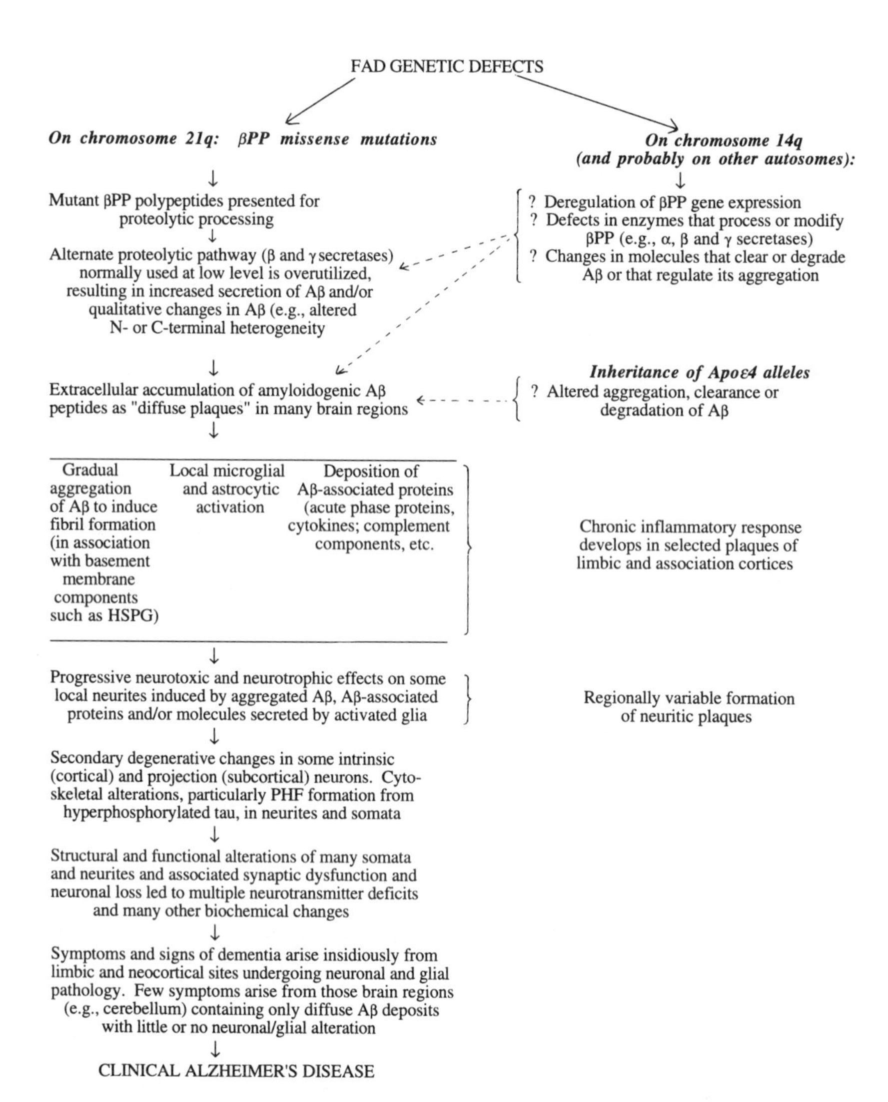

Figure 4 Hypothetical model of the pathogenesis of familial forms of Alzheimer's disease based on currently available information.

Perhaps the genetic form of AD in which such a mechanism is most readily explicable at present is the βPP missense mutation located directly prior to the start site of Aβ (K670N/M671L). This double mutation segregates with the disease phenotype in an extended Swedish pedigree (Mullan et al 1992b) that shows onset of clinical dementia at age ~ 50–55 years and a typical neuropathological phenotype. Studies summarized above have shown (*a*) that expression of this βPP mutation in transfected cells leads to a marked increase in secretion of soluble Aβ; (*b*) that the second substitution (M671L) immediately N-terminal to the β-secretase cleavage site is principally responsible for this elevation; and (*c*) that the enhanced Aβ production can be detected in primary peripheral cells from subjects bearing the mutation many years prior to development of the clinical syndrome. (It will be of interest to learn whether soluble Aβ levels are elevated in the CSF and serum of presymptomatic patients with this mutation.) Therefore, one may speculate that the severalfold increase in Aβ secretion from various βPP-expressing cells in the brain and its vasculature exceeds the ability to degrade and clear the peptide (the mechanisms for which are not yet defined). The subsequent gradual accumulation of Aβ in cerebral tissue and microvessels, perhaps including its binding to and stabilization on basement membrane molecules such as heparan sulfate proteoglycan (Snow et al 1994; Yamaguchi et al 1992), results in the formation of amorphous Aβ deposits (diffuse plaques). Indeed, Snow et al (1990) have shown that diffuse plaques in the cerebral cortex contain heparan sulfate proteoglycan. It is likely that such diffuse plaques can slowly accrue increasing amounts of Aβ, including some Aβ peptide aggregated in polymeric form, which gradually leads to denser, more fibrillar, partially compacted deposits. The latter type of plaque is intimately associated with dystrophic neurites, both within and immediately surrounding the amyloid deposit (Masliah et al 1993).

Another form of AD in which the initiating mechanism appears explicable in part is that which occurs in trisomy 21. Here, increased transcription of βPP in the brain has been demonstrated (see e.g. Neve et al 1988), and neuropathological studies have consistently shown diffuse, non-neuritic deposits of Aβ in limbic and cerebral cortices as the earliest detectable histopathological alteration, sometimes appearing already in childhood (see e.g. Giaccone et al 1989; Mann et al 1992; Royston et al 1994). Whether other gene products encoded on the trisomic portion of chromosome 21 also contribute to the development of AD neuropathology is not known. However, it is clear that diffuse Aβ deposits precede the development of the full-blown neuritic plaques and neurofibrillary tangles of AD in Down's patients by many years. This observation, plus the fact that missense mutations clustered solely in the Aβ region of βPP can cause AD, proves that Aβ deposition does not necessarily require preexisting neuronal injury but rather can precede it. Further, it is noteworthy that humans in their 60's and 70's and monkeys at analogous ages can develop some or many Aβ deposits with the characteristics of diffuse plaques in the virtual absence of the neuronal and glial lesions of AD. This finding again indicates that Aβ accumulation can precede other changes of AD rather than arise secondarily to them. The common observation of diffuse

Aβ deposits in healthy elderly primates, including humans, in the absence of cellular pathology becomes understandable in light of the realization that Aβ is a normal soluble product of cellular metabolism throughout life. A change in the normal balance between production and clearance of Aβ could thus lead to its multifocal cerebral deposition, in a manner analogous to the formation of fatty streaks of cholesterol in arterial walls early in the course of athersclerosis (for review, see Ross 1993).

Yet another genetic basis for AD in which the initial mechanism is partly understood is represented by the missense mutations at codon 717 of βPP_{770}. Early studies of the V717I mutation in transfected cells suggested no major increase in total Aβ secretion (Cai et al 1993). Subsequently, evidence has emerged that mutations at this codon lead to a modest but significant increase in the proportion of Aβ molecules that end at residue 42 rather than at residue 40 (Cai et al 1993; Tamaoka et al 1994). The presence of two additional hydrophobic residues at the C-terminus of Aβ1–42 increases its fibrillogenic potential, in part, by seeding the aggregation of the more common Aβ1–40 peptide and perhaps other shorter peptides (Jarrett et al 1993). It has, therefore, been hypothesized (but not rigorously proven) that the longer Aβ species resulting from the 717 mutations can tip the balance between Aβ production and clearance towards accumulation of insoluble, aggregated Aβ, which is more resistant to degradation and can form the nidus of the diffuse plaque. The concept that diffuse plaques are associated with little local cytopathology is supported by ultrastructural observations (Giaccone et al 1990; Yamaguchi et al 1990) and by the use of quantitative synaptophysin immunocytochemistry to demonstrate that diffuse plaques have synaptic densities indistinguishable from the surrounding Aβ-free cortical neuropil, whereas compacted, neuritic plaques have significant decreases in synaptic density (Masliah et al 1993; Terry et al 1991).

Beyond the examples of the βPP missense mutations and trisomy 21, the earliest pathogenetic events underlying other forms of the AD syndrome remain unknown. It is anticipated that the next form of the disease to be deciphered at the molecular level will be that linked to an unknown gene on chromosome 14q. That the defective gene product is likely to relate in some way—either directly or indirectly—to amyloidogenesis is suggested by the fact that the chromosome 14–linked families show a severe β-amyloidotic histological phenotype that is essentially indistinguishable from that in the βPP-linked cases, but that has a clinical onset about a decade earlier than the latter (Mullan et al 1993). Quantitative studies of βPP expression and metabolism in cultured skin fibroblasts from one chromosome 14–linked family (a Canadian pedigree) suggest an increase in βPP transcription and an associated increase in Aβ production in affected vs unaffected members of the pedigree (H Querfurth et al, submitted). This result raises the possibility that the implicated gene on chromosome 14 is involved in some aspect of the regulation of βPP transcription. Regarding the role of ApoE alleles as a risk factor for the development of late onset AD, a biochemical mechanism has not been established. Neuropathological observations that ApoE immunoreactivity is associated with senile plaques (as are many other proteins) and that the number and density of Aβ plaques is increased in AD patients with one or two ApoE4 alleles (Rebeck et al 1993; Schmechel et al 1993) suggest that ApoE could be involved in some aspect of Aβ transport, clearance, or deposition.

CONCLUSION

The scrutiny to which the cell biology of βPP is currently being subjected bodes well for identifying additional ways in which an imbalance between Aβ production and clearance could occur. Genetically determined alterations in (*a*) the intracellular trafficking of βPP; (*b*) its alternative processing by α- vs β-secretase (each of which may well represent a group of proteases); (*c*) the microheterogeneity of the Aβ peptides; (*d*) the proteolytic systems capable of degrading Aβ once it is secreted; (*e*) the kinetics of Aβ fibrillogenesis; and (*f*) the binding of one of the numerous proteins that associate strongly with Aβ deposits could all be found to underlie different variants of the AD clinico-pathological syndrome. This potentially rich array of molecular defects capable of initiating or accelerating the amyloidogenic cascade (Figure 4) could provide several points for pharmacological intervention in the disease process. As was emphasized above, Aβ deposition appears to be a necessary but not sufficient factor for the development of the AD syndrome. This realization leads to the conclusion that there will also be opportunities for intervening in one or more of the many complex neuronal, astrocytic, and microglial alterations that accompany and follow cerebral Aβ deposition. This multicellular response—which includes the neuronal accumulation of hyperphosphorylated tau protein that forms the principal subunit of the PHF comprising neurofibrillary tangles—is likely to represent the basis for neuronal dysfunction and consequent cognitive impairment in AD. The biochemical and structural characteristics of the profound neuronal cytoskeletal alteration that characterize AD have been recently reviewed (e.g. Lee & Trojanowski 1992), as has the glial activation and inflammatory response occurring in AD cortex (Rogers et al 1992).

Evolving knowledge of the complex pathogenetic cascade of AD suggests a number of possible approaches to prevention or treatment, some of which are already being pursued. A partial list of possible pharmacological targets is provided in Table 1. Among these, inhibition of the proteases that generate Aβ, retardation of the conversion of soluble, monomeric Aβ to insoluble, aggregated Aβ, and interference with the toxic response of neurons and their processes to Aβ

Table 1 Potential therapeutic strategies for preventing or slowing the pathogenetic mechanism of AD

Decrease Aβ production by partial inhibition of the β-secretase and/or γ-secretase proteases
Divert βPP metabolism from amyloidogenic to non-amyloidogenic processing pathways (e.g. by using first messengers that enhance the percentage of βPP molecules cleaved by α-secretase)
Increase the activity of extracellular proteases capable of degrading Aβ and related peptides
Retard the aggregation of soluble Aβ monomers into insoluble, polymeric fibrils
Block the selective neuronal toxicity exerted by aggregated Aβ (e.g. by modulating associated increases in intraneuronal calcium)
Block the activation of microglia and proliferation of astrocytes associated with mature Aβ plaques
Use CNS-specific anti-inflammatory agents to inhibit the cytokine release, complement activation and acute phase protein response that occurs around mature Aβ plaques
Control the imbalance between neuronal kinase and phosphatase activities that results in the accumulation of hyperphosphorylated forms of tau protein and subsequent tangle formation
Administer specific trophic factors that could retard the neuronal loss associated with the development of plaques and tangles

or closely associated proteins are among the strategies most actively explored at present. Numerous other approaches including inhibition of the peri-plaque inflammatory response and interference with altered tau phosphorylation are likely to receive increasing attention in the future. This multipronged attack on the disease process envisions a future in which several distinct pharmacological therapies will be developed to address different stages in its evolution as well as some of the distinct molecular causes of the AD syndrome. If therapeutics are developed that effectively and safely influence the balance between Aβ production and clearance in a way that reduces the cerebral Aβ burden, they have the potential to be applied to slowing the gradual accumulation of Aβ deposits that accompanies normal aging of the human brain.

Literature Cited

Abraham CR, Razzaboni BL, Papastoitsis G, Picard E, Kanemura K, et al. 1992. Purification and cloning of brain proteases capable of degrading the β-amyloid precursor protein. *Ann. NY Acad. Sci.* 674:174–79

Asch AS, Liu I, Bricetti FM, Barnwell JW, Kwakye-Berko F, et al. 1993. Analysis of CD36 binding domains: ligand specificity controlled by dephosphorylation of an ectodomain. *Science* 262:1436–40

Benowitz LI, Rodriguez W, Paskevich P, Muson EJ, Schenk D, Neve RL. 1989. The amyloid precursor protein is concentrated in neuronal lysosomes in normal and Alzheimer disease subjects. *Exp. Neurol* 106:237–50

Breen KC, Bruce M, Anderton BH. 1991. Beta amyloid precursor protein mediates neuronal cell-cell and cell-surface adhesion. *J. Neurosci. Res.* 28:90–100

Busciglio J, Gabuzda DH, Matsudaira P, Yankner BA. 1993. Generation of β-amyloid in the secretory pathway in neuronal and non-neuronal cells. *Proc. Natl. Acad. Sci. USA* 90:2092–96

Buxbaum JD, Gandy SE, Cicchetti P, Ehrlich ME, Czernik AJ, et al. 1990. Processing of Alzheimer β/A4 amyloid precursor protein: Modulation by agents that regulate protein phosphorylation. *Proc. Natl. Acad. Sci. USA* 87:6003–6

Buxbaum JD, Koo EH, Greengard P. 1993. Protein phosphorylation inhibits production of Alzheimer amyloid β/A4 peptide. *Proc. Natl. Acad. Sci. USA* 90:9195–98

Cai X-D, Cheung TT, Younkin SG. 1993. Comparison of the Aβ released from normal amyloid β protein precursor (βAPP) and from mutant $βAPP_S$ linked to familial Alzheimer's disease. *Soc. Neurosci. Abstr.* 19(1):430

Cai X-D, Golde TE, Younkin GS. 1993. Release of excess amyloid β protein from a mutant amyloid β protein precursor. *Science* 259:514–16

Caporaso GL, Gandy SE, Buxbaum JD, Greengard P. 1992. Chloroquine inhibits intracellular degradation but not secretion of Alzheimer β/A4 amyloid precursor protein. *Proc. Natl. Acad. Sci. USA* 89:2252–56

Chartier-Harlin M-C, Crawford F, Houlden H. 1991. Early-onset Alzheimer's disease caused by mutations at codon 717 of the β-amyloid precursor protein gene. *Nature* 353: 844–46

Chen M, Yankner BA. 1991. An antibody to β-amyloid and the amyloid precursor protein inhibits cell-substratum adhesion in many mammalian cell types. *Neurosci. Lett.* 125: 223–26

Chen W-J, Goldstein JL, Brown MS. 1990. NPXY, a sequence often found in cytoplasmic tails, is required for coated-pit mediated internalization of the low density lipoprotein receptor. *J. Biol. Chem.* 265:3116–23

Citron M, Oltersdorf T, Haass C, McConlogue L, Hung AY, et al. 1992. Mutation of the β-amyloid precursor protein in familial Alzheimer's disease increases β-protein production. *Nature* 360:672–74

Citron M, Teplow DB, Schmitt FO, Selkoe DJ. 1993. The N-terminus of amyloid β-peptide is cleaved by a sequence specific protease. *Soc. Neurosci. Abstr.* 19(1):18

Citron M, Vigo-Pelfrey C, Teplow DB, Miller C, Schenk D, et al. 1994. Excessive production of amyloid β-protein by peripheral cells of symptomatic and presymptomatic patients carrying the Swedish familial Alzheimer's disease mutation. *Proc. Natl. Acad. Sci. USA.* Submitted

Cole GM, Huynh TV, Saitoh T. 1989. Evidence for lysosomal processing of beta-amyloid precursor in cultured cells. *Neurochem. Res.* 14:933–39

Corder EH, Saunders AM, Risch NJ, Strittmatter WJ, Schmechel DE, et al. 1994. Protective effect of apolipoprotein E type 2 allele for late onset Alzheimer's disease. *Nature Genet.* 7:180–84

Corder EH, Saunders AM, Strittmatter WJ, Schmechel DE, Gaskell PC, et al. 1993. Gene dose of apolipoprotein E type 4 allele and the risk of Alzheimer's disease in late onset families. *Science* 261:921–23

Davies CA, Mann DMA. 1993. Is the "pre-amyloid" of diffuse plaques in Alzheimer's disease really nonfibrillar? *Am. J. Pathol.* 143:1594–1605

de Sauvage F, Octave JN. 1989. A novel mRNA of the A4 amyloid precursor gene coding for a possibly secreted protein. *Science* 245:651–53

De Strooper B, Umans L, Van Leuven F, Van Den Berghe H. 1993. Study of the synthesis and secretion of normal and artificial mutants of murine amyloid precursor protein: Cleavage of APP occurs in a late compartment of the default secretion pathway. *J. Cell Biol.* 121:295–304

Dovey HF, Suomensaari-Chrysler S, Lieberburg I, Sinha S, Keim PS. 1993. Cells with a familial Alzheimer's disease mutation produce authentic β-peptide. *NeuroReport* 4: 1039–42

Downing JR, Roussel MF, Sherr CJ. 1989. Ligand and protein kinase C down modulate the colony-stimulating factor 1 receptor by independent mechanisms. *Mol. Cell. Biol.* 9: 2890–96

Ehlers MRW, Riordan JF. 1991. Membrane proteins with soluble counterparts: role of proteolysis in the release of transmembrane proteins. *Biochemistry* 30:10065–74

Esch FS, Keim PS, Beattie EC, Blacher RW, Culwell AR, et al. 1990. Cleavage of amyloid β-peptide during constitutive processing of its precursor. *Science* 248:1122–24

Estus S, Golde TE, Kunishita T, Blades D, Lowery D, et al. 1992. Potentially amyloidogenic carboxyl-terminal derivatives of the amyloid protein precursor. *Science* 255: 726–28

Gandy S, Czernik AJ, Greengard P. 1988. Phosphorylation of Alzheimer disease amyloid precursor peptide by protein kinase C and Ca^{2+}/calmodulin-dependent protein kinase II. *Proc. Natl. Acad. Sci. USA* 85:6218–21

Giaccone G, Tagliavini F, Linoli G, Bouras C, Frigerior L, et al. 1989. Down's patients:

extracellular preamyloid deposits precede neuritic degeneration and senile plaques. *Neurosci. Lett.* 97:232–38

Giaccone G, Verga L, Finazzi M, Pollo B, Tagliavini F, et al. 1990. Cerebral preamyloid deposits and congophilic angiopathy in aged dogs. *Neurosci. Lett.* 114:178–83

Glenner GG, Wong CW. 1984. Alzheimer's disease: Initial report of the purification and characterization of a novel cerebrovascular amyloid protein. *Biochem. Biophys. Res. Commun.* 120:885–90

Goate A, Chartier-Harlin M-C, Mullan M, Brown J, Crawford F, et al. 1991. Segregation of a missense mutation in the amyloid precursor protein gene with familial Alzheimer's disease. *Nature* 349:704–6

Golde TE, Estus S, Younkin LH, Selkoe DJ, Younkin SG. 1992. Processing of the amyloid protein precursor to potentially amyloidogenic carboxyl-terminal derivatives. *Science* 255:728–30

Gorevic P, Goni F, Pons-Estel B, Alvarez F, Peress R, Frangione B. 1986. Isolation and partial characterization of neurofibrillary tangles and amyloid plaque cores in Alzheimer's disease: Immunohistological studies. *J. Neuropathol. Exp. Neurol.* 45:647–64

Grundke-Iqbal I, Iqbal K, Quinlan M, Tung Y-C, Zaidi MS, Wisniewski HM. 1986. Microtubule-associated protein tau: a component of Alzheimer paired helical filaments. *J. Biol. Chem.* 261:6084–89

Haass C, Hung AY, Schlossmacher MG, Teplow DB, Selkoe DJ. 1993. β-amyloid peptide and a 3–kDa fragment are derived by distinct cellular mechanisms. *J. Biol. Chem.* 268:3021–24

Haass C, Koo EH, Mellon A, Hung AY, Selkoe DJ. 1992a. Targeting of cell-surface β-amyloid precursor protein to lysosomes: alternative processing into amyloid-bearing fragments. *Nature* 357:500–3

Haass C, Koo EH, Teplow DB, Selkoe DJ. 1994. Polarized secretion of β-amyloid precursor protein and amyloid β-peptide in MDCK cells. *Proc. Natl. Acad. Sci. USA* 91: 1564–68

Haass C, Schlossmacher MG, Hung AY, Vigo-Pelfrey C, Mellon A, et al. 1992b. Amyloid β-peptide is produced by cultured cells during normal metabolism. *Nature* 359:322–25

Haass C, Selkoe DJ. 1993. Cellular processing of β-amyloid precursor protein and the genesis of amyloid β-peptide. *Cell* 75:1039–42

Hung AY, Haass C, Nitsch RM, Qiu WQ, Citron M, et al. 1993. Activation of protein kinase C inhibits cellular production of the amyloid β-protein. *J. Biol. Chem.* 268: 22959–62

Hung AY, Selkoe DJ. 1994. Selective ectodomain phosphorylation and regulated cleavage of β-amyloid precursor protein. *EMBO J.* 13:534–42

Jarrett J, Berger EP, Lansbury PT Jr. 1993. The carboxy terminus of the beta amyloid protein is critical for the seeding of amyloid formation: Implications for the pathogenesis of Alzheimer's disease. *Biochemistry* 32:4693–97

Kang J, Lemaire H, Unterbeck A, Salbaum JM, Masters CL, et al. 1987. The precursor of Alzheimer's disease amyloid A4 protein resembles a cell-surface receptor. *Nature* 325: 733–36

Kitaguchi N, Takahashi Y, Tokushima Y, Shiojiri S, Ito H. 1988. Novel precursor of Alzheimer's disease amyloid protein shows protease inhibitory activity. *Nature* 331:530–32

Klier FG, Cole G, Stallcup W, Schubert D. 1990. Amyloid β-protein precursor is associated with extracellular matrix. *Brain Res.* 515:336–42

Knops J, Gandy S, Greengard P, Lieberburg I, Sinha S. 1993. Serine phosphorylation of the secreted extracellular domain of APP. *Biochem. Biophys. Res. Commun.* 197:380–85

Kondo J, Honda T, Mori H, Hamada Y, Miura R, et al. 1988. The carboxyl third of tau is tightly bound to paired helical filaments. *Neuron* 1:827–34

Koo EH, Haass C, Selkoe DJ. 1993. Familial Alzheimer's disease mutation within transmembrane amyloid precursor protein (APP) sequence results in altered cell surface processing. *Soc. Neurosci. Abstr.* 43:(4):422

Koo EH, Park L, Selkoe DJ. 1993. Amyloid β-protein as a substrate interacts with extracellular matrix to promote neurite outgrowth. *Proc. Natl. Acad. Sci. USA* 90:4748–52

Koo EH, Squazzo S. 1994. Evidence that production and release of amyloid β-protein involves the endocytic pathway. *J. Biol. Chem.* 269:17386–89

Kosik KS, Joachim CL, Selkoe DJ. 1986. Microtubule-associated protein, tau, is a major antigenic component of paired helical filaments in Alzheimer's disease. *Proc. Natl. Acad. Sci. USA* 83:4044–48

Lantz M, Gullberg U, Nilsson E, Olsson I. 1990. Characterization in vitro of a human tumor necrosis factor-binding protein. A soluble form of the tumor necrosis factor receptor. *J. Clin. Invest.* 86:1396–402

Lee VM, Trojanowski JQ. 1992. The disordered neuronal cytoskeleton in Alzheimer's disease. *Curr. Opin. Neurobiol.* 2:653–56

Lee VM-Y, Balin BJ, Otvos L, Trojanowski JQ. 1991. A68. A major subunit of paired helical filaments and derivatized forms of normal tau. *Science* 251:675–78

Levy E, Carman MD, Fernandez-Madrid IJ, Power MD, Lieberburg I, et al. 1990. Mutation of the Alzheimer's disease amyloid gene in hereditary cerebral hemorrhage, Dutch-type. *Science* 248:1124–26

Mann DMA, Esiri MM. 1989. The pattern of acquisition of plaques and tangles in the brains of patients under 50 years of age with Down's syndrome. *J. Neurol. Sci.* 89:169–79

Mann DMA, Yuonis N, Jones D, Stoddart RW. 1992. The time course of pathological events in Down's Syndrome with particular reference to the involvement of microglial cells and deposits of β/A4. *Neurodegeneration* 1: 201–15

Masliah E, Mallory M, Deerinck T, DeTeresa R, Lamont S, et al. 1993. Re-evaluation of the structural organization of neuritic plaques in Alzheimer's disease. *J. Neuropathol. Exp. Neurol.* 52:619–32

Masters CL, Simms G, Weinman NA, Multhaup G, McDonald BL, Beyreuther K. 1985. Amyloid plaque core protein in Alzheimer disease and Down syndrome. *Proc. Natl. Acad. Sci. USA* 82:4245–49

Matlin KS, Simons K. 1984. Sorting of an apical plasma membrane glycoprotein occurs before it reaches the cell surface in cultured epithelial cells. *J. Cell Biol.* 99:2131–39

Matter K, Hunziker W, Mellman I. 1992. Basolateral sorting of LDL receptor in MDCK cells: the cytoplasmic domain contains two tyrosine-dependent targeting determinants. *Cell* 71:741–53

Mattson M, Cheng B, Culwell A, Esch F, Lieberburg I, Rydel R. 1993. Evidence for excitoprotective and intraneuronal calcium-regulating roles for secreted forms of the β-amyloid precursor protein. *Neuron* 10: 243–54

Miller DL, Papayannopoulos IA, Styles J, Bobin SA. 1993. Peptide compositions of the cerebrovascular and senile plaque core amyloid deposits of Alzheimer's disease. *Arch. Biochem. Biophys.* 301:41–52

Milward EA, Papadopoulos R, Fuller SJ, Moir RD, Small D, et al. 1992. The amyloid protein precursor of Alzheimer's disease is a mediator of the effects of nerve growth factor on neurite outgrowth. *Neuron* 9:129–37

Mori H, Takio K, Ogawara M, Selkoe DJ. 1992. Mass spectrometry of purified amyloid β protein in Alzheimer's disease. *J. Biol. Chem.* 267:17082–86

Motte J, Williams RS. 1989. Age-related changes in the density and morphology of plaques and neurofibrillary tangles in Down syndrome brain. *Acta Neuropathol.* 77:535–46

Mullan M, Houlden H, Crawford F, Kennedy A, Rogues P, Rossor M. 1993. Age of onset in familial early onset Alzheimer's disease correlates with genetic etiology. *Am. J. Med. Genet.* 48:129–30

Mullan M, Houlden H, Windelspecht M, Fidani L, Lombardi C, et al. 1992b. A locus for familial early onset Alzheimer's disease on the long arm of chromosome 14, proximal to α_1-antichymotrypsin. *Nature Genet.* 2:340–42

Murrell J, Farlow M, Ghetti B, Benson MD. 1991. A mutation in the amyloid precursor protein associated with hereditary Alzheimer's disease. *Science* 254:97–99

Namba Y, Tomonaga M, Kawasaki H, Otomo E, Ikeda K. 1991. Apolipoprotein E immunoreactivity in cerebral deposits and neurofibrillary tangles in Alzheimer's disease and kuru plaque amyloid in Creutzfeldt-Jacob disease. *Brain Res.* 541:163–66

Neve RL, Finch EA, Dawes LR. 1988. Expression of the Alzheimer amyloid precursor gene transcripts in the human brain. *Neuron* 1:669–77

Ninomiya H, Roch J, Sundsmo MP, Otero DAC, Saitoh T. 1993. Amino acid sequence RERMS represents the active domain of amyloid β/A4 protein precursor that promotes fibroblast growth. *J. Cell Biol.* 121:879–86

Nitsch RM, Farber SA, Growdon JH, Wurtman RJ. 1993. Release of amyloid beta-protein precursor derivatives by electrical depolarization of rat hippocampal slices. *Proc. Natl. Acad. Sci. USA* 90:5191–93

Nukina N, Ihara Y. 1986. One of the antigenic determinants of paired helical filaments is related to tau protein. *J. Biochem.* 99:1541–44

Oltersdorf T, Fritz LC, Schenk DB, Lieberburg I, Johnson-Wood KL, et al. 1989. The secreted form of the Alzheimer's amyloid precursor protein with the Kunitz domain is protease nexin-II. *Nature* 341:144–47

Oltersdorf T, Ward PJ, Henriksson T, Beattie EC, Neve R, et al. 1990. The Alzheimer amyloid precursor protein. Identification of a stable intermediate in the biosynthetic/degradative pathway. *J. Biol. Chem.* 265:4492–97

Pandiella A, Massagué J. 1991. Cleavage of the membrane precursor for transforming growth factor α is a regulated process. *Proc. Natl. Acad. Sci. USA* 88:1726–30

Pericak-Vance MA, Bebout J, Gaskell P. 1991. Linkage studies in familial Alzheimer disease: Evidence for chromosome 19 linkage. *Am. J. Hum. Genet.* 48:1034–50

Pike CJ, Burdick D, Walencewicz AJ, Glabe CG, Cotman CW. 1993. Neurodegeneration induced by beta-amyloid peptides in vitro: the role of peptide assembly state. *J. Neurosci.* 13:1676–87

Ponte P, Gonzalez-DeWhitt P, Schilling J, Miller J, Hsu D, et al. 1988. A new A4 amyloid mRNA contains a domain homologous to serine proteinase inhibitors. *Nature* 331: 525–27

Qiu WQ, Ferreira A, Miller C, Koo EH, Selkoe

DJ. 1994. Cell-surface β-amyloid precursor protein stimulates neurite outgrowth of hippocampal neurons in an isoform-dependent manner. *J. Neurosci.* In press

Querfurth HW, Selkoe DJ. 1994. Calcium ionophore increases amyloid β peptide production by cultured cells. *Biochemistry.* 33: 4450–61

Rebeck GW, Reiter JS, Strickland DK, Hyman BT. 1993. Apolipoprotein E in sporadic Alzheimer's disease: allelic variation and receptor interactions. *Neuron* 11:575–80

Rogers J, Civin WH, Styren SD, McGeer PL. 1992. Immune-related mechanisms of Alzheimer's disease pathogenesis. In *Alzheimer's Disease: New Treatment Strategies,* ed. Z Khachaturian, J Blass, pp. 147–63. New York: Dekker

Roher A, Wolfe D, Palutke M, KuKuruga D. 1986. Purification, ultrastructure, and chemical analyses of Alzheimer's disease amyloid plaque core protein. *Proc. Natl. Acad. Sci. USA* 83:2662–66

Roher AE, Lowenson JD, Clarke S, Woods AS, Cotter RJ, et al. 1993. β-amyloid-(1–42) is a major component of cerebrovascular amyloid deposits: implications for the pathology of Alzheimer's disease. *Proc. Natl. Acad. Sci. USA* 90:10836–40

Ross R. 1993. Atherosclerosis: a defense mechanism gone awry. *Am. J. Pathol.* 143:987–1002

Royston MC, Kodical NS, Mann DMA, Groom K, Landon M, Roberts GW. 1994. Quantitative analysis of β-amyloid deposition in Down's Syndrome using computerized image analysis. *Neurodegeneration* 3:43–51

Saitoh T, Sunsdmo M, Roch J-M, Kimura N, Cole G, et al. 1989. Secreted form of amyloid β protein precursor is involved in the growth regulation of fibroblasts. *Cell* 58:615–22

Sambamurti K, Shioi J, Anderson JA, Pappolla MA, Robakis NK. 1992. The Alzheimer's amyloid precursor is cleaved by APP-secretase intracellularly in either the trans-Golgi network or in post-Golgi vesicles. *Neurobiol. Aging* 13(Suppl. 79):312

Saunders AM, Strittmatter WJ, Schmechel D, George-Hyslop PH, Pericak-Vance MA, et al. 1993. Association of apolipoprotein E allele epsilon 4 with late-onset familial and sporadic Alzheimer's disease. *Neurology* 43: 1467–72

Schellenberg GD, Bird TD, Wijsman EM, Moore DK, Boehnke M, et al. 1988. Absence of linkage of chromosome 21q21 markers to familial Alzheimer's disease. *Science* 241: 1507–10

Schellenberg GD, Bird TD, Wijsman EM, Orr HT, Anderson L, et al. 1992. Genetic linkage evidence for a familial Alzheimer's disease locus on chromosome 14. *Science* 258:668–71

Schmechel DE, Saunders AM, Strittmatter WJ, Crain BJ, Hulette CM, et al. 1993. Increased amyloid β-peptide deposition in cerebral cortex as a consequence of apolipoprotein E gentoype in late-onset Alzheimer disease. *Proc. Natl. Acad. Sci. USA* 90:9649–53

Schubert D, Jin L-W, Saitoh T, Cole G. 1989. The regulation of amyloid β protein precursor secretion and its modulatory role in cell adhesion. *Neuron* 3:689–94

Selkoe DJ, Abraham CR, Podlisny MB, Duffy LK. 1986. Isolation of low-molecular-weight proteins from amyloid plaque fibers in Alzheimer's disease. *J. Neurochem.* 146: 1820–34

Selkoe DJ, Watson D, Hung AY, Teplow DB, Haass C. 1993. Influence of the Aβ C-terminal sequence and of a FAD mutation at βAPP_{692} on the formation of amyloid β-peptide. *J. Neurosci. Abstr.* 19:431

Seubert P, Oltersdorf T, Lee MG, Barbour R, Blomqist C, et al. 1993. Secretion of β-amyloid precursor protein cleaved at the amino-terminus of the β-amyloid peptide. *Nature* 361:260–63

Seubert P, Vigo-Pelfrey C, Esch F, Lee M, Dovey H, et al. 1992. Isolation and quantitation of soluble Alzheimer's β-peptide from biological fluids. *Nature* 359:325–27

Shoji M, Golde TE, Ghiso J, Cheung TT, Estus S, et al. 1992. Production of the Alzheimer amyloid β protein by normal proteolytic processing. *Science* 258:126–29

Sinha S, Dovey HF, Seubert P, Ward PJ, Blacher RW, et al. 1990. The protease inhibitory properties of the Alzheimer's β-amyloid precursor protein. *J. Biol. Chem.* 265: 8983–85

Sisodia SS. 1992. β-amyloid precursor protein cleavage by a membrane-bound protease. *Proc. Natl. Acad. Sci. USA* 89:6075–79

Sisodia SS, Koo EH, Beyreuther K, Unterbeck A, Price DL. 1990. Evidence that β-amyloid protein in Alzheimer's disease is not derived by normal processing. *Science* 248:492–95

Smith RP, Higuchi DA, Broze GJ Jr. 1990. Platelet coagulation factor XIa-inhibitor, a form of Alzheimer amyloid precursor protein. *Science* 248:1126–28

Snow AD, Mar H, Nochlin D, Sakiguchi RT, Kimata K, et al. 1990. Early accumulation of heparan sulfate in neurons and in the β-amyloid protein containing lesions of Alzheimer's disease and Down's syndrome. *Am. J. Pathol.* 137:1253–70

Snow AD, Sekiguchi R, Nochlin D, Fraser P, Kimata K, et al. 1994. An important role of heparan sulfate proteoglycan (Perlecan) in a model system for the deposition and persistence of fibrillar Aβ-amyloid in rat brain. *Neuron* 12:219–34

St. George-Hyslop P, Haines J, Rogaev E, Mortilla M, Vaula G, et al. 1992. Genetic evidence for a novel familial Alzheimer's disease locus on chromosome 14. *Nature Genet.* 2:330–34

St. George-Hyslop PH, Haines JL, Farrer LA, Polinsky R, van Broeckhoven C, et al. 1990. Genetic linkage studies suggest that Alzheimer's disease is not a single homogeneous disorder. *Nature* 347:194–97

Strittmatter WJ, Weisgraber KH, Huand D, Dong L-M, Salvesen GS, et al. 1993. Binding of human apolipoprotein E to synthetic amyloid β peptide: isoform specific effects and implications for late-onset Alzheimer disease. *Proc. Natl. Acad. Sci. USA* 90:8098–8102

Suzuki T, Nairn AC, Gandy SE, Greengard P. 1992. Phosphorylation of Alzheimer amyloid precursor protein by protein kinase C. *Neuroscience* 48:755–61

Tamaoka A, Kalaria RN, Lieberburg I, Selkoe DJ. 1992. Identification of a stable fragment of the Alzheimer amyloid precursor containing the β protein in brain microvessels. *Proc. Natl. Acad. Sci. USA* 89:1345–49

Tamaoka A, Odaka A, Ishibashi Y, Usami M, Suzuki N, et al. 1994. APP717 missense mutation (Val to Ile) selectively increases the amyloidogenic longer form of amyloid β protein (Aβ1–41/43) in brains with familial Alzheimer's disease. *Nature.* Submitted

Tanzi RE, McClatchey AI, Lamperti ED, Villa-Komaroff L, Gusella JF, Neve RL. 1988. Protease inhibitor domain encoded by an amyloid protein precursor mRNA associated with Alzheimer's disease. *Nature* 331:528–32

Terry RD, Masliah E, Salmon DP, Butters N, DeTeresa R, et al. 1991. Physical basis of cognitive alterations in Alzheimer's disease: synapse loss is the major correlate of cognitive impairment. *Ann. Neurol.* 30:572–80

van Broeckhoven C, Backhovens H, Cruts M, DeWinter G, Bruyland M, et al. 1992. Mapping of a gene predisposing to early-onset Alzheimer's disease to chromosome 14q24.3. *Nature Genet.* 2:335–39

van Broeckhoven C, Haan J, Bakker E, Hardy JA, Hul WV, et al. 1990. Amyloid β-protein precursor gene and hereditary cerebral hemorrhage with amyloidosis (Dutch). *Science* 248:1120–22

van Nostrand WE, Cunningham DD. 1987. Purification of protease nexin II from human fibroblasts. *J. Biol. Chem.* 262:8508–14

van Nostrand WE, Schmaier AH, Farrow JS, Cunningham DD. 1990. Protease nexin-II (amyloid β-protein precursor): A platelet α-granule protein. *Science* 248:745–48

van Nostrand WE, Wagner SL, Suzuki M, Choi BH, Farrow JS, et al. 1989. Protease nexin-II, a potent anti-chymotrypsin, shows identity to amyloid β-protein precursor. *Nature* 341: 546–49

Vigo-Pelfrey C, Lee D, Keim PS, Lieberburg I, Schenk D. 1993. Characterization of β-amyloid peptide from human cerebrospinal fluid. *J. Neurochem.* 61:1965–68

Weidemann A, Konig G, Bunke D, Fischer P, Salbaum JM, et al. 1989. Identification, biogenesis and localization of precursors of Alzheimer's disease A4 amyloid protein. *Cell* 57:115–26

Wertkin AM, Turner RS, Pleasure SJ, Golde RE, Younkin SG, et al. 1993. Human neurons derived from a teratocarcinoma cell line express soley the 695–amino acid amyloid precursor protein and produce intracellular β-amyloid of A4 peptides. *Proc. Natl. Acad. Sci. USA* 90:9513–17

Whitson JS, Glabe CG, Shintani E, Abcar A, Cotman CW. 1990. β-Amyloid protein promotes neuritic branching in hippocampal cultures. *Neurosci. Lett.* 110:319–24

Whitson JS, Selkoe DJ, Cotman CW. 1989. Amyloid β protein enhances the survival of hippocampal neurons in vitro. *Science* 243:1488–90

Wischik CM, Novak M, Thogersen HC, Edwards PC, Runswick MJ, et al. 1988. Isolation of a fragment of tau derived from the core of the paired helical filament of Alzheimer's disease. *Proc. Natl. Acad. Sci. USA* 85:4506–10

Wood JG, Mirra SS, Pollock NL, Binder LI. 1986. Neurofibrillary tangles of Alzheimer's disease share antigenic determinants with the axonal microtubule-associated protein tau. *Proc. Natl. Acad. Sci. USA* 83:4040–43

Yamaguchi H, Nakazato Y, Hirai S, Shoji M. 1990. Immunoelectron microscopic localization of amyloid β protein in the diffuse plaques of Alzheimer-type dementia. *Brain Res.* 508:320–24

Yamaguchi H, Yamazaki T, Lemere CA, Frosch MP, Selkoe DJ. 1992. Beta amyloid is focally deposited within the outer basement membrane in the amyloid angiopathy of Alzheimer's disease. *Am. J. Pathol.* 141: 249–59

Yamazaki T, Koo EH, Hedley-Whyte ET, Selkoe DJ. 1993. Intracellular trafficking of cell surface βAPP in living cells. *Soc. Neurosci. Abstr.* 19(1):396

Yankner BA, Duffy LK, Kirschner DA. 1990. Neurotrophic and neurotoxic effects of amyloid β protein: reversal by tachykinin neuropeptides. *Science* 250:279–82

Annu. Rev. Cell Biol. 1994. 10:405–55

STRUCTURE, REGULATION AND FUNCTION OF NF-κB

Ulrich Siebenlist, Guido Franzoso, and Keith Brown
Laboratory of Immunoregulation, National Institute of Allergy and Infectious Diseases, National Institutes of Health, Bethesda, Maryland 20892

KEY WORDS: IκB, Rel, dorsal, ankyrin motif, gene regulation, immune activation, HIV cytokines

---

CONTENTS

0743–4634/94/1115–0405$05.00

INTRODUCTION

Survival of all life forms requires dynamic response to environmental change and challenge. In higher organisms such phenomena include defensive response to stress, injury, viruses, and pathogens. To mount effective responses, an organism requires a sensitive and rapidly acting system to detect and fight such potentially life-threatening circumstances. In the case of viruses and other microbiological pathogens, for example, the immune system is capable of recognizing these agents as foreign by their antigenic properties. To fight them, various immune cells are activated, differentiated, expanded, and summoned to relevant sites. Antibodies are synthesized and cell-mediated defenses activated. To achieve this, recognition of an invading agent by T cells, B cells, macrophages, and endothelial cells must elaborate defined genetic programs, such as the induced synthesis of soluble mediators (cytokines, growth factors, chemokines) and their receptors. The genetic response, i.e. the signal-responsive induction of gene expression, is mediated ultimately by factors that control the transcription of these genes. Among the various factors that contribute to the induction of these rapid-response genes, the transcription factor NF-κB stands out as a central, coordinating regulator. A vast and interrelated array of defense genes is regulated, at least in part, by a family of closely related dimeric complexes collectively referred to as NF-κB. Indeed, the association between induced cellular defense genes and NF-κB transcription factors dates at least as far back as insects in evolutionary terms.

Beyond playing a central role in defensive responses, NF-κB may modulate gene expression in various other situations that signal rapid gene expression. NF-κB and its regulators appear to function in growth control, as evidenced by mis-regulation of NF-κB genes encoding genes in certain tumors. A developmental role is indicated by the dorsal protein, a *Drosophila* homologue of vertebrate NF-κB factors. dorsal controls axial polarity in developing embryos in response to a positional cue.

The diversification of NF-κB during evolution into multiple genetically encoded polypeptides that assemble as homo- and heterodimers, and that bear unique and overlapping functional activities, as well as unique and overlapping expression patterns, portends wide and varied functional roles for this family of transcription factors. In addition to its role in expression of host cell genes, NF-κB has been usurped by various viruses, including the human immunodeficiency virus, to mediate viral gene expression.

NF-κB's most obvious characteristic is its rapid translocation from cytoplasm to nucleus in response to extracellular signals. Under most circumstances, these factors lie dormant in the cytoplasm of unstimulated cells, kept there by an inhibitory protein termed IκB. Many signals inactivate the inhibitor IκB, thereby allowing NF-κB to enter nuclei and rapidly induce coordinate sets of defense-related genes. NF-κB does this by binding to *cis*-acting κB sites in promoters and enhancers of these genes bearing the consensus sequence 5′-GGGPuNNPyPyCC-3′.

In this review we focus on recent developments in the field. For more complete perspectives on NF-κB, we refer the reader to previous reviews that include comprehensive and specifically focused treatments of the subject (Baeuerle 1991; Blank et al 1992; Bose 1992; Bours et al 1992a; Gilmore 1992; Israël 1992; Nolan & Baltimore 1992; Schreck et al 1992a; Beg & Baldwin 1993; Collins 1993; Gilmore & Morin 1993; Grilli et al 1993; Grimm & Baeuerle 1993; Hay 1993; Liou & Baltimore 1993; Kaltschmidt et al 1993; Muller et al 1993; Wasserman 1993; Baeuerle & Henkel 1994; Lenardo & Siebenlist 1994).

MOLECULAR STRUCTURE AND FUNCTION

The Family of Rel/NF-κB Proteins

The active, DNA-binding forms of NF-κB transcription factors are dimeric complexes, composed of various combinations of members of the Rel/NF-κB family of polypeptides. This family is distinguished by the presence of a so-called Rel homology domain (RHD) of about 300 amino acids in length, which displays ~ 35 to 61% identity between various family members (Bours et al 1992a,b; Ryseck et al 1992; Ip et al 1993). The RHD determines DNA binding to κB elements, dimerization with identical or other members of the family, as well as interaction with the IκB family of proteins (see below). The Rel/NF-κB family members can be grouped into two classes; one consists of the p105 and p100 precursor proteins, which are processed proteolytically to the mature p50 and p52 forms, respectively (Bours et al 1990; Ghosh et al 1990; Kieran et al 1990; Meyer et al 1991; Neri et al 1991; Schmid et al 1991; Bours et al 1992b; Mercurio et al 1992), and the other class consists of the Rel (c-Rel), v-Rel, RelA (p65) and RelB proteins (Stephens et al 1983; Wilhelmsen et al 1984; Brownell et al 1989; Grumont & Gerondakis 1989; Nolan et al 1991; Ruben et al 1991,1992; Ballard et al 1992; Ryseck et al 1992), as well as the *Drosophila* proteins dorsal and Dif (Steward 1987; Ip et al 1993) (see Table 1 for nomenclature). A critical distinguishing characteristic between the former (p50 and p52) and latter (p65, Rel, dorsal and Dif) classes is their discrepant ability to activate transcription. Members of the former class, if associated as homodimers, are weak or inert gene activators. In contrast, dimeric forms of NF-κB containing at least onc subunit of the latter category function as potent activators of gene expression. In molecular terms, the activating members of the NF-κB family are understood to contain activation domains that lie outside the RHD.

Table 1 Rel/NF-κB and IκB proteins

Proteins	Other names	Genes	Chromosomal locus
p50 or p105 (NF-κB1)	p110, KBF1, EBP-1	*nfkb1*	NFKB1
p52 or p100 (NF-κB2)	p50B or p97, p49 or p100, p55 or p98, Lyt10, H2TF1	*nfkb2*	NFKB2
Rel	c-Rel	*rel*	REL
v-Rel	—	*v-rel*	—
RelA	p65	*rela*	RELA
RelB	I-Rel	*relb*	RELB
dorsal	—	*dorsal*	—
Dif, Cif	dorsal-related immunity factor cecropia immunoresponsive factor	*dif*	—
IκB-α	MAD-3, pp40, RL/IF-1, ECI-6	*ikba*	IKBA
IκB-β			
IκB-γ	p105/pdI, C-terminal portion of p105	*nfkb1*	NFKB1
Bcl-3		*bcl-3*	BCL3
cactus	—	*cactus*	—

It is unclear whether processing of the p105 or p100 precursor proteins into their mature forms is a regulated process. Evidence indicates that such processing is upregulated during stimulation with extracellular signals (Mellits et al 1993; Mercurio et al 1993). Other studies have indicated that processing is a relatively slow, constitutive process (Fan & Maniatis 1991; Baeuerle & Henkel 1994). p105 is processed at a significantly faster rate than p100 (Mercurio et al 1993; K Brown & U Siebenlist, unpublished observations). The reaction is ATP dependent (Fan & Maniatis 1991). The mature proteins end in a glycine-rich region located directly adjacent to the RHD. The glycine-rich region has been interpreted to function as a flexible hinge in the precursor, also providing potential access to cleavage by endoproteases. The resulting N-terminal halves of the p105 or p100 molecules constitute the mature p50 or p52 molecules, respectively, each containing the intact RHD. Separate, processed C-terminal parts have not been detected in cells, presumably because they are rapidly degraded, either as a consequence of a subsequent, independent processing step or as part of a progressive reaction initiated by the initial processing (Fan & Maniatis 1991; Rice et al 1992). In this regard, precursors with even short, incomplete C-terminal extensions beyond the RHD can be processed into the final mature size, which suggests that an intact C-terminus is not required for processing (Blank et al 1991; Fan & Maniatis 1991). The structure and function of the C-terminal domains removed from p105 and p100 during processing are discussed below.

NF-κB: A Family of Multiple Distinct Dimers

The members of the Rel/NF-κB family of proteins can form almost all theoretically possible homo- and heterodimers, generating many complexes through combinatorial mixing. Dimerization is an obligatory regimen for DNA binding (Logeat et al 1991; Bressler et al 1993), probably because both subunits interact with the κB DNA element (see below). κB sites are dyad symmetric, and each subunit of the NF-κB dimer interacts with a half site of the recognition sequence.

The p50/RelA (p50/p65) complex is present in essentially all cells and is usually the most abundant dimeric complex. Among the other possible dimeric complexes, homodimers of RelA or Rel or RelA/Rel heterodimers are usually detected at only very low levels, but such complexes may, nevertheless, play unique and significant roles in the context of specific promoters (Hansen et al 1992; Nakayama et al 1992; Bakalkin et al 1993; Costello et al 1993; Ganchi et al 1993; Kunsch & Rosen 1993). Complexes that have not been detected, either in vitro and/or in vivo, are RelB homodimers (Ryseck et al 1992; Dobrzanski et al 1993) and RelB/RelA and RelB/Rel heterodimers.

The various dimeric complexes differ in their preference for certain κB sites, transactivation potentials, kinetics of nuclear translocation, and levels of expression in tissue. The nearly ubiquitously expressed p50/RelA tends to be rapidly translocated to the nucleus in response to extracellular activating signals, whereas other complexes such as p50/Rel accumulate in nuclei more slowly, exhibiting a delay of several hours before reaching peak levels (Molitor et al 1990; Doerre et al 1993). The reason for this delay is not understood. Rel complexes have been proposed to function as late-acting competitive downregulators of genes induced by RelA dimers at early times after stimulation (Ballard et al 1992; Doerre et al 1993; La Rosa et al 1994). However, the transfection data on which this conclusion is based may simply reflect a relatively lower (but still significant) transactivation potential of Rel in these cells. In other cells, including embryonal cells, Rel potently transactivated many κB-dependent reporter constructs (Inoue et al 1991; Richardson & Gilmore 1991; Ishikawa et al 1993; Inuzuka et al 1994).

Rel-containing complexes have been observed in many cell lines, but they appear to be preferentially present in hematopoietic cells, particularly in lymphoid cells (Brownell et al 1987, 1989; Feuillard et al 1994). This foreshadows a greater role for Rel in regulating gene expression in such cells, especially in B cells. RelB complexes may be even more restricted, with expression in tissue detectable primarily in interdigitating dendritic cells in thymus and probably in spleen, as judged by in situ hybridization (Carrasco et al 1993). RelB mRNA could not be detected during early mouse embryogenesis, but appears later, at times when the immune system matures and when dendritic cells begin to function to negatively and positively select T cells (Carrasco et al 1993). This correlation portends an important role for RelB in antigen presentation and T cell activation and selection. However, RelB is induced also in fibroblast cell lines subjected to serum stimulation (Ryseck et al 1992). In another study, RelB complexes have also been reported in some B cells (Lernbecher et al 1993). Together, these studies indicate that high expression of Rel and RelB is restricted to certain tissues but that lower and possibly significant levels may be present elsewhere.

Among the various members of the Rel/NF-κB family of proteins, p50 and p52 are most similar in primary amino acid sequence and behave almost identically in transfection experiments when paired with the transactivators RelA, Rel, or RelB, with which they heterodimerize (Bours et al 1992b; G Franzoso et al, unpublished observations). However, p50- and p52-containing complexes may differ functionally in the context of specific promoters where they interact with other transcription factors (Perkins et al 1992).

Protein Domains and their Functions

The highly conserved RHD is responsible for DNA binding, dimerization, nuclear translocation, and binding to IκB inhibitors, and it is the only region shared by all members of this family. Deletions and site-directed mutagenesis have located the DNA-binding region to the N-terminal part of this ~ 300 amino acid domain (Logeat et al 1991; Bressler et al 1993; Coleman et al 1993). A short stretch of amino acids at the beginning of the domain has been found essential for binding (the RXXRXRXXC motif) and is believed to contact DNA directly (Kumar et al 1992; Bressler et al 1993; Toledano et al 1993). The critical cysteine residue within this motif must be in a reduced state because its oxidation interferes with DNA binding (Toledano & Leonard 1991; Kumar et al 1992; Matthews et al 1992, 1993a; Hayashi et al 1993b; Toledano et al 1993). A second region ~ 100 amino acids further C-terminal to the RHD also contributes significantly to binding (Bressler et al 1993). Neither region obviously conforms with known DNA-binding motifs, which suggests that this family of proteins encodes a previously unrecognized DNA-binding motif (Liu et al 1994). Deletions or mutations in the p50 subunit, which interfered with DNA binding but still formed heterodimers with RelA, acted as dominant negative mutants in transfection experiments inhibiting transactivation by RelA (Logeat et al 1991; Bressler et al 1993).

Various studies have indicated that the C-terminal half of the RHD facilitates dimerization (Logeat et al 1991; Bressler et al 1993; Doerre et al 1993; Ganchi et al 1993). The structure of this region is also unknown. Dimerization with different partners may involve distinct amino acids. A potential cAMP-dependent protein kinase (PKA) phosphorylation site or an internal cysteine, when mutated in RelA, severely interferes with homodimerization, but not heterodimerization with p50 (Ganchi et al 1993). It is conceivable that phosphorylation of the PKA site could regulate dimer choices in RelA. The PKA site is conserved among the members of the Rel/NF-κB family, with the exception of RelB and p52 (Bours et al 1992b; Ryseck et al 1992).

The extreme C-terminal end of the RHD contains a conserved cluster of positively charged amino acids that function as a nuclear localization sequence (NLS) and that appear to be important for interaction with the inhibitory IκB proteins (see below) (Blank et al 1991; Beg et al 1992; Ganchi et al 1992; Henkel et al 1992; Matthews et al 1993b; Zabel et al 1993).

RelA (p65), Rel, RelB, and the *Drosophila* protein dorsal (and most likely Dif) contain demonstrable activation domains in polypeptide segments located on the carboxyl terminal side of the RHD (Bull et al 1990; Kamens et al 1990; Ip et al 1991; Richardson & Gilmore 1991; Schmitz & Baeuerle 1991; Ballard et al 1992; Dobrzanski et al 1993) (Figure 1 shows a schematic representation of all members of Rel/NF-κB/dorsal and IκB families). In the case of RelB, an additional activation function has been attributed to its unique N-terminal extension, a short region preceding the RHD that may encode a leucine-zipper-like motif (Dobrzanski et al 1993). According to one report, RelB (termed I-Rel) acted as a dominant inhibitor of NF-κB (Ruben et al 1992), but ample evidence now indicates that RelB functions to potently induce transcription (Bours et al 1992b, 1994; Ryseck et al 1992; Dobrzanski et al 1993). Rel contains two distinct regions that could participate in transactivation (Ishikawa et al 1993; Sarkar & Gilmore 1993). The most C-terminal of these two domains of Rel, as well as the potent transactivation domain present in RelA and the C-terminal domain of RelB, may all belong to the class of "acidic transactivators" (Bull et al 1990; Kamens et al 1990; Schmitz & Baeuerle 1991; Dobrzanski et al 1993), exemplified by the VP16 protein. V-Rel lacks this particular transactivating domain (Kamens et al 1990; Ishikawa et al 1993; Sarkar & Gilmore 1993), and this difference from Rel may be critical to oncogenic potential of the viral form (see below) (Capobianco et al 1990;

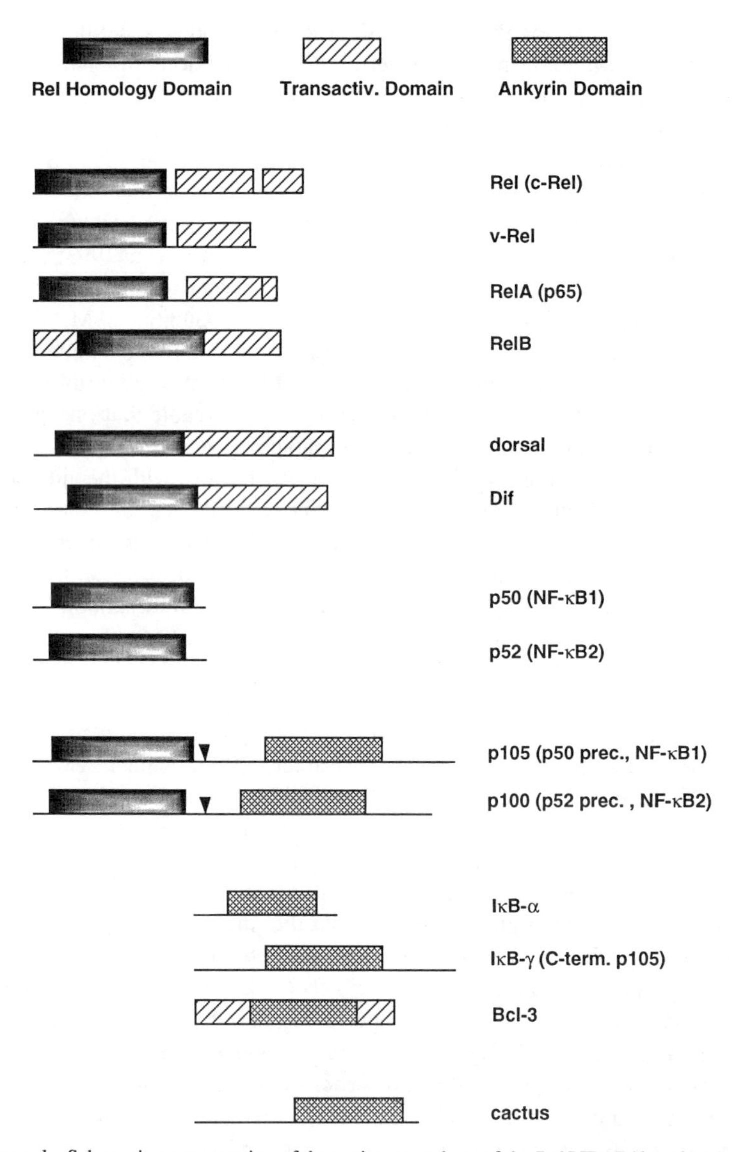

Figure 1 Schematic representation of the various members of the Rel/NF-κB/dorsal and the IκB family of polypeptides. Functional domains are indicated; in most cases the borders of the transactivation domains have not been precisely mapped. The putative transactivation domain of Dif has not been directly demonstrated. See text for details.

Kamens et al 1990; Mosialis et al 1991; Diehl et al 1993; Hrdlickova et al 1994a,b; Kralova et al 1994; Nehyba et al 1994). The residual transactivating function of v-Rel can be observed only in undifferentiated cells (Walker et al 1992; Ishikawa et al 1993; Inuzuka et al 1994; Hrdlickova et al 1994b), whereas Rel can transactivate in a wide range of cells (Kamens et al 1990; Inoue et al 1991; Ishikawa et al 1993). Transactivation by these factors is probably mediated by direct physical interaction with the basal transcription apparatus, possibly allowing the formation of a stable pre-initiation complex, but the precise mechanisms remain to be determined (see Kerr et al 1993; Xu et al 1993).

Different NF-κB dimers display different binding affinities with respect to certain κB sites (Urban & Baeuerle 1990; Neri et al 1991; Bours et al 1992b; Kunsch et al 1992; Perkins et al 1992; Franzoso et al 1993). Dissociation constants of a number of complexes at preferred binding sites were reported in the range of 10^{-12} M (Urban & Baeuerle 1990; Fujita et al 1992; Kretzschmar et al 1992), which indicates strong binding. The transactivation potential of a particular NF-κB dimer in vivo may depend not only on the affinity for a given κB element but also on the DNA sequence-dependent conformation adopted by the bound dimer. Different κB sites appear to allow different protein surfaces to be exposed, and transactivation potential does not necessarily correlate with binding affinity in vitro (Fujita et al 1992; Perkins et al 1992).

Most κB elements are somewhat palindromic, but rarely completely so, which gives these elements a directional sense relative to the transcriptional start site (Baeuerle 1991; Urban & Baeuerle 1990, 1991). This is likely to be important functionally because the p50 subunit of a p50/RelA heterodimer has been shown to interact preferentially with the 5′ half site of the prototypical immunoglobulin (Ig) κB site (5′-GGGAC), while the 3′ half site (5′-GGAAA, opposite strand), which is more variable among different κB sites, interacts preferentially with the RelA (p65) subunit (Urban & Baeuerle 1991; Kunsch et al 1992). Consequently, p50 homodimers prefer to bind palindromic sequences consisting of two Ig κB 5′ half sites, whereas RelA homodimers prefer duplicated Ig κB 3′ half sites. The three G residues in the 5′ half site (top strand) and the two G residues in the 3′ half site (bottom strand) of the IgκB site may be directly contacted, as judged by methylation protection and interference experiments (Baldwin & Sharp 1988; Baeuerle & Baltimore 1989; Israël et al 1989). The more variable bases in between, primarily AT base pairs, may be important for the bending of DNA observed upon binding (Schreck et al 1990). That DNA bending might be critically involved is further supported by work with the IFN-β promoter κB site, where the HMG-I/Y proteins, which are known to facilitate DNA bending, also enhance binding of p50/RelA. The HMG-I/Y proteins recognize the particular AT base pairs of the IFN-β κB site in the minor groove and bind simultaneously with p50/RelA (Thanos & Maniatis 1992).

The Family of IκB-Related Proteins

The IκB family of proteins includes IκB-α, the precursor proteins p105 and p100, IκB-γ, Bcl-3, and the *Drosophila* protein cactus. All members of the family share a partially conserved domain that harbors between six to eight ankyrin motifs, each 33-amino acids in length (Schmitz et al 1991; Blank et al 1992; Nolan & Baltimore 1992). That these motifs are regulators of NF-κB activity was first recognized with the cloning of the p105 precursor (Bours et al 1990; Kieran et al 1990), where they were discovered to inhibit DNA binding of the N-terminal Rel/NF-κB homology domain (Bours et al 1990; Ghosh et al 1990; Kieran et al 1990). The members of this protein family interfere with nuclear translocation and DNA binding of Rel/NF-κB complexes, with several notable exceptions. Although IκB proteins generally interact with dimers (Beg et al 1992), evidence indicates binding of IκB-α to RelA monomers as well (Ganchi et al 1993; P Bressler & U Siebenlist, unpublished observations).

IκB-α (and the ankyrin domain of the p105 precursor) interact with and shield the nuclear localization signal (NLS) located at the end of the RHD domain (Beg et al 1992; Ganchi et al 1992; Henkel et al 1992; Matthews et al 1993b; Zabel et al 1993). IκB-α does so as a monomer with all dimeric NF-κB complexes tested, which suggests that it shields the positively charged NLSs of both subunits of these dimers (Hatada et al 1993). A short, negatively charged region of IκB-α located adjacent to the most C-terminal (sixth) ankyrin motif may be critical (Hatada et al 1992, 1993). An analogous acidic region in the p105 precursor located between the last two ankyrin motifs appears to be necessary for the inhibitory activity of that protein (Blank et al 1991; Hatada et al 1993). In addition, cytoplasmic retention may be assisted by a cooperating domain at the C-termini of Rel and dorsal (Hannink & Temin 1989; Capobianco et al 1990; Isoda et al 1992; Norris & Manley 1992; Diehl et al 1993). Although these regions do not appear to be necessary for binding by IκB-α (see below), they may nevertheless cooperate with IκB-α to help cytoplasmic retention. In the case of RelA, a C-terminal domain was also reported to assist in IκB-α-mediated inhibition of DNA binding (Ganchi et al 1992). Homodimers of p50 or p52 are less efficiently retained in the cytoplasm (Franzoso et al 1992; Kang et al 1992), although they bind IκB-α (Beg et al 1992; Ganchi et al 1992; Franzoso et al 1993). The interaction with the homodimers may not be physiologically relevant. However, IκB-α binds tightly and functionally inhibits all dimeric complexes containing any of the transactivating subunits (RelA, Rel, RelB), in particular those containing RelA.

IκB-α not only prevents DNA binding of the strongly transactivating complexes, but it also dissociates prebound complexes from their cognate DNA sites (Zabel & Baeuerle 1990). While complexes containing either RelA, Rel, or RelB subunits are inhibited in this way (Baeuerle & Baltimore 1988a; Haskill et al 1991; Beg et al 1992; Davis et al 1991; Inoue et al 1992b; Tewari et al 1992), DNA binding by p50 and p52 homodimers is essentially unaffected by IκB-α (Beg et al 1992; Kerr et al 1992; Franzoso et al 1993; Nolan et al 1993). IκB-α is apparently readily displaced from these homodimers in the presence of a κB binding site (Beg et al 1992). To date, no consensus has emerged as to how DNA binding of dimers containing RelA, Rel, or RelB is blocked by IκB. According to one report, IκB-α binding to the N-terminal DNA-binding region of Rel is both necessary and sufficient (Kerr et al 1991), whereas other reports suggest some binding to the N-terminal and a possibly stronger interaction with the NLS (Diehl et al 1993; Kumar & Gelinas 1993). In addition, IκB-α binding to RelA has been localized to the dimerization domain, which also includes the adjoining NLS (Ganchi et al 1992, 1993). According to the model derived from the latter data, DNA binding is blocked indirectly, possibly by changing the interaction between dimer subunits. This mode of action would enable IκB-α to dissociate stably bound complexes from DNA without requiring direct access to the DNA-contacting portion of the bound protein.

Initial purification procedures for IκB activity revealed two distinct fractions termed IκB-α and IκB-β (Zabel & Baeuerle 1990; Link et al 1992). Despite earlier confusion in the literature, direct evidence now demonstrates IκB-α function to be encoded by the cloned human MAD-3 (Haskill et al 1991), chicken pp40 (Davis et al 1991), rat RL/IF-1 (Tewari et al 1992), and porcine ECI-6 (de Martin et al 1993) proteins (Zabel et al 1993). The identity of IκB-β remains unresolved. Based on their charge differences (Link et al 1992), the absence of immunological cross-reactivity (Zabel et al 1993), and difference in molecular weights, (IκB-α is 37 K and IκB-β is reportedly 43 K (Zabel & Baeuerle 1990). IκB-β may be encoded by a separate gene from IκB-α. Also, IκB-β activity is inactivated by phosphatase treatment in vitro, in distinction from the IκB-α-containing fraction (Link et al 1992).

Bcl-3 is an IκB-α-related protein that was first discovered as a translocation of its gene into the immunoglobulin locus in a subset of chronic lymphocytic leukemias (Ohno et al 1990). Whereas IκB-α preferentially interacts with the transactivating complexes, especially those containing RelA, Bcl-3 primarily targets homodimers of p50 and p52 (Franzoso et al 1992, 1993; Hatada et al 1992; Wulczyn et al 1992; Bours et al 1993; Fujita et al 1993; Naumann et al 1993; Nolan et al 1993). Also, Bcl-3 does not interfere with the nuclear translocation of these complexes. Indeed, Bcl-3 can bc readily detected in cell nuclei (Bours et al 1993; Franzoso et al 1993; Nolan et al 1993), although not in all cells (Naumann et al 1993). These differences between IκB-α and Bcl-3 (Hatada et al 1992; Nolan & Baltimore 1992), together with unique features of Bcl-3 encoded outside of the ankyrin domain (Bours et al 1993), indicate a very different function of this protein. Instead of inhibiting transactivation by NF-κB, Bcl-3 promotes κB-dependent transcription (see below). Using in vitro cross-linking experiments, up to two Bcl-3 monomers were reportedly able to interact with a single p50 homodimer (Wulczyn et al 1992). This contrasts with the interaction of only one IκB-α monomer with a p50/RelA dimer (Hatada et al 1993).

The p105 and p100 precursor proteins function as IκB-like proteins (Blank et al 1991; Henkel et al 1992; Rice et al 1992, Mercurio et al 1993; Naumann et al 1993; Scheinman et al 1993; Beraud et al 1994). p105 and p100 are able to dimerize with the various Rel proteins, and they inhibit these proteins from translocating to the nucleus, even in the absence of IκB-α (Henkel et al 1992; Rice et al 1992; Mercurio et al 1993; Naumann et al 1993). Given the relatively long survival of the precursor forms prior to processing, they may function as physiological inhibitors of NF-κB activity. At least part of the pool of Rel proteins is found in association with the p105 and p100 precursors, to the exclusion of IκB-α (Rice et al 1992).

The precursor is speculated possibly to have an IκB-like regulated activity (Mercurio et al 1993). Precursor-containing cytoplasmic complexes not associated with IκB-α could represent an alternative source for nuclear NF-κB activity, which may be activated to translocate to the nucleus by direct signal-induced proteolytic processing. However, no current evidence exists that demonstrates a rapid signal-dependent release from precursor inhibition and subsequent nuclear translocation, although a moderate increase in processing has been reported after activation of some cells (Mercurio et al 1993). The precursor-associated cytoplasmic complexes may not represent a reservoir poised for direct immediate activation but, rather, they may primarily serve as a reservoir to replenish the signal-responsive IκB-α-associated complexes.

IκB-γ contains the C-terminal half of the p105 precursor molecule. It is not a product of processing but rather is synthesized from a separately initiated mRNA present in certain mouse B cells (Inoue et al 1992a; Liou et al 1992), where IκB-γ protein is also found. Nothing is known about control of expression of this mRNA in mouse B cells. Although this protein is capable of inhibiting various NF-κB complexes in vitro in a manner somewhat similar to that of IκB-α (Inoue et al 1992a; Liou et al 1992), its physiological relevance is unclear and no equivalent mRNA or protein has been detected in any human B cells.

The family of IκB proteins is part of a larger family of proteins with more divergent ankyrin motifs (Schmitz et al 1991; Blank et al 1992; Hatada et al 1992; Michaely & Bennett 1992; Nolan & Baltimore 1992). No obvious consensus of biological function exists within the larger family of proteins with ankyrin motifs. The human ankyrin protein, and, in particular its ankyrin domain, links ion channels to cytoskeletal structures (Lux et al 1990). Several transcription factors, including two yeast cell cycle proteins, contain ankyrin repeats, as do various developmentally important regulators in *Drosophila,* nematodes, and frogs (Michaely & Bennett 1992). Most likely, the common denominator between these and other ankyrin motif-containing proteins is that the motifs define structures for protein-protein interactions. The more highly conserved segments within each ankyrin motif are likely to form amphipathic α-helices [although the existence of β-sheets has been hypothesized as well (Michaely & Bennett 1992)], whereas the more divergent segments of the repeats may represent loops that also determine binding specificity. Together the individual ankyrin motifs may form a tertiary structure, possibly by bundling of α-helices (Gay & Ntwasa 1993). For IκB-α, Bcl 3, and p105, the whole ankyrin domain is necessary for inhibitory activity. In the case of IκB-α and p105, a short, negatively charged segment is also required for full inhibition (Hatada et al 1992, 1993; Wulczyn et al 1992; Bours et al 1993, Franzoso et al 1993; see also Blank et al 1991; Inoue et al 1992b).

Historical Perspectives and Nomenclature

NF-κB DNA binding, and transactivation activity was first detected with the κB element of the immunoglobulin enhancer in certain murine B cells (Sen & Baltimore 1986a,b). This discovery was born of prior work that mapped an inducible DNAaseI hypersensitive site to this region in response to differentiation of a mouse preB cell (Chung et al 1983; Parslow & Granner 1983). This κB element was shown to be critical to B cell-specific expression of enhancer function, which led to the conclusion that the bound factor, called NF-κB, was a lineage and differentiation stage-specific factor for B cells (Sen & Baltimore 1986a). It was soon recognized, however, that this binding activity was ubiquitous and could be detected in many cells but usually only after stimulation of such cells by various agents (see Lenardo & Baltimore 1989). Subsequently, the IκB inhibitor was identified as a protein capable of retaining NF-κB in the cytoplasm in the absence of activating signals (Baeuerle & Baltimore 1988a,b). This inhibitor could be dissociated in vitro with detergents (deoxycholate; DOC), which liberated NF-κB (Baeuerle & Baltimore 1988b).

Prior to cloning of the genes, the various NF-κB activities could not be distinguished by the standard electrophoretic mobility shift assay, but it is now known that several different dimeric complexes account for what was earlier thought to be a single NF-κB complex. To be consistent with prior literature, we use the term NF-κB or Rel/NF-κB to collectively denote all κB-binding activity of this family of proteins, and we refer to specific dimers by naming their subunits.

The first cloning of a gene (p50/p105) encoding a component of the NF-κB binding activity was the result of independent and simultaneous efforts in several laboratories (Bours et al 1990; Ghosh et al 1990; Kieran et al 1990; Meyer et al 1991). This development dramatically impacted the field since it soon led to the discovery (and cloning) of a family of Rel/NF-κB and IκB proteins. In addition, it merged this field of research with that of the previously discovered proteins Rel/v-Rel and dorsal.

Table 1 indicates the nomenclature proposed for the individual members of the Rel/NF-κB and IκB families (adapted from a consensus reached in October 1992 at a meeting entitled *NF-κB, Rel and Dorsal: Structure and Function,* held at the Howard Hughes Medical Institute in Bethesda, MD). For simplicity and clarity, we use only the names p50, p52 (or their precursors p105, p100, respectively), RelA, Rel, v-Rel, and RelB throughout this review.

DIRECT MECHANISMS FOR ACTIVATION

Modification and Degradation of IκB-α

Activation of NF-κB from cytoplasmic pools correlates temporally with proteolytic degradation of the inhibitory IκB-α protein. Nearly complete degradation of the inhibitor occurs within minutes after administration of extracellular stimuli such as TNF-α or PMA plus ionomycin, an observation that has now been confirmed in many laboratories (Beg et al 1993; Brown et al 1993; Sun et al 1993, see also Cordle et al 1993; de Martin et al 1993; Henkel et al 1993; Mellits et al 1993; Rice & Ernst 1993; Scott et al 1993; Chiao et al 1994; Read et al 1994). In addition, a post-translational phosphorylation of IκB-α accompanies the degradation (Beg et al 1993; Brown et al 1993; Cordle et al 1993; Sun et al 1994). The modified form of IκB-α does not accumulate but is detected (as a protein with slightly slower electrophoretic mobility) only transiently shortly after stimulation of cells. This is consistent with the idea that phosphorylation tags IκB-α for rapid degradation, although this is still unproven. As IκB-α disappears from cells, NF-κB is freed to enter nuclei and induce genes (Brown et al 1993; Sun et al 1993).

A different model for activation is based on the observation that free, uncomplexed IκB-α is extremely unstable in cells, while complexed IκB is more stable (Beg et al 1993; Brown et al 1993; Rice & Ernst 1993; Scott et al 1993; Sun et al 1993). According to this model (see discussion in Beg et al 1993), phosphorylation of IκB-α releases the inhibitor from NF-κB, thus leaving it unprotected from constitutively active proteases, which then digest the free form. Therefore, proteolytic degradation per se would not be an integral part of activation. This view is also based on early reports indicating that several kinases activate IκB-bound NF-κB in vitro and phosphorylate and inactivate partially purified IκB (Shirakawa & Mizel 1989; Ghosh & Baltimore 1990).

Recently we obtained evidence in support of an essential role for degradation in activating NF-κB, contrary to the above model, which does not envision proteolytic degradation as a critical activating event. IκB-α protein, which is modified by phosphorylation in response to cellular signals, in fact, remains associated with NF-κB, as demonstrated by co-immunoprecipitation with anti-RelA antibodies (YC Lin & U Siebenlist, unpublished observations). This favors the view that proteases are required to liberate NF-κB activity, presumably acting on the NF-κB-IκB-α complex in situ.

Evidence for a critical role of proteases also emerges from studies with inhibitors of chymotrypsin-like proteases, inhibitors that appear to specifically block degradation of IκB-α and activation of NF-κB (Henkel et al 1993; Mellits et al 1993; Chiao et al 1994). However, the interpretation of these data is complicated by the discovery that these protease inhibitors also block the signal-dependent modification of IκB-α (YC Lin et al, unpublished observations). It is therefore unclear how these protease inhibitors function to inhibit activation of NF-κB, but they may have unappreciated activities, including potential effects on signaling kinases.

According to the simplest form of the current model, signal-dependent phosphorylation of IκB-α is sufficient to trigger proteolytic degradation. However, a separate signal-dependent induction of protease activity has not been ruled out. To date the protease(s) involved has not been identified. Intermediate fragments resulting from the degradation of IκB-α have not been reported, which possibly reflects a rapid and complete proteolysis.

The kinase directly responsible for mediating signal-dependent phosphorylation has not been definitively identified either, although some evidence implicates Raf as this kinase. This interpretation is based largely on indirect experiments. For example, cells transfected with active forms of Raf show constitutively active NF-κB. Likewise, cells that express dominant negative forms of Raf fail to activate NF-κB upon exposure to TNF-α, serum, PMA, or UV irradiation (Bruder et al 1993; Devary et al 1993; Finco & Baldwin 1993; Li & Sedivy 1993). These data argue for a role of Raf at some stage during stimulation by these agents (see below). That Raf may be an IκB-α kinase has also been suggested by a presumed direct physical interaction between Raf and IκB-α, as concluded from a yeast two-hybrid analysis in which genes for both proteins were specifically introduced (Li & Sedivy 1993). Furthermore, IκB-α can be phosphorylated by Raf in vitro, although this by itself does not establish IκB-α as a physiological target (Li & Sedivy 1993).

Modification of NF-κB

Although IκB-α appears to be the primary target for activation signals, the associated NF-κB proteins may also be important targets. dorsal is reported to be rapidly phosphorylated by protein kinase A (PKA) in response to activation

(Norris & Manley 1992), and Rel and p105 become tyrosine phosphorylated (Neumann et al 1992; see also Mellits et al 1993). Furthermore, an as yet unidentified kinase has been purified in association with RelA (Hayashi et al 1993a). Despite these observations, there is at present no convincing evidence to indicate a role for modification of NF-κB subunits in triggering activation. In fact, activated nuclear NF-κB, when extracted from cells, remains sensitive to IκB-α inhibition in vitro (Rice & Ernst 1993). Modifications could result following dissociation of IκB-α whereupon new protein surfaces become exposed on NF-κB dimers (e.g. the PKA site located adjacent to the NLS, see above). Likewise, translocation of NF-κB into the nucleus could well result in exposure to new protein kinases. Phosphorylation of NF-κB subunits, if relevant, could affect functions such as transcriptional activation.

Cactus and Dorsal

Considerable genetic data exist concerning the activation of the *Drosophila* protein dorsal during embryogenesis (reviewed by Govind & Steward 1991; St. Johnston & Nüsslein-Volhard 1992). Dorsal is kept in the cytoplasm by cactus, the *Drosophila* homologue of IκB-α, during the early developmental stages. After ten nuclear division cycles, the embryo receives a positional cue through the toll receptor on the ventral side only. In response, dorsal enters nuclei in a gradient fashion. In addition to the negative role played by cactus, which is relieved by the toll-mediated activating signal, at least two further proteins are involved in the signaling cascade linking toll to dorsal, namely tube and pelle. Whereas little is known about the function of tube, pelle is a serine protein kinase that may be a direct regulator of cactus (Letson et al 1991; Shelton & Wasserman 1993). pelle is distantly related to Raf and many other kinases. Another possible parallel between *Drosophila* and mammalian regulation is suggested by the primary amino acid sequence similarity between the cytoplasmic tails of the toll and the mammalian IL-1 receptor (Schneider et al 1991; Wasserman 1993); IL-1 is a potent activator of NF-κB. Since this signaling pathway may have been conserved through evolution, pelle and tube may have direct mammalian counterparts.

SIGNALS AND SIGNALING PATHWAYS FOR NF-κB ACTIVATION

Table 2 lists signals that lead to activation of NF-κB in some cells. Although very different, most signals are directly or indirectly associated with a pathogenic event or physical stress. All signals so far tested target IκB-α, as demonstrated by its ensuing degradation. On the one hand, it is possible that all of these divergent signals converge on an effector molecule upstream of IκB-α,

Table 2 NF-κB-activating stimuli

Cytokines	Tumor necrosis factor-α (TNF-α) Lymphotoxin (LT) (TNF-β) Interleukin-1 α and β (IL-1 α and β) Interleukin-2 (IL-2) Leukemia inhibitory factor (LIF) (Interferon-γ) (Macrophage colony-stimulating factor) (M-CSF) (Granulocyte/macrophage colony-stimulating factor) (GM–CSF)
Mitogens	Antigen Allogeneic stimulation Lectins (PHA, Con A) anti-αβ T cell receptor anti-CD3 anti-CD2 anti-CD28[1]
	Phorbol esters (Diacylglycerol) (DAG) Calcium ionophores (ionomycin, A2837)[1]
	anti-surface IgM (p39) (CD-40 ligand)
	Serum (Platelet-derived growth factor) (PDGF)
Other biological mediators	Leukotriene B4 (Prostaglandin E2) (PGE2)
	(Insulin)
Bacteria and bacterial products	*Shigella flexneri* *Mycobacterium tuberculosis* Cell wall products: Lipopolysaccharide (LPS) Muramyl peptides (G(Anh)MTetra) Toxins: Staphylococcus enterotoxin A and B (SEA and SEB) Toxic shock syndrome toxin-1 (TSST-1) (Cholera toxin)
Viruses and viral products	Human T cell leukemia virus-1 (HTLV-1) Tax Hepatitis B virus (HBV) Hbx MHBs[1] Epstein-Barr virus (EBV) EBNA-2 LMP

Table 2 *(Continued)*

	Cytomegalovirus (CMV)
	(Human immunodeficiency virus-1) (HIV-1)
	Herpes simplex virus-1 (HSV-1)
	Human herpes virus-6 (HHV-6)
	Newcastle disease virus
	Sendai virus
	Adenovirus 5
	ds RNA
Eukaryotic parasite	*Theileria parva*
Physical stress	UV light
	Ionizing radiations (X and γ)
	(Photofrin plus red light)
	(Hypoxia)
	Partial hepatectomy
Oxidative stress	Hydrogen peroxide
	Butyl peroxide
	Oxidized lipids
	(Antimycin A)[1]
Chemical agents	Calyculin A
	Okadaic acid
	(Pervanadate)
	(Ceramide)
	(Dibutyrl c-AMP)
	(Forskolin)
	Protein synthesis inhibitors
	Cycloheximide
	Anisomycin
	Emetine

[1]These agents have been shown to active NF-κB in conbination with other stimuli for references see text and recent reviews (Baeuerle 1991; Grilli et al 1993). Not cited in these reviews are *S. flexneri* (Dyer et al 1993); muramyl peptides (Schreck et al 1992b); G(Anh)MTetra (Dokter et al 1994); SEA, SEB and TSST-1 (Trede et al 1993); cholera toxin and dibutyryl c-AMP (Muroi & Suzuki 1993; Parhami et al 1993); IL-2 (Brach et al 1992b); LIF (Gruss et al 1992); interferon-γ (Narumi et al 1992; Yasumoto et al 1992; Ohmori & Hamilton 1993); PGE2 and forskolin (Muroi & Suzuki 1993); leukotriene B4 (Brach et al 1992a); insulin (Dominguez et al 1993); allogeneic stimulation (Jamieson et al 1991); p39 (Lalmanach-Girard et al 1993); PDGF (Olashaw et al 1992); X-rays (Brach et al 1993a); photofrin and red light (Ryter & Gomer 1993); pervanadate (Schieven et al 1993; Imbert et al 1994); partial hepatectomy (Tewari et al 1992); antimycin (Schulze-Osthoff et al 1993); emetine (Ghersa et al 1992). The activating stimuli listed in parentheses are less well characterized.

for example, an IκB-α kinase. Alternatively, IκB-α itself may be the common target of multiple independent pathways.

It is unclear whether all NF-κB activity that is released into the nucleus originates from a cytoplasmic pool that is bound to and inhibited by IκB-α. As discussed above, there appears to be at least one other pool of cytoplasmic complexes containing of precursor molecules p105 and p100. Despite almost total degradation of IκB-α in response to potent stimuli, a major portion of NF-κB protein usually remains in the cytoplasm (Sun et al 1994), presumably on account of the precursors. In addition, the putative distinct IκB-β protein could contribute to cytoplasmic retention as well. Despite such alternative cytoplasmic, IκB-α-independent pools, there is as yet no evidence for a direct, extracellular signal-responsive translocation of NF-κB complexes from such pools into nuclei.

Activation by TNF-α

Consistent with a role of NF-κB activity in a defensive response, the pro-inflammatory cytokines TNF-α, TNF-β, and IL-1 are potent activators of this transcription factor in a wide variety of cell types (Osborn et al 1989; Messer et al 1990; Hohmann et al 1990a; 1990b). TNF-α, one of the most potent effectors, signals through the 55 kd TNF-I receptor. Although the immediate receptor proximal events are unknown, a phosphatidylcholine-specific phospholipase C is activated to generate 1,2-diacylglycerol (DAG) in response to ligand (Dressler et al 1992; Schutze et al 1992). DAG, in turn, may initiate a signaling cascade that involves, in order, the activation of acidic sphingomyelinases—located in endosomal and lysosomal compartments—the subsequent generation of ceramide from sphingomyelin (Schutze et al 1992), and the ceramide-dependent activation of a serine/threonine kinase (Mathias et al 1991; Joseph et al 1993). In support of this pathway for TNF-α, NF-κB has been activated in permeabilized cells by exogenously added phospholipase C, DAG analogues, and purified acidic sphingomyelinase or ceramide (Schutze et al 1992). Neutral plasma membrane-associated sphingomyelinases have also been proposed to be involved in NF-κB activation, independent of DAG (Yang et al 1993). Addition of sphingomyelinase or cell-permeable ceramide analogues to intact cells induces NF-κB activity (Yang et al 1993).

Raf may lie on this pathway as well, since mitogen-activated protein (MAP) kinase kinase, a downstream effector of Raf function, is activated in response to TNF (Saklatvala et al 1993). Dominant negative Ras and Raf mutants block TNF-α-mediated activation of NF-κB (Devary et al 1993; Finco & Baldwin 1993). Mitochondria could present another downstream effector of ceramide because inhibitors of mitochondrial electron transport function impair TNF-induced activation of NF-κB (Schulze-Osthoff et al 1992). Likewise, depletion of this organelle had similar effects (Schulze-Osthoff et al 1993). As discussed below, mitochondria-generated reactive oxygen intermediates may be critical for activation. Other studies suggest an essential role for protein kinase C (PKC) zeta in TNF-α signaling to NF-κB, possibly as a component of the sphingomyelin pathway. A dominant negative PKC zeta protein inhibits NF-κB activation by TNF-α. Moreover, constitutively active PKC zeta activates NF-κB in NIH3T3 cells (Diaz-Meco et al 1993). Despite the production of DAG (which activates PKC), the activation of NF-κB by TNF-α appears to be independent of PKC. For example, inhibitors of PKC, which block PMA-induced activation of NF-κB, nevertheless fail to block TNF-α-induced activation (Meichle et al 1990).

IL-1 has been reported to signal via production of ceramide as well and it may feed into a pathway similar to that of TNF-α, as evidenced by the largely overlapping biological activities of the two cytokines (Mathias et al 1993). Herbimycin A, an inhibitor of tyrosine phosphorylation, interfers with IL-1α-induced NF-κB activation in T and B cells, which suggests a role for tyrosine kinases during early phases of signaling by this cytokine (Iwasaki et al 1992). Tyrosine kinases are also implicated in signaling by UV, gamma-irradiation, and bacterial lipopolysaccharide (LPS) (Devary et al 1993; Geng et al 1993; Schieven et al 1993).

Redox Regulation

Given the convergence of all signaling paths at or before the IκB-α target, it is possible that a common upstream effector exists, although no such molecule has been identified. However, several indirect lines of evidence suggest a role for reactive oxygen intermediates (ROI) as a common and critical denominator for various activating signals (Schreck et al 1992a,d). This conclusion is based largely on the inhibition of NF-κB activation by a variety of antioxidants. These reagents have been reported to block NF-κB activation in many instances, although the extent of this block appears to vary depending on cell and signal. Inhibitory antioxidants with diverse structural properties include N-acetyl-L-cysteine (NAC, a precursor of glutathione), dithiocarbamates, vitamin E derivatives, and various metal chelators.

Pyrolidinedithiocarbamate (PDTC) in particular has been widely used to inhibit the induced degradation of IκB-α, as shown with TNF-α and PMA stimulation of some Jurkat T cell lines (Beg et al 1993; Henkel et al 1993; Sun et al 1993). Activation of NF-κB by UV light is suppressed by NAC (Devary et al 1993). Support for the involvement of reactive oxygen intermediates as common messenger also derives from evidence showing elevated cellular levels for ROIs in response to TNF-α, IL-1; PMA, LPS, UV light, and gamma-irradiation (Schreck et al 1992a,d; Meyer et al 1993b, Geng et al 1993, Schieven et al 1993). Normal respiration produces basal levels of such intermediates. In the case of TNF-α, mitochondria may be involved in the signal-induced increase of such ROIs, as discussed above (Schulze-Osthoff et al 1992, 1993). Despite these observations, a direct functional role for ROIs in signaling to NF-κB remains to be proven.

Among various ROIs administered to cells in culture, such as hydrogen peroxide, hydroxyl radicals, superoxide, and singlet oxygen, only hydrogen peroxide is an effective activator of NF-κB (Schreck et al 1991, 1992a). It is possible that peroxides may specifically function in the NF-κB signaling cascade. Curiously, tyrosine kinase inhibitors block NF-κB activation following gamma-irradiation, a stimulant that is thought to work primarily through the immediate generation of ROI (Schieven et al 1993). Such inhibitors do not commonly block all NF-κB activation, which suggests that tyrosine kinase functions prior to the generation of ROIs.

Mitogenic Stimuli

Mitogenic stimuli for T and B cells activate NF-κB. In serum-deprived fibroblast cultures, NF-κB is transiently activated by serum stimulation, coincidently with a G_0-G_1 transition (Baldwin et al 1991). NF-κB is not activated in proliferating fibroblasts but may play a role in the initiation of the cell cycle from a resting state. Partial liver hepatectomy results in rapid tissue regeneration associated with immediate activation of NF-κB and the induction of many early response genes seen with other mitogenic systems (Tewari et al 1992). These cases indicate a possible link between NF-κB activation and cell proliferation.

Another connection between NF-κB function and growth is established through the induction of genes encoding growth factors and their receptors, such as IL-2, the IL-2 receptor-α chain, other growth-promoting cytokines, and the c-myc oncogene and IRF-1 and IRF-2 (Baldwin et al 1991; Duyao et al 1990, 1992; Kessler et al 1992; Harada et al 1994) (Table 3). Finally, NF-κB genes can become oncogenic upon genetic changes in structure and/or expression.

Activation by Viruses

The list of activators of NF-κB in Table 2 includes a growing number of viruses and viral products. Such activation is presumably beneficial to their lytic growth cycles; viruses may require NF-κB for transcription of their own genes, or they may require it to regulate various cellular genes. Induction of cellular growth-response genes may be critical to viral integration and replication functions. Viruses may activate NF-κB directly as part of the infection process through virion-associated molecules, or they may do so through viral products synthesized after infection.

Infection of T cells by HTLV I is known to immortalize these cells, a prelude to adult T cell leukemia (reviewed in Smith & Greene 1991). Transformation is mediated by the HTLV I Tax product (Nerenberg et al 1987; Grassman et al 1989), which activates a number of cellular transcription factors, among them NF-κB (Kelly et al 1992). Tax-mediated activation of NF-κB, in turn, contributes to the induced expression of many mitogen-response proteins (Kelly et al 1992) including the IL-2 growth factor and IL-2 receptor-α (Hoyos et al 1989; see also Smith & Greene 1991). The expression of Tax as a

Table 3 NF-γB-responsive genes

Cytokines and growth factors	Interleukin-1β
	Tumor necrosis factor α (TNF-α)
	Lymphotoxin (LT) (TNF-β)
	Interleukin-8
	IP-10
	Gro-α, -β and -γ
	(MIP-1α)
	(MCP-1/JE)
	(RANTES)
	Interleukin-2
	Macrophage colony-stimulating factor (M-CSF)
	Granulocyte/macrophage colony-stimulating factor (GM-CSF)
	Granulocyte colony-stimulating factor (G-CSF)
	(Erythropoietin)
	Interferon-β
	(Interferon-γ)
	Interleukin-6
	Proenkephalin
	(Transforming growth factor-β2) (TGF-β2)
Immunoreceptors	Immunoglobulin κ light chain (lg-κ-LC)
	T cell receptor β chain
	Major histocompatibility complex class I (MHC-I)
	(Major histocompatibility complex II) (MHC-II)
	β2-microglobulin
	MHC-II invariant chain
	Tissue factor-1
	Interleukin-2 receptor α chain
	(CD7)
Adhesion molecules	Endothelial-leukocyte adhesion molecule-1 (ELAM-1)
	Vascular cell adhesion molecule-1 (VCAM-1)
	(Intracellular cell adhesion molecule-1) (ICAM-1)
Acute phase proteins	Angiotensinogen
	Serum amyloid A precursor
	Complement factor B
	Complement factor C4
	(Urokinase-type plasminogen activator)
Viruses	Human immunodeficiency virus 1 (HIV-1)
	(Human immunodeficiency virus 2) (HIV-2)
	Simian immunodeficiency virus (macaques) (SIVmac)
	Cytomegalovirus (CMV)
	Adenovirus
	Herpes simplex virus 1 (HSV-1)
	(Human neurotropic virus) (JCV)
	(Simian virus 40) (SV-40)

Transcription factors and regulators	Rel p105 IκB-α
	Myc Interferon regulatory factor 1 (IRF-1) (Interferon regulatory factor 2) (IRF-2) A-20
Others	NO-synthetase (Apolipoprotein CIII)
	(Perforin)
	Vimentin (Decorin)

For references see text or recent reviews (Baeuerle 1991; Grilli et al 1993). Not cited in these reviews are: HSV-1 (Rong et al 1992); JCV (Ranganathan & Khalili et al 1993); tissue factor-1 (Mackman et al 1991); IP-10 (Ohmori & Hamilton 1993); erythropoietin (Lee-Huang et al 1993); A-20 (Krikos et al 1992); decorin (Santra et al 1994); MCP-1/JE (Rosi et al 1994); M-CSF (Brach et al 1993a; Sater et al 1994); interferon-γ (Sica et al 1992). A definitive involvement of NF-κB in the regulation of the genes listed in parentheses remains to be demonstrated.

transgene in mice caused tumors that regressed when treated with antisense RelA oligonucleotides, which implies a critical growth-associated role for the RelA subunit in these Tax-transformed tumor cells (Kitajima et al 1993).

Several possible mechanisms have been proposed for Tax activation of NF-κB including (*a*) the transcriptional induction of component parts of NF-κB by Tax (Arima et al 1991), (*b*) a post-translational mechanism (Lindholm et al 1992), possibly via generation of ROIs (Schreck et al 1992c), and (*c*) nuclear translocation of NF-κB and transactivation mediated by the reported physical interaction of Tax with the p50/p105 proteins (Hirai et al 1992; Suzuki et al 1993; Watanabe et al 1993). Recent discoveries indicate two potential ways in which Tax may activate NF-κB (T Kanno & U Siebenlist, unpublished observations). Tax was demonstrated to increase proteolytic degradation of the IκB-α protein, although by what means remains to be shown. Despite the continual, loss of IκB-α, high steady state levels of the inhibitor are maintained by NF-κB-induced new synthesis. Increased turnover of IκB-α may interfere with its inhibitory function, thus allowing some NF-κB to escape into the nucleus. It was also demonstrated that Tax can liberate p50/RelA NF-κB complexes held in the cytoplasm by the p100 precursor protein, in a manner analogous to that of IκB-α but independent of it. Tax physically interacts with p100, which may be the mechanism by which Tax antagonizes the cytoplasmic sequestration function encoded in the p100's ankyrin domain. A preferential interaction of Tax with p100 has recently been described elsewhere; its functional significance, however, is reported to be related to viral latency because p100 retains Tax in the cytoplasm, which could preclude nuclear activities of Tax (Beraud et al 1994). The physiological importance of the Tax-mediated release of NF-κB complexes from p100-inhibited cytoplasmic pools remains to be shown. Release of NF-κB from such stores, together with induced proteolysis of IκB-α, may be required to maintain constitutively high levels of nuclear NF-κB.

HIV infection of monocytic cells has been reported to lead to activation of NF-κB after several days in culture (Bachelerie et al 1991; Riviere et al 1991; Paya et al 1992; Roulston et al 1992, 1993). This might ensure continued expression of integrated virus because such expression is dependent on NF-κB-induced transcription through *cis*-acting κB elements in the HIV LTR (Nabel & Baltimore 1987; Ross et al 1991). The mechanism of this activation is unclear. In apparent contradiction to virus-induced activation, the HIV Nef product has been reported as an inhibitor of NF-κB (Niederman et al 1992). Rather than activating NF-κB itself, the HIV-1 virus may depend on environmental cofactors to stimulate its propagation, because resting populations of primary cells fail to replicate HIV (Pantaleo et al 1993b). Sites of inflammation or immune reactivity may represent primary loci for viral spread (Embretson et al 1993; Pantaleo et al 1993a), possibly due to the presence of agents that activate NF-κB, such as the cytokine TNF-α, bacterial toxins, and neutrophil-generated H_2O_2 (see Table 2). Finally, co-infection of patients with the HTLV-I virus can act as a cofactor for HIV replication (Pierik & Murphy 1991), potentially through activation of NF-κB by Tax.

Other Activating Conditions

Okadaic acid and calyculin A, inhibitors of phosphatase 1 and 2A, cause activation of NF-κB (Thevenin et al 1990; Menon et al 1993). Both drugs lead to phosphorylation and proteolytic degradation of IκB-α (Y-C Lin & U Siebenlist, unpublished observations). It is possible that the action of phosphatase inhibitors reflects the direct and regulated involvement of a negatively acting phosphatase in the signal transduction to IκB-α. Alternatively, phosphatases may play an indirect role, for example, by continually counteracting basal, unstimulated activities of kinases that would otherwise activate NF-κB; in this model the phosphatase inhibitors would simply unmask basal activities. According to a recent report, phosphatase inhibitors, by themselves, only function in transformed cells, whereas primary cells require an additional signal such as H_2O_2 to activate NF-κB (Menon et al 1993).

In T cells, the phosphatase calcineurin has been shown to be an important component in signaling NF-κB. T cells typically require two apparently distinct signaling events for IL-2 expression and growth induction, a calcium signal and a signal mediated through PKC (Crabtree 1989). Whereas PMA, acting through PKC, can activate NF-κB weakly in Jurkat T cells, PMA and calcium ionophore act synergistically. Stimulation with PHA is critically dependent on the calcium signal, as first shown by the block to activation of NF-κB by the immunosuppressant cyclosporin A (CsA) (Schmidt et al 1990). This immunosuppressant and the structurally unrelated but functionally equivalent drug FK506 also inhibits the calcium ionophore-dependent synergy with PMA (Mattila et al 1990; Frantz et al 1994). Cyclosporin A and FK506, in combination with cyclophilins and FK-binding proteins (FKBP), respectively, are known to inhibit the calcium-calmodulin-dependent phosphatase calcineurin A (Schreiber & Crabtree 1992). Recently it was shown that a constitutively active calcineurin can replace the synergizing calcium signal for the inactivation of IκB-α (Frantz et al 1994). Calcium-dependent calcineurin apparently functions to augment various signals, and depending on the signal, this may be critical in T cells. The okadaic acid studies imply the existence of negatively acting phosphatase, while calcineurin plays a positive role in activation of NF-κB.

Mature B cells, unlike pre B cells, exhibit constitutive NF-κB activity (Sen & Baltimore 1986a,b). This suggests a possible developmental role for NF-κB factors including the rearrangement and the continuous expression of the immunoglobulin κ light chain locus through its κB element in the intronic enhancer (Lichtenstein et al 1994). Confirming these connections, v-abl, which transforms cells at the preB stage, has been reported to block NF-κB activity as well as κ light chain transcription and rearrangement (Klug et al 1994; Chen et al 1994). Stimulation and differentiation of 70Z/3 preB cells with LPS or IL-1 induces κ light chain gene expression through activation of NF-κB (Sen & Baltimore 1986b). Initially p50/RelA dimers are activated, but later these are replaced by p50/Rel dimers, which are maintained in the nucleus. Mature B cells typically contain activated p50/Rel dimers and sometimes activated RelB-containing dimers. p50/RelA dimers stay in the cytoplasm, where they remain sensitive to stimulation (Liou & Baltimore 1993; Miyamoto et al 1994).

How is NF-κB activity maintained in the nucleus of mature B cells? While the steady-state level of IκB-α is not reduced in these cells relative to unstimulated pre B cells, the turnover of IκB-α is significantly increased in mature B cells (Miyamoto et al 1994). IκB-α half-life is decreased and new synthesis of IκB-α is increased, presumably as a result of induced transcription of the gene encoding IκB-α by NF-κB itself (see Tax-mediated activation above). It is possible that a continuous signaling event in B cells leads to induced degradation of IκB-α and causes long-term activation of Rel complexes. The inhibition of Rel by IκB-α may be less efficient than that of RelA, at least under conditions of increased turnover of IκB-α. Continuous signaling in B cells may be due to weak autocrine or paracrine stimulation by cytokines like IL-1 and TFN-α, cytokines whose expression is positively controlled by NF-κB itself.

MUTUAL REGULATION OF NF-κB AND IκB-α

Induced Synthesis of IκB-α

Cellular stimulation leading to IκB-α proteolysis and nuclear translocation of NF-κB also results in the subsequent rapid induction of IκB-α mRNA and protein, commencing at about 20 min after initial signaling (Brown et al 1993; Scott et al 1993; Sun et al 1993; see also Rice & Ernst 1993; Chiao et al 1994; Read et al 1994). Transfection of cells with transactivating NF-κB subunits induces high levels of endogenous IκB-α mRNA (Brown et al 1993; Scott et al 1993; Sun et al 1993; see also Chiao et al 1994). The cloning and sequencing of the IκB-α promoter has confirmed the predicted presence of κB binding sites, which have been shown to mediate transcriptional stimulation by NF-κB (de Martin et al 1993; LeBail et al 1993; Chiao et al 1994). Therefore, transactivating NF-κB dimers potently induce their own inhibitor, IκB-α, which presumably is destined to restore the inhibited state (Brown et al 1993). This built-in feedback inhibition may assure a transient response once the initiating event fades, an essential feature for the regulation of genes whose functions may be harmful if expressed unchecked.

Removal of the TNF-α stimulus causes rapid inhibition of nuclear NF-κB activity by newly induced IκB-α, as evidenced by continued nuclear NF-κB if IκB-α synthesis is blocked, even in the absence of TNF-α signaling (Brown et al 1993; K Brown & U Siebenlist, unpublished observations). Such observations suggest that newly induced IκB-α can enter nuclei to inhibit the previously activated NF-κB as the stimulating signal fades. Although this has not been demonstrated directly, two arguments further support this contention. First, IκB-α when overexpressed in transfected cells and pp40 in v-rel-transformed lymphoid cells is readily observed in nuclei (Cressman & Taub 1993; Zabel et al 1993; Davis et al 1990). Uncomplexed IκB-α is presumably free to passively diffuse into the nucleus, a situation somewhat analogous to that which exists in untransfected cells during the peak of induced IκB-α synthesis (Brown et al 1993). Complexed IκB-α, on the other hand, is unable to enter the nucleus, possibly because NF-κB dimers and IκB-α mutually inhibit each other. Second, IκB-α can remove bound p50/RelA from its cognate recognition sites on DNA in vitro (Zabel & Baeuerle 1990).

Induced Synthesis of NF-κB Proteins

The precursor proteins p105 and p100 are induced after cellular stimulation with agents that also activate NF-κB (Bours et al 1990; 1992b). Nuclear run-on experiments have demonstrated a transcriptional upregulation in the case of p105 (Gunter et al 1989; Bours et al 1990). Although the induction of the precursors is delayed relative to that of IκB-α, the precursors' genes may nevertheless be directly induced by NF-κB as well (Sun et al 1994). The p105 gene promotor contains functional κB elements (Ten et al 1992; Cogswell et al 1993). Induced expression of the precursors presumably replenishes NF-κB stores after processing. Prior to processing, however, the induced precursors may be critical to limit activation since they would be expected to help retain any newly synthesized transactivating proteins like Rel or RelA in the cytoplasm, which may be an important action if IκB-α is continually degraded in response to stimuli.

In addition to p100 and p105, the expression of the gene encoding Rel is transiently upregulated in response to stimuli that activate NF-κB (Bull et al 1989), and the Rel promoter is known to contain a κB element (Hannink & Temin 1990; Capobianco & Gilmore 1991). RelB may be similarly regulated (see Ryseck et al 1992). The expression of RelA, on the other hand, is essentially constitutive, only small increases, if any, have been noted following various cellular stimulations (Molitor et al 1990), and there are no κB elements in the RelA promoter (Ueberla et al 1993). Induced expression of NF-κB subunits (with the exception of RelA) may be necessary to sustain nuclear NF-κB activity over longer periods of stimulation, as reported for HL60 cells (Hohmann et al 1991). Stimulation of Jurkat T cells over several hours results in increasing amounts of Rel relative to RelA in the nucleus, possibly because of preferentially induced levels of Rel (Molitor et al 1990; Doerre et al 1993). Cells expressing constitutively active NF-κB, such as B cells or HTLV I-transformed T cells, display primarily p50/Rel heterodimers in their nuclei (Liou & Baltimore 1993; Miyamoto et al 1994).

NF-κB-REGULATED GENES: The Promoter Context

Select Groups Of Genes Controlled by NF-κB

Table 3 lists many of the genes whose expression is controlled by NF-κB. In a few instances, this control remains to be demonstrated directly. NF-κB induces, at least in part, many effectors of immune, inflammatory, or acute phase responses. In particular, numerous genes encoding cytokines/growth factors/chemokines, biological messengers that control and coordinate the functions of many cells, are responsive to NF-κB. Some of these messengers can use NF-κB to amplify their signal by a positive autoregulatory loop; the pro-inflammatory cytokines TNF-α, TNF-β, and IL-1 not only activate NF-κB, but they themselves are induced by NF-κB (Messer et al 1990; Hiscott et al 1993; see also Baeuerle 1991). In the case of the TNF-α gene, the observed importance of NF-κB for its expression may depend on the cell type and the segment of the enhancer/promoter region analyzed (Goldfeld et al 1993; Ziegler-Heitbrock et al 1993). Other critical genes induced by NF-κB include the acute phase regulator IL-6, the antiviral IFN-β, and various chemokines that summon cells to sites of inflammation such as IL-8, Gro, and possibly also MCP-1, Mip1-α, and Rantes (Anisowicz et al 1991; Grove & Plumb 1993; Joshi-Barve et al 1993; Danoff et al 1994; Shattuck et al 1994; see also Baeuerle 1991; Grilli et al 1993).

Migration of immune cells into inflamed tissue involves adhesion to the blood vessel endothelium and subsequent extravasation. This process requires NF-κB-mediated induced expression of the cell surface adhesion proteins, endothelial leukocyte adhesion molecule 1 (E-selectin or ELAM-1), and vascular cell adhesion molecule 1 (VCAM-1) (Iademarco et al 1992; Neish et al 1992; Kaszubska et al 1993; Shu et al 1993). The intercellular cell adhesion molecule 1 (ICAM-1) may also be under the control of NF-κB (Voraberger et al 1991). VCAM-1 and ICAM-1 are important in monocyte adherence to endothelial cells. Atherosclerotic lesions, which begin with monocyte depositions on blood vessel walls, may be caused by the NF-κB-induced inappropriate expression of cytokines and adhesion receptors on endothelial cells, possibly in response to oxidative stress signals, such as oxidized lipoproteins (Andalibi et al 1993; Collins 1993; Liao et al 1993; Parhami et al 1993). The association of NF-κB with adhesion is also suggested by experiments in which antisense RelA oligonucleotides (and in some instances antisense p50 oligonucleotides) administered to various cells, including embryonal stem cells, cause complete detachment from the substratum (Narayanan et al 1993; Sokoloski et al 1993). Also, PMA-induced adhesion of HL-60 cells could be inhibited by competitive binding of NF-κB in vivo (Eck et al 1993).

A number of viruses use the NF-κB transcription factor to regulate their own expression. Activation of NF-κB may be mediated by the virus or viral products, or it may depend on environmental signals (see above). CMV activates and is transcriptionally stimulated by NF-κB (Sambucetti et al 1989; Boldogh et al 1993; Kowalik et al 1993). Expression of HIV-1 is critically dependent on tandem κB sites in its LTR (Nabel & Baltimore 1987; Pierce et al 1988). This has been clearly demonstrated for the simian immunodeficiency virus (SIV) (Bellas et al 1993) in monocytes/macrophages but may also apply to HIV in these cells, primary T cells, and some T cell lines (Ross et al 1991). HIV-2 possesses only one κB element but contains sites for several other factors (Leiden et al 1992; Hannibal et al 1993).

In almost no cases does NF-κB act alone to regulate its target genes. Promoters or enhancers typically depend on multiple transcription factors for their activity. In some cases, the non-NF-κB factors may be constitutively active in cells, which makes NF-κB the sole mediator of signals. A recently emerging mechanism by which NF-κB and other DNA-binding factors may potently combine to induce transcription involves the direct physical association between these factors. Rather than simply combining the actions of their respective transactivation domains, cooperatively acting transcription factors can synergize to stimulate the process of transcription initiation. One consequence of physical interaction is cooperative DNA-binding on adjoining sites, essential for engagement of a weak binding site. Another consequence of physical interactions of bound factors may be to allow higher-order structures to form in which DNA is bent to bring upstream binding factors into close proximity with the basal transcription apparatus at the start site of transcription. Physical association of some factors with NF-κB may also result in negative regulation. A less obvious aspect of transcriptional control by association may be the selection of a particular transcription factor from among several that may otherwise compete for the same DNA site. Thus the particular NF-κB complex that binds a given κB site may be selected from among several by physical interaction with a transcription factor for an adjoining site.

Cooperative and Antagonistic Interactions With Multiple Transcription Factors

One of the best studied promoters/enhancers is that of the IFN-β gene, which is potently induced by viruses, in part through the activation of NF-κB (Hiscott et al 1989; Thanos & Maniatis 1992; Du et al 1993). Induction of IFN-β is an intracellular defense response against viral infection. Induction appears to be initiated by double-stranded RNA (dsRNA) (Visvanathan & Goodbourn 1989; Thanos & Maniatis 1992; Du et al 1993), probably through the action of the dsRNA-sensitive kinase (Hovanessian 1991). The IFN-β promoter contains a binding site for an ATF-2 homodimer and/or ATF-2/c-jun heterodimer, two sites for IRF-1 and a site for NF-κB (Miyamoto et al 1988; Harada et al 1989; Du & Maniatis 1992; Reis et al 1992; Thanos & Maniatis 1992; Du et al 1993); these sites lie next to each other in the order listed. In addition, the ATF-2 site is flanked by sequences recognized by the ubiquitous HMG I/Y, and the NF-κB site also harbors such a sequence within it (Thanos & Maniatis 1992; Du et al 1993). The current model envisions a cooperative interaction between HMG I/Y with NF-κB as well as with ATF-2. HMG I/Y facilitates binding by both transcription factors, most likely by protein-protein interactions, which have been demonstrated, and possibly also by this small protein's ability to bend DNA. Finally, ATF-2 and NF-κB interact with each other; this may bring the distal part of the enhancer bearing the ATF-2 complex next to the basal transcription factor apparatus, which lies close to the NF-κB complex, to facilitate transactivation by ATF-2 (Du et al 1993). The combination of factor interactions has one further critical consequence, namely the generation of a very stable protein-DNA complex at the promoter, perhaps a requirement for strong transactivation.

The potent induction of IFN-β by virus is most likely a consequence of the activation of all component parts, ATF-2, NF-κB, and IRF-1. Induction of IFN-β by other stimuli, for example by cytokines, may involve synergy between NF-κB and IRF-1 only (Fujita et al 1989; Abdollahi et al 1991). An analogous situation may exist in the MHC class I promoter, where adjoining binding sites for these two factors mediate the synergy of TNF-α (through NF-κB) and interferon-γ (through IRF-1) (Johnson & Pober 1994; Ten et al 1993) possibly by physical association (P Drew & K Ozato, unpublished observations).

The interplay of NF-κB with ATF factors appears to be critical for the induction of the E-selectin gene (ELAM-1) as well (Kaszubska et al 1993). This cell adhesion protein can be induced, for example, by cytokines, on endothelial cells where it acts to facilitate the binding and extravasation of neutrophils and a subset of leukocytes from the bloodstream into sites of inflammation (Collins 1993).

The interaction between NF-κB and ATF-2 may be analogous to the interaction between NF-κB and c-Jun and c-Fos, since all of these proteins that interact with NF-κB contain basic region plus leucine zipper (bZIP) domains (Stein et al 1993b) but are otherwise unrelated. In the case of c-Jun and c-Fos, it has been reported that their bZIP domain interacts with the RHD of p65 (Stein et al 1993a). Interaction between DNA binding/dimerization domains (such as bZIP and RHD) rather than between transactivation domains may explain cooperative DNA-binding. The functional synergy of NF-κB with c-Jun and/or c-Fos may be seen with either a κB binding site or an AP-1 binding site alone, i.e. only one of the factors is required to contact DNA, while the other appears to act as an accessory factor that does not bind DNA (Stein et al 1993a). It is not clear if such interactions occur in normal untransfected cells, or if these data reflect only what occurs when binding sites for these factors are appropriately juxtaposed on a given promoter element.

A functional interaction of NF-κB with basic helix-loop-helix transcription factors (bHLH) in the immunoglobulin κ enhancer may be expected due to the close proximity of their respective binding sites (Staudt & Lenardo 1991; Grilli et al 1993). This idea is supported by the demonstrated physical interaction of the *Drosophila* protein dorsal with several bHLH proteins (see below). Placement of the κ enhancer into *Drosophila* embryos results in striped patterns of expression determined by the interaction of dorsal and *Drosophila* bHLH proteins with the κB element and E boxes (mammalian bHLH-binding sites), respectively, on the mammalian enhancer (Gonzalez-Crespo & Levine 1994).

The Sp1 transcription factor acts in synergy with the NF-κB transcription factor to induce transcription of the HIV-1 LTR (Perkins et al 1993). Given that the spatial arrangement of the two respective binding sites is critical to this synergy, a direct interaction between these factors has been postulated. Sp1 is ubiquitous and is thought to act mostly in a constitutive manner.

The cytokine IL-6 and the chemokine IL-8 are synergistically induced by

NF-κB and the C/EBP family of transcription factors through adjoining binding sites in their promoters (NF-IL6 is one member of the C/EBP family) (Kunsch & Rosen 1993; Matsusaka et al 1993; Stein & Baldwin 1993). C/EBP proteins, like CREB/ATF and Jun/Fos, are bZIP DNA-binding factors. The bZIP motifs of several C/EBP proteins have been demonstrated to interact with several RHDs (Le Clair et al 1992; Matsusaka et al 1993; Stein et al 1993b). Binding by C/EBP family members to cognate sites and the resulting transactivation are augmented in the presence of NF-κB, even in the absence of a κB element, while the reverse is true at κB sites; high exogenous expression of C/EBP proteins inhibits transactivation through κB sites (Stein & Baldwin 1993; Stein et al 1993b). In the context of the IL-8 promoter, NF-κB binding at the κB site allows C/EBP to cooperatively engage a relatively weak adjoining C/EBP binding site, which leads to synergistic transactivation (Stein & Baldwin 1993). The NF-κB complex acting on the IL-8 promoter may not be the classic p50/RelA complex but rather a RelA homodimer (and possibly a RelA/Rel heterodimer) (Kunsch & Rosen 1993; Stein & Baldwin 1993). A RelA/Rel complex has been implicated in the control of the urokinase gene (Hansen et al 1992, 1994). Synergistic activation by NF-κB and C/EBP proteins may also be important in the regulation of angiotensinogen (Ron et al 1990; 1991), serum amyloid A (Betts et al 1993), and nitric oxide synthetase (Lowenstein et al 1993; Xie et al 1994), all of which are expressed during acute phase reactions. The interaction of NF-κB and C/EBP transcription factors could be the basis for the more global synergy between the cytokines IL-1 and IL-6 in acute phase responses (Betts et al 1993), T cell activation, and immunoglobulin secretion because IL-1 activates NF-κB and IL-6 potently activates C/EBP.

The anti-inflammatory properties of steroids may be mediated in part by direct interaction with NF-κB. The induced expression of the cytokine IL-6, which is dependent on functional synergy between NF-κB and C/EBP-β (NF-IL6), is specifically downregulated by dexamethasone-activated glucocorticoid receptors (Ray & Prefontaine 1994). This inhibition is mediated through a direct physical association of RelA (p65) with the ligand-bound receptor. In support of such a complex, the dexamethasone-induced activation of the mouse mammary tumor virus is inhibited by overexpression of RelA (Ray & Prefontaine 1994). Thus the global antagonistic activities of steroids and inflammatory agents could be explained by the physical complexing of steroid receptors and NF-κB.

Dorsal and Dif

Dorsal is an essential morphogen that determines dorsoventral polarity in the developing *Drosophila* embryos (Govind & Steward 1991; Ip & Levine, 1992; St. Johnston & Nüsslein-Volhard 1992). Thus far no equivalent process or protein has been identified in development of vertebrates. However, NTera-2 embryonal carcinoma cells contain little or no NF-κB proteins but can be induced with the differentiating agent retinoic acid to express these proteins and their activities (Segars et al 1993). Also, early developing thymocytes contain activated NF-κB (Zuniga-Pflucker et al 1993). These observations raise the possibility that NF-κB may act in early mammalian development as well. Dorsal is activated to enter nuclei prior to cellularization in a gradient fashion, dependent on a ventrally generated positional cue. This protein is lacking in dorsal-most nuclei, and peak levels of nuclear dorsal exist on the ventral-most side. The diverse patterns generated by dorsal are infinitely more complex than the dorsal gradient might predict. Dorsal accomplishes distinct and sharp patterns of expression of various regulatory genes by a number of mechanisms including (*a*) cooperative interactions with other proteins, in particular bHLH proteins, (*b*) differential engagement of cognate binding sites, depending on the affinity of the site and the concentration of dorsal, and (*c*) direct repressive effects of dorsal in combination with unknown proteins or indirectly through a specifically induced repressor.

Ventral-most areas give rise to embryonal mesoderm and the dorsal-induced proteins Twist and Snail are essential to this process (Ip et al 1992b, Jiang et al 1992; Jiang & Levine 1993). Twist, a bHLH protein, requires high levels of dorsal for expression because of relatively weak dorsal binding sites in its promoter (Thisse et al 1988; Jiang et al 1992). This bHLH protein then cooperates with dorsal to assure strong and uniform expression of certain target genes, especially if they contain weak binding sites for dorsal and depend on both proteins, as does Snail. Snail is a zinc finger DNA-binding protein that functions as a repressor (Kosman et al 1991; Leptin 1991; Rao et al 1991; Ip et al 1992a,b; Jiang & Levine 1993). Since Snail depends on both dorsal and Twist for expression, it is expressed with a sharp border as a result of the multiple effects of a shallow dorsal gradient and a somewhat steeper Twist gradient (Kosman et al 1991; Leptin 1991; Ip et al 1992b). Rhomboid, a putative transmembrane receptor essential for differentiation of the ventral epidermis, is expressed only in ventral-lateral regions, the presumptive neuroectoderm (Bier et al 1990). Low levels of dorsal can induce Rhomboid expression through synergistic effects with juxtaposed binding sites for various bHLH proteins that include Daughterless, Achaete, and Scute (Ip et al 1992a; Gonzalez-Crespo & Levine 1993). Expression in the mesoderm is specifically repressed by Snail, which may physically overlap some of the binding sites for the bHLH proteins (Ip et al 1992a; Kasai et al 1992). Finally, dorsal directly represses ventral and ventrolateral expression of Zerknullt and Decapentaplegic, two proteins detected only in dorsal regions. This effect appears to be mediated by an unknown co-repressor that cooperatively acts with dorsal to repress gene expression across long distances on the chromosome (Jiang et al

1992; Kirov et al 1993). Whereas dorsal itself is intrinsically an activator, here it acts as a repressor. The mechanism for this is not known, but is likely to involve cooperative DNA binding interactions, as seen with bHLH proteins.

Recent discoveries in *Drosophila* have shown that the connection between regulation of defense systems and NF-κB is evolutionarily conserved. Insects display a rapid response to bacterial infection, including phagocytosis and elaboration of several diverse anti-bacterial proteins, like cecropins, attacins, diptericins, defensins, and lysozymes (reviewed by Boman et al 1991; Hultmark 1993). Many of the genes for these proteins are regulated by an inducible NF-κB-like factor through a κB-like element in their promoters. This was first discovered through studies of the cecropin and diptericin genes of the giant moth Cecropia (Sun et al 1991; Sun & Faye 1992; Engstrom et al 1993; Kappler et al 1993). This factor has now been cloned in *Drosophila* by homology to RHD proteins, and is termed Dif, for dorsal-related immunity factor (Ip et al 1993). The RHD of Dif is nearly as related to vertebrate Rel/NF-κB proteins as it is to dorsal. Dif is not expressed during early embryogenesis but rather during later stages of development. Nuclear Dif activity is inducible from cytoplasmic stores in cells of the fat body and in hemolymph. It is likely that cactus, the IκB inhibitor of dorsal, also functions as the inhibitor of Dif, given cactus's continued expression in adult tissues.

An ancient acute phase-like response is speculated to be the original function for the Rel/NF-κB/dorsal/Dif proteins (Ip et al 1993; Hultmark 1994).

Bcl-3: A Transcriptional Activator

Bcl-3 is a member of the IκB family of proteins, but recent evidence suggests that unlike IκB-α, this protein activates transcription through κB sites. Two mechanisms for activation have been proposed, an indirect one, in which Bcl-3 antagonizes inhibitory p50 homodimers (Franzoso et al 1992, 1993), and a direct mechanism in which Bcl-3 acts as an accessory factor, coupling with otherwise inert p52 homodimers to form competent transactivators (Bours et al 1993).

What is the evidence for an inhibitory role for p50 homodimers? Numerous transfection studies in a number of cells indicate that p50 homodimers are unable, by themselves, to significantly transactivate a variety of κB-dependent reporter constructs, and this is consistent with the lack of an identifiable transactivation domain on p50 (Schmid et al 1991; Schmitz & Baeuerle 1991; Ballard et al 1992; Bours et al 1992b, 1993; Franzoso et al 1992, 1993; Kunsch et al 1992; Mercurio et al 1992; Ryseck et al 1992; see also Fujita et al 1992, 1993; Kretzschmar et al 1992; Moore et al 1993 for opposing views). Indeed, p50 homodimers usually behave as transcriptional inhibitors in vivo, counteracting potent transactivation by p50/RelA heterodimers by direct competition for κB sites (Schmitz & Baeuerle 1991; Franzoso et al 1992, 1993). This inhibition is particularly noticeable for κB sites that display a preference for the p50 homodimers (Franzoso et al 1993). In resting, unstimulated cells, the ubiquitous p50 homodimers localize largely to nuclei where they may help to protect some genes against inadvertent induction by low levels of the potently transactivating NF-κB dimers (Franzoso et al 1993).

In vivo data further support an inhibitory role for p50 homodimers. The expression of IL-2 in resting, non-transformed mouse T cell clones is under negative control by p50 homodimers. Only a complete stimulation protocol with peptide and antigen presenting cells results in expression of IL-2 and correlates with the specific removal of p50 homodimers from DNA (Kang et al 1992). p50 homodimers also appear to inhibit expression of MHC class I genes. Highly metastatic cell variants, expressing few class I genes, display high levels of these homodimers, whereas the reverse is true for less tumorigenic variants (Plaksin et al 1993). It is conceivable, of course, that p50 homodimers, although apparently unable to transactivate by themselves, could do so indirectly by cooperating with other transcription factors in the context of a particular promoter.

Bcl-3 preferentially targets p50 homodimers and p52 homodimers. It not only interferes with the DNA binding of p50 homodimers, but also dissociates already bound homodimers from their cognate binding sites (Hatada et al 1992; Franzoso et al 1993; Nolan et al 1993). Thus Bcl-3 can promote transcription by removing tightly bound inhibitory p50 homodimers from DNA and freeing such sites for binding by potently transactivating heterodimers such as p50/ RelA (Franzoso et al 1992, 1993). Such a situation may exist during mitogenic stimulation of primary T cells, where prebound p50 homodimers would be expected to block NF-κB heterodimers from engaging these κB sites, at least until the newly induced Bcl-3 protein enters the nuclei to relieve inhibition (Franzoso et al 1993). Preliminary evidence suggests that normal antigenic signaling of non-transformed T cells induces Bcl-3 (M Lenardo & U Siebenlist, 1994). Thus Bcl-3 could be the protein that removes p50 homodimers from the IL-2 promoter to allow expression of this cytokine. This form of regulation is very different from that of IκB-α in that it is positive and occurs in the nucleus.

Bcl-3 can also participate directly in transactivation through targeting of p52 homodimers (Bours et al 1993). p52 is structurally and functionally highly similar to p50, although its expression is generally lower, and homodimers are rarely observed, except in a few highly differentiated cells such as myelomas (Bours et al 1993; Chang et al 1994; G Franzoso & U Siebenlist, unpublished observations). Bcl-3 tightly associates with both homodimers, but unlike p50 homodimers, p52 homodimers are not efficiently dissociated from κB binding sites. Rather, Bcl-3 and p52 homodimers form ternary complexes with the κB sites. This enables these two proteins to induce transcription through the κB site, owing to transactivation domains residing on Bcl-3. Bcl-3 harbors two such domains located outside the ankyrin domain on the N- and C-terminal parts of the protein (Bours et al 1993). Significant transactivation by Bcl-3 and p50 homodimers is not observed, regardless of the p50 construct used for transfections, probably because of efficient Bcl-3-mediated dissociation of p50 homodimers from DNA (Bours et al 1993; G Franzoso & U Siebenlist, unpublished observations). However, others have suggested that Bcl-3 and p50 homodimers together can transactivate, albeit weakly (Fujita et al 1993). While direct transactivation by Bcl-3 through p52 homodimers is potent by comparison with p50 homodimers, it is conceivable that cell- or signal-dependent modifications of p50 or Bcl-3 (such as phosphorylation; see Nolan et al 1993) might also allow these molecules to transactivate.

It is interesting to speculate that the nuclear presence of p52 homodimers in some highly differentiated cells, together with constitutive expression of Bcl-3 (Bhatia et al 1991; Bours et al 1993; G Franzoso & U Siebenlist, unpublished observations), could result in transactivation of a limited set of genes in such cells, even in the absence of any stimulating signal. In this way, transactivation of select genes could occur by an IκB-α-independent pathway, a scenario that might be attractive to short-lived terminally differentiated cells.

NF-κB, TUMORIGENESIS AND APOPTOSIS

The v-Rel oncogene was isolated from the acutely transforming and replication-defective avian retrovirus REV-T. Infection of young chicks typically results in fatal tumors within 7 to 10 days. These tumors are primarily of hematopoietic origin, often with markers present for lymphoid or myeloid cells. Chicken embryo fibroblasts, splenocytes, and bone marrow cells can also be transformed in vitro (Rice & Gilden 1988; Boehmelt et al 1992; Bose 1992; Morrison et al 1992). In addition to various internal amino acid differences, v-Rel lacks the most C-terminal transactivation domain present on its cellular homologue Rel. Several conflicting theories have been advanced in the past regarding the mechanism by which v-Rel transforms cells and also regarding the potential of Rel to transform (Gilmore 1992; Boehmelt et al 1992; Bose 1992; Walker et al 1992). Recently, however, a new consensus has emerged: (*a*) Rel, when overexpressed in sufficient quantities, can transform some cells, albeit much more weakly than v-Rel (Hrdlickova et al 1994a; Kralova et al 1994). (*b*) C-terminal deletions of Rel, as well as particular RHD mutations in Rel, all enhance the tumorigenic potential. The C-terminal deletions eliminate a transactivation domain, as well as a cytoplasmic retention function; it is not known how this domain contributes to cytoplasmic localization. A similarly acting domain has also been noted at the C-terminus of dorsal; these domains may assist IκB-α-mediated inhibition (see above) (Hrdlickova et al 1994a; Kralova et al 1994; Nehyba et al 1994). (*c*) The amino acid changes in the most N-terminal part of the v-Rel RHD, as compared to Rel, may subtly affect the DNA binding properties relative to Rel, whereas mutations located within a region, including the C-terminal part of the RHD, may change the interaction with the IκB-α inhibitor (Diehl et al 1993). IκB-α was significantly less efficient in inhibiting DNA binding by Rel proteins that carry these internal mutations. This may help to establish an increased nuclear presence of v-Rel. (*d*) The transactivation domain that remains in v-Rel is necessary for transformation, which suggests that transcriptional stimulation of at least some target genes is required (Sarkar & Gilmore 1993). This refutes earlier scenarios in which the oncogenic properties of v-Rel were attributed to its apparent inhibition of p50/RelA-mediated transactivation, a phenomenon that could simply reflect a generally weaker transactivation potential of v-Rel, at least in most cells (see above) (Ballard et al 1990, 1992; Inoue et al 1991; Richardson & Gilmore 1991; McDonnell et al 1992). In support of a critical role for gene transactivation by v-Rel, this protein was reported to be a particularly potent activator of specific genes such as those coding for MHC class I and II proteins and IL-2 receptor-α (Boehmelt et al 1992; Hrdlickova et al 1994a, 1994b; Nehyba et al 1994).

The C-terminal transactivation domain of Rel, which is deleted in v-Rel, may normally play a role in the induction of anti-oncogenic functions such as apoptosis. Expression of high amounts of Rel was correlated with apoptosis and autophagocytosis in a large variety of cells in developing chick embryos. Furthermore, avian bone marrow cells expressing high amounts of Rel undergo a form of programmed cell death (Abbadie et al 1993). These effects appear to be cell specific because high expression in primary chick embryo fibroblasts transforms them and extends their life span (Abbadie et al 1993; Kralova et al 1994). The C-terminal deletion in v-Rel may prevent the induction of cell death in lymphoid cells because v-Rel transformed bursal lymphocytes are resistant to apoptotic stimuli like radiation, calcium ionophore, or dexamethasone (Neiman et al 1991).

Consistent with a transforming potential of Rel in avian cells, disruptions of the human Rel locus have been observed in several tumors, although no conforming pattern has been noted (Lu et al 1991). In addition to an association of v-Rel and Rel with avian tumors, the p52/p100 gene is translocated in a number of human B and T cell tumors and in particular in cutaneous lymphomas (Neri et al 1991; Fracchiolia et al 1993). In all cases, the translocations physically and functionally disrupt the ankyrin domain, which leads to proteins with an intact RHD and a remaining variably sized ankyrin repeat domain. These proteins are further processed into mature p52 proteins, and both forms are present and can be observed in nuclei, unlike the p100 precursor (Chang et al 1994; G Franzoso & U Siebenlist, unpublished observations) [although complexes of the precursor have been reported to have some DNA-binding activity (Potter et al 1993; Scheinman et al 1993)]. It is not known how the truncated p100 proteins contribute to tumorigenesis. It is possible that loss of a functional ankyrin domain significantly weakens cytoplasmic sequestration, thus allowing transactivating Rel/NF-κB dimers to more readily translocate to the nucleus.

As noted, Bcl-3 is another member of the NF-κB/IκB family that has been associated with rare forms of chronic lymphocytic leukemias (Ohno et al 1990). Bcl-3 is overexpressed in these tumor cells by its juxtaposition to the immunoglobulin enhancer. To date, no cell lines have been established from these leukemias, and it is not known how Bcl-3 contributes to tumorigenesis. However, based on the research cited above, one may predict a role for overexpressed Bcl-3 in inappropriate transactivation of specific κB-dependent target genes.

NF-κB AS A TARGET FOR THERAPIES

NF-κB in its various forms has many wide-ranging effects that are controlled by a complex regulatory network of inhibitors and co-activators. Given the intimate connection between defense reactions and NF-κB, this transcription factor and its regulators could provide central targets for therapeutic intervention in various diseases or pathological conditions such as toxic/septic shock, graft vs host reactions, acute inflammatory conditions, acute phase responses, radiation damage, atherosclerosis, and cancer. NF-κB could also be a therapeutic target against infections by viruses such as HIV-1. Specific targets for small synthetic inhibitors could be the transcription factor(s) itself or any of the essential signaling proteins including specific protein kinases and the protease(s) that digest IκB-α. In contrast to short-term treatments, a long-term systemic block of NF-κB activity, if desired, would likely severely suppress many important host functions. In these cases, a less broadly acting target for blocking may be advisable, such as that provided by the combination of NF-κB with certain other transcription factors in the context of a given promoter. In this way, limited sets of relevant genes could be inhibited.

ACKNOWLEDGMENTS

We thank all members of our laboratory for their many contributions to this review. We are grateful to Dr. Anthony S. Fauci for his continuing support. We apologize for being unable to cite many important contributions in this field due to the extremely large number of relevant publications. We thank M. Rust for her skilled assistance in the preparation of this review.

Literature Cited

Abbadie C, Kabrun N, Bouali F, Smardova J, Stehelin D, et al. 1993. High levels of c-rel expression are associated with programmed cell death in the developing avian embryo and in bone marrow cells in vitro. *Cell* 75:899–912

Abdollahi A, Lord KA, Hoffman-Liebermann B, Liebermann D. 1991. Interferon regulatory factor 1 is a myeloid differentiation primary response gene induced by interleukin 6 and leukemia inhibitory factor: role in growth inhibition. *Cell Growth Differ.* 2:401–7

Andalibi A, Liao F, Imes S, Fogelman AM, Lusis AJ. 1993. Oxidized lipoproteins influence gene expression by causing oxidative stress and activating the transcription factor NF-κB. *Biochem. Soc. Trans.* 21:651–55

Anisowicz A, Messineo M, Lee SW, Sager R. 1991. An NF-κB-like transcription factor mediates IL-1/TNF-α induction of gro in human fibroblasts. *J. Immunol.* 147:520–27

Arima N, Molitor JA, Smith MR, Kim JH. 1991. Human T-cell leukemia virus type I Tax induces expression of the Rel-related family of κB enhancer-binding proteins: evidence for a pretranslational component of regulation. *J. Virol.* 65:6892–99

Bachelerie F, Alcami J, Arenzana-Seisdedos F, Virelizier J-L. 1991. HIV enhancer activity perpetuated by NF-κB induction on infection of monocytes. *Nature* 350:709–12

Baeuerle PA. 1991. The inducible transcription activator NF-κB: regulation by distinct protein subunits. *Biochim. Biophys. Acta* 1072:63–80

Baeuerle PA, Baltimore D. 1988a. IκB: a specific inhibitor of the NF-κB transcription factor. *Science* 242:540–46

Baeuerle PA, Baltimore D. 1988b. Activation of DNA-binding activity in an apparently cytoplasmic precursor of the NF-κB transcription factor. *Cell* 53:211–17

Baeuerle PA, Baltimore D. 1989. A 65-kD subunit of active NF-κB is required for inhibition of NF-κB by IκB. *Genes Dev.* 3: 1689–98

Baeuerle PA, Henkel T. 1994. Function and activation of NF-κB in the immune system. *Annu. Rev. Immunol.* 12:141–79

Bakalkin GYA, Yakovleva T, Terenius L. 1993. NF-κB-like factors in the murine brain. Developmentally-regulated and tissue-specific expression. *Mol. Brain Res.* 20:137–46

Baldwin AS, Azizkhan JC, Jensen DE, Beg AA, Coodly LR. 1991. Induction of NF-κB DNA-binding activity during the G_0 to G_1 transition in mouse fibroblasts. *Mol. Cell. Biol.* 11:4943–51

Baldwin AS, Sharp PA. 1988. Two transcription factors, NF-κB and H2TF1, interact with a single regulatory sequence in the class I major histocompatibility complex promoter. *Proc. Natl. Acad. Sci. USA* 85:723–27

Ballard DW, Dixon EP, Peffer NJ, Bogerd H, Doerre S, et al. 1992. The 65-kDa subunit of human NF-κB functions as a potent transcriptional activator and a target for v-Rel-mediated repression. *Proc. Natl. Acad. Sci. USA* 89:1875–79

Ballard DW, Walker WH, Doerre S, Sista P, Molitor JA, et al. 1990. The V-REL oncogene encodes a κB enhancer binding protein that inhibits NF-κB function. *Cell* 63:803–14

Beg AA, Baldwin AS. 1993. The IκB proteins: multifunctional regulators of Rel/NF-κB transcription factors. *Genes Dev.* 7:2064–70

Beg AA, Finco TS, Nantermet PV, Baldwin AS. 1993. Tumor necrosis factor and interleukin-1 lead to phosphorylation and loss of IκB α: a mechanism for NF-κB activation. *Mol. Cell. Biol.* 13:3301–10

Beg AA, Ruben SM, Scheinman RI, Haskill S, Rosen CA, et al. 1992. IκB interacts with the nuclear localization sequences of the subunits of NF-κB: a mechanisms for cytoplasmic retention. *Genes Dev.* 6:1899–913

Bellas RE, Hopkins N, Li Y. 1993. The NF-κB binding site is necessary for efficient replication of simian immunodeficiency virus of macaques in primary macrophages but not in T cells in vitro. *J. Virol.* 67:2908–13

Beraud C, Sun S-C, Ganchi P, Ballard DW, Green WC. 1994. Human T-cell leukemia virus type I tax associates with and is negatively regulated by the NF-κB2 p100 gene product: implications for viral latency. *Mol. Cell. Biol.* 14:1374–82

Betts JC, Cheshire JK, Akira S, Kishimoto T, Woo P. 1993. The role of NF-κB and NF-IL-6 transactivating factors in the synergistic activation of human serum amyloid A gene expression by interleukin-1 and interleukin-6. *J. Biol. Chem.* 268:25624–31

Bhatia K, Huppi K, McKeithan T, Siwarski D. 1991. Mouse bcl-3: cDNA structure, mapping and stage-dependent expression in B lymphocytes. *Oncogene* 6:1569–73

Bier E, Jan LY, Jan Y. 1990. Rhomboid, a gene required for dorsoventral axis establishment and peripheral nervous system development in *Drosophila melanogaster*. *Genes Dev.* 4: 190–203

Blank V, Kourilsky P, Israël A. 1991. Cytoplasmic retention, DNA binding and processing of the NF-κB p50 precursor are controlled by a small region in its C-terminus. *EMBO J.* 10:4159–67

Blank V, Kourilsky P, Israël A. 1992. NF-κB

and related proteins: Rel/dorsal homologies meet ankyrin-like repeats. *Trends Biochem. Sci.* 17:135–40

Boehmelt G, Walker A, Kabrun N, Melitzer G, Beug H, et al. 1992. Hormone-regulated v-rel estrogen receptor fusion protein: reversible induction of cell transformation and cellular gene expression. *EMBO J.* 11:4641–52

Boldogh I, Fons MP, Albrecht T. 1993. Increased levels of sequence-specific DNA-binding proteins in human cytomegalovirus-infected cells. *Biochem. Biophys. Res. Commun.* 197:1505–10

Boman HG, Faye I, Gudmundsson GH, Lee J-Y, Lidholm DA. 1991. Cell free immunity in cecropia—a model system for antibacterial proteins. *Eur. J. Biochem.* 201:23–31

Bose HJ. 1992. The Rel family: models for transcriptional regulation and oncogenic transformation. *Biochim. Biophys. Acta* 1114:1–17

Bours V, Azarenko V, Dejardin E, Siebenlist U. 1994. Human RelB (I-Rel) functions as a κB site-dependent transactivating member of the family of Rel-related proteins. *Oncogene.* In press

Bours V, Burd PR, Brown K, Villalobos J, Park S, et al. 1992b. A novel mitogen-inducible gene product related to p50-p105-NF-κB participates in transactivation through a κB site. *Mol. Cell. Biol.* 12:685–95

Bours V, Franzoso G, Azarenko V, Park S, Kanno T, et al. 1993. The oncoprotein Bcl-3 directly transactivates through κB motifs via association with DNA-binding p50B homodimers. *Cell* 72:729–39

Bours V, Franzoso G, Brown K, Park S, Azarenko V, et al. 1992a. Lymphocyte activation and the family of NF-κB transcription factor complexes. *Curr. Top. Microbiol. Immunol.* 182:411–20

Bours V, Villalobos J, Burd PR, Kelly K, Siebenlist U. 1990. Cloning of a mitogen-inducible gene encoding a κB DNA-binding protein with homology to the rel oncogene and to cell-cycle motifs. *Nature* 348:76–80

Brach MA, Arnold C, Kiehntopf M, Gruss HJ, Herrman F. 1993a. Transcriptional activation of the macrophage colony-stimulating factor gene by IL-2 is associated with secretion of bioactive macrophage colony-stimulating factor protein by monocytes and involves activation of the transcription factor *J. Immunol.* 150:5535–43

Brach MA, deVos S, Arnold C, Gruss H-J, Mertelsmann R, et al. 1992a. Leukotriene B4 transcriptionally activates interleukin-6 expression involving NK-κB and NF-IL-6. *Eur. J. Immunol.* 22:2705–11

Brach MA, Gruss HJ, Kaisho T, Asano Y. 1993b. Ionizing radiation induces expression of interleukin 6 by human fibroblasts involving activation of nuclear factor-kappa B. *J. Biol. Chem.* 268:8466–72

Brach MA, Gruss HJ, Riedel D, Mertelsmann R, Hermann F. 1992b. Activation of NF-κB by IL-2 in human blood monocytes. *Cell Growth Differ.* 3:421–27

Brach MA, Henschler R, Mertelsmann RH, Herrman F. 1991. Regulation of M-CSF expression by M-CSF: role of protein kinase C and transcription factor. *Pathobiology* 59: 284–88

Bressler P, Brown K, Timmer W, Bours V, Siebenlist U, et al. 1993. Mutational analysis of the p50 subunit of NF-κB and inhibition of NF-κB activity by *trans*-dominant p50 mutants. *J. Virol.* 67:288–93

Brown K, Park S, Kanno T, Franzoso G, Siebenlist U. 1993. Mutual regulation of the transcriptional activator NF-κB and its inhibitor, I kappa B-α. *Proc. Natl. Acad. Sci. USA* 90:2532–36

Brownell E, Mathieson B, Young HA, Keller J, Ihle JN, et al. 1987. Detection of c-rel-related transcripts in mouse hematopoietic tissues, fractionated lymphocyte populations, and cell lines. *Mol. Cell. Biol.* 7:1304–9

Brownell E, Mittereder N, Rice NR. 1989. A human rel protooncogene cDNA containing an Alu fragment as a potential coding exon. *Oncogene* 4:935–42

Bruder JT, Heidecker G, Tan T-H, Weske JC, Derse D, et al. 1993. Oncogene activation of HIV-LTR-driven expression via the NF-κB binding sites. *Nucleic Acids Res.* 21:5229–34

Bull P, Hunter T, Verma IM. 1989. Transcriptional induction of the murine c-rel gene with serum and phorbol-12-myristate-13-acetate in fibroblasts. *Mol. Cell. Biol.* 9:5239–43

Bull P, Morley KL, Hoekstra MF, Hunter T, Verma IM. 1990. The mouse c-rel protein has an N-terminal regulatory domain and a C-terminal transcriptional *trans*-activation domain. *Mol. Cell. Biol.* 10:5473–85

Capobianco AJ, Gilmore TD. 1991. Repression of the chicken c-rel promoter by v-Rel in chicken-embryo fibroblasts is not mediated through a consensus NF-κ-B binding site. *Oncogene* 6:2203–10

Capobianco AJ, Simmons DL, Gilmore TD. 1990. Cloning and expression of a chicken c-rel cDNA: unlike p69v-rel, p68c-rel is a cytoplasmic protein in chicken embryonic fibroblasts. *Oncogene* 5:257–65

Carrasco D, Ryseck R-P, Bravo R. 1993. Expression of rel/B transcripts during lymphoid organ development: specific expression in dendritic antigen-presenting cells. *Development* 118:1221–31

Chang CC, Zhang J, Lombardi L, Neri A, Dalla-Favera R. 1994. Mechanism of expression and role in transcriptional control of the proto-oncogene NF-κB-2/lyt-10. *Oncogene* 9:923–33

Chen Y-Y, Wang LC, Huang MS, Rosenberg N. 1994. An active v-able protein tyrosine kinase blocks immunoglobulin light-chain gene rearrangement. *Genes Dev.* 8:688–97

Chiao PJ, Miyamoto S, Verma IM. 1994. Autoregulation of IκBα activity. *Proc. Natl. Acad. Sci. USA* 91:28–32

Chung SY, Folsom V, Wooley J. 1983. DNase I-hypersensitive sites in the chromatin of immunoglobulin κ light chain genes. *Proc. Natl. Acad. Sci. USA* 80:2427–31

Cogswell PC, Scheinman RI, Baldwin AS. 1993. Promoter of the human NF-κB p50/p105 gene: regulation by NF-κB subunits and by c-Rel. *J. Immunol.* 150:2794–804

Coleman TA, Kunsch C, Maher M, Ruben SM, Rosen CA. 1993. Acquisition of NF-κB1-selective DNA binding by substitution of four amino acid residues from NF-κB1 into RelA. *Mol. Cell. Biol.* 13:3850–59

Collins T. 1993. Biology of disease. Endothelial nuclear factor-κB and the initiation of the atherosclerotic lesion. *Lab. Invest.* 68:499–506

Cordle SR, Donald R, Read MA, Hawiger J. 1993. Lipopolysaccharide induces phosphorylation of MAD3 and activation of c-Rel and related NF-κB proteins in human monocytic THP-1 cells. *J. Biol. Chem.* 268:11803–10

Costello R, Lipcey C, Algarte M, Cerdan C, Baeuerle PA, et al. 1993. Activation of primary human T-lymphocytes through CD2 plus CD28 adhesion molecules induces long-term nuclear expression of NF-κB. *Cell Growth Differ.* 4:329–39

Crabtree GR. 1989. Contingent genetic regulatory events in T lymphocyte activation. *Science* 243:355–61

Cressman DE, Taub R. 1993. IκB alpha can localize in the nucleus but shows no direct transactivation potential. *Oncogene* 8:2567–73

Danoff TM, Lalley PA, Chang YS, Heeger PS, Neilson EG. 1994. Cloning, genomic organization, and chromosomal localization of the *scya5* gene encoding the murine chemokine RANTES. *J. Immunol.* 152:1182

Davis N, Bargmann W, Lim M-Y, Bose H. 1990. Avian reticuloendotheliosis virus-transformed lymphoid cells contain multiple pp59v-rel complexes. *J. Virol.* 64:584–91

Davis N, Ghosh S, Simmons DL, Tempst P, Liou H, et al. 1991. Rel-associated pp40: an inhibitor of the rel family of transcription factors. *Science* 253:1268–71

de Martin R, Vanhove B, Cheng Q, Hofer E, Csizmadia V, et al. 1993. Cytokine-inducible expression in endothelial cells of an IκBα-like gene is regulated by NF-κB. *EMBO J.* 12:2773–79

Devary Y, Rosette C, DiDonato JA, Karin M. 1993. NF-κB activation by ultraviolet light not dependent on a nuclear signal. *Science* 261:1442–45

Diaz-Meco MT, Berra E, Municio MM, Sanz L, Lozano J, et al. 1993. A dominant negative protein kinase c zeta subspecies blocks NF-κB activation. *Mol. Cell. Biol.* 13:4770–75

Diehl JA, McKinsey TA, Hannink M. 1993. Differential pp40/IκB-β inhibition of DNA binding by rel proteins. *Mol. Cell. Biol.* 13:1769–78

Dobrzanski P, Ryseck R-P, Bravo R. 1993. Both N- and C-terminal domains of RelB are required for full transactivation: role of the N-terminal leucine zipper-like motif. *Mol. Cell. Biol.* 13:1572–82

Doerre S, Sista P, Sun S-C, Ballard DW, Greene WC. 1993. The c-rel protooncogene product represses NF-κB p65-mediated transcriptional activation of the long terminal repeat of type 1 human immunodeficiency virus. *Proc. Natl. Acad. Sci. USA* 90:1023–27

Dokter WHA, Dijkstra AJ, Koopmans SB, Stulp BK, Keck W, et al. 1994. G(Anh) MTetra, a natural bacterial cell wall breakdown product, induces interleukin-1β and interleukin-6 expression in human monocytes. *J. Biol. Chem.* 269:4201–6

Dominguez I, Sanz L, Arenzana-Seisdedos F, Diaz-Meco MT. 1993. Inhibition of protein kinase C zeta subspecies blocks the activation of an NF-κB-like activity in *Xenopus laevis* oocytes. *Mol. Cell. Biol.* 13:1290–95

Dressler KA, Mathias S, Kolesnick RN. 1992. Tumor necrosis factor-alpha activates the sphingomyelin signal transduction pathway in a cell-free system. *Science* 255:1715–18

Du W, Maniatis T. 1992. An ATF/CREB binding site is required for virus induction of the human interferon beta gene. *Proc. Natl. Acad. Sci. USA* 89:2150–54

Du W, Thanos D, Maniatis T. 1993. Mechanisms of transcriptional synergism between distinct virus-inducible enhancer elements. *Cell* 74:887–98

Duyao MP, Buckler AJ, Sonenshein GE. 1990. Interaction of an NF-κB-like factor with a site upstream of the c-myc promoter. *Proc. Natl. Acad. Sci. USA* 87:4727–31

Duyao MP, Kessler DJ, Spicer DB, Sonenshein GE. 1992. Transactivation of the c-myc gene by HTLV-1 tax is mediated by NF-κB. *Curr. Top. Microbiol. Immunol.* 182:421–24

Dyer RB, Collaco CR, Niesel DW, Herzog NK. 1993. *Shigella flexneri* invasion of HeLa cells induces NF-κB DNA-binding activity. *Infect. Immunol.* 61:4427–33

Eck SL, Perkins ND, Carr DP, Nabel GJ. 1993. Inhibition of phorbol ester-induced cellular adhesion by competitive binding of NF-κB B in vivo. *Mol. Cell. Biol.* 13:6530–36

Embretson J, Zupancic M, Ribas JL, Burke A. 1993. Massive covert infection of helper T lymphocytes and macrophages by HIV during the incubation period of AIDS. *Nature* 362:359–62

Engstrom Y, Kadalayil L, Sun S-C, Samakovlis C, Hultmark D, et al. 1993. κB-like motifs regulate the induction of immune genes in *Drosophila. J. Mol. Biol.* 232:327–33

Fan CM, Maniatis T. 1991. Generation of p50 subunit of NF-κB by processing of p105 through an ATP-dependent pathway. *Nature* 354:395–98

Feuillard J, Korner M, Fourcade C, Costa A. 1994. Visualization of the endogenous NF-κB p50 subunit in the nucleus of follicular dendritic cells in germinal centers. *J. Immunol.* 152:12–21

Finco TS, Baldwin AS. 1993. κB site-dependent induction of gene expression by diverse inducers of nuclear factor κB requires Raf-1. *J. Biol. Chem.* 24:17676–79

Fracchiolia NS, Lombardi L, Slaina M, Migliazzi A, Baldini L, et al. 1993. Structural alterations of the NF-κB transcription factor lyt-10 in lymphoid malignancies. *Oncogene* 8:2839–45

Frantz B, Nordby EC, Bren G, Steffan N, Paya CV, et al. 1994. Calcineurin acts in synergy with PMA to inactivate IκB/MAD3, an inhibitor of NF-κB. *EMBO J.* 13:861–70

Franzoso G, Bours V, Azarenko V, Park S, Tomita-Yamaguchi M, et al. 1993. The oncoprotein Bcl-3 can facilitate NF-κB-mediated transactivation by removing inhibiting p50 homodimers from select κB sites. *EMBO J.* 12:3893–901

Franzoso G, Bours V, Park S, Tomita-Yamaguchi M, Kelly K, et al. 1992. The candidate oncoprotein Bcl-3 is an antagonist of p50/NF-κB-mediated inhibition. *Nature* 359: 339–42

Fujita T, Nolan GP, Ghosh S, Baltimore D. 1992. Independent modes of transcriptional activation by the p50 and p65 subunits of NF-κB. *Genes Dev.* 6:775–87

Fujita T, Nolan GP, Liou H-C, Scott ML, Baltimore D. 1993. The candidate proto-oncogene bcl-3 encodes a transcriptional coactivator that activates through NF-κB p50 homodimers. *Genes Dev.* 7:1354–63

Fujita T, Reis LFL, Watanabe N, Kimura Y, Taniguchi T, et al. 1989. Induction of the transcription factor IRF-1 and IFN-κ mRNAs by cytokines and activators of second-messenger pathways. *Proc. Natl. Acad. Sci. USA* 86:9936–40

Ganchi PA, Sun S-C, Greene WC, Ballard DW. 1992. IκB/MAD-3 masks the nuclear localization signal of NF-κB p65 and requires the transactivation domain to inhibit NF-κB p65 DNA binding. *Mol. Cell. Biol.* 3:1339–52

Ganchi PA, Sun S-C, Greene WC, Ballard DW. 1993. A novel NF-κB complex containing p65 homodimers: implications for transcriptional control at the level of subunit dimerization. *Mol. Cell. Biol.* 13:7826–35

Gay NJ, Ntwasa M. 1993. The *Drosophila* ankyrin repeat protein cactus has a predominantly alpha-helical secondary structure. *FEBS Lett.* 355:155–60

Geng Y, Zhang B, Lotz M. 1993. Protein tyrosine kinase activation is required for lipopolysaccaride induction of cytokines in human blood monocytes. *J. Immunol.* 151: 6692–700

Ghersa P, van Huijsduijnen RH, Whelan J, DeLamarter JF. 1992. Labile proteins play a dual role in the control of endothelial leukocyte adhesion molecule-1 (ELAM-1) gene regulation. *J. Biol. Chem.* 267:19226–32

Ghosh S, Baltimore D. 1990. Activation in vitro of NF-κB by phosphorylation of its inhibitor IκB. *Nature* 344:678–82

Ghosh S, Gifford AM, Riviere LR, Tempst P, Nolan GP, et al. 1990. Cloning of the p50 DNA binding subunit of NF-κB: homology to rel and dorsal. *Cell* 62:1019–29

Gilmore TD. 1992. Role of rel family genes in normal and malignant lymphoid cell growth. *Cancer Surveys* 15:69–87

Gilmore TD, Morin PJ. 1993. The IκB proteins: members of a multifunctional family. *Trends Genet.* 9:427–33

Goldfeld AE, McCaffrey PG, Strominger JL, Rao A. 1993. Identification of a novel cyclosporin-sensitive element in the human tumor necrosis factor alpha gene promoter. *J. Exp. Med.* 178:1365–79

Gonzalez-Crespo S, Levine M. 1993. Interactions between dorsal and helix-loop-helix proteins initiate the differentiation of the embryonic mesoderm and neuroectoderm in *Drosophila. Genes Dev.* 7:1703–13

Gonzalez-Crespo S, Levine M. 1994. Related target enhancers for dorsal and NF-κB signaling pathways. *Science* 264:255–58

Govind S, Steward R. 1991. Dorsovental pattern formation in *Drosophila:* signal transduction and nuclear targeting. *Trends Genet.* 7:119–25

Grassman R, Dengler C, Muller-Fleckenstein I, Fleckenstein B, McGuire K, et al. 1989. Transformation of continuous growth of primary human T lymphocytes by human T-cell leukemia virus type I X-region genes transduced by a Herpes virus saimiri vector. *Proc. Natl. Acad. Sci. USA* 86:3351–55

Grilli M, Jason J-S, Lenardo MJ. 1993. NF-κB and rel-participants in a multiform transcriptional regulatory system. *Int. Rev. Cytol.* 143: 1–62

Grimm S, Baeuerle PA. 1993. The inducible transcription factor NF-κB: structure-function relationship of its protein subunits. *Biochem. J.* 290:297–308

Grove M, Plumb M. 1993. C/EBP, NF-κB, and c-Ets family members and transcriptional regulation of the cell-specific and inducible macrophage inflammatory protein 1 alpha immediate-early gene. *Mol. Cell. Biol.* 13: 5276–89

Grumont RJ, Gerondakis S. 1989. Structure of a mammalian c-rel protein deduced from the nucleotide sequence of murine cDNA clones. *Oncogene Res.* 4:1–8

Gruss HJ, Brach MA, Herrmann F. 1992. Involvement of nuclear factor-κB in induction of the interleukin-6 gene by leukemia inhibitory factor. *Blood* 80:2563–70

Gunter KC, Irving SG, Zipfel PF, Siebenlist U, Kelly K. 1989. Cyclosporin A-mediated inhibition of mitogen-induced gene transcription is specific for the mitogenic stimulus and cell type. *J. Immunol.* 142:3286–91

Hammarskjold M-L, Simurda MC. 1992. Epstein-Barr virus latent membrane protein transactivates the human immunodeficiency virus type 1 long terminal repeat through induction of NF-κB activity. *J. Virol.* 66:6496–501

Hannibal MC, Markovitz DM, Clark N, Nabel GJ. 1993. Differential activation of human immunodeficiency virus type 1 and 2 transcription by specific T-cell activation signals. *J. Virol.* 67:5035–40

Hannink M, Temin HM. 1989. Transactivation of gene expression by nuclear and cytoplasmic rel proteins. *Mol. Cell. Biol.* 9:4323–36

Hannink M, Temin HM. 1990. Structure and autoregulation of the c-rel promoter. *Oncogene* 5:1843–50

Hansen SK, Baeuerle PA, Blasi F. 1994. Purification, reconstitution, and IκB association of the c-Rel-p65 (RelA) complex, a strong activator of transcription. *Mol. Cell. Biol.* 14: 2593–603

Hansen SK, Nerlov C, Zabel U, Verde P, Johnsen M, et al. 1992. A novel complex between the p65 subunit of NF-κB and c-Rel binds to a DNA element involved in the phorbol ester induction of the human urokinase gene. *EMBO J.* 11:205–13

Harada H, Fujita T, Miyamoto M, Kimura Y, Maruyama M, et al. 1989. Structurally similar but functionally distinct factors, IRF-1 and IRF-2, bind to the same regulatory elements of IFN and IFN-inducible genes. *Cell* 58:729–39

Harada H, Takahashi E-I, Itoh S, Harada K, Hori T-A. 1994. Structure and regulation of the human interferon regulatory factor (IRF-1) and IRF-2 genes: implications for a gene network in the interferon system. *Mol. Cell. Biol.* 14:1500–9

Haskill S, Beg AA, Tompkins SM, Morris JS, Yurochko AD, et al. 1991. Characterization of an immediate-early gene induced in adherent monocytes that encodes IκB-like activity. *Cell* 65:1281–89

Hatada EN, Naumann M, Scheidereit C. 1993. Common structural constituents confer IκB activity to NF-κB p105 and IκB/MAD-3. *EMBO J.* 12:2781–88

Hatada EN, Nieters N, Wulczyn FG, Naumann M, Meyer R, et al. 1992. The ankyrin repeat domains of the NF-κB precursor p105 and the prootoncogene bcl-3 act as specific inhibitors of NF-κB DNA binding. *Proc. Natl. Acad. Sci. USA* 89:2489–93

Hayashi T, Sekine T, Okamoto T. 1993a. Identification of a new serine kinase that activates NF-κB by direct phosphorylation. *J. Biol. Chem.* 268:26790–95

Hayashi T, Ueno Y, Okamoto T. 1993b. Oxidoreductive regulation of NF-κB. Involvement of a cellular reducing catalyst thioredoxin. *J. Biol. Chem.* 268:11380–88

Henkel T, Alkalay I, Machleidt T, Kronke M, Ben-Neriah Y, et al. 1993. Rapid proteolytic degradation of IκB-α induced by stimulation of cells with phorbol ester, cytokines and lipopolysaccharide is a necessary step in the activation of NF-κB. *Nature* 365:182–85

Henkel T, Zabel U, vanZee K, Muller JM, Fanning E, et al. 1992. Intramolecular masking of the nuclear localization signal and dimerization domain in the precursor for the p50 NF-κB subunit. *Cell* 68:1121–33

Hirai H, Fujisawa J, Suzuki T, Ueda K, Muramatsu M, et al. 1992. Transcriptional activator tax of HTLV-1 binds to the NF-κB precursor p105. *Oncogene* 7:1737–42

Hiscott J, Alper D, Cohen L, LeBlanc JF, Sportza L, et al. 1989. Induction of human interferon gene expression is associated with a nuclear factor that interacts with the NF-κB site of the human immunodeficiency virus enhancer. *J. Virol.* 63:2557–66

Hiscott J, Marois J, Garoufalis J, D'Addario M. 1993. Characterization of a functional NF-κB site in the human interleukin 1 beta promoter: evidence for a positive autoregulatory loop. *Mol. Cell. Biol.* 13:6231–40

Hohmann H-P, Brockhaus M, Baeuerle PA, Remy R. 1990a. Expression of the types A and B tumor necrosis factor (TNF) receptors is independently regulated, and both receptors mediate activation of the transcription factor NF-κB. *J. Biol. Chem.* 265:22409–17

Hohmann H-P, Remy R, Poschl B, vanLoon APG. 1990b. Tumor necrosis factors-α and -β bind to the same two types of tumor necrosis factor receptors and maximally activate the transcription factor NF-κB at low receptor occupancy and within minutes after receptor binding. *J. Biol. Chem.* 265:15183–88

Hohmann H-P, Remy R, Scheidereit C, vanLoon APGM. 1991. Maintenance of NF-κB activity is dependent on protein synthesis and the continuous presence of external stimuli. *Mol. Cell. Biol.* 11:259–66

Hovanessian AG. 1991. Interferon-induced and

double-stranded RNA-activated enzymes: a specific protein kinase and 2′, 5′-oligoadenylate synthetases. *J. Interferon Res.* 11: 199–205

Hoyos B, Ballard DW, Bohnlein E, Siekevitz M, Greene WC. 1989. κB-specific DNA binding proteins: role in the regulation of human interleukin-2 gene expression. *Science* 244:457–60

Hrdlickova R, Nehyba J, Humphries EH. 1994a. In vivo evolution of c-rel oncogenic potential. *J. Virol.* 68:2371–82

Hrdlickova R, Nehyba J, Humphries EH. 1994b. v-rel induces expression of three avian immunoregulatory surface receptors more efficiently than c-rel. *J. Virol.* 68:308–19

Hultmark D. 1993. Immune reactions in *Drosophila* and other insects: a model for innate immunity. *Trends Genet.* 9:178–93

Hultmark D. 1994. Ancient relationships. *Nature* 367:116–17

Iademarco MF, McQuillan JJ, Rosen GD, Dean DC. 1992. Characterization of the promoter for vascular cell adhesion molecule-1 (VCAM-1). *J. Biol. Chem.* 267:16323–29

Imbert V, Peyron JF, Farahifar D, Mari B. 1994. Induction of tyrosine phosphorylation and T-cell activation by vanadate peroxide, an inhibitor of protein tyrosine phosphatases. *Biochem. J.* 297:163–73

Inoue J-I, Kerr LD, Kakizuka A, Verma IM. 1992a. IκB-γ, a 70 kd protein identical to the C-terminal half of p110 NF-κB: a new member of the IκB family. *Cell* 68:1109–20

Inoue J-I, Kerr LD, Ransone LJ, Bengal E, Hunter T, et al. 1991. c-rel activates but v-rel suppresses transcription from κB sites. *Proc. Natl. Acad. Sci. USA* 88:3715–19

Inoue J-I, Kerr LD, Rashid D, Davis N, Bose HR, et al. 1992b. Direct association of pp40/IκB-β with rel/NF-κB transcription factors: role of ankyrin repeats in the inhibition of DNA binding activity. *Proc. Natl. Acad. Sci. USA* 89:4333–37

Inuzuka M, Ishikawa H, Kumar S, Gelinas C, Ito Y. 1994. The viral and cellular Rel oncoproteins induce the differentiation of P19 embryonal carcinoma cells. *Oncogene* 9:133–40

Ip YT, Kraut R, Levine M, Rushlow C. 1991. The dorsal morphogen is a sequence-specific DNA-binding protein that interacts with a long-range repression element in Drosophila. *Cell* 64:439–46

Ip YT, Levine M. 1992. The role of the dorsal morphogen gradient in *Drosophila* embryogenesis. *Semin. Dev. Biol.* 3:15–23

Ip YT, Park RE, Kosman D, Bier E, Levine M. 1992a. The dorsal gradient morphogen regulates stripes of rhomboid expression in the presumptive neuroectoderm of the *Drosophila* embryo. *Genes Dev.* 6:1728–39

Ip YT, Park RE, Kosman D, Yazdanbakhsh K, Levine M. 1992b. dorsal-twist interactions establish Snail expression in the presumptive mesoderm of the *Drosophila* embryo. *Genes Dev.* 6:1518–30

Ip YT, Reach M, Engstrom Y, Kadalayil L, Cai H, et al. 1993. *Dif*, a dorsal-related gene that mediates an immune response in Drosophila. *Cell* 75:753–63

Ishikawa H, Asano M, Kanda T, Kumar S, Gelinas C, et al. 1993. Two novel functions associated with the Rel oncoproteins: DNA replication and cell-specific transcriptional activation. *Oncogene* 8:2889–96

Isoda K, Roth S, Nüsslein-Volhard C. 1992. The functional domains of the *Drosophila* morphogen dorsal: evidence from the analysis of mutants. *Genes Dev.* 6:619–30

Israël A. 1992. The rel/NF-κB family of transcription factors: a novel mechanism to control gene expression. *Pathol. Biol.* 40:212–14

Israël A, LeBail O, Hatat D, Piette J, Kieran M, et al. 1989. TNF stimulates expression of mouse MHC class I genes by inducing an NF-κB-like enhancer binding activity which displaces constitutive factors. *EMBO J.* 8: 3793–800

Iwasaki T, Uehara Y, Graves L, Rachie N, Bomsztyk K. 1992. Herbimycin A blocks IL-1-induced NF-κB DNA-binding activity in lymphoid cell lines. *FEBS Lett.* 298:240–44

Jamieson C, McCaffrey PG, Rao A, Sen R. 1991. Physiologic activation of T cells via the T cell receptor induces NF-κB. *J. Immunol.* 147:416–20

Jiang J, Levine M. 1993. Binding affinities and cooperative interactions with bHLH activators delimit threshold responses to the dorsal gradient morphogen. *Cell* 72:741–52

Jiang J, Rushlow CA, Zhou Q, Small S, Levine M. 1992. Individual dorsal morphogen binding sites mediate activation and repression in the *Drosophila* embryo. *EMBO J.* 11:3147–54

Johnson DR, Pober JS. 1994. HLA class I heavy-chain gene promoter elements mediating synergy between tumor necrosis factor and interferons. *Mol. Cell. Biol.* 14:1322–32

Joseph CK, Byun HS, Bittman R, Kolesnick RN. 1993. Substrate recognition by ceramide-activated protein kinase. Evidence that kinase activity is proline-directed. *J. Biol. Chem.* 268:20002–6

Joshi-Barve SS, Rangnekar VV, Sells SF, Rangnekar VM. 1993. Interleukin-1-inducible expression of gro-β via NF-κB activation is dependent upon tyrosine kinase signaling. *J. Biol. Chem.* 268:18018–29

Kaltschmidt B, Baeuerle PA, Kaltschmidt C. 1993. Potential involvement of the transcription factor NF-κB in neurological disorders. *Mol. Aspects Med.* 14:171–90

Kamens J, Richardson P, Mosialos G, Brent R,

Gilmore TD. 1990. Oncogenic transformation by v-Rel requires an amino-terminal activation domain. *Mol. Cell. Biol.* 10:2840–47

Kang SM, Tran AC, Grilli M, Lenardo MJ. 1992. NF-κB subunit regulation in non-transformed CD4+ T lymphocytes. *Science* 256:1452–56

Kappler C, Meister M, Lagueux M, Gateff E, Hoffman JA, et al. 1993. Insect immunity. Two 17 bp repeats nesting a κB-related sequence confer inducibility to the diptericin gene and bind a polypeptide in bacteria-challenged *Drosophila. EMBO J.* 12:1561–68

Kasai Y, Nambu JR, Lieberman PM, Crews ST. 1992. Dorsal-ventral patterning in *Drosophila:* DNA binding of snail protein to the *single-minded* gene. *Proc. Natl. Acad. Sci. USA* 89:3414–18

Kaszubska W, van Huijsduijnen RH, Ghersa P, DeRaemy-Schenk AM. 1993. Cyclic AMP-independent ATF family members interact with NF-κB and function in the activation of the E-selectin promoter in response to cytokines. *Mol. Cell. Biol.* 13:7180–90

Kelly K, Davis P, Mitsuya H, Irving S, Wright J, et al. 1992. A high proportion of early response genes are constitutively activated in T cells by HTLV-1. *Oncogene* 7:1463–70

Kerr LD, Duckett CS, Wamsley P, Zhang Q, Chiao P, et al. 1992. The protooncogene BCL-3 encodes an IκB protein. *Genes Dev.* 6:2352–63

Kerr LD, Inoue J-I, Davis N, Link E, Baeuerle PA, et al. 1991. The Rel-associated pp40 protein prevents DNA binding of Rel and NF-κB: relationship with IκB-β and regulation by phosphorylation. *Genes Dev.* 5:1464–76

Kerr LD, Ransone LJ, Wamsley P, Schmitt MJ, Boyer TG, et al. 1993. Association between proto-oncoprotein Rel and TATA-binding protein mediates transcriptional activation by NF-κB. *Nature* 365:412–19

Kessler DJ, Duyao M, Spicer DB. Sonenshein GE. 1992. NF-κB-like factors mediate interleukin-1 induction of c-myc gene transcription in fibroblasts. *J. Exp. Med.* 176:787–92

Kieran M, Blank V, Logeat F, Vandekerckhove J, Lottspeich F, et al. 1990. The DNA binding subunit of NF-κB is identical to factor KBF1 and homologous to the rel oncogene product. *Cell* 62:1007–18

Kirov N, Zhelnin L, Shah J, Rushlow C. 1003. Conversion of a silencer into an enhancer: evidence for a co-repressor in dorsal-mediated repression in *Drosophila. EMBO J.* 12:3193–99

Kitajima I, Shinohara T, Bilakovics J, Brown DA. 1993. Ablation of transplanted HTLV-I tax-transformed tumors in mice by antisense inhibition of NF-κB. *Science* 259:1523–26

Klug CA, Gerety SJ, Shah PC, Chen Y-Y, Rice NR, et al. 1994. The v-abl tyrosine kinase negatively regulates NF-κB/Rel factors and blocks kappa gene transcription in pre-B lymphocytes. *Genes Dev.* 8:678–87

Koong AC, Chen EY, Giaccia AJ. 1994. Hypoxia causes the activation of nuclear factor κB through the phosphorylation of I κBα on tyrosine residues. *Cancer Res.* 54:1425–30

Kosman D, Ip YT, Levine M, Arora K. 1991. Establishment of the mesoderm-neuroectoderm boundary in the *Drosophila* embryo. *Science* 254:118–22

Kowalik TF, Wing B, Haskill JS, Azizkhan JC, Baldwin AS, et al. 1993. Multiple mechanisms are implicated in the regulation of NF-κB activity during human cytomegalovirus infection. *Proc. Natl. Acad. Sci. USA* 90:1107–11

Kralova J, Schatzle JD, Bargmann W, Bose HR. 1994. Transformation of avian fibroblasts overexpressing the c-rel proto-oncogene and a variant of c-rel lacking 40 C-terminal amino acids. *J. Virol.* 68:2073–83

Kretzschmar M, Meisterernst M, Scheidereit C, Li G, Roeder RG. 1992. Transcriptional regulation of the HIV-1 promoter by NF-κB in vitro. *Genes Dev.* 6:761–74

Krikos A, Laherty CD, Dixit VM. 1992. Transcriptional activation of the tumor necrosis factor alpha-inducible zinc finger protein, A20, is mediated by kappa B elements. *J. Biol. Chem.* 267:17971–76

Kumar S, Gelinas C. 1993. IκB α-mediated inhibition of v-Rel DNA binding requires direct interaction with the RXXRXRXXC Rel/κB DNA-binding motif. *Proc. Natl. Acad. Sci. USA* 90:8962–66

Kumar S, Rabson AB, Gelinas C. 1992. The RXXRXRXXC motif conserved in all Rel/κB proteins is essential for the DNA-binding activity and redox regulation of the v-Rel oncoprotein. *Mol. Cell. Biol.* 12:3094–106

Kunsch C, Rosen CA. 1993. NF-κB subunit-specific regulation of the interleukin-8 promoter. *Mol. Cell. Biol.* 13:6137–46

Kunsch C, Ruben SM, Rosen CA. 1992. Selection of optimal κB Rel DNA-binding motifs: interaction of both subunits of NF-κB with DNA is required for transcriptional activation. *Mol. Cell. Biol.* 12:4412–21

Laherty CD, Hu HM, Opipari AW, Wang F, Dixit VM. 1992. The Epstein-Barr virus LMP1 gene product induces A20 zinc finger protein expression by activating NF-κB. *J. Biol. Chem.* 267:24157–60

Lalmanach-Girard AC, Chiles TC, Parker DC, Rothstein TL. 1993. T cell-dependent induction of NF-κB in B cells. *J. Exp. Med.* 177:1215–19

LaRosa FA, Pierce JW, Sonenshein GE. 1994. Differential regulation of the c-myc oncogene promoter by the NF-κB rel family of transcription factors. *Mol. Cell. Biol.* 14:1039–44

Lauer U, Weiss L, Lipp M, Hofschneider PH, Kekulé A. 1994. The hepatitis B virus *PreS2/S[t]* transactivator utilizes AP-1 and other transcription factors for transactivation. *Hepatology* 19:23–31

LeBail O, Schmidt-Ullrich R, Israël A. 1993. Promoter analysis of the gene encoding the IκB-α/MAD3 inhibitor of NF-κB: positive regulation by members of the rel/NF-κB family. *EMBO J.* 12:5043–49

LeClair KP, Blanar MA, Sharp PA. 1992. The p50 subunit of NF-κB associates with the NF-IL6 transcription factor. *Proc. Natl. Acad. Sci. USA* 89:8145–49

Lee-Huang S, Lin JJ, Kung HF, Huang PL. 1993. The human erythropoietin-encoding gene contains a CAAT box, TATA boxes and other transcriptional regulatory elements in this 5′ flanking region. *Gene* 128:227–36

Leiden JM, Wang CY, Petryniak B, Markovitz DM. 1992. A novel Ets-related transcription factor, Elf-1, binds to human immunodeficiency virus type 2 regulatory elements that are required for inducible *trans* activation in T cells. *J. Virol.* 66:5890–97

Lenardo MJ, Baltimore D. 1989. NF-κB: a pleiotropic mediator of inducible and tissue-specific gene control. *Cell* 58:227–29

Lenardo MJ Siebenlist U. 1994. Bcl-3-mediated nuclear regulation of the NF-κB *trans-activating* factor. *Immunol. Today* 15:145–46

Leptin M. 1991. Twist and snail as positive and negative regulators during *Drosophila* mesoderm development. *Genes Dev.* 5:1568–76

Lernbecher T, Muller U, Wirth T. 1993. Distinct NF-κB/Rel transcription factors are responsible for tissue-specific and inducible gene activation. *Nature* 365:767–70

Letson A, Alexander S, Orth K, Wasserman SA. 1991. Genetic and molecular characterization of *tube*, a *Drosophila* gene maternally required for embryonic dorsoventral polarity. *Proc. Natl. Acad. Sci. USA* 88:810–14

Li S, Sedivy JM. 1993. Raf-1 protein kinase activates the NF-κB transcription factor by dissociating the cytoplasmic NF-κB-IκB complex. *Proc. Natl. Acad. Sci. USA* 90: 9247–51

Liao F, Andalibi A, deBeer FC, Fogelman AM, Lusis AJ. 1993. Genetic control of inflammatory gene induction and NF-κB-like transcription factor activation in response to an atherogenic diet in mice. *J. Clin. Invest.* 91: 2572–79

Lichtenstein M, Keini G, Cedar H, Bergman Y. 1994. B cell-specific demethylation: a novel role for the intronic kappa chain enhancer sequence. *Cell* 76:913–23

Lindholm PF, Reid RL, Brady JN. 1992. Extracellular Tax-1 protein stimulates tumor necrosis factor-β and immunoglobulin κ light chain expression in lymphoid cells. *J. Virol.* 66:1294–302

Link E, Kerr LD, Schreck R, Zabel U, Verma I, et al. 1992. Purified IκB-β is inactivated upon dephosphorylation. *J. Biol. Chem.* 267: 239–46

Liou HC, Baltimore D. 1993. Regulation of the NF-κB/rel transcription factor and IκB inhibitor system. *Curr. Opin. Cell Biol.* 5:477–87

Liou HC, Nolan GP, Ghosh S, Fujita T, Baltimore D. 1992. The NF-κB p50 precursor, p105, contains an internal IκB-like inhibitor that preferentially inhibits p50. *EMBO J.* 11: 3003–9

Liu J, Sodeoka M, Lane WS, Verdine GL. 1994. Evidence for a non-alpha-helical DNA-binding motif in the Rel homology region. *Proc. Natl. Acad. Sci. USA* 91:908–12

Logeat F, Israël N, Ten R, Blank V, LeBail O, et al. 1991. Inhibition of transcription factors belonging to the rel/NF-κB family by a trans-dominant negative mutant. *EMBO J.* 10: 1827–32

Lowenstein CJ, Alley EW, Raval P, Snowman AM, Snyder SH, et al. 1993. Macrophage nitric oxide synthase gene: two upstream regions mediate induction by interferon gamma and lipopolysaccharide. *Proc. Natl. Acad. Sci. USA* 90:9730–34

Lu D, Thompson JD, Gorski GK, Rice NR, Mayer MG, et al. 1991. Alterations at the rel locus in human lymphoma. *Oncogene* 6: 1235–41

Lux SE, John KM, Bennett V. 1990. Analysis of cDNA human erythrocyte ankyrin indicates a repeated structure with homology to tissue-differentiation and cell cycle control proteins. *Nature* 344:36–42

Mackman N, Brand K, Edgington TS. 1991. Lipopolysaccharide-mediated transcriptional activation of the human tissue factor gene in THP-1 monocytic cells requires both activator protein 1 and nuclear factor κB binding sites. *J. Exp. Med.* 174:1517–26

Mathias S, Dressler KA, Kolesnick RN. 1991. Characterization of a ceramide-activated protein kinase: stimulation by tumor necrosis-factor-α. *J. Biol. Chem.* 88:10009–13

Mathias S, Younes A, Kan CC, Orlow I, Joseph C, et al. 1993. Activation of the spingomyelin signaling pathway in intact EL4 cells and in a cell-free system by IL-1β. *Science* 259: 519–22

Matsusaka T, Fujikawa K, Nishio Y, Mukaida N. 1993. Transcription factors NF-IL6 and NF-κB synergistically activate transcription of the inflammatory cytokines, interleukin 6 and interleukin 8. *Proc. Natl. Acad. Sci. USA* 90:10193–97

Matthews JR, Kaszubska W, Turcatti G, Wells TN, Hay RT. 1993a. Role of cysteine 62 in DNA recognition by the P50 subunit of NF-kappa B. *Nucleic Acids Res.* 21:1727–34

Matthews JR, Wakasugi N, Virelizier J-L, Yodoi J, Hay RT. 1992. Thioredoxin regulates the DNA binding activity of NF-κB by reduction of a disulfide bond involving cysteine 62. *Nucleic Acids Res.* 30:3821–30

Matthews JR, Watson E, Buckley S, Hay RT. 1993b. Interaction of the C-terminal region of p105 with the nuclear localisation signal of p50 is required for inhibition of NF-kappa B DNA binding activity. *Nucleic Acids Res.* 21:4516–23

Mattila P, Ullman KS, Fiering S, Emmel EA, McCutcheon M, et al. 1990. The action of cyclosporin A and FK506 suggest a novel step in the activation of T lymphocytes. *EMBO J.* 9:4425–33

McDonnell PC, Kumar S, Rabson AB, Gelinas C. 1992. Transcriptional activity of rel family proteins. *Oncogene* 7:163–70

Meichle A, Schutze S, Hensel G, Brunsing D, Krönke M. 1990. Protein kinase C-independent activation of nuclear factor κB by tumor necrosis factor. *J. Biol. Chem.* 265:8339–43

Mellits KH, Hay RT, Goodburn S. 1993. Proteolytic degradation of MAD3 (IκB alpha) and enhanced processing of the NF-κB precursor p105 are obligatory steps in the activation of NF-κB. *Nucleic Acids Res.* 21: 5059–66

Menon SD, Qin S, Guy GR, Tan YH. 1993. Differential induction of nuclear NF-κB by protein phosphatase inhibitors in primary and transformed human cells. *J. Biol. Chem.* 268: 26805–12

Mercurio F, Didonato J, Rosette C, Karin M. 1992. Molecular cloning and characterization of a novel Rel/NF-κB family member displaying structural and functional homology to NF-κB p50-p105. *DNA Cell Biol.* 11: 523–37

Mercurio F, Didonato JA, Rosette C, Karin M. 1993. p105 and p98 precursor proteins play an active role in NF-κB-mediated signal transduction. *Genes Dev.* 7:705–18

Messer G, Weiss EH, Baeuerle PA. 1990. Tumor necrosis factor beta (TNF-β) induces binding of the NF-κB transcription factor to a high-affinity κB element in the TNF-β promoter. *Cytokine* 2:389–97

Meyer M, Caselmann WH, Schluter V, Schreck R, et al. 1992. Hepatitis B virus transactivator MHBst: activation of NF-kappa B, selective inhibition by antioxidants and integral membrane localization. *EMBO* J. 11:2992–3001

Meyer R, Hatada EN, Hohmann HP, Haiker M, Bartsch C, et al. 1991. Cloning of the DNA-binding subunit of human nuclear factor κB: the level of its mRNA is strongly regulated by phorbol ester or tumor necrosis factor α. *Proc. Natl. Acad. Sci. USA* 88:966–70

Michaely P, Bennett V. 1992. The ANK repeat: a ubiquitous motif involved in macromolecular recognition. *Trends Cell Biol.* 2:127–29

Miyamoto M, Fujita T, Kimura Y, Maruyama M, Harada H, et al. 1988. Regulated expression of a gene encoding a nuclear factor, IRF-1, that specifically binds to IFN-κ gene regulatory elements. *Cell* 54:903–13

Miyamoto S, Chiao PJ, Verma IM. 1994. Enhanced IκBα degradation is responsible for constitutive NF-κB activity in mature murine B-cell lines. *Mol. Cell. Biol.* 14:3276–82

Molitor JA, Walker WH, Doerre S, Ballard DW, Greene WC. 1990. NF-κB: a family of inducible and differentially expressed enhancer-binding proteins in human T cells. *Proc. Natl. Acad. Sci. USA* 87:10028–32

Moore PA, Ruben SM, Rosen CA. 1993. Conservation of transcriptional activation functions of the NF-κB p50 and p65 subunits in mammalian cells and *Saccharomyces cerevisiae. Mol. Cell. Biol.* 13:1666–74

Morrison LE, Boehmelt G, Beug H, Enrietto PJ. 1992. Expression of v-rel by a replication-competent virus in chicken embryo fibroblasts. *Oncogene* 6:1657–66

Mosialos G, Hamer P, Capobianco AJ, Laursen RA, Gilmore TD. 1991. A protein kinase A recognition sequence is structurally linked to transformation by p59v-rel and cytoplasmic retention of p68c-rel. *Mol. Cell. Biol.* 11: 5867–77

Muller JM, Ziegler-Heitbrock HWL, Baeuerle PA. 1993. Nuclear factor κB, a mediator of lipopolysaccharide effects. *Immunobiology* 187:233–56

Muroi M, Suzuki T. 1993. Role of protein kinase A in LPS-induced activation of NF-κB proteins of a mouse macrophage-like cell line, J774. *Cell. Signal.* 5:289–98

Nabel G, Baltimore D. 1987. An inducible transcription factor activates expression of human immunodeficiency virus in T cells. *Nature* 326:711–13

Nakayama K, Shimizu H, Mitomo K, Watanabe T. 1992. A lymphoid cell-specific nuclear factor containing c-Rel-like proteins preferentially interacts with interleukin-6 κB-related motifs whose activities are repressed in lymphoid cells. *Mol. Cell. Biol.* 12:1736–46

Narayanan R, Higgins KA, Perez JR, Coleman TA, Rosen CA. 1993. Evidence for differential functions of the p50 and p65 subunits of NF-κB with a cell adhesion model. *Mol. Cell. Biol.* 13:3802–10

Narumi S, Tebo JM, Finke JH, Hamilton TA. 1992. IFN-gamma and IL-2 cooperatively activate NF-κB in murine peritoneal macrophages. *J. Immunol.* 149:529–34

Natoli G, Avantaggiati ML, Balsano C, De Marzio E, et al. 1992. Characterization of the hepatitis B virus preS/S region encoded transcriptional transactivator. *Virology* 187:663–70

Naumann M, Wulczyn FG, Scheidereit C. 1993. The NF-κB precursor p105 and the proto-oncogene product Bcl-3 are IκB molecules and control nuclear translocation of NF-κB. *EMBO J.* 12:213–22

Nehyba J, Hrdlickova R, Humphries EH. 1994. Evolution of the oncogenic potential of v-rel: rel-induced expression of immunoregulatory receptors correlates with tumor development and in vitro transformation. *J. Virol.* 68: 2039–50

Neiman PE, Thomas SJ, Loring G. 1991. Induction of apoptosis during normal and neoplastic B-cell development in the bursa of *Fabricius. Proc. Natl. Acad. Sci. USA* 88: 5857–61

Neish AS, Williams AJ, Palmer HJ, Whitley MZ, Collins T. 1992. Functional analysis of the human vascular cell adhesion molecule 1 promoter. *J. Exp. Med.* 176:1583–93

Nerenberg M, Hinrichs SH, Reynolds RK, Khoury G, Jay G. 1987. The *tat* gene of human T-lymphotropic virus type 1 induces mesenchymal tumors in transgenic mice. *Science* 237:1324–29

Neri A, Chang CC, Lombardi L, Salina M, Corradini P, et al. 1991. B cell lymphoma-associated chromosomal translocation involves candidate oncogene lyt-10, homologous to NF-κB p50. *Cell* 67:1075–87

Neumann M, Tsapos K, Schoppler JA, Ross J, Franza BR. 1992. Identification of complex formation between two intracellular tyrosine kinase substrates: human c-Rel and the p105 precursor of p50 NF-κB. *Oncogene* 7:2095–104

Niederman TMJ, Garcia JV, Hastings WR, Luria S, Ratner L. 1992. Human immunodeficiency virus type 1 nef protein inhibits NF-κB induction in human T cells. *J. Virol.* 66:6213–19

Nielsch U, Zimmer SG, Babiss LE. 1991. Changes in NF-kappa B and ISGF3 DNA binding activities are responsible for differences in MHC and beta-IFN gene expression in Ad5- versus Ad12-transformed cells. *EMBO* J. 10:4169–75

Nolan GP, Baltimore D. 1992. The inhibitory ankyrin and activator Rel proteins. *Curr. Opin. Genet. Dev.* 2:211–20

Nolan GP, Fujita T, Bhatia K, Huppi K, Liou H-C, et al. 1993. The bcl-3 proto-oncogene encodes a nuclear IκB-like molecule that preferentially interacts with NF-κB p50 in a phosphorylation-dependent manner. *Mol. Cell. Biol.* 13:3557–66

Nolan GP, Ghosh S, Liou H-C, Tempst P, Baltimore D. 1991. DNA binding and IκB inhibition of the cloned p65 subunit of NF-κB, a rel-related polypeptide. *Cell* 64: 961–69

Norris JL, Manley JL. 1992. Selective nuclear transport of the *Drosophila* morphogen dorsal can be established by a signaling pathway involving the transmembrane protein Toll and protein kinase A. *Genes Dev.* 6:1654–67

Ohmori Y, Hamilton TA. 1993. Cooperative interaction between interferon (IFN) stimulus response element and κB sequence motifs controls IFN-γ- and lipopolysaccharide-stimulated transcription from the murine IP-10 promoter. *J. Biol. Chem.* 268:6677–88

Ohno H, Takimoto G, McKeithan TW. 1990. The candidate proto-oncogene bcl-3 is related to genes implicated in cell lineage determination and cell cycle control. *Cell* 60: 991–97

Olashaw NE, Kowalik TF, Huang ES, Pledger WJ. 1992. Induction of NF-κB-like activity by platelet-derived growth factor in mouse fibroblasts. *Mol. Cell. Biol.* 3:1131–39

Osborn L, Kunkel S, Nabel GJ. 1989. TNF-α and interleukin 1 stimulate the human immunodeficiency virus enhancer by activation of the NF-κB. *Proc. Natl. Acad. Sci. USA* 86:2336–40

Oster W, Brach MA, Gruss HJ, Mertelsmann R, Herrmann F. 1992. Interleukin-1 beta (IL-1 beta) expression in human mononuclear phagocytes is differentially regulated by granulocyte-macrophage colony-stimulating factor (GM-CSF), M-CSF, and IL-3. *Blood* 79:1260–65

Pantaleo G, Graziosi C, Demarest JF, Butini L. 1993a. HIV infection is active and progressive in lymphoid tissue during the clinically latent stage of disease. *Nature* 362:355–58

Pantaleo G, Graziosi C, Fauci AS. 1993b. New concepts in the immunopathogenesis of human immunodeficiency virus infection. *N. Engl. J. Med.* 328:327–35

Parhami F, Fang ZT, Fogelman AM, Andalibi A. 1993. Minimally modified low density lipoprotein-induced inflammatory responses in endothelial cells are mediated by cyclic adenosine monophosphate. *J. Clin. Invest.* 92:471–78

Parslow TG, Granner DK. 1983. Structure of a nuclease-sensitive region inside the immunoglobin kappa gene: evidence for a role in gene regulation. *Nucleic Acids Res.* 11:4775–92

Paya CV, Ten RM, Bessia C, Alcami J, Hay RT, et al. 1992. NF-κB-dependent induction of the NF-κB p50 subunit gene promoter underlies self-perpetuation of human immunodeficiency virus transcription in monocytic cells. *Proc. Natl. Acad. Sci. USA* 89:7826–30

Perkins ND, Edwards NL, Duckett CS, Agranoff AB. 1993. A cooperative interaction between NF-κB and Sp1 is required for HIV-1 enhancer activation. *EMBO J.* 12: 3551–58

Perkins ND, Schmid RM, Duckett CS, Leung K, Rice NR, et al. 1992. Distinct combina-

tions of NF-κB subunits determine the specificity of transcriptional activation. *Proc. Natl. Acad. Sci. USA* 89:1529–33
Pierce JW, Lenardo M, Baltimore D. 1988. An oligonucleotide that binds nuclear factor NF-κB acts as a lymphoid-specific and inducible enhancer element. *Proc. Natl. Acad. Sci. USA* 85:1482–86
Pierik LT, Murphy EL. 1991. The clinical significance of HTLV-I and HTLV-II infection in the AIDS epidemic. *AIDS Clin. Rev.* pp. 39–57
Plaksin D, Baeuerle P, Eisenbach L. 1993. KBF-1 (p50 NF-κB homodimer) acts as a repressor of H-2Kb gene expression in metastatic tumor cells. *J. Exp. Med.* 177:1651–62
Potter DA, Larson CJ, Eckes P, Schmid RM, Nabel GJ, et al. 1993. Purification of the major histocompatibility complex class I transcription factor H2TF1. *J. Biol. Chem.* 268:18882–90
Ranganathan PN, Khalili K. 1993. The transcriptional enhancer element, κB, regulates promoter activity of the human neurotropic virus, JCV, in cells derived from the CNS. *Nucleic Acids Res.* 21:1959–64
Rao Y, Vaessin H, Jan LY, Jan Y-N. 1991. Neuroectoderm in *Drosophila* embryos is dependent on the mesoderm for positioning but not for formation. *Genes Dev.* 5:1577–88
Ray A, Prefontaine KE. 1994. Physical association and functional antagonism between the p65 subunit of transcription factor NF-κB and the glucocorticoid receptor. *Proc. Natl. Acad. Sci. USA* 91:752–56
Read MA, Whitley MZ, Williams AJ, Collins T. 1994. NF-κB and IκB-α: an inducible regulatory system in endothelial activation. *J. Exp. Med.* 179:503–412
Reis LFL, Harada H, Wolchok JD, Taniguchi T, Vilcek J. 1992. Critical role of a common transcription factor, IRF-1, in the regulation of IFN-κ and IFN-inducible genes. *EMBO J.* 11:185–93
Rice NR, Ernst MK. 1993. In vivo control of NF-κB activation by IκB-α. *EMBO J.* 12:4685–95
Rice NR, Gilden RV. 1988. The Rel oncogene. In *The Oncogene Handbook,* ed. EP Reddy, AM Skalka, T Curran, pp. 495–562. Amsterdam: Elsevier Science
Rice NR, MacKichan ML, Israël A. 1992. The precursor of NF-κp50 has IkB-like functions. *Cell* 71:243–53
Richardson PM, Gilmore TD. 1991. vRel is an inactive member of the Rel family of transcriptional activating proteins. *J. Virol.* 65:3122–30
Riviere Y, Blank V, Kourilsky P, Israël A. 1991. Processing of the precursor of NF-κB by the HIV-1 protease during acute infection. *Nature* 350:622–25
Ron D, Brasier AR, Habener JF. 1990. Transcriptional regulation of hepatic angiotensinogen gene expression by the acute-phase response. *Mol. Cell. Endocrinol.* 74: C97–104
Ron D, Brasier AR, Habener JF. 1991. A new family of large nuclear proteins that recognize nuclear factor kappa B-binding sites through a zinc finger motif. *Mol. Cell. Biol.* 11:2887–95
Rong BL, Libermann TA, Kogawa K, Ghosh S, Cao L-X, et al. 1992. HSV-1-inducible proteins bind to NF-κB-like sites in the HSV-1 genome. *Virology* 189:750–56
Rosl F, Lengert N, Albrecht J, Kleine K. 1994. Differential regulation of the JE gene encoding the monocyte chemoattractant protein (MCP-1) in cervical carcinoma cells and derived hybrids. *J. Virol.* 68:2142–50
Ross I, Buckler-White AJ, Rabson AB, Ingland G, Martin MA. 1991. Contribution of NF-κB and SP1 binding motifs to the replicative capacity of human immunodeficiency virus type 1: distinct patterns of viral growth are determined by T cell types. *J. Virol.* 65: 4350–58
Roulston A, Beauparlant P, Rice N, Hiscott J. 1993. Chronic human immunodeficiency virus type 1 infection stimulates distinct NF-κB B/rel DNA binding activities in myelomonoblastic cells. *J. Virol.* 67:5235–46
Roulston A, D'Addario M, Boulerice F, Caplan S, Wainberg MA, et al. 1992. Induction of monocytic differentiation and NF-κB-like activities by human immunodeficiency virus 1 infection of myelomonoblastic cells. *J. Exp. Med.* 175:751–63
Ruben SM, Dillon PJ, Schreck R, Henkel T, Chen CH, et al. 1991. Isolation of a rel-related human cDNA that potentially encodes the 65-kD subunit of NF-κB. *Science* 251: 1490–93
Ruben SM, Klement JF, Coleman TA, Maher M, Chen CH, et al. 1992. I-Rel: a novel rel-related protein that inhibits NF-κB transcriptional activity. *Genes Dev.* 6:745–60
Ryseck RP, Bull P, Takamiya M, Bours V, Siebenlist U, et al. 1992. RelB, a new rel family transcription activator that can interact with p50-NF-κB. *Mol. Cell. Biol.* 12: 674–84
Ryter SW, Gomer CJ. 1993. Nuclear factor κB binding activity in mouse L1210 cells following photofrin II-mediated photosensitization. *Photochem. Photobiol.* 58:753–56
Saklatvala J, Rawlinson LM, Marshall CJ, Kracht M. 1993. Interleukin-1 and tumour necrosis factor activate the mitogen activated protein (MAP) kinase kinase in cultured cells. *FEBS Lett.* 334:189–92
Sambucetti LC, Cherrington JM, Wilkinson GWG, Mocarski ES. 1989. NF-κB activation of the cytomegalovirus enhancer is mediated by a viral transactivator and by T cell stimulation. *EMBO J.* 8:4251–58

Santra M, Danielson KG, Iozzo RV. 1994. Structural and functional characterization of the human decorin gene promoter. *J. Biol. Chem.* 269:579–87

Sarkar S, Gilmore TD. 1993. Transformation by the vRel oncoprotein requires sequences carboxy-terminal to the Rel homology domain. *Oncogene* 8:2245–52

Sater RA. 1994. Basal expression of the human macrophage colony-stimulating factor (M-CSF) gene in K562 cells. *Leuk. Res.* 18:133–43

Scala G, Quinto I, Ruocco MR, Mallardo M, Ambrosino C, et al. 1993. Epstein-Barr virus nuclear antigen 2 transactivates the long terminal repeat of human immunodeficiency virus type 1. *J. Virol.* 67:2853–61

Scheinman RI, Beg AA, Baldwin AS. 1993. NF-κB p100 (lyt-10) is a component of H2TF1 and can function as an IκB-like molecule. *Mol. Cell. Biol.* 13:6089–101

Schieven GL, Kirihara JM, Myers DE, Ledbetter JA, Uckun FM. 1993. Reactive oxygen intermediates activate NF-κB in a tyrosine-kinase dependent mechanism and in combination with vanadate activate the p56 lck and p59 fyn tyrosine kinase in human lymphocytes. *Blood* 82:1212–20

Schmid RM, Perkins ND, Duckett CS, Andrews PC, Nabel GJ. 1991. Cloning of an NF-κB subunit which stimulates HIV transcription in synergy with p65. *Nature* 352: 733–36

Schmidt A, Hennighausen L, Siebenlist U. 1990. Inducible nuclear factor binding to the kappa B elements of the human immunodeficiency virus enhancer in T cells can be blocked by cyclosporin A in a signal-dependent manner. *J. Virol.* 64:4037–41

Schmitz ML, Baeuerle PA. 1991. The p65 subunit is responsible for the strong transcription activation potential of NF-κB. *EMBO J.* 10: 3805–17

Schmitz ML, Henkel T, Baeuerle PA. 1991. Proteins controlling the nuclear uptake of the NF-κB, rel and dorsal. *Trends Cell Biol.* 1: 130–37

Schneider DS, Hudson KL, Lin T, Anderson KV. 1991. Dominant and recessive mutations define functional domains of Toll, a transmembrane protein required for dorsal-ventral polarity in the *Drosophila* embryo. *Genes Dev.* 5:797–807

Schreck R, Albermann K, Baeuerle PA. 1992a. NF-κB: an oxidative stress-responsive transcription factor of eukaryotic cells (a review). *Free Radical Res. Commun.* 17: 221–37

Schreck R, Bevec D, Dukor R, Baeuerle PA, Chedid L, Bahr GM. 1992b. Selection of a muramyl peptide based on its lack of activation of nuclear factor-κB as a potential adjuvant for AIDS vaccines. *Clin. Exp. Immunol.* 90:188–93

Schreck R, Grassman R, Fleckenstein B, Baeuerle PA. 1992c. Antioxidants selectively suppress activation of NF-κB by human T-cell leukemia virus type I tax protein. *J. Virol.* 66:6288–93

Schreck R, Meier B, Maennel DN, Droge W, Baeuerle A. 1992d. Dithiocarbamates as potent inhibitors of nuclear factor κB activation in intact cells. J. *Exp. Med.* 175:1181–94

Schreck R, Rieber P, Baeuerle PA. 1991. Reactive oxygen intermediates as apparently widely used messengers in the activation of the NF-κB transcription factor and HIV-1. *EMBO J.* 10:2247–58

Schreck R, Zorbas H, Winnacker EL, Baeuerle PA. 1990. The NF-κB transcription factor induces DNA binding which is modulated by its 65-kD subunit. *Nucleic Acids Res.* 18: 6497–502

Schreiber SL, Crabtree GR. 1992. The mechanisms of action of cyclosporin A and FK506. *Immunol. Today* 13:136–42

Schulze-Osthoff K, Bakker AC, Van-Haesebroeck B, Beyaert R, Jacob WA, et al. 1992. Cytotoxic activity of tumor necrosis factor is mediated by early damage to mitochondrial functions—evidence for the involvement of mitochondrial radical generation. *J. Biol. Chem.* 267:5317–22

Schulze-Osthoff K, Beyaert R, Van Dervoorde V, Haegeman G, Fiers W. 1993. Depletion of the mitochondrial electron transport abrogates the cytotoxic and gene induction effects of toxic and gene induction effects of TNF. *EMBO J.* 12:3095–104

Schutze S, Pothoff K, Machleidt T, Bercovic D, Wiegmann K, et al. 1992. TNF activates NF-κB by phosphatidylcholine-specific phospholipase C-induced "acidic" sphingomyelin breakdown. *Cell* 71:765–76

Scott ML, Fujita T, Liou H-C, Nolan GP, Baltimore D. 1993. The p65 subunit of NF-κB regulates IκB by two distinct mechanisms. *Genes Dev.* 7:1266–76

Segars JH, Nagata T, Bours V, Medin IA, Franzoso G, et al. 1993. Retinoic acid induction of major histocompatibility class complex I genes in NTera-2 embryonal carcinoma cells involves induction of NF-κB (p50-p-65) and retinoic acid beta-retinoid x receptor beta heterodimers. *Mol. Cell. Biol.* 13:6157–69

Sen R, Baltimore D. 1986a. Multiple nuclear factors interact with the immunoglobulin enhancer sequences. *Cell* 46:705–16

Sen R, Baltimore D. 1986b. Inducibility of κ immunoglobulin enhancer-binding protein NF-κB by a postranslational mechanism. *Cell* 47:921–28

Shattuck RL, Wood LD, Jaffe GJ, Richmond A. 1994. MGSA/GRO transcription is differentially regulated in normal retinal pigment epithelial and melanoma cells. *Mol. Cell. Biol.* 14:791–802

Shelton CA, Wasserman SA. 1993. Pelle encodes a protein kinase required to establish dorsoventral polarity in the Drosophila embryo. *Cell* 72:515–25

Shirakawa F, Mizel SB. 1989. In vitro activation and nuclear translocation of NF-κB catalyzed by cyclic AMP-dependent protein kinase and protein kinase C. *Mol. Cell. Biol.* 9:2424–30

Shu HB, Agranoff AB, Nabel EG, Leung K. 1993. Differential regulation of vascular cell adhesion molecule 1 gene expression by specific NF-κB subunits in endothelial and epithelial cells. *Mol. Cell. Biol.* 13:6283–89

Sica A, Tan TH, Rice N, Kretzschmar M, et al. 1992. The c-rel protooncogene product c-Rel but not NF-kappa B binds to the intronic region of the human interferon-gamma gene at a site related to an interferon-stimulable response element. *Proc. Natl. Acad. Sci. USA* 89:1740–44

Smith MR, Greene WC. 1991. Molecular biology of type I human T-cell leukemia virus (HTLV-I) and adult T-cell leukemia. *J. Clin. Invest.* 87:761–66

Sokoloski JA, Sartorelli AC, Rosen CA, Narayanan R. 1993. Antisense oligonucleotides to the p65 subunit of NF-κB block CD11b expression and alter adhesion properties of differentiated HL-60 granulocytes. *Blood* 82:625–32

St. Johnston D, Nüsslein-Volhard C. 1992. The origin of pattern and polarity in the Drosophila embryo. *Cell* 68:201–19

Staudt LM, Lenardo MJ. 1991. Immunoglobulin gene transcription. *Annu. Rev. Immunol.* 9:373–98

Stein B, Baldwin AS. 1993. Distinct mechanisms for regulation of the interleukin-8 gene involve synergism and cooperativity between C/EBP and NF-κB. *Mol. Cell. Biol.* 13:7191–98

Stein B, Baldwin AS, Ballard DW, Greene WC, Angel P, et al. 1993a. Cross-coupling of the NF-κB p65 and Fos/Jun transcription factors produces potentiated biological function. *EMBO J.* 12:3879–91

Stein B, Cogswell PC, Baldwin AS. 1993b. Functional and physical associations between NF-κB and C/EBP family members: a rel domain b-ZIP interaction. *Mol. Cell. Biol.* 13:3964–74

Stephens RM, Rice NR, Hiebsch RR, Bose HR, Gilden RV. 1983. Nucleotide sequence of v-rel: the oncogene of the reticuloendotheliosis virus. *Proc. Natl. Acad. Sci. USA* 80:6229–32

Steward R. 1987. *Dorsal,* an embryonic polarity gene in *Drosophila* is homologous to the vertebrate proto-oncogene, c-rel. *Science* 238:692–94

Sun S-C, Faye I. 1992. Affinity purification and characterization of CIF, an insect immunoresponsive factor with NF-κB-like properties. *Comp. Biochem. Physiol. B* 103:225–33

Sun S-C, Ganchi PA, Ballard DW, Greene WC. 1993. NF-κB controls expression of inhibitor IκBα: evidence for an inducible autoregulatory pathway. *Science* 259:1912–15

Sun S-C, Ganchi PA, Beraud C, Ballard DW, Greene WC. 1994. Autoregulation of the NF-κB transactivator RelA (p65) by multiple cytoplasmic inhibitors containing ankyrin motifs. *Proc. Natl. Acad. Sci. USA* 91:1346–50

Sun S-C, Lindstrom I, Lee J-Y, Faye I. 1991. Structure and expression of the attacin genes in *Hyalophora cecropria. Eur. J. Biochem.* 196:247–54

Suzuki T, Hirai H, Fujisawa J, Fujita T, Yoshida M. 1993. A *trans*-activator Tax of human T-cell leukemia virus type 1 binds to NF-κB p50 and serum response factor (SRF) and associates with enhancer DNAs of the NF-κB site and CArG box. *Oncogene* 8:2391–97

Ten RM, Blank V, LeBail O, Kourilsky P, Israël A. 1993. Two factors, IRF1 and KBF1/NF-κB, cooperate during induction of MHC class I gene expression by interferon α/β or Newcastle disease virus. *CR Acad. Sci.* 316:496–501

Ten RM, Paya CV, Israël N, LeBail O, Mattei M-G, et al. 1992. The characterization of the promoter of the gene encoding the p50 subunit of NF-κB indicates that it participates in its own regulation. *EMBO J.* 11:195–203

Tewari M, Dobrzanski P, Mohn KL, Cressman DE, Hsu J-C, et al. 1992. Rapid induction in regenerating liver of RL/IF-1 (an IκB that inhibits NF-κB, RelB-p50, and c-Rel-p50) and PHF, a novel κB site-binding complex. *Mol. Cell. Biol.* 12:2898–908

Thanos D, Maniatis T. 1992. The high mobility group protein HMG I(Y) is required for NF-κB-dependent virus induction of the human IFN-κ gene. *Cell* 71:777–89

Thevenin C, Kim S-C, Rieckmann P, Fujiki H, Norcross MA, et al. 1990. Induction of nuclear factor-κB and the human immunodeficiency virus long terminal repeat by okadaic acid, a specific inibitor of phosphatases I and 2A. *New Biol.* 2:793–800

Thisse B, Stoetzel C, Gorostiza TC, Perrin-Schmitt F. 1988. Sequence of the twist gene and nuclear localization of its protein in endomesodermal cells of early *Drosophila* embryos. *EMBO J.* 7:2175–83

Toledano MB, Ghosh D, Trinh F, Leonard WJ. 1993. N-terminal DNA-binding domains contribute to differential DNA-binding

specificities of NF-κB p50 and p65. *Mol. Cell. Biol.* 13:852–60
Toledano MB, Leonard WJ. 1991. Modulation of transcription factor NF-κB binding activity by oxidation-reduction in vitro. *Proc. Natl. Acad. Sci. USA* 88:4328–32
Trede NS, Castigli E, Geha RS, Chatila T. 1993. Microbial superantigens induce NF-κB in the human monocytic cell line THP-1. *J. Immunol.* 150:5604–13
Ueberla K, Lu YC, Chung E, Haseltine WA. 1993. The NF-κB p65 promoter. *J. AIDS Res.* 6:227–30
Urban MB, Baeuerle PA. 1990. The 65-kD subunit of NF-κB is a receptor for IκB and a modulator of DNA-binding specificity. *Genes Dev.* 4:1975–84
Urban MB, Baeuerle PA. 1991. The role of the p50 and p65 subunits of NF-κB in the recognition of cognate sequences. *New Biol.* 3: 279–88
Visvanathan KV, Goodbourn S. 1989. Double-stranded RNA activates binding of NF-κB to an inducible element in the human κ-interferon promoter. *EMBO J.* 8:1129–38
Voraberger G, Schafer R, Stratowa C. 1991. Cloning of the human gene for intercellular adhesion molecule 1 and analysis of its 5′ regulatory region. Induction by cytokines and phorbol ester. *J. Immunol.* 147:2777–86
Walker WH, Stein B, Ganchi PA, Hoffman JA, Kaufman PA, et al. 1992. The v-rel oncogene: insights into the mechanism of transcriptional activation, repression, and transformation. *J. Virol.* 66:5018–29
Wasserman SA. 1993. A conserved signal transduction pathway regulating the activity of the rel-like proteins dorsal and NF-κB. *Mol. Cell. Biol.* 4:767–71
Watanabe M, Muramatsu M, Hirai H, Suzuki T. 1993. HTLV-I encoded Tax in association with NF-κκ precursor p105 enhances nuclear localization of NF-κB p50 and p65 in transfected cells. *Oncogene* 8:2949–58
Wilhelmsen KC, Eggleton K, Temin HM. 1984. Nucleic acid sequence of the oncogene v-rel in reticuloendotheliosis virus strain T and its cellular homolog, the protooncogene c-rel. *J. Virol.* 52:172–82
Wulczyn FG, Naumann M, Scheidereit C. 1992. Candidate proto-oncogene bcl-3 encodes a subunit-specific inhibitor of transcription factor NF-κB. *Nature* 358:597–99
Xie QW, Kashiwabara Y, Nathan C. 1994. Role of transcription factor NF-κB/Rel in induction of nitric oxide synthase. *J. Biol. Chem.* 269:4705–5708
Xu X, Prorock C, Ishikawa H, Maldonado E. 1993. Functional interaction of the v-Rel and c-Rel oncoproteins with the TATA-binding protein and association with transcription factor IIB. *Mol. Cell. Biol.* 13:6733–41
Yang Z, Costanzo M, Golde W, Kolesnick RN. 1993. Tumor necrosis factor activation of the sphingomyelin pathway signals nuclear factor kappa B translocation in intact HL-60 cells. *J. Biol. Chem.* 268:20520–23
Yasumoto K, Okamoto S, Mukaida N, Murakami S. 1992. Tumor necrosis factor alpha and interferon gamma synergistically induce interleukin 8 production in a human gastric cancer cell line through acting concurrently on AP-1 and NF-κB-like binding sites of the interleukin 8 gene. *J. Biol. Chem.* 267: 22506–11
Zabel U, Baeuerle PA. 1990. Purified human IκB can rapidly dissociate the complex of the NF-κB transcription factor with its cognate DNA. *Cell* 61:255–65
Zabel U, Henkel T, dosSantosSilva M, Baeuerle P. 1993. Nuclear uptake control of NF-κB by MAD-3, and IκB protein present in the nucleus. *EMBO J.* 12:201–11
Zhang Y, Broser M, Rom WN. 1994. Activation of the interleukin 6 gene by *Mycobacterium tuberculosis* or lipopolysaccharide is mediated by nuclear factors NF-IL6 and NF-kappa B. *Proc. Natl. Acad. Sci. USA* 91:2225–29
Ziegler-Heitbrock HWL, Sternsdorf T, Liese J, Belohradsky B, Weber C, et al. 1993. Pyrrolidine dithiocarbamate inhibits NF-κB mobilization and TNF production in human monocytes. *J. Immunol.* 151:6986–93
Zuniga-Pflucker JC, Schwartz ML, Lenardo MY. 1993. Gene transcription in differentiating immature T cell receptor (neg) thymocytes resembles antigen-activated mature T cells. *Medicine* 178:1139–49

SUBJECT INDEX

CUMULATIVE INDEXES

CONTRIBUTING AUTHORS, VOLUMES 6–10

CUMULATIVE INDEXES

CHAPTER TITLES, VOLUMES 6–10

ANNUAL REVIEWS

a nonprofit scientific publisher
4139 El Camino Way
P.O. Box 10139
Palo Alto, CA 94303-0139 • USA

ORDER FORM

ORDER TOLL FREE
1.800.523.8635
from USA and Canada

Fax: 1.415.855.9815

Annual Reviews publications may be ordered directly from our office; through booksellers and subscription agents, worldwide; and through participating professional societies. **Prices are subject to change without notice. We do not ship on approval.**

- **Individuals:** Prepayment required on new accounts. in US dollars, checks drawn on a US bank.
- **Institutional Buyers:** Include purchase order. Calif. Corp. #161041 • ARI Fed. I.D. #94-1156476
- **Students / Recent Graduates: $10.00 discount from retail price, per volume.** ***Requirements:*** **1.** be a degree candidate at, or a graduate within the past three years from, an accredited institution; **2.** present proof of status (photocopy of your student I.D. or proof of date of graduation); **3.** Order direct from Annual Reviews; **4.** prepay. This discount **does not** apply to standing orders, *Index on Diskette,* Special Publications, ARPR, or institutional buyers.
- **Professional Society Members:** Many Societies offer *Annual Reviews* to members at reduced rates. Check with your society or contact our office for a list of participating societies.
- **California orders** add applicable sales tax. • **Canadian orders** add 7% GST. Registration #R 121 449-029.
- **Postage paid** by Annual Reviews (4th class bookrate/surface mail). UPS ground service is available at $2.00 extra per book within the contiguous 48 states only. UPS air service or US airmail is available to any location at actual cost. UPS requires a street address. P.O. Box, APO, FPO, not acceptable.
- **Standing Orders:** Set up a standing order and the new volume in series is sent automatically each year upon publication. Each year you can save 10% by prepayment of prerelease invoices sent 90 days prior to the publication date. Cancellation may be made at any time.
- **Prepublication Orders:** Advance orders may be placed for any volume and will be charged to your account upon receipt. Volumes not yet published will be shipped during month of publication indicated.

NOTE: For copies of individual articles from any *Annual Review,* or copies of any article cited in an *Annual Review,* call **Annual Reviews Preprints and Reprints (ARPR)** toll free 1-800-347-8007 (fax toll free 1-800-347-8008) from the USA or Canada. From elsewhere call 1-415-259-5017.

ANNUAL REVIEWS SERIES *Volumes not listed are no longer in print*		**Prices, postpaid, per volume. USA/other countries**	Regular Order Please send Volume(s):	Standing Order Begin with Volume:
❑ *Annual Review of*	**ANTHROPOLOGY**			
Vols. 1-20	(1972-91)	$41 / $46		
Vols. 21-22	(1992-93)	$44 / $49		
Vol. 23	(avail. Oct. 1994)	$47 / $52	Vol(s). ______	Vol. ______
❑ *Annual Review of*	**ASTRONOMY AND ASTROPHYSICS**			
Vols. 1, 5-14, 16-29	(1963, 67-76, 78-91)	$53 / $58		
Vols. 30-31	(1992-93)	$57 / $62		
Vol. 32	(avail. Sept. 1994)	$60 / $65	Vol(s). ______	Vol. ______
❑ *Annual Review of*	**BIOCHEMISTRY**			
Vols. 31-34, 36-60	(1962-65,67-91)	$41 / $47		
Vols. 61-62	(1992-93)	$46 / $52		
Vol. 63	(avail. July 1994)	$49 / $55	Vol(s). ______	Vol. ______
❑ *Annual Review of*	**BIOPHYSICS AND BIOMOLECULAR STRUCTURE**			
Vols. 1-20	(1972-91)	$55 / $60		
Vols. 21-22	(1992-93)	$59 / $64		
Vol. 23	(avail. June 1994)	$62 / $67	Vol(s). ______	Vol. ______